Emerging Dietary Bioactives in Health and Disease

Emerging Dietary Bioactives in Health and Disease

Editor

Elad Tako

MDPI

Basel • Beijing • Wuhan • Barcelona • Belgrade • Novi Sad • Cluj • Manchester

Editor
Elad Tako
Department of Food Science
Cornell University
Ithaca, NY, USA

Editorial Office
MDPI
St. Alban-Anlage 66
4052 Basel, Switzerland

This is a reprint of articles from the Special Issue published online in the open access journal *Nutrients* (ISSN 2072-6643) (available at: https://www.mdpi.com/journal/nutrients/special_issues/Bioactives_Health_Disease).

For citation purposes, cite each article independently as indicated on the article page online and as indicated below:

Lastname, A.A.; Lastname, B.B. Article Title. *Journal Name* **Year**, *Volume Number*, Page Range.

ISBN 978-3-0365-8632-8 (Hbk)
ISBN 978-3-0365-8633-5 (PDF)
doi.org/10.3390/books978-3-0365-8633-5

Cover image courtesy of Elad Tako

Contents

About the Editor

Elad Tako

Elad Tako holds degrees in animal science (B.S.), endocrinology (M.S.), and physiology/nutrigenomics (Ph.D.), with previous appointments at the Hebrew University of Jerusalem, North Carolina State University, and Cornell University. As an Associate Professor with the Department of Food Science at Cornell University, Dr. Tako's research focuses on various aspects of trace mineral deficiencies, emphasizing molecular, physiological and nutritional factors and practices that influence intestinal micronutrient absorption. With over 150 peer-reviewed publications and presentations, he leads a research team focused on understanding the interactions between dietary factors, physiological and molecular biomarkers, the microbiome, and intestinal functionality. His research accomplishments include the development of the *Gallus gallus* intra-amniotic administration procedure and establishing recognized approaches for the use of animal models within mineral bioavailability and intestinal absorption screening processes. He has also developed a zinc status physiological blood biomarker (red blood cell linoleic acid: dihomo–linolenic acid ratio) and molecular tissue biomarkers to assess the effect of dietary mineral deficiencies on intestinal functionality and determine how micronutrient dietary deficiencies alter gut microbiota composition and function.

nutrients MDPI

Editorial

Emerging Dietary Bioactives in Health and Disease

Elad Tako

Department of Food Science, Cornell University, Stocking Hall, Ithaca, NY 14853-7201, USA; et79@cornell.edu

This monograph, based on a Special Issue of Nutrients, contains 16 manuscripts—2 review manuscripts and 14 original research manuscripts—that reflect the wide spectrum of currently conducted research in the field of Emerging Dietary Bioactives in Health and Disease. The manuscripts in this Special Issue collection include contributions from researchers from multiple countries, including the USA, Brazil, United Kingdom, Italy, Norway, China, and Israel. The presented manuscripts cover a wide variety and range of topics in the field of dietary bioactives in health and disease, with emphasis on the investigation of the effects hydrolyzed protein of chia (*Salvia hispanica* L.) and *Lacticaseibacillus paracasei* on intestinal functionality, morphology, and bacterial populations, in vivo (*Gallus gallus*) [1]; the demonstration of the capacity of macauba (*Acrocomia aculeata*) pulp oil to prevent adipogenesis, inflammation and oxidative stress in mice fed a high-fat diet [2]; the demonstration of the capacity of kombuchas from green and black tea to modulate the gut microbiota and improve the intestinal health of wistar rats fed a high-fat high-fructose diet [3]; the methodological preparation, characterization, wound healing, and cytotoxicity assay of pegylated nanophytosomes loaded with 6-gingerol [4]; the discussion of empire apple (*Malus domestica*) juice, pomace, and pulp and the modulation of intestinal functionality, morphology, and bacterial populations in vivo (*Gallus gallus*) [5]; the effect of chia (*Salvia hispanica* L.) associated with high-fat diet on the intestinal health of wistar rats [6]; the demonstration of the capacity of curcumin-added whey protein to positively modulate skeletal muscle inflammation and oxidative damage after exhaustive exercise [7]; the introduction of the novel intra-amniotic administration—an emerging method with which to investigate necrotizing enterocolitis in vivo (*Gallus gallus*) [8]; the investigation of the effect of black corn anthocyanin-rich extract (*Zea mays* L.) on cecal microbial populations in vivo (*Gallus gallus*) [9]; the demonstration of BRD9 inhibition by natural polyphenols which target DNA damage/repair and apoptosis in human colon cancer cells [10]; the alterations in intestinal brush border membrane functionality and bacterial populations following the intra-amniotic administration (*Gallus gallus*) of catechin and its derivatives [11]; the performance of a comparison of the effects of concord grape (*Vitis labrusca* L.) puree, juice, and pomace on intestinal morphology and functionality, and on bacterial populations in vivo (*Gallus gallus*) [12]; the demonstration of how the intra-amniotic administration (*Gallus gallus*) of genistein alters mineral transport, intestinal morphology, and gut microbiota [13]; the occurrence of the alterations in intestinal brush border membrane functionality and bacterial populations following intra-amniotic administration (*Gallus gallus*) of nicotinamide riboside and its derivatives [14]; a literature review and discussion of dietary trehalose as a bioactive nutrient [15]; and an analysis of how resistant maltodextrin consumption in a double-blind, randomized, crossover clinical trial induces specific changes in potentially beneficial gut bacteria [16]. These wide spectra of topics further demonstrate the importance and relevance of dietary bioactive compounds, as well as their relevance in health and disease circumstances.

Plant-based diets contain wide varieties of metabolites that may impact on health and disease prevention. Most are focused on the potential bioactivity and nutritional relevance of several classes of phytochemicals, such as polyphenols, flavonoids, carotenoids, phyto-oestrogens, and frucrooligo-saccharides [17]. These compounds are found in fruit,

Citation: Tako, E. Emerging Dietary Bioactives in Health and Disease. *Nutrients* **2023**, *15*, 1956. https://doi.org/10.3390/nu15081956

Received: 11 April 2023
Accepted: 17 April 2023
Published: 19 April 2023

vegetables, and herbs [18]. The permitted daily intakes of some of these compounds may exceed 100 mg. Moreover, intestinal bacterial activity may transform complex compounds such as anthocyanins, procyanidins, and isoflavones into simple phenolic metabolites [19]. The colon is thus a rich source of potentially active phenolic acids that may impact both locally and systemically on gut health. Further, nondigestible fibers (prebiotics) are dietary substrates that selectively promote proliferation and/or activity of health-promoting bacterial populations in the colon [20]. Prebiotics, such as inulin, raffinose, and stachyose, have a proven ability to promote the abundance of intestinal bacterial populations, which may provide additional health benefits to the host [21–26]. Further, various pulse seed soluble (fiber) extracts are responsible for improving gastrointestinal motility, intestinal functionality and morphology, and mineral absorption [25,27]. Studies have indicated that the consumption of seed origin soluble extracts can upregulate the expression of BBM proteins that contribute to the digestion and absorption of nutrients. Soluble extracts can positively affect intestinal health by increasing the mucus production, goblet cells number/diameter, villus surface area, and crypt depth [25,27]. These functional and morphological effects appear to occur due to the increased motility of the digestive tract, leading to hyperplasia and/or hypertrophy of muscle cells. Plant-origin soluble extracts may act, directly or indirectly, to increase mineral solubility and, therefore, dietary bioavailability. This occurs due to fiber fermentation and the bacterial production of SCFA, a process that reduces intestinal pH, inhibits the growth of potentially pathogenic bacterial population and increases the solubility and, therefore, absorption of minerals. The SCFA can increase the proliferation of epithelial cells, which, in turn, increases the absorptive surface area, an occurrence which contributes to the absorption of nutrients [28–30]. Several phenolic acids and other phytochemicals affect the expression and activity of enzymes involved in the production of inflammatory mediators of pathways thought to be important in the development of gut disorders, including colon cancer. However, it is still unclear as to which of these compounds are beneficial to gut health. Hence, the aim of the current Special Issue is to further explore the interactions between dietary bioactive compounds and overall health, and to further expand and add research knowledge on the vital role of dietary bioactive compounds in various nutrition-related physiological and metabolic pathways, and beyond.

This Special Issue and collection of manuscripts constitutes a useful summary of progress in various areas related to emerging dietary bioactives in health and disease.

Conflicts of Interest: The author declares no conflict of interest.

References

1. Mishima, M.D.V.; Martino, H.S.D.; Kolba, N.; Shah, D.D.; Grancieri, M.; Dos Santos, K.M.O.; Lima, J.P.; Da Silva, B.P.; Gonzalez de Mejia, E.; Tako, E. Effects of Intra-Amniotic Administration of the Hydrolyzed Protein of Chia (*Salvia hispanica* L.) and Lacticaseibacillus paracasei on Intestinal Functionality, Morphology, and Bacterial Populations, In Vivo (*Gallus gallus*). *Nutrients* **2023**, *15*, 1831. [CrossRef]
2. Sant' Ana, C.T.; Agrizzi Verediano, T.; Grancieri, M.; Toledo, R.C.L.; Tako, E.; Costa, N.M.B.; Martino, H.S.D.; de Barros, F.A.R. Macauba (*Acrocomia aculeata*) Pulp Oil Prevents Adipogenesis, Inflammation and Oxidative Stress in Mice Fed a High-Fat Diet. *Nutrients* **2023**, *15*, 1252. [CrossRef] [PubMed]
3. Costa, M.A.d.C.; Dias Moreira, L.d.P.; Duarte, V.d.S.; Cardoso, R.R.; São José, V.P.B.d.; Silva, B.P.d.; Grancieri, M.; Corich, V.; Giacomini, A.; Bressan, J.; et al. Kombuchas from Green and Black Tea Modulate the Gut Microbiota and Improve the Intestinal Health of Wistar Rats Fed a High-Fat High-Fructose Diet. *Nutrients* **2022**, *14*, 5234. [CrossRef] [PubMed]
4. Al-Samydai, A.; Qaraleh, M.A.; Alshaer, W.; Al-Halaseh, L.K.; Issa, R.; Alshaikh, F.; Abu-Rumman, A.; Al-Ali, H.; Al-Dujaili, E.A.S. Preparation, Characterization, Wound Healing, and Cytotoxicity Assay of PEGylated Nanophytosomes Loaded with 6-Gingerol. *Nutrients* **2022**, *14*, 5170. [CrossRef]
5. Jackson, C.; Shukla, V.; Kolba, N.; Agarwal, N.; Padilla-Zakour, O.I.; Tako, E. Empire Apple (*Malus domestica*) Juice, Pomace, and Pulp Modulate Intestinal Functionality, Morphology, and Bacterial Populations In Vivo (*Gallus gallus*). *Nutrients* **2022**, *14*, 4955. [CrossRef]
6. Mishima, M.D.V.; Da Silva, B.P.; Gomes, M.J.C.; Toledo, R.C.L.; Mantovani, H.C.; José, V.P.B.d.S.; Costa, N.M.B.; Tako, E.; Martino, H.S.D. Effect of Chia (*Salvia hispanica* L.) Associated with High-Fat Diet on the Intestinal Health of Wistar Rats. *Nutrients* **2022**, *14*, 4924. [CrossRef]

7. Dias, K.A.; da Conceição, A.R.; Pereira, S.M.S.; Oliveira, L.A.; da Silva Rodrigues, J.V.; Dias, R.S.; de Paula, S.O.; Natali, A.J.; da Matta, S.L.P.; Gonçalves, R.V.; et al. Curcumin-Added Whey Protein Positively Modulates Skeletal Muscle Inflammation and Oxidative Damage after Exhaustive Exercise. *Nutrients* **2022**, *14*, 4905. [CrossRef]
8. Kolba, N.; Cheng, J.; Jackson, C.D.; Tako, E. Intra-Amniotic Administration—An Emerging Method to Investigate Necrotizing Enterocolitis, In Vivo (*Gallus gallus*). *Nutrients* **2022**, *14*, 4795. [CrossRef]
9. Agrizzi Verediano, T.; Agarwal, N.; Stampini Duarte Martino, H.; Kolba, N.; Grancieri, M.; Dias Paes, M.C.; Tako, E. Effect of Black Corn Anthocyanin-Rich Extract (*Zea mays* L.) on Cecal Microbial Populations In Vivo (*Gallus gallus*). *Nutrients* **2022**, *14*, 4679. [CrossRef]
10. Kapoor, S.; Damiani, E.; Wang, S.; Dharmanand, R.; Tripathi, C.; Tovar Perez, J.E.; Dashwood, W.M.; Rajendran, P.; Dashwood, R.H. BRD9 Inhibition by Natural Polyphenols Targets DNA Damage/Repair and Apoptosis in Human Colon Cancer Cells. *Nutrients* **2022**, *14*, 4317. [CrossRef]
11. Kolba, N.; Zarei, A.; Cheng, J.; Agarwal, N.; Dadmohammadi, Y.; Khazdooz, L.; Abbaspourrad, A.; Tako, E. Alterations in Intestinal Brush Border Membrane Functionality and Bacterial Populations Following Intra-Amniotic Administration (*Gallus gallus*) of Catechin and Its Derivatives. *Nutrients* **2022**, *14*, 3924. [CrossRef] [PubMed]
12. Agarwal, N.; Shukla, V.; Kolba, N.; Jackson, C.; Cheng, J.; Padilla-Zakour, O.I.; Tako, E. Comparing the Effects of Concord Grape (*Vitis labrusca* L.) Puree, Juice, and Pomace on Intestinal Morphology, Functionality, and Bacterial Populations In Vivo (*Gallus gallus*). *Nutrients* **2022**, *14*, 3539. [CrossRef]
13. Cheng, J.; Kolba, N.; Sisser, P.; Turjeman, S.; Even, C.; Koren, O.; Tako, E. Intraamniotic Administration (*Gallus gallus*) of Genistein Alters Mineral Transport, Intestinal Morphology, and Gut Microbiota. *Nutrients* **2022**, *14*, 3473. [CrossRef]
14. Kolba, N.; Zarei, A.; Cheng, J.; Agarwal, N.; Dadmohammadi, Y.; Khazdooz, L.; Abbaspourrad, A.; Tako, E. Alterations in Intestinal Brush Border Membrane Functionality and Bacterial Populations Following Intra-Amniotic Administration (*Gallus gallus*) of Nicotinamide Riboside and Its Derivatives. *Nutrients* **2022**, *14*, 3130. [CrossRef] [PubMed]
15. Chen, A.; Gibney, P.A. Dietary Trehalose as a Bioactive Nutrient. *Nutrients* **2023**, *15*, 1393. [CrossRef] [PubMed]
16. Mai, V.; Burns, A.M.; Solch, R.J.; Dennis-Wall, J.C.; Ukhanova, M.; Langkamp-Henken, B. Resistant Maltodextrin Consumption in a Double-Blind, Randomized, Crossover Clinical Trial Induces Specific Changes in Potentially Beneficial Gut Bacteria. *Nutrients* **2022**, *14*, 2192. [CrossRef]
17. Valero-Cases, E.; Cerdá-Bernad, D.; Pastor, J.J.; Frutos, M.J. Non-Dairy Fermented Beverages as Potential Carriers to Ensure Probiotics, Prebiotics, and Bioactive Compounds Arrival to the Gut and Their Health Benefits. *Nutrients* **2020**, *12*, 1666. [CrossRef]
18. Upadhyay, S.; Dixit, M. Role of Polyphenols and Other Phytochemicals on Molecular Signaling. *Oxid. Med. Cell Longev.* **2015**, *2015*, 504253. [CrossRef]
19. Shinwari, K.J.; Rao, P.S. Trends in Food Science & Technology. Stability of bioactive compounds in fruit jam and jelly during processing and storage: A review. *Trends Food Sci. Technol.* **2018**, *75*, 181–193.
20. Hartono, K.; Reed, S.; Ayikarkor Ankrah, N.; Tako, T. Alterations in gut microflora populations and brush border functionality following intra-amniotic daidzein administration. *RSC Adv.* **2015**, *5*, 6407–6412. [CrossRef]
21. Slavin, J. Fiber and Prebiotics: Mechanisms and Health Benefits. *Nutrients* **2013**, *5*, 1417–1435. [CrossRef] [PubMed]
22. Tako, E.; Rutzke, M.A.; Glahn, R.P. Using the domestic chicken (*Gallus gallus*) as an in vivo model for Fe bioavailability. *J. Poult. Sci.* **2010**, *89*, 514–521. [CrossRef] [PubMed]
23. Pacifici, S.; Song, J.; Zhang, C.; Wang, Q.; Kolba, N.; Tako, E. Intra Amniotic Administration of Raffinose and Stachyose Affects the Intestinal Brush Border Functionality and Alters Gut Microflora Populations. *Nutrients* **2017**, *9*, 304. [CrossRef]
24. Hou, T.; Kolba, N.; Tako, E. Intra-Amniotic Administration (*Gallus gallus*) of Cicer arietinum and Lens culinaris Prebiotics Extracts and Duck Egg White Peptides Affects Calcium Status and Intestinal Functionality. *Nutrients* **2017**, *9*, 785. [CrossRef] [PubMed]
25. Hou, T.; Tako, E. The In Ovo Feeding Administration (*Gallus Gallus*)—An Emerging In Vivo Approach to Assess Bioactive Compounds with Potential Nutritional Benefits. *Nutrients* **2018**, *10*, 418. [CrossRef]
26. Morais Dias, D.; Kolba, N.; Hart, J.; Ma, M.; Sha, S.; Lakshmanan, N.; Nutti Regini, M.; Duarte Martino, H.; Tako, E. Soluble extracts from carioca beans (*Phaseolus vulgaris* L.) affect the gut microbiota and iron related brush border membrane protein expression in vivo (*Gallus gallus*). *Food Res. Int.* **2019**, *120*, 172–180. [CrossRef]
27. Beasley, J.T.; Johnson, A.A.T.; Kolba, N.; Bonneau, J.P.; Ohayon, M.N.; Koren, O.; Tako, E. Nicotianamine-chelated iron improves micronutrients physiological status and gastrointestinal health in vivo (*Gallus gallus*). *Sci. Rep.* **2020**, *10*, 2297. [CrossRef]
28. Wang, X.; Kolba, N.; Tako, E. Alterations in gut microflora populations and brush border functionality following intra-amniotic administration (*Gallus gallus*) of wheat bran prebiotics extracts. *Food Funct.* **2019**, *10*, 4834–4843. [CrossRef]
29. Reed, S.; Neuman, H.; Moscovich, S.R.; Koren, O.; Tako, E. Chronic Zinc Deficiency AltersChick Gut Microbiota Composition and Function. *Nutrients* **2015**, *7*, 9768–9784. [CrossRef]
30. Reed, S.; Knez, M.; Uzan, A.; Stangoulis, J.; Koren, O.; Tako, E. Alterations in the gut (*Gallus gallus*) microbiota following the consumption of zinc biofortified wheat (*Triticum aestivum*) -based diet. *J. Agric. Food Chem.* **2018**, *66*, 6291–6299. [CrossRef]

nutrients

MDPI

Article

BRD9 Inhibition by Natural Polyphenols Targets DNA Damage/Repair and Apoptosis in Human Colon Cancer Cells

Sabeeta Kapoor [1], Elisabetta Damiani [2], Shan Wang [3], Ravirajan Dharmanand [4], Chakrapani Tripathi [1], Jorge Enrique Tovar Perez [1], Wan Mohaiza Dashwood [1], Praveen Rajendran [1,5,*] and Roderick Hugh Dashwood [1,5,*]

1 Center for Epigenetics & Disease Prevention, Texas A&M Health, Houston, TX 77030, USA
2 Department of Life and Environmental Sciences, Università Politecnica delle Marche, 60121 Ancona, Italy
3 Stomatology Hospital, School of Stomatology, Zhejiang University School of Medicine, Hangzhou 310030, China
4 Center for Infectious & Inflammatory Diseases, Texas A&M Health, Houston, TX 77030, USA
5 Department of Translational Medical Sciences, Texas A&M College of Medicine, Houston, TX 77030, USA
* Correspondence: prajendran@tamu.edu (P.R.); rdashwood@tamu.edu (R.H.D.); Tel.: +1-713-677-7803 (P.R.); +1-713-677-7806 (R.H.D.)

Abstract: Epigenetic mechanisms play an important role in the etiology of colorectal cancer (CRC) and other malignancies due, in part, to deregulated bromodomain (BRD) functions. Inhibitors of the bromodomain and extraterminal (BET) family have entered into clinical trials as anticancer agents, and interest has grown in other acetyl 'reader' proteins as therapeutic targets, including non-BET member bromodomain-containing protein 9 (BRD9). We report here that overexpression of BRD9 is associated with poor prognosis in CRC patients, and that siRNA-mediated knockdown of BRD9 decreased cell viability and activated apoptosis in human colon cancer cells, coincident with increased DNA damage. Seeking natural compounds as BRD9 antagonists, molecular docking in silico identified several polyphenols such as Epigallocatechin-3-gallate (EGCG), Equol, Quercetin, and Aspalathin, with favorable binding energies, supported by BROMOscan® (DiscoverX) and isothermal titration calorimetry experiments. Polyphenols mimicked BRD9 knockdown and iBRD9 treatment in reducing colon cancer cell viability, inhibiting colony formation, and enhancing DNA damage and apoptosis. Normal colonic epithelial cells were unaffected, signifying cancer-specific effects. These findings suggest that natural polyphenols recognize and target BRD9 for inhibition, and might serve as useful lead compounds for bromodomain therapeutics in the clinical setting.

Keywords: apoptosis; bromodomain inhibition; colorectal cancer; molecular docking

Citation: Kapoor, S.; Damiani, E.; Wang, S.; Dharmanand, R.; Tripathi, C.; Tovar Perez, J.E.; Dashwood, W.M.; Rajendran, P.; Dashwood, R.H. BRD9 Inhibition by Natural Polyphenols Targets DNA Damage/Repair and Apoptosis in Human Colon Cancer Cells. *Nutrients* **2022**, *14*, 4317. https://doi.org/10.3390/nu14204317

Academic Editor: Elad Tako

Received: 20 September 2022
Accepted: 14 October 2022
Published: 15 October 2022

1. Introduction

CRC is a leading cause of cancer-related death worldwide, characterized by the accumulation of genetic and epigenetic alterations, including deregulated bromodomain (BRD) functions [1–3]. BRDs are evolutionarily conserved epigenetic 'reader' modules that recognize acetylated lysine residues on target proteins [4,5]. BRD-containing factors link acetylated histone and nonhistone proteins to the activation of RNA polymerase II, with critical roles in gene transcription, splicing, chromatin remodeling, protein scaffolding, and signal transduction [6]. Bromodomain-containing protein 9 (BRD9) is a recently identified subunit of the ATP-dependent chromatin remodeling switch/sucrose non-fermentable (SWI/SNF) complex that interacts with chromatin partners and transcription factors implicated in cell proliferation, apoptosis, differentiation, and cancer [7,8]. BRD9 also is involved in DNA replication, DNA repair, and cellular responses to DNA damage [9]; thus, targeting BRD9 provides opportunities for new avenues in cancer treatment. Bromodomain and extraterminal (BET) antagonists, such as JQ1, exhibit anticancer efficacy in clinical trials, but not without associated toxicity and drug resistance [10]. These observations have given impetus to the targeting of other BRD-containing proteins with a view to improved safety and efficacy. Nutrients from diet have an important role in the development of CRC, and more research is needed on specific foods and dietary compounds that can combat the development and progression of CRC. Dietary phytochemicals have been studied as modulators of epigenetic readers/writers/erasers [3,11,12], and the metabolites are untapped sources of novel bioactives for 'drugging' chromatin [3,13]. This report focuses on natural polyphenols as BRD9 inhibitors in CRC cells and the use of nutrients in targeting an epigenetic 'reader' protein deregulated in CRC.

2. Materials and Methods

2.1. Cells and Treatments

Human colon cancer cells (SW620, SW480) and non-transformed colonic epithelial cells (CCD841) were purchased from ATCC and used within 10–15 passages. Each cell line was confirmed independently to be of human origin, with no mammalian interspecies contamination, and with the correct genetic profile based on allele-specific markers (Idexx Radil, Columbia, MO, USA). Cells were cultured in McCoy's 5A media (Invitrogen) or EMEM (Invitrogen), supplemented with 10% FBS and 1% penicillin/streptomycin, at 37 °C in a humidified chamber with 5% CO2. Chemicals, reagents, and cell culture media were from reputable commercial vendors, including supplies for pooled siRNAs, as reported [14–16].

2.2. Molecular Docking

Iterative docking of the BRD9 bromodomain with energy-minimized ligand was performed using reported methodologies [13–17], including SWISS-MODEL, AutoDock Vina, PDBePISA and LPC/CSU [18–22].

2.3. BROMOscan® Screening

BRD9 inhibitor screening was performed using the BROMOscan® Technology Platform (DiscoverX, Eurofins, Dallas, TX, USA) [23], with iBRD9 as a positive control and all compounds at a 10 μM final concentration. Test agents were blinded at the time of the assay.

2.4. Isothermal Titration Calorimetry

Experiments were conducted in a microcalorimeter (MicroCal PEAQ-ITC, Malvern Panalytical) using normal and reverse titration methods. The BRD9 protein was expressed in chemically competent *Escherichia coli* cells and purified by affinity chromatography. A microsyringe was loaded with 40-μL protein sample in isothermal titration calorimetry (ITC) buffer, in the concentration range 150−300 μM, and inserted into the calorimetric cell (0.2 mL cell volume), which was filled with test compound (10−30 μM) or protein (15−50 μM). Titrations involved 20–30 repeated 2-μL injections, with a duration of 4-s per injection. Titration experiments were optimized to ensure complete saturation of the protein or ligand before the final injection. Data were corrected by subtracting the heat of dilution, determined from independent titrations. MicroCal Origin 7.0 software was used to determine enthalpies of binding (ΔH) and binding constants (Kb), as reported [24].

2.5. Cell Viability

Cell Counting Kit-8 (CCK8, APExBIO, Boston, MA, USA) was used as reported [25]. Cells in the exponential growth phase were plated at a cell density of 5000 cells per well in 96-well tissue culture plates. After attachment overnight, cells were treated with compounds for 72 h across a broad range of concentrations, as shown in the figures. The colony formation assay also was performed, as described [26]. Cells were trypsinized and plated in 6-well dishes (500 cells/well), allowed to attach overnight, and then treated with test agents for 72 h. Seven days later, cells were fixed and stained with crystal violet. Colonies were counted and the surviving fraction was calculated as the ratio of the number of colonies in the treated sample to the number of colonies in the untreated control. Triplicate wells were established for each condition.

2.6. Immunoblotting

As reported [25], whole cell lysates (20 µg protein/lane) were separated by SDS-PAGE on 4–12% Bis-Tris gel (NuPAGE, Invitrogen, CA, USA) and transferred to nitrocellulose membranes (Invitrogen, CA, USA). Membranes were saturated with 2% Bovine Serum Albumin for 1 h, followed by overnight incubation at 4 °C with primary antibodies for BRD9 (1:1000, Active Motif, Carlsbad, CA, USA, #61537), pRPA32 S4/S8 (1:500, Bethyl Laboratories, Montgomery, TX, USA, #A300-245A), β-Actin (1:5000, Sigma-Aldrich, St. Louis, MO, USA, #A5441), c-Myc (1:1000, #D3N8F), pH2AX (1:500, #2577S), poly(ADP-ribose)polymerase (PARP) (1:1000, #9542), and cleaved caspase-3 (1:1000, #9661) from Cell Signaling (Danvers, MA, USA). After washing, membranes were incubated with horseradish peroxidase-conjugated secondary antibodies for 1 h. Bands were visualized using Western Lightning Plus-ECL Enhanced Chemiluminescence Substrate (Perkin Elmer, Waltham, MA, USA) and detected using a ChemiDoc MP Imaging System (Bio-Rad, Hercules, CA, USA).

2.7. Apoptosis

Colon cancer cells were treated with test compounds for 72 h in 6-well plates, and apoptosis assays were conducted as reported [27] using a BD Pharmingen PE Annexin V Apoptosis Detection Kit I (BD Biosciences, San Jose, CA, USA). Briefly, cells were collected, washed with PBS, and incubated in binding buffer with 5 µL of PE Annexin V for 5 min and 5 µL of 7-AAD for 15 min in the dark at 37 °C. Percent of apoptotic cells was determined using LSR II Flow cytometer (BD Biosciences) and FlowJo 10.8.1 software.

2.8. Statistics

Unless stated otherwise, findings are representative outcomes from three or more biological and technical replicates. Paired comparisons were made between test agent and vehicle control using Student's *t*-test in GraphPad Prism 9.4.

3. Results

3.1. BRD9 Overexpression Is Associated with Reduced Survival in CRC Patients

Publicly available data revealed that Survival Probability was significantly worse in CRC cancer patients for whom BRD9 transcript expression was high in adenocarcinomas as compared with those harboring relatively low BRD9 levels (Figure 1A). In tissue microarrays, BRD9 protein was immunolocalized mainly to the nucleus, and was strongly positive in colon cancers as compared to normal colon (Figure 1B). These observations implicated an oncogenic role in CRC, and efforts to inhibit or downregulate BRD9.

3.2. BRD9 Regulates DNA Damage/Repair and Apoptosis in Human Colon Cancer Cells

In metastasis-lineage SW620 cells, transient transfection of BRD9 siRNAs for 72 h followed by immunoblotting revealed concentration-dependent reduction of BRD9 and c-Myc protein compared with vehicle-treated and scrambled siRNA-treated controls (Figure 2A). A concomitant increase was observed in pH2AX and pRPA32, which are well-characterized readouts of DNA damage and repair [16,25,28], along with cleaved PARP and caspase-3 apoptosis markers (Figure 2A, arrows). Morphological examination and quantification in the CCK8 assay revealed that siRNA-mediated BRD9 knockdown reduced cell viability significantly (Figure 2B), and there was a concentration-dependent decrease in the colony formation assay (Figure 2C). These findings indicated that BRD9 regulates DNA damage/repair, cell viability, and apoptosis induction in human colon cancer cells.

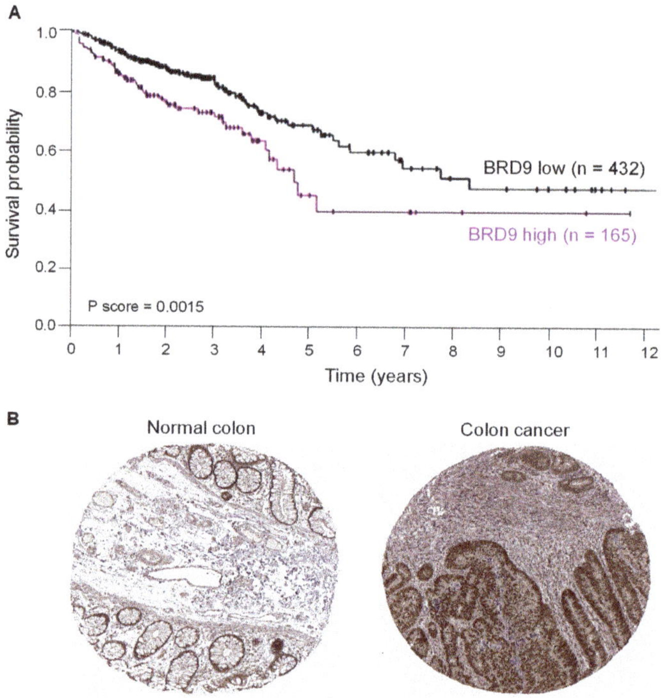

Figure 1. BRD9 overexpression is associated with reduced survival in CRC patients. (**A**) Overall survival probability in colorectal cancer (CRC) patients with high vs. low BRD9 mRNA expression in tumors (Human Protein Atlas at https://www.proteinatlas.org/ENSG00000028310-BRD9/pathology/colorectal+cancer, accessed on 13 June 2022). (**B**) Immunohistochemistry images of BRD9 protein expression in normal colon (https://www.proteinatlas.org/ENSG00000028310-BRD9/tissue/colon#img, accessed on 13 June 2022) and colon cancer (https://www.proteinatlas.org/ENSG00000028310-BRD9/pathology/colorectal+cancer#img, accessed on 13 June 2022).

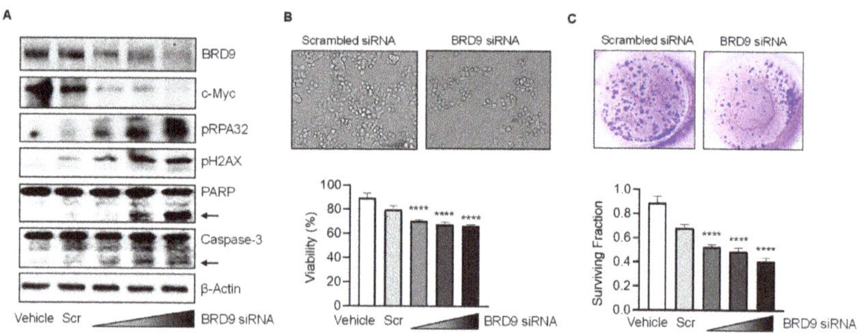

Figure 2. BRD9 depletion alters cell viability, DNA damage and apoptosis in human colon cancer cells. (**A**) Immunoblotting of SW620 cells 72 h after vehicle, scrambled (Scr) or BRD9 siRNA treatment. Knockdown experiments employed a pool of two different hairpin loop siRNAs; wedge symbol = 18.7, 37.5, 75 pmol siRNA transfected. β-Actin, loading control; arrows, positions of cleaved PARP and Caspase-3. (**B**) Representative morphology of SW620 cells and quantification of cell viability in the CCK-8 assay at 48 h. (**C**) Representative colony formation images and quantification of surviving fraction at 48 h. Data = mean ± SD, $n = 3$, ****$p < 0.0001$, by Student's t-test vs. vehicle control.

3.3. BRD9 Interacts with Natural Polyphenols

Extending prior molecular docking in silico [13,14,16], interactions of BRD9 were examined with various natural polyphenols (Figure 3A,B). Test agents bound to the acetyl lysine binding site with favorable orientations and binding energies in the range −8.1 to −9.9 kcal/mol, comparable to the synthetic chemical probe iBRD9 [29]. However, whereas 10 μM iBRD9 exhibited 99% inhibition in BROMOscan® screens, agents such as Orientin and Aspalathin at the same concentration had 45% and 22% inhibitory activity, respectively (Figure 3B, right column). Additional positive controls corroborated the selectivity and specificity of BROMOscan® screens, including JQ1 towards BRD4 and BI7273 for BRD7 (Supplementary Table S1). In ITC experiments, an apparent Kd value of 20 μM supported moderate reversible interactions with purified BRD9 protein, and predicted 1:1 or 3:1 molar ratios for Aspalathin and Orientin, respectively (Figure 3C).

A

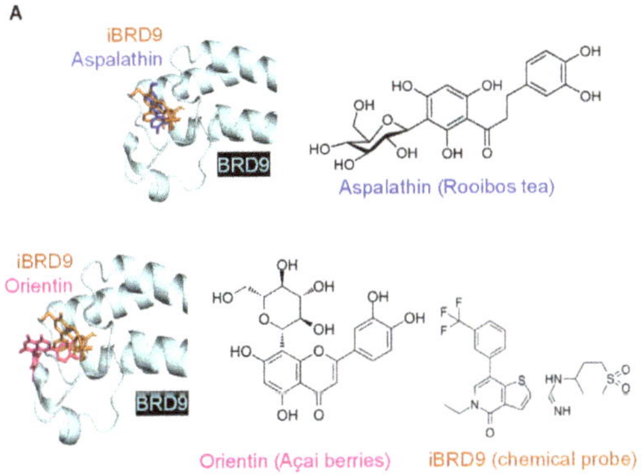

B

Compound	Binding affinity (kcal/mol) with BRD9 *in silico*	Inhibition (%) of BRD9 in Bromoscan™
Orientin	−9.1	45
EGCG	−8.5	49
Equol	−8.1	40
Hyperoside	−9.9	50
Quercetin	−8.8	41
Isoorientin	−9.2	46
Aspalathin	−9.2	22
iBRD9	−9.3	99

C

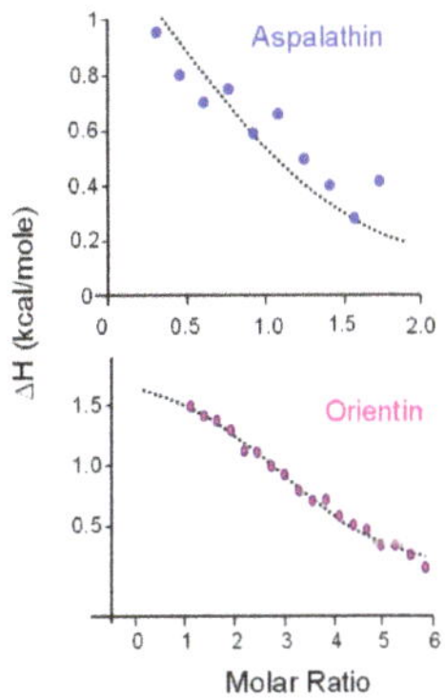

Figure 3. Interaction of natural polyphenols with BRD9. (**A**) Chemical structures of test agents and representative docking orientations in silico. (**B**) Binding affinities in silico and percent inhibition of BRD9 in the Bromoscan™ assay, with iBRD9 as a blinded positive control and 10 μM final concentration of each test agent. EGCG (Epigallocatechin-3-gallate), iBRD9 (Inhibitior of Bromodomain-containing protein 9) (**C**) Isothermal titration calorimetry (ITC) data for purified BRD9 protein with Aspalathin (Kd 20 μM, molar ratio N = 1) and Orientin (Kd 20 μM, molar ratio N = 3).

3.4. BRD9 Inhibition Reduces Colon Cancer Cell Viability and Increases DNA Damage and Apoptosis

In metastasis lineage SW620 colon cancer cells, iBRD9 and predicted BRD9-inhibitory natural polyphenols exhibited concentration-dependent inhibition of cell viability and colony formation (Figure 4A–C), with a significant increase in the percentage of cells undergoing apoptosis (Figure 4D,E). The non-metastatic cell line SW480 obtained from the same patient as SW620 cells responded similarly in cell viability and colony formation assays, whereas CCD841 non-transformed colonic epithelial cells were resistant. For example, IC_{50} values for Aspalathin were in the range 20–31 μM in SW620 and SW480 cells vs. >120 μM in CCD841 cells (Supplementary Table S2). These findings implied the preferential targeting of colon cancer vs. normal colonic epithelial cells by the test agents. Immunoblotting of whole cell lysates revealed that with reduced BRD9 expression, test agents produced a marked loss of oncogenic c-Myc protein, induction of pRPA32 and pH2AX DNA damage/repair markers (Figure 4F, Supplementary Figure S1), and increased cleaved PARP and caspase-3 indicative of apoptosis (Figure 4F). The latter experiments implicated quercetin as the most effective polyphenol for apoptosis induction, although this was not the case in FACS-based assays (Figure 4E). This apparent discrepancy can be due to differences between the appearance of a cell surface apoptosis marker versus the cleavage of a nuclear enzyme by caspases during apoptosis, that might be resolved by additional time-course studies.

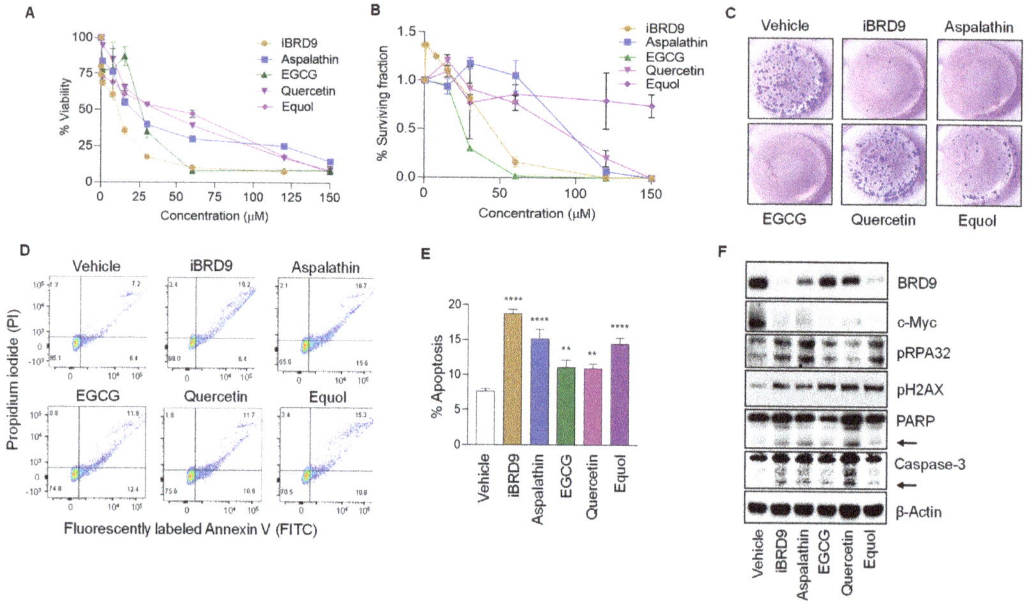

Figure 4. Polyphenols recapitulate BRD9 inhibitor phenotypes in colon cancer cells. (**A**) SW620 cell viability data from CCK-8 assay at 72 h. (**B**) Colony formation data showing surviving fraction (%). (**C**) Representative images of crystal violet-stained colonies after treatment with test compounds. (**D**) Apoptosis as detected by flow cytometry after treatment with 30 μM test compounds and (**E**) quantification of three biological replicates. Data = mean ± SD, $n = 3$. ** $p < 0.01$, **** $p < 0.0001$ by Student's t-test vs. vehicle. (**F**) Immunoblotting of SW620 cell lysates after treatment with 30 μM of compounds. β-Actin, loading control; arrows, cleaved PARP and Caspase-3.

4. Discussion

BRD-containing proteins serve as scaffolds, transcription factors, and chromatin remodeling agents with critical roles in physiology and pathophysiology [2]. The interaction of BRD9 with SWI/SNF complexes represents an area of much current interest for cancer treatment [30–32]. BRD9 is overexpressed in many malignancies, including CRC (Figure 1) and Gastrointestinal Stromal Tumors (GISTs) [33]. We observed that siRNA-mediated loss of BRD9 expression in human colon cancer cells produced marked changes in phenotypic and molecular readouts (Figure 2), consistent with the overarching goal of targeting oncogenic functions arising from BRD9 overexpression.

Previous studies demonstrated that several residues in the ZA and ZB loops of the BRD9 bromodomain, such as Asp144, Ile53, Lys91, Thr104, Pro82, Asn140, Asn100, and Phe44, are important for optimal BRD9 inhibition [34]. Based on this information, natural compounds were screened via molecular docking in silico, which identified favorable interactions for natural polyphenols such as Epigallocatechin-3-gallate (EGCG), Equol, quercetin, and Aspalathin with the bromodomain of BRD9 (Figure 3A,B), and corroborated via BROMOscan® and ITC experiments. Notably, an apparent Kd of 20 μM (Figure 3C) and IC_{50} values of 20–30 μM (Supplementary Table S2) fall within the physiological range for polyphenols and their bioactive metabolites [35,36].

Natural compounds with BRD9-targeting capability (Figure 3) had enhanced DNA damage/repair and apoptosis endpoints in cell-based assays, comparable to iBRD9 (Figure 4). Natural polyphenols lowered BRD9 protein expression in the relative order Equol > Aspalathin > Quercetin > EGCG (Figure 4F). The imperfect association between loss of BRD9 and c-

Myc expression implicated additional factors in downregulating the oncoprotein in colon cancer cells. Multiple BRD9-binding sites reside within *MYC* enhancer elements [33,37], with various multiprotein complexes differentially regulating apoptosis outcomes in CRC and other malignancies [33,38–40]. Similarly, the lack of a good association between BRD9 expression and the phenotypic markers (Figure 4) can be attributed to additional effects of polyphenols on other molecular targets and signaling pathways that regulate antioxidant, anti-inflammatory, anti-proliferative, and cell cycle inhibitory actions [41]. Nonetheless, BRD9 expression and pRPA32 were interrelated (Figure 4F), and up to a 6-fold increase in pH2AX and pRPA32 expression in colon cancer cells treated with iBRD9 or natural polyphenols (Supplementary Figure S1) suggests that DNA damage/repair pathways are worthy of further investigation. This would include BRD9 knockdown and recovery experiments to discern the precise mechanisms.

Polyphenols occur in many foods and beverages, including green tea, curry spices, grapes, soy, and berries. They hold considerable promise for cancer treatment [42], including the bioactive metabolites [43], due to the targeting of epigenetic mechanisms, DNA damage responses and oncogenic signaling networks in cancer vs. normal cells [11]. Dietary polyphenols such as green tea-derived EGCG are known to activate cell cycle arrest and apoptosis in cancer cells [42,44]. Aspalathin from Rooibos tea also exerted anti-proliferative and proapoptotic effects in liver and colon cancer cells [45,46]. Quercetin from citrus fruits exhibited anticancer activity in colon cancer models [47] and enhanced the actions of BET inhibitors [48]. Equol is a microbial metabolite derived from the soy isoflavone daidzein, with anticancer activity in the breast and prostate [49]. Repurposing of natural agents and their metabolites as lead compounds for new BRD9-targeting antagonists would provide an avenue for future intervention in the clinical setting, and a strong rationale to continue investigation into dietary strategies for CRC prevention.

5. Conclusions

BRD9 overexpression in CRC is synonymous with oncogenic activity and provides an avenue for precision medicine. In human colon cancer cells, knockdown of BRD9, pharmacologic inhibition with iBRD9, and targeting of BRD9 overexpression via natural polyphenols resulted in marked changes in phenotype and molecular readouts, whereas normal colonic epithelial cells were resistant. Future studies will repurpose other nutrients as lead compounds for improved BRD9 inhibition and antitumor activity.

Supplementary Materials: The following supporting information can be downloaded at: https://www.mdpi.com/article/10.3390/nu14204317/s1. Table S1: Bromodomain inhibition. Bromoscan™ screening at 10 µM of test agent; – not determined; Table S2: Concentration for 50% inhibition (IC50) values in colon cancer (SW620, SW480) and normal colonic epithelial cells (CCD841) after 72 h of treatment, based on cell viability and colony formation assays as in Figure 4A,B; Figure S1: Densitometry data for (A) pH2AX and (B) pRPA32 and (C) calculated fold change compared to vehicle, from the immunoblotting of SW620 cell lysates shown in Fig 4F.

Author Contributions: Conceptualization, S.K., E.D.,W.M.D., P.R. and R.H.D.; Data curation, S.K., E.D., S.W., R.D. and P.R.; Formal analysis, S.K., E.D., S.W., R.D., C.T., J.E.T.P. and P.R.; Investigation, S.K., P.R. and R.H.D.; Methodology, S.K., E.D., S.W., R.D., C.T. and P.R. Project administration, W.M.D., P.R. and R.H.D.; Resources, W.M.D., P.R. and R.H.D.; Supervision, P.R. and R.H.D.; Validation, S.K., E.D. and P.R.; Visualization, P.R. and R.H.D.; Writing—review and editing, S.K., P.R. and R.H.D. All authors have read and agreed to the published version of the manuscript.

Funding: Research supported by grants CA090890 and CA122959 from the NCI, by the John S. Dunn Foundation, and by a Chancellor's Research Initiative from Texas A&M University.

Institutional Review Board Statement: Not applicable.

Informed Consent Statement: Not applicable.

Data Availability Statement: See text for online sources and accessibility.

Acknowledgments: The authors thank the members of Dashwood laboratory, High-throughput Screening Core, and Antibody & Biopharmaceuticals Core (ABC) for technical assistance.

Conflicts of Interest: The authors declare no conflict of interest.

References

1. Jung, G.; Hernández-Illán, E.; Moreira, L.; Balaguer, F.; Goel, A. Epigenetics of colorectal cancer: Biomarker and therapeutic potential. *Nat. Rev. Gastroenterol. Hepatol.* **2020**, *17*, 111–130. [CrossRef]
2. Fujisawa, T.; Filippakopoulos, P. Functions of bromodomain-containing proteins and their roles in homeostasis and cancer. *Nat. Rev. Mol. Cell Biol.* **2017**, *18*, 246–262. [CrossRef] [PubMed]
3. Damiani, E.; Duran, M.N.; Mohan, N.; Rajendran, P.; Dashwood, R.H. Targeting epigenetic 'Readers' with natural compounds for cancer interception. *J. Cancer Prev.* **2020**, *25*, 189. [CrossRef]
4. Filippakopoulos, P.; Picaud, S.; Mangos, M.; Keates, T.; Lambert, J.P.; Barsyte-Lovejoy, D.; Felletar, I.; Volkmer, R.; Müller, S.; Pawson, T.; et al. Histone recognition and large-scale structural analysis of the human bromodomain family. *Cell* **2012**, *149*, 214–231. [CrossRef] [PubMed]
5. Zou, Z.; Huang, B.; Wu, X.; Zhang, H.; Qi, J.; Bradner, J.; Nair, S.; Chen, L.F. Brd4 maintains constitutively active NF-κB in cancer cells by binding to acetylated RelA. *Oncogene* **2014**, *33*, 2395–2404. [CrossRef]
6. Sanchez, R.; Zhou, M.M. The role of human bromodomains in chromatin biology and gene transcription. *Curr. Opin. Drug Discov. Devel.* **2009**, *12*, 659–665. [PubMed]
7. Wilson, B.G.; Roberts, C.W. SWI/SNF nucleosome remodellers and cancer. *Nat. Rev. Cancer* **2011**, *11*, 481–492. [CrossRef] [PubMed]
8. Zhu, X.; Liao, Y.; Tang, L. Targeting BRD9 for Cancer Treatment: A New Strategy. *Onco Targets Ther.* **2020**, *13*, 13191–13200. [CrossRef]
9. Campos, C.; Fragoso, S.; Luís, R.; Pinto, F.; Brito, C.; Esteves, S.; Pataco, M.; Santos, S.; Machado, P.; Vicente, J.B.; et al. High-throughput sequencing identifies 3 novel susceptibility genes for hereditary melanoma. *Genes* **2020**, *11*, 403. [CrossRef]
10. Xu, Y.; Vakoc, C.R. Targeting Cancer Cells with BET Bromodomain Inhibitors. *Cold Spring Harb. Perspect. Med.* **2017**, *7*, a026674. [CrossRef] [PubMed]
11. Rajendran, P.; Ho, E.; Williams, D.E.; Dashwood, R.H. Dietary phytochemicals, HDAC inhibition, and DNA damage/repair defects in cancer cells. *Clin. Epigenetics* **2011**, *3*, 4. [CrossRef] [PubMed]
12. Yang, A.Y.; Kim, H.; Li, W.; Kong, A.-N.T. Natural compound-derived epigenetic regulators targeting epigenetic readers, writers and erasers. *Curr. Top. Med. Chem.* **2016**, *16*, 697–713. [CrossRef]
13. Chen, Y.S.; Li, J.; Neja, S.; Kapoor, S.; Tovar Perez, J.E.; Tripathi, C.; Menon, R.; Jayaraman, A.; Lee, K.; Dashwood, W.M.; et al. Metabolomics of Acute vs. Chronic Spinach Intake in an Apc–Mutant Genetic Background: Linoleate and Butanoate Metabolites Targeting HDAC Activity and IFN-γ Signaling. *Cells* **2022**, *11*, 573. [CrossRef]
14. Rajendran, P.; Johnson, G.; Li, L.; Chen, Y.S.; Dashwood, M.; Nguyen, N.; Ulusan, A.; Ertem, F.; Zhang, M.; Li, J.; et al. Acetylation of CCAR2 Establishes a BET/BRD9 Acetyl Switch in Response to Combined Deacetylase and Bromodomain Inhibition. *Cancer Res.* **2019**, *79*, 918–927. [CrossRef] [PubMed]
15. Rajendran, P.; Delage, B.; Dashwood, W.M.; Yu, T.W.; Wuth, B.; Williams, D.E.; Ho, E.; Dashwood, R.H. Histone deacetylase turnover and recovery in sulforaphane treated colon cancer cells: Competing actions of 14-3-3 and Pin1 in HDAC3/SMRT corepressor complex dissociation/reassembly. *Mol. Cancer* **2011**, *10*, 68. [CrossRef] [PubMed]
16. Rajendran, P.; Kidane, A.I.; Yu, T.W.; Dashwood, W.M.; Bisson, W.H.; Löhr, C.V.; Ho, E.; Williams, D.E.; Dashwood, R.H. HDAC turnover, CtIP acetylation and dysregulated DNA damage signaling in colon cancer cells treated with sulforaphane and related dietary isothiocyanates. *Epigenetics* **2013**, *8*, 612–623. [CrossRef] [PubMed]
17. Gille, C.; Fähling, M.; Weyand, B.; Wieland, T.; Gille, A. Alignment-Annotator web server: Rendering and annotating sequence alignments. *Nucleic Acids Res.* **2014**, *42*, W3–W6. [CrossRef] [PubMed]
18. Trott, O.; Olson, A.J. AutoDock Vina: Improving the speed and accuracy of docking with a new scoring function, efficient optimization, and multithreading. *J. Comput. Chem.* **2010**, *31*, 455–461. [CrossRef]
19. Biasini, M.; Bienert, S.; Waterhouse, A.; Arnold, K.; Studer, G.; Schmidt, T.; Kiefer, F.; Cassarino, T.G.; Bertoni, M.; Bordoli, L.; et al. SWISS-MODEL: Modelling protein tertiary and quaternary structure using evolutionary information. *Nucleic Acids Res.* **2014**, *42*, W252–W258. [CrossRef]
20. Paxman, J.J.; Heras, B. Bioinformatics tools and resources for analyzing protein structures. In *Proteome Bioinformatics*; Humana Press: New York, NY, USA, 2017; pp. 209–220.
21. Krissinel, E.; Henrick, K. Inference of macromolecular assemblies from crystalline state. *J. Mol. Biol.* **2007**, *372*, 774–797. [CrossRef]
22. Sobolev, V.; Sorokine, A.; Prilusky, J.; Abola, E.E.; Edelman, M. Automated analysis of interatomic contacts in proteins. *Bioinformatics* **1999**, *15*, 327–332. [CrossRef]
23. Jacoby, E.; Tresadern, G.; Bembenek, S.; Wroblowski, B.; Buyck, C.; Neefs, J.M.; Rassokhin, D.; Poncelet, A.; Hunt, J.; van Vlijmen, H. Extending kinome coverage by analysis of kinase inhibitor broad profiling data. *Drug Discov. Today* **2015**, *20*, 652–658. [CrossRef] [PubMed]
24. Wiseman, T.; Williston, S.; Brandts, J.F.; Lin, L.N. Rapid measurement of binding constants and heats of binding using a new titration calorimeter. *Anal. Biochem.* **1989**, *179*, 131–137. [CrossRef]

25. Kapoor, S.; Gustafson, T.; Zhang, M.; Chen, Y.S.; Li, J.; Nguyen, N.; Perez, J.E.T.; Dashwood, W.M.; Rajendran, P.; Dashwood, R.H. Deacetylase Plus Bromodomain Inhibition Downregulates ERCC2 and Suppresses the Growth of Metastatic Colon Cancer Cells. *Cancers* **2021**, *13*, 1438. [CrossRef] [PubMed]
26. Katz, D.; Ito, E.; Lau, K.S.; Mocanu, J.D.; Bastianutto, C.; Schimmer, A.D.; Liu, F.F. Increased efficiency for performing colony formation assays in 96-well plates: Novel applications to combination therapies and high-throughput screening. *Biotechniques* **2008**, *44*, ix–xiv. [CrossRef] [PubMed]
27. Nian, H.; Bisson, W.H.; Dashwood, W.M.; Pinto, J.T.; Dashwood, R.H. Alpha-keto acid metabolites of organoselenium compounds inhibit histone deacetylase activity in human colon cancer cells. *Carcinogenesis* **2009**, *30*, 1416–1423. [CrossRef]
28. Okonkwo, A.; Mitra, J.; Johnson, G.S.; Li, L.; Dashwood, W.M.; Hegde, M.L.; Yue, C.; Dashwood, R.H.; Rajendran, P. Heterocyclic Analogs of Sulforaphane Trigger DNA Damage and Impede DNA Repair in Colon Cancer Cells: Interplay of HATs and HDACs. *Mol. Nutr. Food Res.* **2018**, *62*, e1800228. [CrossRef]
29. Theodoulou, N.H.; Bamborough, P.; Bannister, A.J.; Becher, I.; Bit, R.A.; Che, K.H.; Chung, C.W.; Dittmann, A.; Drewes, G.; Drewry, D.H.; et al. Discovery of I-BRD9, a Selective Cell Active Chemical Probe for Bromodomain Containing Protein 9 Inhibition. *J. Med. Chem.* **2016**, *59*, 1425–1439. [CrossRef]
30. Chung, C.W.; Dean, A.W.; Woolven, J.M.; Bamborough, P. Fragment-based discovery of bromodomain inhibitors part 1: Inhibitor binding modes and implications for lead discovery. *J. Med. Chem.* **2012**, *55*, 576–586. [CrossRef]
31. Prieto-Martínez, F.D.; Medina-Franco, J.L. Flavonoids as putative epi-modulators: Insight into their binding mode with BRD4 bromodomains using molecular docking and dynamics. *Biomolecules* **2018**, *8*, 61. [CrossRef]
32. Dutra, L.A.; Heidenreich, D.; Silva, G.D.; Chin, C.M.; Knapp, S.; Santos, J.L. Dietary compound resveratrol is a pan-BET bromodomain inhibitor. *Nutrients* **2017**, *9*, 1172. [CrossRef] [PubMed]
33. Mu, J.; Sun, X.; Zhao, Z.; Sun, H.; Sun, P. BRD9 inhibition promotes PUMA-dependent apoptosis and augments the effect of imatinib in gastrointestinal stromal tumors. *Cell Death Dis.* **2021**, *12*, 962. [CrossRef] [PubMed]
34. Su, J.; Liu, X.; Zhang, S.; Yan, F.; Zhang, Q.; Chen, J. Insight into selective mechanism of class of I-BRD9 inhibitors toward BRD9 based on molecular dynamics simulations. *Chem. Biol. Drug Des.* **2019**, *93*, 163–176. [CrossRef]
35. Kay, C.D.; Pereira-Caro, G.; Ludwig, I.A.; Clifford, M.N.; Crozier, A. Anthocyanins and flavanones are more bioavailable than previously perceived: A review of recent evidence. *Annu. Rev. Food Sci. Technol.* **2017**, *8*, 155–180. [CrossRef] [PubMed]
36. Luca, S.V.; Macovei, I.; Bujor, A.; Miron, A.; Skalicka-Woźniak, K.; Aprotosoaie, A.C.; Trifan, A. Bioactivity of dietary polyphenols: The role of metabolites. *Crit. Rev. Food Sci. Nutr.* **2020**, *60*, 626–659. [CrossRef] [PubMed]
37. Hohmann, A.F.; Martin, L.J.; Minder, J.L.; Roe, J.S.; Shi, J.; Steurer, S.; Bader, G.; McConnell, D.; Pearson, M.; Gerstberger, T.; et al. Sensitivity and engineered resistance of myeloid leukemia cells to BRD9 inhibition. *Nat. Chem. Biol.* **2016**, *12*, 672–679. [CrossRef] [PubMed]
38. Zhao, R.; Liu, Y.; Wu, C.; Li, M.; Wei, Y.; Niu, W.; Yang, J.; Fan, S.; Xie, Y.; Li, H.; et al. BRD7 Promotes Cell Proliferation and Tumor Growth Through Stabilization of c-Myc in Colorectal Cancer. *Front. Cell Dev. Biol.* **2021**, *9*, 955. [CrossRef]
39. Bell, C.M.; Raffeiner, P.; Hart, J.R.; Vogt, P.K. PIK3CA cooperates with KRAS to promote MYC activity and tumorigenesis via the bromodomain protein BRD9. *Cancers* **2019**, *11*, 1634. [CrossRef] [PubMed]
40. Greco, C.; Alvino, S.; Buglioni, S.; Assisi, D.; Lapenta, R.; Grassi, A.; Stigliano, V.; Mottolese, M.; Casale, V. Activation of c-MYC and c-MYB proto-oncogenes is associated with decreased apoptosis in tumor colon progression. *Anticancer Res.* **2001**, *21*, 3185–3192.
41. Tresserra-Rimbau, A.; Lamuela-Raventos, R.M.; Moreno, J.J. Polyphenols, food and pharma. Current knowledge and directions for future research. *Biochem. Pharmacol.* **2018**, *156*, 186–195. [CrossRef]
42. Kim, E.; Bisson, W.H.; Löhr, C.V.; Williams, D.E.; Ho, E.; Dashwood, R.H.; Rajendran, P. Histone and Non-Histone Targets of Dietary Deacetylase Inhibitors. *Curr. Top. Med. Chem.* **2016**, *16*, 714–731. [CrossRef] [PubMed]
43. Rajendran, P.; Williams, D.E.; Ho, E.; Dashwood, R.H. Metabolism as a key to histone deacetylase inhibition. *Crit. Rev. Biochem. Mol. Biol.* **2011**, *46*, 181–199. [CrossRef] [PubMed]
44. Zhou, Y.; Zheng, J.; Li, Y.; Xu, D.P.; Li, S.; Chen, Y.M.; Li, H.B. Natural Polyphenols for Prevention and Treatment of Cancer. *Nutrients* **2016**, *8*, 515. [CrossRef] [PubMed]
45. Samodien, S.; Kock, M.; Joubert, E.; Swanevelder, S.; Gelderblom, W.C.A. Differential Cytotoxicity of Rooibos and Green Tea Extracts against Primary Rat Hepatocytes and Human Liver and Colon Cancer Cells—Causal Role of Major Flavonoids. *Nutr. Cancer* **2021**, *73*, 2050–2064. [CrossRef]
46. Wang, B.J.; Huang, S.H.; Kao, C.L.; Muller, C.J.F.; Wang, Y.P.; Chang, K.H.; Wen, H.C.; Yeh, C.C.; Shih, L.J.; Kao, Y.H.; et al. Aspalathus linearis suppresses cell survival and proliferation of enzalutamide-resistant prostate cancer cells via inhibition of c-Myc and stability of androgen receptor. *PLoS ONE* **2022**, *17*, e0270803. [CrossRef] [PubMed]
47. Esmeeta, A.; Adhikary, S.; Dharshnaa, V.; Swarnamughi, P.; Ummul Maqsummiya, Z.; Banerjee, A.; Pathak, S.; Duttaroy, A.K. Plant-derived bioactive compounds in colon cancer treatment: An updated review. *Biomed. Pharmacother.* **2022**, *153*, 113384. [CrossRef]
48. Pham, T.N.D.; Stempel, S.; Shields, M.A.; Spaulding, C.; Kumar, K.; Bentrem, D.J.; Matsangou, M.; Munshi, H.G. Quercetin Enhances the Anti-Tumor Effects of BET Inhibitors by Suppressing hnRNPA1. *Int. J. Mol. Sci.* **2019**, *20*, 4293. [CrossRef] [PubMed]
49. Mayo, B.; Vázquez, L.; Flórez, A.B. Equol: A Bacterial Metabolite from The Daidzein Isoflavone and Its Presumed Beneficial Health Effects. *Nutrients* **2019**, *11*, 2231. [CrossRef]

nutrients

MDPI

Article

Alterations in Intestinal Brush Border Membrane Functionality and Bacterial Populations Following Intra-Amniotic Administration (*Gallus gallus*) of Catechin and Its Derivatives

Nikolai Kolba, Amin Zarei, Jacquelyn Cheng, Nikita Agarwal, Younas Dadmohammadi, Leila Khazdooz, Alireza Abbaspourrad * and Elad Tako *

Department of Food Science, Cornell University, Ithaca, NY 14853, USA
* Correspondence: alireza@cornell.edu (A.A.); et79@cornell.edu (E.T.); Tel.: +1-607-255-2923 (A.A.); +1-607-255-0884 (E.T.)

Abstract: Catechin is a flavonoid naturally present in numerous dietary products and fruits (e.g., apples, berries, grape seeds, kiwis, green tea, red wine, etc.) and has previously been shown to be an antioxidant and beneficial for the gut microbiome. To further enhance the health benefits, bioavailability, and stability of catechin, we synthesized and characterized catechin pentaacetate and catechin pentabutanoate as two new ester derivatives of catechin. Catechin and its derivatives were assessed in vivo via intra-amniotic administration (*Gallus gallus*), with the following treatment groups: (1) non-injected (control); (2) deionized H_2O (control); (3) Tween (0.004 mg/mL dose); (4) inulin (50 mg/mL dose); (5) Catechin (6.2 mg/mL dose); (6) Catechin pentaacetate (10 mg/mL dose); and (7) Catechin pentabutanoate (12.8 mg/mL dose). The effects on physiological markers associated with brush border membrane morphology, intestinal bacterial populations, and duodenal gene expression of key proteins were investigated. Compared to the controls, our results demonstrated a significant ($p < 0.05$) decrease in *Clostridium* genera and *E. coli* species density with catechin and its synthetic derivative exposure. Furthermore, catechin and its derivatives decreased iron and zinc transporter (Ferroportin and ZnT1, respectively) gene expression in the duodenum compared to the controls. In conclusion, catechin and its synthetic derivatives have the potential to improve intestinal morphology and functionality and positively modulate the microbiome.

Keywords: intra-amniotic administration; brush border membrane; catechin derivatives; microbiome

Citation: Kolba, N.; Zarei, A.; Cheng, J.; Agarwal, N.; Dadmohammadi, Y.; Khazdooz, L.; Abbaspourrad, A.; Tako, E. Alterations in Intestinal Brush Border Membrane Functionality and Bacterial Populations Following Intra-Amniotic Administration (*Gallus gallus*) of Catechin and Its Derivatives. *Nutrients* **2022**, *14*, 3924. https://doi.org/10.3390/nu14193924

Academic Editor: Marloes Dekker Nitert

Received: 30 August 2022
Accepted: 20 September 2022
Published: 22 September 2022

1. Introduction

Catechins are potent polyphenolic compounds naturally present in plant-based foods such as vegetables, beverages, green tea, and fruits such as cherries, grapes, apples, and pears [1]. Like other flavonoids, catechins possess a multitude of biological and pharmaceutical properties that can improve human health and prevent various diseases. For example, they can inhibit the activities of various reactive oxygen species (ROS) by preventing damage to proteins, DNA, and cell membranes [2,3]. Compared to antioxidants such as ascorbic acid, α-tocopherol, curcumin, and butylated hydroxyanisole, catechins are more effective at scavenging free radicals [1,4]. The antioxidant capacity of catechins has previously been shown to be powerful inhibitors against various cancers and tumors in humans and animals [4–7]. Due to high antioxidant activity, these compounds have been reported to inhibit the oxidation of low-density lipoproteins (LDL), preventing plaque formation and decreasing cardiovascular disease incidence [8–10]. Catechins have also been shown to possess other beneficial and protective properties, such as anti-inflammatory, anti-HIV, anti-aging, anti-allergic, anti-diabetic, and anti-bacterial activities [11–18]. Moreover, these compounds exhibit noteworthy neuroprotection against neurodegenerative diseases such as Alzheimer's disease, Parkinson's disease, and ischemic damage [19,20].

Recent studies have demonstrated that polyphenolic compounds can be used as prebiotics to improve the structure and function of the gut microbiota [21,22]. Dietary polyphenolic compounds can be transformed via intestinal microbiota fermentation into short-chain fatty acids (SCFAs) [23–26]. SCFAs are a source of energy for colonic cells, where increased SCFA production has been associated with increased populations of beneficial bacteria, acidification of the intestinal lumen, and, consequently, the inhibition of harmful pathogen growth [27]. Acetate and butyrate are two key SCFAs utilized as a source of energy for epithelial cells to improve intestinal barrier integrity and prevent the translocation of antigens and pathogens into circulation [28]. Moreover, increased SCFA production has been associated with increases in villus surface area, positive modulations in gut microbiota composition, and enhanced functionality of the duodenal brush border membrane (BBM) [29].

While catechin consumption has been found to possess a myriad of beneficial physiological effects, low stability (i.e., pH, thermal, storage) and permeability of catechins in body membranes limit their bioavailability [1]. Further, catechins can be significantly degraded in the key site of human digestion, the small intestine, where ROS, digestive secretions, and changes in pH exacerbate catechin autoxidation and degradation [30]. Thus, microencapsulation and chemical modification can improve catechin stability and bioavailability during digestion [1,31,32]. In this vein, a purposeful chemical modification of natural polyphenolic compounds, such as adding short-chain ester groups, can increase compound stability. Moreover, probiotic bacterial populations can hydrolyze and metabolize these short-chain ester groups to produce beneficial SCFAs.

In this work, we reported the modification of catechin by converting it to catechin pentabutanoate (a new molecule that has not been previously reported) and catechin pentaacetate as two types of catechin short-chain ester derivatives (Scheme 1). Further, to characterize the physiological effects of catechin and its short-chain ester derivatives on brush border membrane (BBM) functionality, intestinal morphology, and intestinal microbial populations, we utilized the in vivo *Gallus gallus* model. The *Gallus gallus* has been used as a novel and cost-effective animal model to elucidate the physiological effects of plant bioactives and nutritional solutions relevant to human nutrition [33–40]. Recently, the positive impact of catechins, as food additives, on poultry growth performance and egg quality has been studied and found to improve poultry antioxidant status [41,42]. Our present study utilizes the intra-amniotic administration approach (*in ovo* feeding) in the *Gallus gallus*, where the amniotic fluid, which is naturally and orally consumed by the embryo starting at day 17, is entirely consumed by hatch, which allows for testing the effects of the solution administered into the amniotic fluid on the different systems of interest [29,39,43–45]. In the current study, the impact of intra-amniotic administration of catechin and its derivatives at a dosage range of 6.2–12.8 mg/mL on BBM functionality was assessed by evaluating duodenal gene expression of biomarkers of mineral status, BBM digestive and absorptive ability, immune function, and inflammation in vivo in the *Gallus gallus*. A secondary objective was to evaluate the effects of the intra-amniotic administration of catechin and its derivatives on cecal bacterial populations by quantifying the relative abundances of health-promoting populations (*Bifidobacterium* spp. and *Lactobacillus* spp.) versus those of potentially pathogenic bacteria (*E. coli* and *Clostridium* spp.). We hypothesize that catechin and its more stable derivatives, when administered intra-amniotically, will cause favorable alterations in BBM functionality and development and positively modulate the gut microbial populations, where increased effects will be seen with catechin derivatives due to increased stability within the gastrointestinal tract as was demonstrated in our previous work [46].

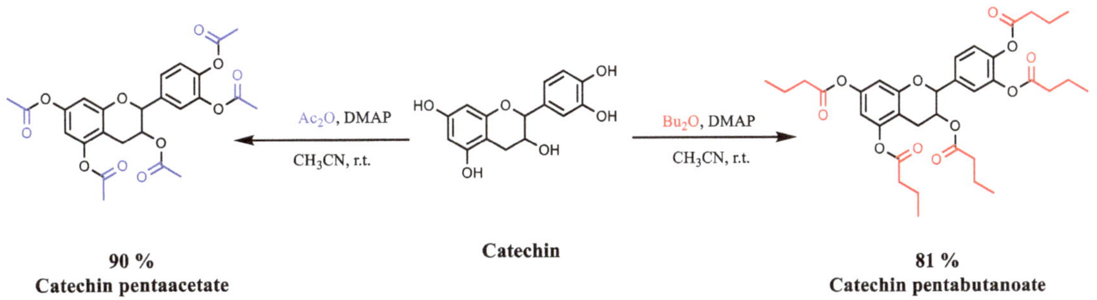

Scheme 1. Synthesis of catechin pentaacetate and catechin pentabutanoate. The reported percentages are the reaction yield.

2. Materials and Methods

2.1. Animals

Cornish cross-fertile broiler chicken eggs (n = 41) were obtained from a commercial hatchery (Moyer's chicks, Quakertown, PA, USA). The eggs were incubated at the Cornell University Animal Science poultry farm incubator under optimal conditions. All animal protocols were approved by Cornell University Institutional Animal Care and Use Committee (IACUC #2020-0077).

2.2. Materials

(+)-Catechin and butyric anhydride with 98% purity and 4-dimethylamino pyridine with 99% purity were purchased from Sigma Aldrich (St. Louis, MO, USA). Acetic anhydride with 99% purity was purchased from Acros (Morris Plains, NJ, USA). Silica gel (P60, 40–63 μm, 60 Å) was purchased from SiliCycle (Quebec, QC, Canada), and Silica Gel 60 F254 Coated Aluminum-Backed TLC (thin layer chromatography) sheets were purchased from EMD Millipore (Billerica, MA, USA).

2.3. General Procedure for the Synthesis of Catechin Pentabutanoate

To a round bottom flask with a magnetic stir bar, septa, and nitrogen inlet, 200 mg (0.65 mmol) of catechin.H_2O, 2.1 mL (12.6 mmol) of butyric anhydride, 35 mg (0.28 mmol) of 4-dimethylamino pyridine, and 5 mL of acetonitrile were added. The reaction mixture was stirred at room temperature for 24 h under a nitrogen atmosphere. The progress of the reaction was followed by thin-layer chromatography (TLC). After the reaction, the excess amount of butyric anhydride was evaporated, and the crude product was extracted with 30 mL of ethyl acetate. Then, the organic phase was washed with (4 × 20 mL) of HCl (2 M). After that, the organic phase was washed with (3 × 20 mL) of $NaHCO_3$ (0.4 M). Finally, the organic solvent was dried with sodium sulfate and evaporated by a rotary evaporator to obtain the product in 81% yield (337 mg) as a light-yellow viscose liquid (λ_{max} in ethanol was 271 nm).

2.4. General Procedure for the Synthesis of Catechin Pentaacetate

To a round bottom flask with a magnetic stir bar, septa, and nitrogen inlet, 200 mg (0.65 mmol) of catechin.H_2O, 1.2 mL (12.6 mmol) of acetic anhydride, 35 mg (0.28 mmol) of 4-dimethylamino pyridine, and 5 mL of acetonitrile were added. The reaction mixture was stirred at room temperature for 24 h under a nitrogen atmosphere. TLC followed the progress of the reaction. After the reaction, the excess acetic anhydride was evaporated, and the crude product was extracted with 30 mL of ethyl acetate. Then, the organic phase was washed with (4 × 20 mL) of HCl (2 M). After that, the organic phase was washed with (3 × 20 mL) of $NaHCO_3$ (0.4 M). Finally, the organic solvent was dried with sodium sulfate and evaporated by a rotary evaporator to obtain the product in 90% yield (293 mg) as a light-yellow viscose liquid (λ_{max} in ethanol was 270 nm).

2.5. Characterization of Catechin Pentabutanoate and Catechin Pentaacetate

2.5.1. ^{1}H NMR (500 MHz) and ^{13}C NMR

A 500 MHz NMR (Bruker AVANCE) spectrometer was used for ^{1}H NMR (500 MHz) and ^{13}C NMR (125 MHz) spectra in $CDCl_3$. The chemical shifts were expressed in δ (ppm) relative to Tetramethylsilane (TMS) as the internal standard, and coupling constants (J) were measured in Hz. Spin multiplicities were described as singlet (s), doublet (d), triplet (t), quartet (q), sextet (sext), and multiplet (m).

2.5.2. Fourier Transform Infrared Spectroscopy

Fourier transform infrared spectra (ATR-FTIR) were recorded on a Shimadzu IRAffinity-1S spectrophotometer.

2.5.3. Ultraviolet–Visible Spectroscopy (UV-Vis)

UV-vis was recorded on a Shimadzu UV-2600 spectrophotometer, as was previously described [32,46].

2.5.4. Liquid Chromatography-Mass Spectrometry Analysis

For LC-MS analysis, we used an LC (Agilent 1100 series) coupled with a mass spectrometer. All samples were passed through a 13 mm nylon syringe filter with a 0.22 µm pore size before injection to ensure the removal of the solid contaminants. Reverse-phase chromatography was used with a Phenomenex Luna Omega (Phenomenex) LC column with the following specifications: 100 × 4.6 mm, 3 µm, polar C18, 100 Å pore size with a flow rate of 0.3 mL min^{-1}. LC eluents include MiliQ-H_2O containing 0.1 % formic acid (15%) and acetonitrile (85%) using isocratic elution. The time of analysis was 20 min. The mass spectrometer (Finnigan LTQ mass spectrometer) was equipped with an electrospray interface (ESI) set in positive electrospray ionization mode for analysis and with 15 kV collision energy.

2.5.5. Particle Size Measurement

To ensure nanosize particles for intra-amniotic administration, the particle size distribution and mean particle diameter (Zeta average size) of catechin pentabutanoate and catechin pentaacetate in DI H_2O containing 0.4% (w/v) tween 80 were measured using a commercial dynamic light-scattering device (Nano-ZS, Malvern Instruments, Worcestershire, UK).

2.6. Preparation of Catechin Solution (0.02 M)

0.31 g of catechin was dissolved in 50 mL of DI H_2O containing 2% (v/v) EtOH to prepare a 0.02 mol/L catechin solution.

2.7. Dispersion of Catechin Pentabutanoate and Catechin Pentaacetate in DI H_2O

One mmol of catechin pentabutanoate or catechin pentaacetate was dissolved in 2 mL of pure EtOH, and then this phase was added to the 50 mL of DI H_2O containing 0.4% (w/v) tween 80 and homogenized at room temperature at 15,000 rpm. The total concentration of catechin pentaacetate in this emulsion was approximately 0.02 mol/L.

2.8. Intra-Amniotic Administration

The intra-amniotic administration procedure was previously described by Tako et al. [47]. On Day 17 of embryonic incubation, eggs with viable embryos were weighed and divided into seven groups (n = 10) with approximately equal weight distribution. The seven groups were assigned: (1) non-injected (control); (2) deionized H_2O (control); (3) Tween (0.004 mg/mL dose); (4) inulin (50 mg/mL dose); (5) catechin (6.2 mg/mL dose); (6) catechin pentaacetate (10 mg/mL dose); and (7) catechin pentabutanoate (12.8 mg/mL dose). The intra-amniotic injection solution (1 mL per egg) was injected with a 21-gauge needle into the amniotic fluid, identified by candling. Following the injection, the injection sites were sealed with cellophane tape. Eggs were then placed in hatching baskets, with each treatment equally represented at each incubator location.

2.9. Blood and Tissue Collection

As previously described [48], immediately following hatching (Day 21), birds were weighed and euthanized with CO_2 exposure. Blood was collected using micro-hematocrit heparinized capillary tubes (Thermo Fisher Scientific, Waltham, MA, USA). The small intestines, ceca, and livers were quickly removed from the carcasses and placed in separate sterile cryovials (Simport, Beloeil, QC, Canada) for storage. Ceca were weighed before storage. The samples were immediately frozen in liquid nitrogen and stored at −80 °C until analysis [49].

2.10. Blood Hemoglobin Measurements

Blood Hemoglobin (Hb) concentrations were determined spectrophotometrically using the QuantiChromTM Hemoglobin Assay (DIHB-250, BioAssay Systems, Hayward, CA, USA) following the kit manufacturer's instructions.

2.11. Isolation of Total RNA From Chicken Duodenum

Total RNA was extracted from 30 mg of the proximal duodenal tissue using a Qiagen RNeasy Mini Kit (Qiagen Inc., Germantown, MD) according to the manufacturer's protocol. Total RNA was eluted in 50 μL of RNase-free H_2O. All steps were carried out under RNase-free conditions. RNA was quantified with a NanoDrop 2000 (ThermoFisher Scientific, Waltham, MA, USA) at $A_{260/280}$. RNA was stored at −80 °C until use.

2.12. Real-Time Polymerase Chain Reaction

As previously described [34,43,44,46], the primers used in the real-time polymerase chain reactions (RT-PCR) were designed using Real-Time Primer Design Tool software (IDT DNA, Coralvilla, IA, USA) based on 13 gene sequences from the GenBank database. The sequences are shown in Table S1. The amplicon length was limited to 90 to 150 bp, the size of the primers was from 17-to 25-mer, and the GC content was between 41 and 55%. The specificity of the primers was tested by performing a BLAST search against the genomic NCBI database.

cDNA was generated using a C1000 Touch thermocycler (Biorad, Hercules, CA, USA) and a Promega-Improm-II Reverse Transcriptase Kit (Catalog #A1250) 20 μL reverse transcriptase reaction. The reverse transcriptase reaction consisted of 1 μg total RNA template, ten μM random hexamer primers, and two mM of oligo-dT primers. All reactions were performed under the following conditions: 94 °C for 5 min, 60 min at 42 °C, 70 °C for 15 min, and hold at 4 °C. The concentration of cDNA obtained was determined with a NanoDrop 2000 at $A_{260/280}$ with an extinction coefficient of 33 for single-stranded DNA.

RT-PCR was performed with a Bio-RadCFX96 Touch (Hercules, CA, USA). The ten μL RT-PCR mixtures consisted of cDNA (2 μg), 2X BioRad SSO Advanced Universal SYBR Green Supermix (Cat #1725274, Hercules, CA, USA), forward and reversed primers, and nuclease-free H_2O (for the no template control). The no-template control of nuclease-free H_2O was included to exclude DNA contamination in the PCR mix. All reactions were performed in duplicates and under the following conditions: initial denaturing at 95 °C for 30 s, 40 cycles of denaturing at 95 °C for 15 s, various annealing temperatures according to IDT for 30 s and elongating at 60 °C for 30 s. After the cycling process was completed, melting curves were determined from 65.0 °C to 95.0 °C with increments of 0.5 °C for 5 s to ensure the amplification of a single product. RT-PCR efficiency values for the eleven genes were as follows: DcytB, 1.046; DMT1, 0.998; Ferroportin, 1.109; ZIP9, 1.035; ZnT1, 1.09; ZnT7, 1.013; SGLT-1, 0.994; SI, 1.032; AP, 1.015; Muc2, 1.102; NF-κβ1, 1.113; TNF-α, 1.046; IL8, 0.998; 18s rRNA, 0.994. Gene expression levels were obtained from Ct values based on the 'second derivative maximum' computed by the Bio-Rad CFX Maestro Software Version

2.2 (Bio-Rad, Hercules, CA, USA). Gene expression was normalized to the expression of 18S rRNA.

2.13. Collection of Microbial Samples and Intestinal Contents DNA Isolation

The contents of the ceca were placed into a sterile 15 mL tube (Corning, Corning, NY, USA) containing 9 mL of sterile 1X phosphate-buffered saline (PBS) and homogenized by vortexing with glass beads (3 mm diameter) for 3 min. Debris was removed by centrifugation at 700× *g* for 1 min, and the supernatant was collected and centrifuged at 12,000× *g* for 5 min. The pellet was washed twice with 1X PBS and stored at −20 °C until DNA extraction.

The pellet was re-suspended in 50 mM EDTA to extract DNA and treated with 10 mg/mL lysozyme (Sigma Aldrich CO., St. Louis, MO, USA) for 45 min at 37 °C. The bacterial genomic DNA was then isolated using a Wizard Genomic DNA purification kit (Promega Corp., Madison, WI, USA), following the manufacturer's instructions.

2.14. PCR Amplification of Bacterial 16s rDNA

Primers for *Lactobacillus, Bifidobacterium, Clostridium,* and *E. coli* were designed according to previously published data by Zhu et al. [50]. The universal primers, which identify all known strains of bacteria in the intestine, were prepared with the invariant region in the 16S rRNA of bacteria and used as an internal standard to normalize the results. PCR products were separated by electrophoresis on 2% agarose gel, stained with ethidium bromide, and quantified using the Quantity One 1-D analysis software (BioRad, Hercules, CA, USA).

2.15. Glycogen Analysis

Glycogen analysis was obtained from the pectoralis muscle, and liver previously described [51]. Briefly, the frozen pectoral muscle or liver samples were homogenized in 8% perchloric acid. Pectoral muscle samples were centrifuged at 12,000× *g* for 15 min, and liver samples were centrifuged at 4000× *g* for 15 min at 4 °C. The supernatant was removed, and 1.0 mL of petroleum ether was added to each tube. After mixing, the petroleum ether fraction was removed, and samples from the bottom layer were transferred to a 96-well plate containing 300 μL of an iodine color reagent. All samples were read at a wavelength of 450 nm in a microplate spectrophotometer (Epoch, BioTek, VT, USA), and the amount of glycogen was calculated according to a standard curve. The amount of glycogen present in the pectoral sample was determined by multiplying the weight of the tissue by the amount of glycogen per 1 g of wet tissue.

2.16. Morphometric Examination of Duodenal Tissue

Villus epithelium analysis was conducted as previously published [43,44,52,53]. The duodenal samples were soaked in buffered formaldehyde (4% (v/v)), dehydrated, cleared, and embedded in paraffin. Numerous sections were cut with a thickness of 5 μm and put on slides. The sections were deparaffinized in xylene, after which they were rehydrated in a series of graded alcohol. Finally, the slides were stained with Alcian Blue–periodic acid-Schiff and investigated under light microscopy. The variables were assessed (light microscope, EPIX XCAP software, standard version, Olympus, Waltham, MA, USA) for the following: villus length, villus diameter, depth of crypts, goblet cell diameter, crypt goblet cell number, and villus goblet cells type number (acidic, neutral, or mixed). Four segments were examined for each biological sample and five biological samples per treatment group. The goblet cells were enumerated at ten villi/samples, and the means were calculated for statistical analysis.

2.17. Statistical Analysis

Experimental treatments for the intraamniotic administration assay were arranged in a completely randomized design. Initially, data was investigated for normality utilizing the Kolmogorov–Smirnov test. Results were analyzed by one-way multiple analysis of variance (ANOVA). Statistical analyses were carried out using SPSS version 27.0 software. Differences between treatment groups were compared with a post hoc Duncan test, with statistically different results at $p < 0.05$. Results are expressed as mean ± standard error, $n \geq 8$.

3. Results

After chemical modification of catechin-to-catechin pentabutanoate and catechin pentaacetate, the structures of these compounds were characterized by FTIR, ^{1}H NMR, ^{13}C NMR, and LC-MS in the following orders:

3.1. Fourier-Transform Infrared (FTIR) Spectroscopy of Catechin Pentabutanoate

In the structure of catechin pentabutanoate, there are four butanoate chains bonded to aromatic rings. A strong band at 1761 cm^{-1} confirms the carbonyl of these ester groups. Another band at 1739 cm^{-1} demonstrates the presence of the aliphatic ester group in this molecule. FTIR results show that two bands at 2966 and 2875 cm^{-1} are attributed to asymmetric and symmetric stretching vibrations of aliphatic C-H, respectively (Figure S1). The stretching vibrations of C=C bonds in aromatic rings appear at 1622 and 1595, 1506, and 1489 cm^{-1}. The bands at 1458 and 1361 cm^{-1} are the aliphatic C-H bending vibrations of methylene and methyl groups, respectively. The stretching vibrations of C-O bonds appear at 1242 and 1122 cm^{-1} as strong and broadband. Finally, the bands at 916, 837, 800, and 748 cm^{-1} are evidence of the out-of-plane C-H bending vibrations of the aromatic rings (Figure S1).

3.2. ^{1}H NMR of Catechin Pentabutanoate

The ^{1}H NMR (500 MHz) of catechin pentabutanoate was performed in $CDCl_3$ at room temperature (Figure S2). The expanded ^{1}H NMR of this compound showed that the most deshielded proton (H5′) at 7.25 ppm that appeared as a doublet of doublets peak (J_1 = 8.5 Hz, J_2 = 2.5 Hz) was attributed to the aromatic hydrogen located on the aromatic ring at carbon 5′ (Figure S3). The chemical shifts of two aromatic hydrogens (H4′ and H1′) were so close together that they overlapped and appeared as a doublet peak at 7.19 ppm. The other aromatic hydrogens (H8 and H6) were observed in this compound as two doublet peaks (J = 2 Hz) at 6.66 and 6.59 ppm, respectively. A quartet peak (J = 7 Hz) at 5.28 ppm was attributed to H3 located on the aliphatic ring in the structure of catechin pentabutanoate. Another hydrogen on the aliphatic ring is H2 which appeared at 5.14 ppm as a doublet peak (J = 6.5 Hz). In this molecule, there were two diastereotopic hydrogens (H4) at 2.90 ppm (J_1 = 16.75 Hz, J_2 = 5.5 Hz) and 2.66 ppm (J_1 = 16.75 Hz, J_2 = 6.5 Hz) as two doublets of doublets patterns (J_1 = 16.5 Hz, J_2 = 6 Hz).

In the five chain ester groups of catechin pentabutanoate, four CH_2 groups near the ester carbonyl groups bonded to aromatic rings, which appeared as multiplets between 2.55–2.51 ppm with integral 8 (Figure S4). The chemical shift of another CH_2 group in the vicinity of an ester carbonyl group bonded to an aliphatic ring was 2.23 ppm as a triplet peak (J = 7.5 Hz). In those ester chains bonded to the aromatic rings, a multiplet peak at 1.78 ppm was attributed to the four methylene groups near the methyl groups. It should be mentioned that in another ester chain bonded to the aliphatic ring, methylene is in the neighborhood of a methyl group that appeared at 1.55 ppm as a sextet (J = 7.5 Hz). A multiplet peak at 1.05 ppm with an integral of 12 confirmed the existence of four methyl groups at the end of the butanoate esters arms bonded to the aromatic rings. Finally, a triplet peak (J = 7.5 Hz) at 0.85 ppm was attributed to a CH_3 group at the end of the butanoate ester chain bonded to the aliphatic ring.

3.3. ^{13}C NMR of Catechin Pentabutanoate

The ^{13}C NMR (125 MHz) of catechin pentabutanoate in $CDCl_3$ was also studied at room temperature (Figure S5). The ^{13}C NMR of this compound revealed five peaks at

172.76, 171.66, 171.07, 170.74, and 170.72 ppm attributed to the five different carbonyl carbons of the ester groups in the structure of this molecule. There are twelve distinct peaks at 154.46, 149.92, 149.47, 142.24, 142.20, 136.03, 124.41, 123.70, 121.87, 110.18, 108.79, and 107.60 ppm for the carbons of the aromatic rings. In the aliphatic ring of catechin pentabutanoate, two different carbons bonded to ether and ester oxygens (C2 and C3) appear at 77.82 and 68.14 ppm, respectively. Another carbon in the aliphatic ring was a benzylic methylene carbon (C4) with a chemical shift of 24.18 ppm. In the five short ester chains of catechin pentabutanoate, five dissimilar peaks at 36.19, 36.03, 36.00, 35.89, and 35.84 ppm were attributed to the five CH_2 groups near the carbonyl carbons of the ester groups (Figure S5). The rest of the methylene groups in these chains appeared at 18.44, 18.41, 18.39, 18.37, and 18.25 ppm. In this compound, five methyl groups were at the end of ester chains. Because the chemical shifts of these methyl groups were close, one of the methyl groups overlapped with another, resulting in these five methyl groups appearing as four peaks at 13.67, 13.64, 13.63, and 13.42 ppm (Figures S5 and S6).

3.4. LC-MS of Catechin Pentabutanoate

For further assurance of the synthesis and purity of catechin pentabutanoate, LC-MS was performed to determine the molecular weight of catechin pentabutanoate (Figure S7). The results of selected reaction monitoring (SRM) show a single peak with 641.22 m/z (M + H), which agrees with the structure and molecular weight of catechin pentabutanoate. A fragment with 553.04 m/z is attributed to eliminating a butanoic acid molecule from the catechin pentabutanoate.

3.5. Particle Sizes of Catechin Pentabutanoate

Catechin pentabutanoate was dispersed in DI H_2O using 2% (v/v) ethanol as a co-solvent and 0.4 % (w/v) of tween 80 as a surfactant. The average size of the catechin pentabutanoate particles was 232 nm, and the polydispersity index (PDI) was 0.185.

3.6. FTIR of Catechin Pentaacetate

FTIR (cm^{-1}): 3024 (stretching vibration of aromatic C-H), 2939 (stretching vibration of aliphatic C-H), 2852 (symmetric stretching vibration of aliphatic C-H), 1764 (stretching vibration of C=O in aryl ester groups), 1739 (stretching vibration of C=O in aliphatic ester group), 1622, 1593, and 1489 (stretching vibrations of aromatic C=C bonds), 1369 (out-of-plane C-H bending vibrations of CH_3), 1174, 1120, and 1028 (stretching vibrations of C-O bonds), 898, 839, 750, and 665 ((out-of-plane C-H bending vibrations of aromatic rings) Supporting information, Figure S8).

3.7. ^1H NMR of Catechin Pentaacetate

^1H NMR (500 MHz, $CDCl_3$) δ (ppm): 7.25 (dd, J_1 = 8.5, J_2 = 2 Hz, 1 H, aromatic ring), 7.19 (d, J = 8.5 Hz, 1 H, aromatic ring), 7.17 (d, J = 2.5 Hz, 1 H, aromatic ring), 6.66 (d, J = 2.5 Hz, 1 H, aromatic ring), 6.59 (d, J = 2.5 Hz, 1 H, aromatic ring), 6.75 (d, J = 2.5 Hz, 1 H, aromatic ring), 5.25 (q, J = 6.0 Hz, 1 H, aliphatic ring), 5.14 (d, J = 6.5 Hz, 1 H, aliphatic ring), 2.86 (dd, J_1 = 16.75, J_2 = 5.5 Hz, 1 H, diastereotopic methylene in aliphatic ring), 2.67 (dd, J_1 = 16.5, J_2 = 6.5 Hz, 1 H, diastereotopic methylene in aliphatic ring), 2.27–2.26 (m, 12 H, four CH_3 groups), 1.99 (s, 3 H, CH_3). (Supporting information, Figures S9–S11).

3.8. ^{13}C NMR of Catechin Pentaacetate

^{13}C NMR (125 MHz, $CDCl_3$) δ (ppm): 170.13, 168.98, 168.37, 168.07, and 168.06 (carbonyl of ester groups), 154.37, 149.86, 149.42, 142.11, 142.09, 136.14, 124.41, 123.71, 121.78, 110.18, 108.78, and 107.69 (aromatic rings), 77.64, 68.27, and 23.91 (aliphatic ring), 21.10, 20.96, 20.79, 20.65, and 20.63 (methyl groups). (Supporting information, Figure S12).

3.9. Particle Sizes of Catechin Pentaacetate

Catechin pentaacetate was dispersed in DI H_2O using 2 % (v/v) ethanol as a cosolvent and 0.4 % (w/v) of tween 80 as a surfactant. The average size of the catechin pentaacetate particles was 509 nm, and the PDI was 0.503.

3.10. Gross Physiological Parameters

There were no significant differences in body weight, cecum weight, and cecum-to-body weight ratios across all treatment groups (Table 1).

Table 1. Gross physiological parameters measured on the day of hatch [1].

Group Name	Average Body Weight (g)	Average Cecum Weight (g)	CW: BW
NI	49.7 ± 1.9 [a]	0.39 ± 0.06 [a]	0.007 ± 0.001 [a]
H_2O	48.0 ± 2.49 [a]	0.43 ± 0.07 [a]	0.009 ± 0.002 [a]
5 % Inulin	48.3 ± 1.12 [a]	0.42 ± 0.11 [a]	0.009 ± 0.002 [a]
0.4% Tween 80	47.8 ± 1.2 [a]	0.40 ± 0.04 [a]	0.008 ± 0.001 [a]
Catechin	48.6 ± 1.66 [a]	0.30 ± 0.03 [a]	0.006 ± 0.001 [a]
Catechin-P-A	48.1 ± 2.36 [a]	0.35 ± 0.06 [a]	0.008 ± 0.001 [a]
Catechin-P-B	47.5 ± 1.74 [a]	0.28 ± 0.05 [a]	0.006 ± 0.001 [a]

[1] Values are the means ± SEM, n = 10. [a] Treatment groups not indicated by the same letter are significantly different ($p < 0.05$). NI, non-injected; CW, cecum weight; BW, body weight; Catechin-P-A, Catechin pentaacetate; Catechin-P-B, Catechin pentabutanoate.

3.11. Hemoglobin Concentration and Glycogen Concentrations of the Pectoral and Hepatic Tissues

There were no significant differences between experimental groups in hemoglobin levels and pectoral and hepatic glycogen content (Table 2).

Table 2. Hemoglobin, concentrations (g/dL) [1].

Group Name	Average Hemoglobin (g/dL)	Average Pectoral Glycogen (mg/mL)	Average Hepatic Glycogen (mg/mL)
NI	9.2 ± 1.2 [a]	0.011 ± 0.005 [a]	0.3 ± 0.08 [a]
H_2O	9.1 ± 1.3 [a]	0.013 ± 0.007 [a]	0.56 ± 0.09 [a]
5 % Inulin	9.8 ± 1.0 [a]	0.008 ± 0.002 [a]	0.33 ± 0.13 [a]
0.4% Tween 80	12.0 ± 2.1 [a]	0.004 ± 0.002 [a]	0.35 ± 0.07 [a]
Catechin	9.1 ± 1.4 [a]	0.005 ± 0.002 [a]	0.38 ± 0.07 [a]
Catechin-P-A	9.2 ± 1.8 [a]	0.006 ± 0.002 [a]	0.29 ± 0.05 [a]
Catechin-P-B	9.3 ± 0.5 [a]	0.007 ± 0.002 [a]	0.41 ± 0.11 [a]

[1] Values are the means ± SEM, n = 10. [a] Treatment group not indicated by the same letter is significantly different ($p < 0.05$). NI = non-injected. Catechin-P-A, Catechin pentaacetate; Catechin-P-B, Catechin pentabutanoate.

3.12. Cecal Microbiota Analysis

As demonstrated in Figure 1, *Bifidobacterium* populations were significantly increased ($p < 0.05$) in the catechin pentaacetate (catechin-P-A) and catechin pentabutanoate (catechin-P-B) groups when compared with all other treatment groups. Additionally, there were significant differences between 0.4% Tween 80 and catechin groups compared to the no injection, H_2O injection, and 5% inulin control groups. It was significantly increased ($p < 0.05$). *Lactobacillus* populations were found in the Catechin-P-A, Catechin-P-B, and 5% inulin groups compared with the catechin, 0.4% Tween 80, H_2O, and NI groups. The highest *Lactobacillus* density levels were seen in 5% inulin and catechin-P-A administration compared to all other groups ($p < 0.05$). Within *E. coli*, the highest density of bacteria was found within the no injection, H_2O, and 5% inulin groups compared to 0.4% Tween 80, Catechin, Catechin-P-A, and Catechin-P-B groups ($p < 0.05$). *Clostridium* populations were significantly elevated in the 5% inulin group compared to all the other groups.

	Bifidobacterium	*Lactobacillus*	*E. coli*	*Clostridium*
NI	c 1.155 ± 0.028	b 0.932 ± 0.079	ab 0.889 ± 0.03	b 0.816 ± 0.025
H_2O	c 1.208 ± 0.088	a 1.145 ± 0.025	b 0.844 ± 0.034	b 0.77 ± 0.058
0.4% Tween 80	b 2.503 ± 0.146	d 0.546 ± 0.011	c 0.67 ± 0.008	b 0.933 ± 0.012
5% Inulin	c 1.101 ± 0.104	a 1.19 ± 0.039	a 0.917 ± 0.027	a 2.389 ± 0.156
Catechin	b 2.578 ± 0.059	c 0.679 ± 0.01	c 0.719 ± 0.015	b 0.927 ± 0.002
Catechin-P-A	a 3.326 ± 0.16	cd 0.647 ± 0.008	c 0.702 ± 0.004	b 0.947 ± 0.006
Catechin-P-B	a 3.407 ± 0.312	d 0.561 ± 0.008	c 0.679 ± 0.007	b 0.924 ± 0.003

Low AU — High AU

Figure 1. Effects of intra-amniotic injections on cecal genera and species-level bacterial populations (day of hatch). Values are means ± SEM, n = 10. [a–d] Per bacterial category (in the same column), treatment groups that do not share any letters are significantly different according to a one-way ANOVA with a post hoc Duncan test ($p < 0.05$). NI, No Injection; Catechin-P-A, Catechin pentaacetate; Catechin-P-B, Catechin pentabutanoate.

3.13. Duodenal Gene Expression

3.13.1. Fe-Related Proteins

For the proteins directly responsible for iron uptake, the gene expression for Duodenal cytochrome b (DcytB) and Divalent Metal Transporter 1 was not significantly altered between groups. Ferroportin expression was significantly downregulated ($p < 0.05$, Figure 2) in the 0.4% Tween 80, 5% Inulin, Catechin-P-A, and Catechin-P-B compared to the NI control group.

	DcytB	DMT1	Ferroportin	ZIP9	ZnT1	ZnT7	SGLT1
NI	a 5.074 ± 0.034	a 1.007 ± 0.002	a 14.964 ± 0.926	a 2.708 ± 0.064	a 11.561 ± 0.638	a 1.49 ± 0.015	ab 0.93 ± 0.002
H_2O	a 4.852 ± 0.162	a 1.009 ± 0.004	ab 12.45 ± 1.195	ab 2.551 ± 0.086	b 9.842 ± 0.767	ab 1.454 ± 0.015	ab 0.937 ± 0.005
0.4% Tween 80	a 4.728 ± 0.082	a 1.01 ± 0.002	b 11.718 ± 0.588	b 2.488 ± 0.049	bc 8.27 ± 0.43	ab 1.463 ± 0.007	b 0.923 ± 0.005
5% Inulin	a 4.995 ± 0.174	a 1.012 ± 0.003	b 10.652 ± 0.846	b 2.397 ± 0.077	bc 8.307 ± 0.755	b 1.433 ± 0.014	a 0.939 ± 0.005
Catechin	a 5.136 ± 0.082	a 1.002 ± 0.002	b 10.728 ± 0.666	ab 2.509 ± 0.077	bc 8.413 ± 0.367	ab 1.452 ± 0.015	ab 0.928 ± 0.002
Catechin-P-A	a 4.913 ± 0.152	a 1.007 ± 0.005	b 11.06 ± 0.625	b 2.464 ± 0.041	bc 8.658 ± 0.426	ab 1.454 ± 0.011	ab 0.929 ± 0.006
Catechin-P-B	a 4.797 ± 0.127	a 1.009 ± 0.002	b 10.448 ± 0.37	b 2.408 ± 0.038	c 7.866 ± 0.3	b 1.431 ± 0.008	ab 0.931 ± 0.006

	SI	AP	MUC2	NF-κB1	TNF-α	IL8
NI	ab 2.713 ± 0.037	a 1.535 ± 0.019	a 1.92 ± 0.026	a 19.409 ± 1.36	a 3.61 ± 0.126	a 1.006 ± 0.002
H_2O	b 2.612 ± 0.065	ab 1.488 ± 0.023	ab 1.838 ± 0.036	ab 16.146 ± 1.428	ab 3.313 ± 0.146	a 1.006 ± 0.004
0.4% Tween 80	ab 2.717 ± 0.03	b 1.482 ± 0.007	b 1.819 ± 0.017	ab 17.583 ± 0.982	a 3.578 ± 0.065	a 0.777 ± 0.221
5% Inulin	ab 2.641 ± 0.069	b 1.471 ± 0.012	b 1.773 ± 0.032	b 13.742 ± 1.372	b 3.035 ± 0.143	a 1.01 ± 0.003
Catechin	a 2.792 ± 0.017	b 1.449 ± 0.012	b 1.808 ± 0.025	ab 16.235 ± 1.404	ab 3.406 ± 0.125	a 0.994 ± 0.002
Catechin-P-A	ab 2.745 ± 0.056	b 1.47 ± 0.014	b 1.802 ± 0.021	ab 16.74 ± 0.679	a 3.468 ± 0.134	a 0.997 ± 0.003
Catechin-P-B	ab 2.7 ± 0.038	b 1.454 ± 0.009	b 1.775 ± 0.011	ab 15.324 ± 0.879	ab 3.251 ± 0.079	a 0.827 ± 0.175

Low AU — High AU

Figure 2. Effect of the intraamniotic administration of experimental solutions on intestinal gene expression. Values are the means ± SEM, n = 10. [a–c] Per gene (in the same column), treatments groups not indicated by the same letter are significantly different ($p < 0.05$). NI, No Injection; Catechin-P-A, Catechin pentaacetate; Catechin-P-B, Catechin pentabutanoate. DcytB, Duodenal cytochrome B; DMT1, Divalent metal transporter 1; ZIP9, Zinc Transporter 9; ZnT1, Zinc transporter 1; ZnT7, Zinc transporter 7; SGLT-1, Sodium-glucose cotransporter 1; SI, Sucrose isomaltase; AP, amino peptidase; MUC2, Mucin 2; NF-κβ1, Nuclear factor-κβ1; TNF-α, Tumor necrosis factor-alpha; IL8, Interleukin 8; 18S rRNA, 18S Ribosomal subunit.

3.13.2. Zn-Related Proteins

The expression of zinc-related proteins related to cellular uptake, transport, and storage; the gene expression zinc transporter 9 (ZIP9) was significantly down-regulated ($p < 0.05$) with 0.4% Tween 80, 5% Inulin, Catechin, Catechin-P-A, and Catechin-P-B compared with the NI group. Zinc transporter 1 (ZnT1) gene expression was significantly down-regulated ($p < 0.05$, Figure 2) with Catechin-P-B compared to the NI and H_2O groups. Additionally, zinc transporter 7 (ZnT7) was downregulated considerably ($p < 0.05$, Figure 2) with 5% inulin and catechin-P-B compared to the NI group.

3.13.3. BBM Functionality and Mucin Production Proteins

Sodium-glucose cotransporter 1 (SGLT-1) expression was significantly ($p < 0.05$, Figure 2) upregulated with 5% inulin compared to the 0.4% Tween 80 group. There was a significant upregulation with Catechin compared to the H_2O injection with Sucrose isomaltase (SI). Amino peptidase (AP) and Mucin 2 (MUC2) were significantly downregulated in the 5% inulin, Catechin, Catechin-P-A and Catechin-P-B groups compared with the NI group.

3.13.4. Pro-Inflammatory Proteins

There was significant downregulation ($p < 0.05$, Figure 2) with 5% inulin compared to the NI groups in nuclear transcription factor-κβ 1(NF-κβ1) expression, and there were no significant differences with the catechin groups compared to the controls. Tumor necrosis factor (TNF-α) had significant downregulation ($p < 0.05$, Figure 2) in 5% inulin when compared to the NI, 0.4% Tween 80, and catechin-P-A groups. There were no significant differences in gene expression of interleukin-8 (IL8) between all the groups.

3.14. Duodenal Morphometric Parameters

The catechin and synthetic compounds treatment groups presented elevated ($p < 0.05$) villus surface area and crypt depth versus the controls (NI and H_2O). Paneth cell diameter and number values were significantly ($p < 0.05$) smaller in catechin, catechin-P-A, and catechin-P-B compared to NI and H_2O (Table 3).

Table 3. Effect of the intra-amniotic administration of experimental solutions on the duodenal small intestinal villus, crypt depth, and Paneth cells [1].

Treatment	Villus Surface Area (μm^2)	Crypt Depth (μm)	Paneth Cell/Crypt	Paneth Cell Diameter (μm)
NI	35,338.95 ± 814.06 ^{d}	22.04 ± 0.66 ^{cd}	1.81 ± 0.07 ^{b}	1.88 ± 0.10 ^{a}
H_2O	36,111.84 ± 735.90 ^{d}	21.80 ± 0.54 ^{d}	2.02 ± 0.08 ^{a}	1.70 ± 0.04 ^{b}
0.4% Tween 80	34,700.16 ± 1443.64 ^{d}	21.77 ± 1.17 ^{d}	1.61 ± 0.05 ^{c}	1.46 ± 0.03 ^{d}
5% Inulin	38,457.03 ± 1257.71 ^{cd}	21.49 ± 0.64 ^{d}	2.29 ± 0.09 ^{a}	1.64 ± 0.04 ^{bc}
Catechin	40,686.07 ± 1337.05 ^{bc}	23.58 ± 0.54 ^{bcd}	1.65 ± 0.05 ^{bc}	1.52 ± 0.03 ^{cd}
Catechin-P-A	41,642.87 ± 1189.86 ^{abc}	35.04 ± 0.85 ^{a}	1.46 ± 0.04 ^{c}	1.39 ± 0.03 ^{d}
Catechin-P-B	43,200.19 ± 1177.55 ^{ab}	25.34 ± 0.98 ^{b}	1.51 ± 0.05 ^{c}	1.41 ± 0.03 ^{d}

[1] Values are the means ± SEM, $n = 5$. $^{a–d}$ Treatment groups not indicated by the same letter are significantly different ($p < 0.05$). NI, No Injection; Catechin-P-A, Catechin pentaacetate; Catechin-P-B, Catechin pentabutanoate.

The villi goblet cell diameter, neutral villi goblet cell, mixed villi goblet cell, and a total number of villi and crypt goblet cells, where catechin and catechin-P-B were significantly smaller than the NI control group. However, the acidic villi goblet cell number of catechin and synthetic catechin compounds were significantly more numerous than the NI and 5% inulin treatment groups. Further, 0.4% tween was significantly higher than all other treatment groups in crypt goblet cell diameter. There was no significant difference between any treatment group in acidic crypt goblet cells per crypt. However, there was a significantly lower number of mixed and neutral goblet cells per crypt of catechin than NI and H_2O. Furthermore, there was a significantly lower ($p < 0.05$) amount of total goblet cells per crypt in the catechin and synthetic compounds (Catechin-P-A and Catechin-P-B) groups relative to the controls (NI and H_2O) (Table 4).

Table 4. Effect of the intra-amniotic administration of experimental solutions on the goblet cells [1].

Treatment	Villi Goblet Diameter (μm)	Villi Goblet Acid/Villi (#)	Villi Goblet Neutral/ Villi (#)	Villi Goblet Mixed/Villi (#)	Villi Goblet Cell (#)	Crypt Goblet Diameter (μm)	Crypt Goblet Acid/Crypt (#)	Crypt Goblet Neutral/ Crypt (#)	Crypt Goblet Mixed/Crypt (#)	Crypt Goblet Cell (#)
NI	3.45 ± 0.07 d	31.89 ± 0.89 a	1.85 ± 0.19 a	5.89 ± 0.28 a	39.63 ± 0.93 a	3.24 ± 0.04 c	7.74 ± 0.24 a	1.56 ± 0.24 b	0.86 ± 0.11 c	10.15 ± 0.41 ab
H_2O	3.43 ± 0.06 d	16.27 ± 0.67 c	1.42 ± 0.17 b	5.29 ± 0.72 a	28.94 ± 0.76 b	2.74 ± 0.04 e	7.66 ± 0.22 a	2.62 ± 0.21 a	0.86 ± 0.08 c	11.14 ± 0.35 a
0.4% Tween 80	4.93 ± 0.09 a	34.26 ± 0.95 a	0.04 ± 0.02 c	0.35 ± 0.08 d	30.20 ± 1.24 b	3.95 ± 0.09 a	7.92 ± 0.32 a	0.01 ± 0.01 d	1.32 ± 0.12 b	8.68 ± 0.38 cd
5% Inulin	3.20 ± 0.06 e	23.96 ± 0.68 b	0.08 ± 0.02 c	1.46 ± 0.14 c	25.50 ± 0.75 c	2.18 ± 0.04 f	8.45 ± 0.36 a	0.63 ± 0.11 c	1.70 ± 0.15 a	10.78 ± 0.43 a
Catechin	4.15 ± 0.07 b	32.14 ± 0.68 a	0.01 ± 0.01 c	4.29 ± 0.24 b	29.16 ± 1.19 b	3.72 ± 0.06 b	7.8 ± 0.22 a	0.24 ± 0.04 d	0.38 ± 0.06 d	8.41 ± 0.24 d
Catechin-P-A	3.50 ± 0.05 d	32.36 ± 0.94 a	0.05 ± 0.03 c	3.69 ± 0.28 b	28.89 ± 1.32 b	3.02 ± 0.05 d	7.94 ± 0.26 a	0.07 ± 0.02 d	0.96 ± 0.10 c	8.96 ± 0.30 cd
Catechin-P-B	3.91 ± 0.07 c	33.99 ± 1.02 a	0.10 ± 0.03 c	1.60 ± 0.37 c	24.80 ± 1.41 c	3.27 ± 0.06 c	8.26 ± 0.32 a	0.02 ± 0.01 d	1.25 ± 0.09 b	9.52 ± 0.35 bc

[1] Values are the means ± SEM, $n = 5$. [a–f] Treatment groups not indicated by the same letter are significantly different ($p < 0.05$). NI, No Injection; Catechin-P-A, Catechin pentaacetate; Catechin-P-B, Catechin pentabutanoate. [#] Number of cells.

4. Discussion

Catechin, a flavan-3-ol, is a type of flavonoid and a secondary metabolite of plants that provides antioxidant roles in plants and humans. Although catechin ingestion has been associated with significant beneficial physiological changes related to catechin antioxidant activity [1,4–7], further understanding of duodenal tissue-level effects associated with catechin exposure is needed. Additionally, the instability of catechin within the digestive tract severely hinders its bioavailability. This study focuses on novel and more stable catechin derivatives, catechin-P-A, and catechin-P-B, synthesized using acetic anhydride and butyric anhydride, respectively. The studied esterification reactions were executed under mild conditions and resulted in acceptable yields, as shown in our previous work [46]. The final purified products were characterized using FTIR, NMR, and LC-MS to confirm their structures, demonstrating the synthesis of catechin pentabutanoate and catechin pentaacetate with high purity. Moreover, we could disperse these compounds in H_2O for intra-amniotic administration studies.

The effects of synthesized and characterized catechin, catechin-P-A, and catechin-P-B on duodenal BBM development, functionality, and cecal microbial populations were investigated in vivo. These compounds did not significantly affect the body weight, cecum, or to-body weight ratio (Table 1). A potential explanation for this finding is the previous short exposure period in other intra-amniotic administration studies investigating the effects of plant-origin bioactive compounds [44,45,48,52,54].

We studied the effects of the investigated catechin compounds on cecal bacteria populations and found significant differences between genera population densities (Figure 1). Catechin consumption has previously been shown to alter intestinal bacterial populations in vivo, including increasing the populations of beneficial SCFA-producing bacteria [21,22]. We found significant ($p < 0.05$) increases in the *Bifidobacterium* population with catechin derivative exposure (catechin-P-A and catechin-P-B) compared to all other groups. As previously seen by Janiak et al. (2018), catechin (from green tea extract) administration resulted in increased *Bifidobacterium* spp. (increase in colony-forming units/mL) [55]. The increases in *Bifidobacterium* have previously been associated with catechin consumption, where colonic microbiota are hypothesized to metabolize catechin to γ-valerolactones and hippuric acids, which undergo further biotransformation, glucuronidation, sulfation, and methylation within the liver [56–58]. Therefore, beneficial bacteria can utilize the catechin and bacterial metabolites as substrates to obtain energy, similar to the effects of growth stimulation observed with inulin and galactooligosaccharides [59,60]. Within the potentially pathogenic bacteria, *E. coli* and *Clostridium* spp., there was a significant ($p < 0.05$) decrease in bacterial density associated with catechin compound administration relative to the controls. Ma et al. (2019) found that certain flavanols (e.g., catechin) can have antimicrobial activity against pathogenic bacteria such as *E. coli* and *Clostridium* [61]. These findings can be attributed to increased SCFA production associated with catechin metabolism by microorganisms within the intestine [62–64]. Further, dietary supplementation with catechins and similar compounds could have promoted the expression of serum leptin and induced carbohydrate degradation, which may regulate the presence and abundance of certain bacterial strains [65]. Additionally, the production of beneficial bacteria, *Bifidobacterium*, can improve intestinal barrier function, maintain intestinal homeostasis, and aid in lowering immune response from the host (Paneth cell number and diameter, Table 3).Moreover, we hypothesize that exposure to the SCFA (acetate and butyrate) groups esterified to the catechin derivatives (treatment groups catechin-P-A and catechin-P-B) increased villus surface area and crypt depth (Table 3). Liao et al. (2020) previously investigated the relationship between gut microbiota, SCFAs, and intestinal morphology of growing broilers, where increased acetate exposure was found to increase *Lactobacillus* genera and villus height, which is congruent with our findings [66]. This possibility could explain the significant differences in villi acidic goblet cells between the catechin synthetic compounds and controls in Table 4. It has been shown previously that acetate and butyrate SCFAs mediate intestinal mucosa [67–70] and epithelial barrier cell proliferation [71–74].

This maintenance of intestinal homeostasis and immune response can be seen within genotypical responses from key BBM digestive and absorptive proteins, as previously demonstrated through intra-amniotic administration of polyphenols and other plant-origin dietary bioactive compounds [33,36,52,54,75]. In the current study, we investigated catechin and its derivatives on BBM gene expression (Figure 2), and the results demonstrated that catechin, catechin-P-A, and catechin-P-B exposure decreased the expression of ferroportin (transporter of Fe into circulation) and zinc transporters (ZIP9, ZnT7). This decrease in gene expression of ferroportin may be due to the beneficial effects of flavonoids exhibiting antioxidant activity by chelating the redox-active transition metals (e.g., iron) that may act as ROS generators [76–78]. Further, catechin and other flavonoids have also been demonstrated to be complex with redox-inactive metals (e.g., zinc), causing alterations in glucidic and lipid metabolisms [76,78–83]. Quesada et al. (2011) demonstrated that catechin and other procyanidins displayed an affinity for zinc cations in a solution high enough to dissociate zinc from Zinquin (zinc-specific chelator) both intracellularly (within organelles) and within the cytoplasm [78].

NF-κβ1 is one of the pathways used in the small intestine to respond to microbial dysbiosis because it is a central mediator of immune response found in innate and adaptive immunity [84,85]. When NF-κB1 is stimulated, it translocates within the cell nucleus and is involved in various biological functions, including the release of proinflammatory mediators, such as tumor necrosis factor-alpha (TNF-α) [86,87]. The downregulation of pro-inflammatory proteins (NF-κβ1 and TNF-α) was found with 5% inulin, catechin, catechin-P-A and catechin-P-B exposure compared to the NI control group. This observation agrees with our hypothesis that the intra-amniotic administration of catechin and its synthetic compounds beneficially modulates BBM functionality through positive alterations in cecal microbiota populations, as described above (Figure 1). These findings are similar to the trends in Surh et al. (2001), where catechin and epigallocatechin gallate were shown to inhibit cyclooxygenase-2 and nitric oxide synthase expression, blocking NF-κβ1 activation [88,89]. Paneth cells are key cells within the small intestine that defend the host during innate immunity, and our findings with decreased NF-κβ1 expression are supported by decreases in Paneth cell number and diameter within the crypt (Table 3). Additionally, we can associate the downregulation of these pro-inflammatory proteins with the increased beneficial bacterial populations (Figure 1), which increases overall SCFA production, where SCFA exerts anti-inflammatory effects in the intestinal mucosa (Table 4) because of inhibition by histone deacetylases and activating G-protein coupled receptors as described by Parada Venegas et al. (2019) [90]. Taken together, additional studies are warranted to assess shifts in intestinal functionality, development and microbiota posthatch and during a long-term feeding trial associated with the consumption of catechin and its more stable derivatives to confirm results within this study. Overall, current results demonstrate a unique approach to evaluate the impact of catechin and its derivatives on BBM functionality markers, the intestinal microbiome and other physiological parameters, as previously demonstrated

Finally, it is important to emphasize that further assessments are necessary to confirm the potential effects of catechin and its derivatives on intestinal morphology and functionality due to potential SCFA production by cecal and/or small intestinal microbiome, as discussed. Overall, current results introduce an innovative approach to evaluating the impact of catechin and its derivatives on BBM functional biomarkers, and intestinal microbial populations, in vivo [46].

5. Conclusions

The intra-amniotic administration of physiologically relevant dosages of catechin and its synthetic derivative (catechin pentaacetate and catechin pentabutanoate) on intestinal microbiota populations, brush border membrane morphology, and physiological parameters such as average body weight, cecum weight, hemoglobin concentrations, pectoral and hepatic glycogen concentrations, and duodenal gene expression of newborn chickens are reported. The data suggests that each compound can positively alter duodenal brush border membrane functionality, morphology, and cecal microbial populations (*Bifidobacterium* and *Lactobacillus*). This is one of the first studies to synthesize two synthetically derived catechin esters with increased bioavailability to reveal potential health benefits. Given these findings, these novel catechin esters should be evaluated in additional long-term studies to elucidate the potential health benefits and mechanisms of action.

Supplementary Materials: The following are available online at https://www.mdpi.com/article/10.3390/nu14193924/s1, Figure S1: FTIR of catechin pentabutanoate; Figure S2: ^{1}H NMR of catechin pentabutanoate in $CDCl_3$; Figure S3: Expanded (7.27–2.64 ppm) ^{1}H NMR of catechin pentabutanoate in $CDCl_3$; Figure S4: Expanded (2.55–0.84 ppm) 1H NMR of catechin pentabutanoate in $CDCl_3$; Figure S5: ^{13}C NMR of catechin pentabutanoate in $CDCl_3$; Figure S6: Expanded ^{13}C NMR of catechin pentabutanoate in $CDCl_3$; Figure S7: SRM LC-MS of catechin pentabutanoate. (a) SRM LC of catechin pentabutanoate. (b) Mass spectrum of catechin pentabutanoate; Figure S8: FTIR of catechin pentaacetate; Figure S9: ^{1}H NMR of catechin pentaacetate in $CDCl_3$, Figure S10: Expanded 1H NMR of catechin pentaacetate in $CDCl_3$; Figure S11: Expanded ^{1}H NMR of catechin pentaacetate in $CDCl_3$; Figure S12: ^{13}C NMR of catechin pentaacetate in $CDCl_3$. Table S1: Real-time polymerase chain reaction (RT-PCR) primer sequence.

Author Contributions: Chemical synthesis and analysis: A.Z., L.K., Y.D. and A.A.; In vivo experiment, data analysis, collection, and methodology: N.K., J.C., N.A. and E.T.; N.K., J.C. and E.T. wrote the manuscript draft; E.T. and A.A. finalized the manuscript and were the primary investigators who led the research conducted and presented in this manuscript. All authors have read and agreed to the published version of the manuscript.

Funding: This research received no external funding.

Institutional Review Board Statement: The animal protocol used in this study was conducted according to the guidelines of the Declaration of Helsinki and approved by the Cornell University Institutional Animal Care and Use Committee by the ethic approval code 2020-0077.

Informed Consent Statement: Not applicable.

Conflicts of Interest: The authors declare no conflict of interest.

References

1. Ahmad, M.; Mudgil, P.; Gani, A.; Hamed, F.; Masoodi, F.A.; Maqsood, S. Nano-Encapsulation of Catechin in Starch Nanoparticles: Characterization, Release Behavior and Bioactivity Retention during Simulated in-Vitro Digestion. *Food Chem.* **2019**, *270*, 95–104. [CrossRef] [PubMed]
2. Grzesik, M.; Naparło, K.; Bartosz, G.; Sadowska-Bartosz, I. Antioxidant Properties of Catechins: Comparison with Other Antioxidants. *Food Chem.* **2018**, *241*, 480–492. [CrossRef] [PubMed]
3. Takahashi, H.; Kosaka, M.; Watanabe, Y.; Nakade, K.; Fukuyama, Y. Synthesis and Neuroprotective Activity of Bergenin Derivatives with Antioxidant Activity. *Bioorg. Med. Chem.* **2003**, *11*, 1781–1788. [CrossRef]
4. Gopal, J.; Muthu, M.; Paul, D.; Kim, D.-H.; Chun, S. Bactericidal Activity of Green Tea Extracts: The Importance of Catechin Containing Nano Particles. *Sci. Rep.* **2016**, *6*, 19710. [CrossRef]
5. Coșarcă, S.; Tanase, C.; Muntean, D.L. Therapeutic Aspects of Catechin and Its Derivatives–An Update. *Acta Biol. Marisiensis* **2019**, *2*, 21–29. [CrossRef]
6. Sharma, P.; Goyal, P.K. Anti-Oxidative and Anti-Metalotoxic Properties of Green Tea Catechin: A Preliminary Study. *Am. J. Ethnomed.* **2015**, *2*, 21–38.
7. Yang, C.S.; Maliakal, P.; Meng, X. Inhibition of Carcinogenesis by Tea. *Annu. Rev. Pharmacol. Toxicol.* **2002**, *42*, 25–54. [CrossRef] [PubMed]
8. Jeong, W.-S.; Kong, A.-N.T. Biological Properties of Monomeric and Polymeric Catechins: Green Tea Catechins and Procyanidins. *Pharm. Biol.* **2004**, *42*, 84–93. [CrossRef]
9. Mukhtar, H.; Ahmad, N. Tea Polyphenols: Prevention of Cancer and Optimizing Health. *Am. J. Clin. Nutr.* **2000**, *71*, 1698S–1702S. [CrossRef]
10. Chyu, K.-Y.; Babbidge, S.M.; Zhao, X.; Dandillaya, R.; Rietveld, A.G.; Yano, J.; Dimayuga, P.; Cercek, B.; Shah, P.K. Differential Effects of Green Tea–Derived Catechin on Developing Versus Established Atherosclerosis in Apolipoprotein E–Null Mice. *Circulation* **2004**, *109*, 2448–2453. [CrossRef]
11. Donà, M.; Dell'Aica, I.; Calabrese, F.; Benelli, R.; Morini, M.; Albini, A.; Garbisa, S. Neutrophil Restraint by Green Tea: Inhibition of Inflammation, Associated Angiogenesis, and Pulmonary Fibrosis. *J. Immunol* **2003**, *170*, 4335–4341. [CrossRef]
12. Nance, C.L.; Shearer, W.T. Is Green Tea Good for HIV-1 Infection? *J. Allergy Clin. Immunol.* **2003**, *112*, 851–853. [CrossRef]
13. Esposito, E.; Rotilio, D.; Dimatteo, V.; Digiulio, C.; Cacchio, M.; Algeri, S. A Review of Specific Dietary Antioxidants and the Effects on Biochemical Mechanisms Related to Neurodegenerative Processes. *Neurobiol. Aging* **2002**, *23*, 719–735. [CrossRef]
14. Matsui, T. Condensed Catechins and Their Potential Health-Benefits. *Eur. J. Pharmacol.* **2015**, *765*, 495–502. [CrossRef] [PubMed]
15. Pan, Z.; Zhou, Y.; Luo, X.; Ruan, Y.; Zhou, L.; Wang, Q.; Yan, Y.J.; Liu, Q.; Chen, J. Against NF-KB/Thymic Stromal Lymphopoietin Signaling Pathway, Catechin Alleviates the Inflammation in Allergic Rhinitis. *Int. Immunopharmacol.* **2018**, *61*, 241–248. [CrossRef] [PubMed]
16. Addepalli, V.; Suryavanshi, S.V. Catechin Attenuates Diabetic Autonomic Neuropathy in Streptozotocin Induced Diabetic Rats. *Biomed. Pharmacother.* **2018**, *108*, 1517–1523. [CrossRef] [PubMed]
17. Ganeshpurkar, A.; Saluja, A.K. Protective Effect of Catechin on Humoral and Cell Mediated Immunity in Rat Model. *Int. Immunopharmacol.* **2018**, *54*, 261–266. [CrossRef]
18. Stapleton, P.D.; Shah, S.; Anderson, J.C.; Hara, Y.; Hamilton-Miller, J.M.T.; Taylor, P.W. Modulation of β-Lactam Resistance in Staphylococcus Aureus by Catechins and Gallates. *Int. J. Antimicrob. Agents* **2004**, *23*, 462–467. [CrossRef]
19. Mandel, S.; Youdim, M.B.H. Catechin Polyphenols: Neurodegeneration and Neuroprotection in Neurodegenerative Diseases. *Free Radic. Biol. Med.* **2004**, *37*, 304–317. [CrossRef]
20. Mandel, S.A.; Avramovich-Tirosh, Y.; Reznichenko, L.; Zheng, H.; Weinreb, O.; Amit, T.; Youdim, M.B.H. Multifunctional Activities of Green Tea Catechins in Neuroprotection. *Neurosignals* **2005**, *14*, 46–60. [CrossRef]
21. Agrizzi Verediano, T.; Agarwal, N.; Gomes, M.J.C.; Duarte Martino, H.S.; Tako, E. Effects of Dietary Fiber on Intestinal Iron Absorption, and Physiological Status: A Systematic Review of in Vivo and Clinical Studies. *Crit. Rev. Food Sci. Nutr.* **2022**, *1*, 1–18. [CrossRef] [PubMed]
22. Tian, L.; Tan, Y.; Chen, G.; Wang, G.; Sun, J.; Ou, S.; Chen, W.; Bai, W. Metabolism of Anthocyanins and Consequent Effects on the Gut Microbiota. *Crit. Rev. Food Sci. Nutr.* **2019**, *59*, 982–991. [CrossRef]
23. Gibson, G.R.; Scott, K.P.; Rastall, R.A.; Tuohy, K.M.; Hotchkiss, A.; Dubert-Ferrandon, A.; Gareau, M.; Murphy, E.F.; Saulnier, D.; Loh, G.; et al. Dietary Prebiotics: Current Status and New Definition. *Food Sci. Technol. Bull. Funct. Foods* **2010**, *7*, 1–19. [CrossRef]
24. van der Beek, C.M.; Canfora, E.E.; Kip, A.M.; Gorissen, S.H.M.; Olde Damink, S.W.M.; van Eijk, H.M.; Holst, J.J.; Blaak, E.E.; Dejong, C.H.C.; Lenaerts, K. The Prebiotic Inulin Improves Substrate Metabolism and Promotes Short-Chain Fatty Acid Production in Overweight to Obese Men. *Metabolism* **2018**, *87*, 25–35. [CrossRef] [PubMed]
25. McLoughlin, R.F.; Berthon, B.S.; Jensen, M.E.; Baines, K.J.; Wood, L.G. Short-Chain Fatty Acids, Prebiotics, Synbiotics, and Systemic Inflammation: A Systematic Review and Meta-Analysis. *Am. J. Clin. Nutr.* **2017**, *106*, 930–945. [CrossRef] [PubMed]
26. Preidis, G.A.; Versalovic, J. Targeting the Human Microbiome With Antibiotics, Probiotics, and Prebiotics: Gastroenterology Enters the Metagenomics Era. *Gastroenterology* **2009**, *136*, 2015–2031. [CrossRef] [PubMed]
27. Tian, B.; Zhang, Z.; Zhao, J.; Ma, Q.; Liu, H.; Nie, C.; Ma, Z.; An, W.; Li, J. Dietary Whole Goji Berry (*Lycium barbarum*) Intake Improves Colonic Barrier Function by Altering Gut Microbiota Composition in Mice. *Int. J. Food Sci. Technol.* **2021**, *56*, 103–114. [CrossRef]

28. Morais, C.A.; de Rosso, V.V.; Estadella, D.; Pisani, L.P. Anthocyanins as Inflammatory Modulators and the Role of the Gut Microbiota. *J. Nutr. Biochem.* **2016**, *33*, 1–7. [CrossRef]
29. Hou, T.; Tako, E. The In Ovo Feeding Administration (*Gallus Gallus*)—An Emerging In Vivo Approach to Assess Bioactive Compounds with Potential Nutritional Benefits. *Nutrients* **2018**, *10*, 418. [CrossRef]
30. Neilson, A.P.; Hopf, A.S.; Cooper, B.R.; Pereira, M.A.; Bomser, J.A.; Ferruzzi, M.G. Catechin Degradation with Concurrent Formation of Homo- and Heterocatechin Dimers during in Vitro Digestion. *J. Agric. Food Chem.* **2007**, *55*, 8941–8949. [CrossRef]
31. Bridson, J.H.; Grigsby, W.J.; Main, L. Synthesis and Characterization of Flavonoid Laurate Esters by Transesterification. *J. Appl. Polym. Sci.* **2013**, *129*, 181–186. [CrossRef]
32. Zarei, A.; Khazdooz, L.; Madarshahian, S.; Enayati, M.; Mosleh, I.; Lin, T.; Yan, B.; Ufheil, G.; Wooster, T.J.; Abbaspourrad, A. Synthesis, Stability, and Bioavailability of Nicotinamide Riboside Trioleate Chloride. *Nutrients* **2021**, *14*, 113. [CrossRef]
33. Martino, H.S.D.; Kolba, N.; Tako, E. Yacon (*Smallanthus sonchifolius*) Flour Soluble Extract Improve Intestinal Bacterial Populations, Brush Border Membrane Functionality and Morphology in Vivo (*Gallus gallus*). *Food Res. Int.* **2020**, *137*, 109705. [CrossRef] [PubMed]
34. Gomes, M.J.C.; Martino, H.S.D.; Kolba, N.; Cheng, J.; Agarwal, N.; De Moura Rocha, M.; Tako, E. Zinc Biofortified Cowpea (*Vigna unguiculata* L. Walp.) Soluble Extracts Modulate Assessed Cecal Bacterial Populations and Gut Morphology in Vivo (*Gallus gallus*). *Front. Biosci. Landmark* **2022**, *27*, 140–153. [CrossRef] [PubMed]
35. Warkentin, T.; Kolba, N.; Tako, E. Low Phytate Peas (*Pisum sativum* L.) Improve Iron Status, Gut Microbiome, and Brush Border Membrane Functionality In Vivo (*Gallus gallus*). *Nutrients* **2020**, *12*, 2563. [CrossRef]
36. Carboni, J.; Reed, S.; Kolba, N.; Eshel, A.; Koren, O.; Tako, E. Alterations in the Intestinal Morphology, Gut Microbiota, and Trace Mineral Status Following Intra-Amniotic Administration (*Gallus gallus*) of Teff (Eragrostis Tef) Seed Extracts. *Nutrients* **2020**, *12*, 3020. [CrossRef] [PubMed]
37. Dias, D.M.; Kolba, N.; Hart, J.J.; Ma, M.; Sha, S.T.; Lakshmanan, N.; Nutti, M.R.; Martino, H.S.D.; Glahn, R.P.; Tako, E. Soluble Extracts from Carioca Beans (*Phaseolus Vulgaris* L.) Affect the Gut Microbiota and Iron Related Brush Border Membrane Protein Expression in Vivo (*Gallus gallus*). *Food Res. Int.* **2019**, *123*, 172–180. [CrossRef] [PubMed]
38. Wang, Q. *The Effect of Intra-Amnitotic Administration of Plant Origin Prebiotics (Raffinose and Stachyose) on the Intestinal Bacterial Populations, Fe Status and Brush Border Membrane Functionality*; Cornell University: Ithaca, NY, USA, 2016.
39. Givisiez, P.E.N.; Moreira Filho, A.L.B.; Santos, M.R.B.; Oliveira, H.B.; Ferket, P.R.; Oliveira, C.J.B.; Malheiros, R.D. Chicken Embryo Development: Metabolic and Morphological Basis for in Ovo Feeding Technology. *Poult. Sci.* **2020**, *99*, 6774–6782. [CrossRef]
40. Dolgorukova, A.M.; Titov Fisinin, V.I.; Zotov, A.A. Prenatal nutrition of poultry and its postnatal effects (review). *Sel'skokhozyaistvennaya Biol. (Agric. Biol.)* **2020**, *55*, 1061–1072. [CrossRef]
41. Abd El-Hack, M.E.; Elnesr, S.S.; Alagawany, M.; Gado, A.; Noreldin, A.E.; Gabr, A.A. Impact of Green Tea (Camellia Sinensis) and Epigallocatechin Gallate on Poultry. *World's Poult. Sci. J.* **2020**, *76*, 49–63. [CrossRef]
42. Yuan, Z.H.; Zhang, K.Y.; Ding, X.M.; Luo, Y.H.; Bai, S.P.; Zeng, Q.F.; Wang, J.P. Effect of Tea Polyphenols on Production Performance, Egg Quality, and Hepatic Antioxidant Status of Laying Hens in Vanadium-Containing Diets. *Poult. Sci.* **2016**, *95*, 1709–1717. [CrossRef] [PubMed]
43. Agarwal, N.; Kolba, N.; Jung, Y.; Cheng, J.; Tako, E. Saffron (*Crocus sativus* L.) Flower Water Extract Disrupts the Cecal Microbiome, Brush Border Membrane Functionality, and Morphology In Vivo (*Gallus gallus*). *Nutrients* **2022**, *141*, 220. [CrossRef] [PubMed]
44. Agarwal, N.; Kolba, N.; Khen, N.; Even, C.; Turjeman, S.; Koren, O.; Tako, E. Quinoa Soluble Fiber and Quercetin Alter the Composition of the Gut Microbiome and Improve Brush Border Membrane Morphology In Vivo (*Gallus gallus*). *Nutrients* **2022**, *14*, 448. [CrossRef] [PubMed]
45. Agrizzi Verediano, T.; Stampini Duarte Martino, H.; Kolba, N.; Fu, Y.; Cristina Dias Paes, M.; Tako, E. Black Corn (*Zea mays* L.) Soluble Extract Showed Anti-Inflammatory Effects and Improved the Intestinal Barrier Integrity in Vivo (*Gallus gallus*). *Food Res. Int.* **2022**, *157*, 111227. [CrossRef] [PubMed]
46. Kolba, N.; Zarei, A.; Cheng, J.; Agarwal, N.; Dadmohammadi, Y.; Khazdooz, L.; Abbaspourrad, A.; Tako, E. Alterations in Intestinal Brush Border Membrane Functionality and Bacterial Populations Following Intra-Amniotic Administration (*Gallus gallus*) of Nicotinamide Riboside and Its Derivatives. *Nutrients* **2022**, *14*, 3130. [CrossRef]
47. Tako, E. Dietary Plant-Origin Bio-Active Compounds, Intestinal Functionality, and Microbiome. *Nutrients* **2020**, *12*, 3223. [CrossRef]
48. Kolba, N.; Guo, Z.; Olivas, F.M.; Mahler, G.J.; Tako, E. Intra-Amniotic Administration (*Gallus gallus*) of TiO_2, SiO_2, and ZnO Nanoparticles Affect Brush Border Membrane Functionality and Alters Gut Microflora Populations. *Food Chem. Toxicol.* **2019**, *135*, 110896. [CrossRef]
49. Tako, E.; Glahn, R.P.; Knez, M.; Stangoulis, J.C. The Effect of Wheat Prebiotics on the Gut Bacterial Population and Iron Status of Iron Deficient Broiler Chickens. *Nutr. J.* **2014**, *13*, 58. [CrossRef]
50. Zhu, X.Y.; Zhong, T.; Pandya, Y.; Joerger, R.D. 16S RRNA-Based Analysis of Microbiota from the Cecum of Broiler Chickens. *Appl. Environ. Microbiol.* **2002**, *68*, 124–137. [CrossRef]
51. Da Pereira, S.B.; Kolba, N.; Duarte Martino, H.S.; Hart, J.J.; Tako, E. Soluble Extracts from Chia Seed (*Salvia Hispanica* L.) Affect Brush Border Membrane Functionality, Morphology and Intestinal Bacterial Populations In Vivo (*Gallus gallus*). *Nutrients* **2019**, *11*, 2457. [CrossRef]

52. Gomes, M.J.C.; Kolba, N.; Agarwal, N.; Kim, D.; Eshel, A.; Koren, O.; Tako, E. Modifications in the Intestinal Functionality, Morphology and Microbiome Following Intra-Amniotic Administration (*Gallus gallus*) of Grape (Vitis Vinifera) Stilbenes (Resveratrol and Pterostilbene). *Nutrients* **2021**, *13*, 3247. [CrossRef] [PubMed]
53. Beasley, J.T.; Johnson, A.A.T.; Kolba, N.; Bonneau, J.P.; Glahn, R.P.; Ozeri, L.; Koren, O.; Tako, E. Nicotianamine-Chelated Iron Positively Affects Iron Status, Intestinal Morphology and Microbial Populations in Vivo (*Gallus gallus*). *Sci. Rep.* **2020**, *10*, 2297. [CrossRef] [PubMed]
54. Pacifici, S.; Song, J.; Zhang, C.; Wang, Q.; Glahn, R.; Kolba, N.; Tako, E. Intra Amniotic Administration of Raffinose and Stachyose Affects the Intestinal Brush Border Functionality and Alters Gut Microflora Populations. *Nutrients* **2017**, *9*, 304. [CrossRef]
55. Janiak, M.A.; Amarowicz, R.; Rostek, D. Influence of Catechin Fraction and High Molecular Fraction from Green Tea Extract on *Lactobacillus, Bifidobacterium* and *Streptococcus* Strains. *Nat. Prod. Commun.* **2018**, *13*, 1–4. [CrossRef]
56. Shabbir, U.; Rubab, M.; Daliri, E.B.-M.; Chelliah, R.; Javed, A.; Oh, D.-H. Curcumin, Quercetin, Catechins and Metabolic Diseases: The Role of Gut Microbiota. *Nutrients* **2021**, *13*, 206. [CrossRef]
57. Marhuenda-Muñoz, M.; Laveriano-Santos, E.P.; Tresserra-Rimbau, A.; Lamuela-Raventós, R.M.; Martínez-Huélamo, M.; Vallverdú-Queralt, A. Microbial Phenolic Metabolites: Which Molecules Actually Have an Effect on Human Health? *Nutrients* **2019**, *11*, 2725. [CrossRef]
58. Luo, J.; Han, L.; Liu, L.; Gao, L.; Xue, B.; Wang, Y.; Ou, S.; Miller, M.; Peng, X. Catechin Supplemented in a FOS Diet Induces Weight Loss by Altering Cecal Microbiota and Gene Expression of Colonic Epithelial Cells. *Food Funct.* **2018**, *9*, 2962–2969. [CrossRef]
59. Rodríguez, H.; de las Rivas, B.; Gómez-Cordovés, C.; Muñoz, R. Degradation of Tannic Acid by Cell-Free Extracts of Lactobacillus Plantarum. *Food Chem.* **2008**, *107*, 664–670. [CrossRef]
60. Liu, Z.; Bruins, M.E.; Ni, L.; Vincken, J.-P. Green and Black Tea Phenolics: Bioavailability, Transformation by Colonic Microbiota, and Modulation of Colonic Microbiota. *J. Agric. Food Chem.* **2018**, *66*, 8469–8477. [CrossRef]
61. Ma, Y.; Ding, S.; Fei, Y.; Liu, G.; Jang, H.; Fang, J. Antimicrobial Activity of Anthocyanins and Catechins against Foodborne Pathogens *Escherichia coli* and Salmonella. *Food Control.* **2019**, *106*, 106712. [CrossRef]
62. Li, Q.; Van Herreweghen, F.; Onyango, S.O.; De Mey, M.; Van de Wiele, T. In Vitro Microbial Metabolism of (+)-Catechin Reveals Fast and Slow Converters with Individual-Specific Microbial and Metabolite Markers. *J. Agric. Food Chem.* **2022**, acs.jafc.2c00551. [CrossRef] [PubMed]
63. Hodges, J.K.; Sasaki, G.Y.; Bruno, R.S. Anti-Inflammatory Activities of Green Tea Catechins along the Gut–Liver Axis in Nonalcoholic Fatty Liver Disease: Lessons Learned from Preclinical and Human Studies. *J. Nutr. Biochem.* **2020**, *85*, 108478. [CrossRef] [PubMed]
64. Guo, T.; Song, D.; Cheng, L.; Zhang, X. Interactions of Tea Catechins with Intestinal Microbiota and Their Implication for Human Health. *Food Sci. Biotechnol.* **2019**, *28*, 1617–1625. [CrossRef] [PubMed]
65. Zheng, C.; Liu, R.; Xue, B.; Luo, J.; Gao, L.; Wang, Y.; Ou, S.; Li, S.; Peng, X. Impact and Consequences of Polyphenols and Fructooligosaccharide Interplay on Gut Microbiota in Rats. *Food Funct.* **2017**, *8*, 1925–1932. [CrossRef]
66. Liao, X.; Shao, Y.; Sun, G.; Yang, Y.; Zhang, L.; Guo, Y.; Luo, X.; Lu, L. The Relationship among Gut Microbiota, Short-Chain Fatty Acids, and Intestinal Morphology of Growing and Healthy Broilers. *Poult. Sci.* **2020**, *99*, 5883–5895. [CrossRef]
67. Rechkemmer, G.; von Engelhardt, W. Concentration- and PH-Dependence of Short-Chain Fatty Acid Absorption in the Proximal and Distal Colon of Guinea Pig (*Cavia porcellus*). *Comp. Biochem. Physiol. Part A Physiol.* **1988**, *91*, 659–663. [CrossRef]
68. Brown, A.J.; Goldsworthy, S.M.; Barnes, A.A.; Eilert, M.M.; Tcheang, L.; Daniels, D.; Muir, A.I.; Wigglesworth, M.J.; Kinghorn, I.; Fraser, N.J.; et al. The Orphan G Protein-Coupled Receptors GPR41 and GPR43 Are Activated by Propionate and Other Short Chain Carboxylic Acids. *J. Biol. Chem.* **2003**, *278*, 11312–11319. [CrossRef]
69. Miyauchi, S.; Gopal, E.; Fei, Y.-J.; Ganapathy, V. Functional Identification of SLC5A8, a Tumor Suppressor Down-Regulated in Colon Cancer, as a Na+-Coupled Transporter for Short-Chain Fatty Acids. *J. Biol. Chem.* **2004**, *279*, 13293–13296. [CrossRef]
70. Ritzhaupt, A.; Wood, I.S.; Ellis, A.; Hosie, K.B.; Shirazi-Beechey, S.P. Identification and Characterization of a Monocarboxylate Transporter (MCT1) in Pig and Human Colon: Its Potential to Transport L -Lactate as Well as Butyrate. *J. Physiol.* **1998**, *513*, 719–732. [CrossRef]
71. Park, J.; Kotani, T.; Konno, T.; Setiawan, J.; Kitamura, Y.; Imada, S.; Usui, Y.; Hatano, N.; Shinohara, M.; Saito, Y.; et al. Promotion of Intestinal Epithelial Cell Turnover by Commensal Bacteria: Role of Short-Chain Fatty Acids. *PLoS ONE* **2016**, *11*, e0156334. [CrossRef]
72. Lukovac, S.; Belzer, C.; Pellis, L.; Keijser, B.J.; de Vos, W.M.; Montijn, R.C.; Roeselers, G. Differential Modulation by Akkermansia Muciniphila and Faecalibacterium Prausnitzii of Host Peripheral Lipid Metabolism and Histone Acetylation in Mouse Gut Organoids. *mBio* **2014**, *5*, e01438-14. [CrossRef]
73. Kaiko, G.E.; Ryu, S.H.; Koues, O.I.; Collins, P.L.; Solnica-Krezel, L.; Pearce, E.J.; Pearce, E.L.; Oltz, E.M.; Stappenbeck, T.S. The Colonic Crypt Protects Stem Cells from Microbiota-Derived Metabolites. *Cell* **2016**, *165*, 1708–1720. [CrossRef]
74. Zhang, J.; Yi, M.; Zha, L.; Chen, S.; Li, Z.; Li, C.; Gong, M.; Deng, H.; Chu, X.; Chen, J.; et al. Sodium Butyrate Induces Endoplasmic Reticulum Stress and Autophagy in Colorectal Cells: Implications for Apoptosis. *PLoS ONE* **2016**, *11*, e0147218. [CrossRef] [PubMed]

75. Limage, R.; Tako, E.; Kolba, N.; Guo, Z.; García-Rodríguez, A.; Marques, C.N.H.; Mahler, G.J. TiO_2 Nanoparticles and Commensal Bacteria Alter Mucus Layer Thickness and Composition in a Gastrointestinal Tract Model. *Small* **2020**, *16*, 2000601. [CrossRef] [PubMed]
76. Hider, R.C.; Liu, Z.D.; Khodr, H.H. Metal Chelation of Polyphenols. In *Methods in Enzymology*; Elsevier: Amsterdam, The Netherlands, 2001; Volume 335, pp. 190–203, ISBN 978-0-12-182236-1.
77. Scalbert, A.; Morand, C.; Manach, C.; Rémésy, C. Absorption and Metabolism of Polyphenols in the Gut and Impact on Health. *Biomed. Pharmacother.* **2002**, *56*, 276–282. [CrossRef]
78. Quesada, I.M.; Bustos, M.; Blay, M.; Pujadas, G.; Ardèvol, A.; Salvadó, M.J.; Bladé, C.; Arola, L.; Fernández-Larrea, J. Dietary Catechins and Procyanidins Modulate Zinc Homeostasis in Human HepG2 Cells. *J. Nutr. Biochem.* **2011**, *22*, 153–163. [CrossRef] [PubMed]
79. Kagaya, N.; Kawase, M.; Maeda, H.; Tagawa, Y.; Nagashima, H.; Ohmori, H.; Yagi, K. Enhancing Effect of Zinc on Hepatoprotectivity of Epigallocatechin Gallate in Isolated Rat Hepatocytes. *Biol. Pharm. Bull.* **2002**, *25*, 1156–1160. [CrossRef]
80. Sun, S.; He, G.; Yu, H.; Yang, J.; Borthakur, D.; Zhang, L.; Shen, S.; Das, U.N. Free Zn2+ Enhances Inhibitory Effects of EGCG on the Growth of PC-3 Cells. *Mol. Nutr. Food Res.* **2008**, *52*, 465–471. [CrossRef]
81. Esparza, I.; Salinas, Í.; Santamaría, C.; García-Mina, J.M.; Fernández, J.M. Electrochemical and Theoretical Complexation Studies for Zn and Cu with Individual Polyphenols. *Anal. Chim. Acta* **2005**, *543*, 267–274. [CrossRef]
82. Daniel, H.; tom Dieck, H. Nutrient-Gene Interactions: A Single Nutrient and Hundreds of Target Genes. *Biol. Chem.* **2004**, *385*, 571–583. [CrossRef]
83. Dieck, H.T.; Döring, F.; Fuchs, D.; Roth, H.-P.; Daniel, H. Transcriptome and Proteome Analysis Identifies the Pathways That Increase Hepatic Lipid Accumulation in Zinc-Deficient Rats. *J. Nutr.* **2005**, *135*, 199–205. [CrossRef] [PubMed]
84. Lagha, A.B.; Grenier, D. Tea Polyphenols Inhibit the Activation of NF-KB and the Secretion of Cytokines and Matrix Metalloproteinases by Macrophages Stimulated with Fusobacterium Nucleatum. *Sci. Rep.* **2016**, *6*, 34520. [CrossRef] [PubMed]
85. Elphick, D.A. Paneth Cells: Their Role in Innate Immunity and Inflammatory Disease. *Gut* **2005**, *54*, 1802–1809. [CrossRef] [PubMed]
86. Gutsmann, T.; Schromm, A.; Brandenburg, K. The Physicochemistry of Endotoxins in Relation to Bioactivity. *Int. J. Med. Microbiol.* **2007**, *297*, 341–352. [CrossRef] [PubMed]
87. Ghareeb, K.; Awad, W.A.; Soodoi, C.; Sasgary, S.; Strasser, A.; Böhm, J. Effects of Feed Contaminant Deoxynivalenol on Plasma Cytokines and MRNA Expression of Immune Genes in the Intestine of Broiler Chickens. *PLoS ONE* **2013**, *8*, e71492. [CrossRef] [PubMed]
88. Surh, Y.-J.; Chun, K.-S.; Cha, H.-H.; Han, S.S.; Keum, Y.-S.; Park, K.-K.; Lee, S.S. Molecular Mechanisms Underlying Chemopreventive Activities of Anti-Inflammatory Phytochemicals: Down-Regulation of COX-2 and INOS through Suppression of NF-KB Activation. *Mutat. Res./Fundam. Mol. Mech. Mutagenesis* **2001**, *480–481*, 243–268. [CrossRef]
89. Hamer, M. The Beneficial Effects of Tea on Immune Function and Inflammation: A Review of Evidence from in Vitro, Animal, and Human Research. *Nutr. Res.* **2007**, *27*, 373–379. [CrossRef]
90. Parada Venegas, D.; De la Fuente, M.K.; Landskron, G.; González, M.J.; Quera, R.; Dijkstra, G.; Harmsen, H.J.M.; Faber, K.N.; Hermoso, M.A. Short Chain Fatty Acids (SCFAs)-Mediated Gut Epithelial and Immune Regulation and Its Relevance for Inflammatory Bowel Diseases. *Front. Immunol.* **2019**, *10*, 277. [CrossRef]

nutrients MDPI

Article

Comparing the Effects of Concord Grape (*Vitis labrusca* L.) Puree, Juice, and Pomace on Intestinal Morphology, Functionality, and Bacterial Populations In Vivo (*Gallus gallus*)

Nikita Agarwal, Viral Shukla, Nikolai Kolba, Cydney Jackson, Jacquelyn Cheng, Olga I. Padilla-Zakour and Elad Tako *

Department of Food Science, Cornell University, Stocking Hall, Ithaca, NY 14850, USA
* Correspondence: et79@cornell.edu; Tel.: +1-607-255-0884

Abstract: This is a preliminary study evaluating the effect of different fractions of Concord grapes (*Vitis labrusca* L.) on the brush border membrane (BBM) morphology, duodenal gene expression, and specific gut bacterial populations. For this study, we utilized a unique intraamniotic approach, wherein, the test substances are administered into the amnion of the *Gallus gallus* egg (on day 17). The embryo orally consumes the amniotic fluid along with the injected test substance before the hatch. We randomly divided ~50 fertilized eggs into 5 groups including 6% grape (juice, puree, and pomace) along with controls (no injection and diluent—H_2O). The grape juice was prepared by crushing the grapes; the grape residues were used as pomace. The grape puree included the grape skin, endocarp, mesocarp, and juice but not the seeds. On day 21, the hatch day, the blood, pectoral muscle, liver, duodenum, and large intestine were harvested. Our results showed no significant differences in blood glucose, pectoral glycogen level, or body weight. However, significant ($p < 0.05$) differences in duodenal and liver gene expression were observed between the treatment groups. The grape puree treatment resulted in higher *Clostridium* numbers and lower *Bifidobacterium* numbers when compared to all other groups. In summary, the dietary consumption of grape polyphenols has the potential to beneficially modulate aspects of intestinal health provided their concentration is limited.

Keywords: grape pomace; stilbenes; intra-amniotic administration; poultry feed; gut microbiome; brush border membrane; in-ovo; nutrition

Citation: Agarwal, N.; Shukla, V.; Kolba, N.; Jackson, C.; Cheng, J.; Padilla-Zakour, O.I.; Tako, E. Comparing the Effects of Concord Grape (*Vitis labrusca* L.) Puree, Juice, and Pomace on Intestinal Morphology, Functionality, and Bacterial Populations In Vivo (*Gallus gallus*). *Nutrients* **2022**, *14*, 3539. https://doi.org/10.3390/nu14173539

Academic Editor: Max Petrov

Received: 1 August 2022
Accepted: 25 August 2022
Published: 27 August 2022

1. Introduction

The Concord grape is native to North America and was first propagated in New York State in the 1870s [1]. The state currently is the second-largest producer (after Washington) and houses the largest Concord grape industry in the country [2]. The demand for grapes and their products was 53.8 million in 2020 in the United States alone [3]. The grape industry squeezes out the juice from the fruit leaving behind the skin, seeds, and flesh. These by-products, collectively referred to as grape pomace (GP), account for 25% of the harvested grape by weight [4]. Grape pomace has a high concentration of health-promoting polyphenols with a low pH making GP resistant to biological degradation [5]. Currently, the use of GP is inadequate, with most of it being dumped in landfills causing economic and environmental problems [6]. Grape puree (GPR), consisting of skin, flesh, and juice, is a by-product of grape seed extracts [7]. The objective of this study was to compare the different grape fractions—puree, juice, and pomace for their effects on intestinal morphology (villi surface area, goblet and Paneth cell number, and diameter), functionality (duodenal gene expression), and gut microbial populations, in-vivo (*Gallus gallus*).

The bioactives and micronutrients are concentrated in different fractions/matrices of the grape [8]. Another red grape variety (Bordo) was found to have the majority of its hydroxybenzoic acids (including gallic, syringic, and ellagic), hydroxycinnamic acids, flavonol- epicatechin, proanthocyanidin—B2, flavonols (myricetin, quercetin), and anthocyanins (cyanidin, delphinidin, malvidin, and peonidin) concentrated in the peel compared to its pulp or seed [8]. As the pulp fraction is constituted of ~80% water, most of the micronutrients and crude fiber are found in seed and peel [9]. Grape seeds are said to have 11% fiber, 3% minerals, 35% fiber, and only 7% water [10]. Given the nutritional composition, low cost, and sustainability aspects of utilizing GP, numerous studies have been conducted to valorize GP [5,6,11].

Pharmacologically, grape skin has been said to have antioxidant [12,13], cardio-protectant [14], anticancer [15,16], anti-inflammatory [17,18], and anti-microbial activity [19,20]. Whereas grape seeds and grape seed extracts have been extensively studied and said to be anti-diabetic, antioxidant, anti-platelet, anti-cholesterol, anti-inflammatory, anti-aging, anti-neurodegenerative, and anti-microbial agents [21–24]. In the present study, however, we were focused on comparing grape fractions for their effects on molecular, morphological, and microbial aspects of intestinal health, in-vivo (*Gallus gallus*).

Diet is one of the major factors that influence gut microbiota [25,26]. The microbiota in turn affects digestion and nutrient absorption, among others [27]. The microbes ferment the food and produce a range of metabolites as by-products including amino acids, short-chain fatty acids, and enzymes, among others [28]. These metabolites are in direct contact with the cells of the brush border membrane (BBM) including goblet cells, Paneth cells, and enterocytes affecting their proliferation [29]. The increased proliferation of enterocytes leads to an increase in absorptive (villus) surface area which in turn aids in efficient digestion and nutrient absorption, among others. On the other hand, goblet and Paneth cells are secretory cells producing mucin and anti-microbial peptides, respectively. Mucin forms the mucus layer that houses the microbiome, and the peptides regulate the relative abundance of the resident bacterial populations [30].

The *Gallus gallus* in-vivo model is a physiologically relevant model for assessing the effects of bioactive compounds on humans as the two have similar intestinal morphology, microbiota (phylum-level), and certain metabolism-related genes [31–40]. Here we utilize the embryonic phase of the bird, closed off from the external environment (in the egg) with the only variable being the injected treatment. Followed by multi-level analysis to potentially uncover the mechanism by which these grape fractions may benefit health.

2. Materials and Methods

2.1. Sample Preparation

Concord grapes (*Vitis labrusca* L.) were picked locally and processed at the Cornell Food Venture Center Pilot Plant (Geneva, NY, USA). Grapes were manually destemmed and washed. The grape puree (GPR) was made by grinding whole grapes in a bench-scale processor (Robo Coupe; Jackson, Mississippi) for 30 s at 1500 RPM. Seeds were manually removed, and the resulting pulp was then freeze-dried (Max53, Millrock Technology, Kingston, NY, USA) for 48 h. The dried mass was then ground into a fine powder using a bench-scale processor. Grape juice (GJ) was made to mimic typical industry standards. GJ was made by pre-crushing and enzymatically treating (Rapidase® added at 0.2 mL/kg, DSM Food Specialties USA, Inc., South Bend, IN, USA) the grapes at 50 °C for 1 h. They were then pressed using a pilot-scale hydraulic press (Orchard Equipment Co., Conway, MA, USA) at 8.27×10^6 —9.64×10^6 Pa. The resulting juice was cold stabilized at 2 °C for 48 h. The juice was commercially sterilized by hot-packing into bottles at 85 °C for 2 min (Microthermics; Raleigh, North Carolina, USA). Grape pomace (GP) was made by taking the resulting pomace from the juice pressing and freeze-drying for 48 h. The dried mass was ground into a fine powder using a bench-scale grinder. The powders were vacuumed sealed, and all samples were kept frozen until use. Figure 1 depicts the processes carried out.

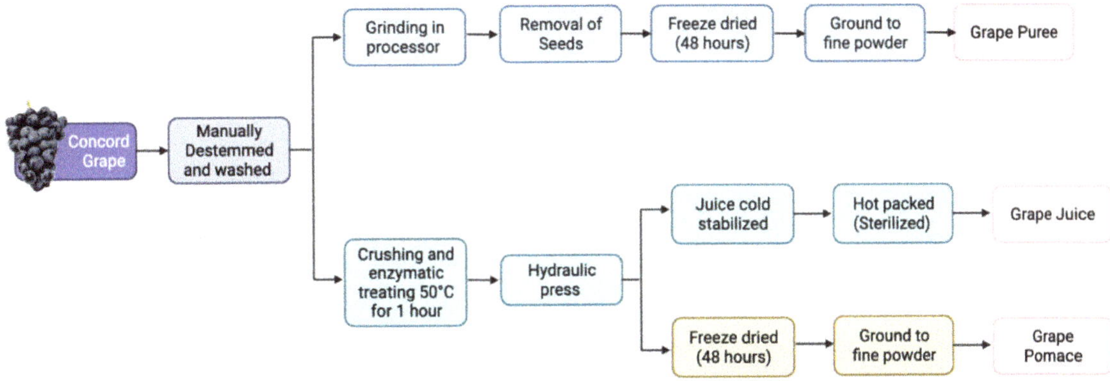

Figure 1. Depicts the processes carried out to prepare the grape puree, pomace, and juice. The Concord grapes were hand-picked, manually destemmed, and washed. The juice was cold pressed followed by sterilization. The grape puree and pomace were freeze-dried. The resultant fractions included: puree (skin + juice + flesh), juice, and pomace (skin + seeds + flesh).

2.2. Polyphenols and Carbohydrate Analysis

2.2.1. Grape Sample Preparation

Grape samples were extracted using absolute methanol under dark and constant agitation for 2 h. The resulting slurry was centrifuged and decanted to obtain the supernatant. The resulting extract and washings were diluted in distilled water to achieve a 15% w/v extract which was used for further analysis as below.

2.2.2. Polyphenol Analysis

Total polyphenol content (TPC) was determined using the Folin-Ciocalteu method as described by Waterhouse [41]. Briefly, the extract was reacted with Folin-Ciocalteu reagent and allowed to incubate at room temperature. The reaction was then quenched with sodium carbonate solution. The samples were then immediately measured for absorbance at 765 nm using a UV-visible spectrophotometer (Thermo Fisher; Waltham, MA, USA). TPC was calculated as gallic equivalents (GE) using a standard curve prepared under the same conditions.

Total monomeric anthocyanin (MA) content was determined using the pH differential method [42]. Briefly, extracts were diluted with pH 1.0 (0.025 M potassium chloride) and pH 4.5 (0.4 M sodium acetate) buffers and allowed to incubate at room temperature for 20 min. Absorbance was measured at 520 nm and 700 nm using a UV-visible spectrophotometer. Total MA content was calculated as cyanidin-3-glucoside equivalents (CE) using the equation below:

$$((\text{A520, pH1} - \text{A700, pH1}) - (\text{A520, pH4.5} - \text{A700, pH4.5})) \times 529 \times \text{dilute factor} \times 1000/28{,}000$$

2.2.3. Fibrous and Non-Fibrous Carbohydrate Analysis

The non-fibrous carbohydrate analysis (NFC) was conducted according to AOAC 962.09. Acid detergent fiber (ADF) and neutral detergent fiber (NDF) analyses were conducted according to 973.18. The analysis was performed by Dairy One Co-Op Inc (Ithaca, NY, USA).

2.3. Animals and Study Design

Fertile Cornish Cross broiler eggs (n = 60) were purchased from a commercial hatchery (Moyer's Chicks, Quakertown, PA, USA). The eggs were incubated under optimal conditions at the Cornell University Animals Science poultry farm incubator. All animal protocols were approved by the Cornell University Institutional Animal Care and Use Committee (ethic approval code: 2020-0077).

2.3.1. Water Extract Preparation

The grape pomace, puree, and juice were further extracted and diluted (in water) before they were administered into the amniotic fluid of the egg. The osmolarity of 4,6,8, and 10% solutions were tested to ensure the value was below 320 Osm/kg H_2O and 6% was selected. The day before injection (i.e., day 16 of embryonic incubation) the respective 6% solutions were prepared in distilled water. The solutions were immersed in a water bath at 60 °C for 60 min. The solutions were centrifuged, and the supernatant was collected and stored at −20 °C until the next day.

2.3.2. Intra-Amniotic Administration and Sample Collection

On day 17 of embryonic incubation, the prepared water extracts (Section 2.3.1) were thawed at 21 °C for 1 h and then placed alongside the eggs in the incubator. Candling was used to distinguish the viable eggs, and the non-viable eggs were appropriately discarded. The viable eggs were weighed and randomly distributed into 5 groups with approximately 10 eggs each. The amniotic fluid, i.e., the injection spot, was determined using candling and was marked. A 1 mL aliquot of the water extracts was then injected, respectively, using a 21-gauge needle and covered with cellophane tape. The 5 groups were assigned as follows: controls, (1) no injection, (2) H_2O and treatments all 6% solutions of (3) GJ (4) GP, (5) GPR. The injected eggs were placed in hatching baskets in the incubator until hatch day, i.e., day 21. Soon after hatching, the chicks were euthanized in a CO_2 chamber. The blood, pectoral muscle, duodenum, cecum, and liver were collected. The blood and pectoral muscle were placed on ice, whereas the other tissues were immediately kept in liquid nitrogen. The samples were shifted to −20 °C until further analysis.

2.4. Blood Glucose Measurements

Blood was collected in 1.5 mL Eppendorf tubes and stored at 4 °C until analysis. The Accu-Chek® blood glucose monitor (Roche, Indianapolis, IN, USA) was used to determine blood glucose levels. Exactly 0.6 μL of blood was placed on the disposable electrochemical test stripe. The digitally displayed reading corresponding to each sample was recorded.

2.5. Pectoral Muscle–Glycogen Content

Pectoral muscle samples were weighed (20 ± 1 mg) and homogenized in 8% perchloric acid following the method described by Dreiling et al. [43]. Glycogen was estimated using spectroscopy measuring the wavelength at 450 nm and calculated against a standard curve [36].

2.6. Gene Expression Analysis

2.6.1. Isolation of Total RNA from Duodenum and Liver Tissue Samples

Total RNA was isolated using Qiagen RNeasy Mini Kit (RNeasy Mini Kit, Qiagen Inc., Valencia, CA, USA) following the manufacturer's protocol. Briefly, 30 ± 2 mg of the proximal duodenal tissue and liver tissue (n = 5) were weighed in 2 mL tubes. RLT® (plus β-mercaptoethanol) was added to each tube and a rotor-stator homogenizer (Omni International, Inc, Kennesaw, GA, USA) was used to disrupt the tissue samples. The lysate so obtained was centrifuged at 8000× g for 3 min at 20 °C. The supernatant was transferred to a new tube and 700 μL of 70% ethanol was added to each tube. The sample was then run through the RNeasy kit's mini-column and centrifuged at 8000× g for 15 s. The flow-through was discarded. To this, 500 μL of RPE® buffer was added and centrifuged at 8000 × g for 15 s. This step with RPE® was repeated for 2 min. The filter part of the column alone was transferred to a new 1.5 mL collection tube and 50 μL of RNase-free water was used to elute RNA from the filter into the tube. Eluted RNA was stored at −80 °C until analysis. All steps were carried out in an RNase-free environment. RNA was quantified using a spectrophotometer at 260/280 nm. Gel electrophoresis (EtBr stain) was used to verify the integrity of the RNA obtained. DNA contamination was removed using TURBO DNase removal kit (AMBION, Austin, TX, USA).

2.6.2. Real-Time Polymerase Chain Reaction (RT-PCR)

cDNA was created using the 20 μL reverse transcriptase (RT) reaction, performed in a BioRad C1000 Touch Thermocycler using the Improm-II Reverse Transcriptase Kit (Promega, Madison, WI, USA). The concentration of cDNA was assessed using Nanodrop (Thermo Fisher Scientific, Waltham, MA, USA). Further details can be found in a previous publication [31].

2.6.3. Primer Design

The primers were designed based on gene sequences from the GenBank database, and the Real-Time Primer Design Tool software (IDT DNA, Coralville, IA, USA) was used [32]. The primer sequences related to iron, zinc, Vitamin A metabolism, immune response, and brush border membrane functionality that was used in this study are summarized in Table 1. The specificity of the primers was verified using the BLAST search against the genomic National Center for Biotechnology Information (NCBI) database. The reference gene used was the 18S rRNA specific for the *Gallus gallus* model.

Table 1. The sequences of the primers (both forward and reserve) used in this study are displayed. GenInfo Identifier number and base-pair lengths have also been specified. All were assessed in the duodenum except for those represented by * these were assessed in the liver.

Analyte	Forward Primer (5′-3′)	Reverse Primer (5′-3′)	Base Pairs Length	GI Number
Iron Metabolism				
DMT1	TTGATTCAGAGCCTCCCATTAG	GCGAGGAGTAGGCTTGTATTT	101	751817
Ferroportin	CTCAGCAATCACTGGCATCA	ACTGGGCAACTCCAGAAATAAG	98	423984
DcytB	CATGTGCATTCTCTTCCAAAGTC	CTCCTTGGTGACCGCATTAT	103	20380692
Hepcidin *	AGACGACAATGCAGACTAACC	CTGCACCAATCCCACATTTC	132	SAMN08056490
Immune Response				
NF-κB	CACAGCTGGAGGGAAGTAAAT	TTGAGTAAGGAAGTGAGGTTGAG	100	396033
IL-6	ACCTCATCCTCCGAGACTTTA	GCACTGAAACTCCTGGTCTT	105	395337
TNF-α	GACAGCCTATGCCAACAAGTA	TTACAGGAAGGGCAACTCATC	109	374125
Zinc Metabolism				
ZnT1	GGTAACAGAGCTGCCTTAACT	GGTAACAGAGCTGCCTTAACT	105	423089
ZnT7	GGAAGATGTCAGGATGGTTCA	CGAAGGACAAATTGAGGCAAAG	87	424464
ZIP4	TCTCCTTAGCAGACAATTGAG	GTGACAAACAAGTAGGCGAAAC	95	107050877
ZIP1	TGCCTCACTTTCCCTCAC	GGCTCTTAAGGGCACTTCT	144	121112053
Vitamin A Metabolism				
CRBP2	GGCTACATGGTTGCACTAGACA	AACCACCCGGTTATCGAGTC	195	NM_001277417.1
LRAT	GATTTTGCCTATGGCGGCAG	TTGTCGGTCTGGAAGCTGAC	197	22403
STRA6 *	GTGCGCTGAACTTTGTCTGC	TTCTTCCTGCTCCCGACCT	116	415301
RBP4 *	TGCCACCAACACAGAACTCTC	CTTTGAAGCTGCTCACACGG	149	396454
BBM Functionality				
VDAC2	CAGCACTCGCTTTGGAATTG	GTGTAACCCACTCCAACTAGAC	99	395498
SI	CCAGCAATGCCAGCATATTG	CGGTTTCTCCTTACCACTTCTT	95	425007
OCLN	GTCTGTGGGTTCCTCATCGT	GTTCTTCACCCACTCCTCCA	124	396026
MUC6	CCAACTTGCAGTGTTCCAAAG	CTGACAGTGTAGAGCAAGTACAG	106	414878
18s rRNA	GCAAGACGAACTAAAGCGAAAG	TCGGAACTACGACGGTATCT	100	7262899

DMT1: Divalent metal transporter 1; Dcytb: Duodenal cytochrome b; NF-κB: Nuclear factor-kappa beta; IL6: Interleukin 6; TNF-α: Tumor Necrosis Factor Alpha; ZIP: Zrt-, Irt-like proteins; ZnT: zinc transporter; CRBP2: Cellular retinol-binding protein; LRAT: Lecithin retinol acyltransferase; STRA6: Signaling receptor and transporter of retinol; RBP: Retinol-binding protein; VDAC: Voltage-dependent anion channel; SI: Sucrose isomaltase; OCLN: Occludin; MUC6: Mucin.

2.6.4. Real-Time qPCR Design

All procedures were conducted as previously described [36,38,39,44]. Briefly, cDNA was used for each 10 μL reaction together with 2×BioRad SSO Advanced Universal SYBR Green Supermix (Hercules, CA, USA) which included buffer, Taq DNA polymerase, dNTPs, and SYBR green dye. Specific primers (forward and reverse) (Table 1) and cDNA or water (for no template control) were added to each PCR reaction. For each gene, the optimal $MgCl_2$ concentration produced the amplification plot with the lowest cycle product (Cp), the highest fluorescence intensity, and the steepest amplification slope. Master mix (8 μL) was pipetted into the 96-well plate and 2 µL cDNA was added as a PCR template. Each run contained 7 standard curve points in duplicate. A no-template control of nuclease-free water was included to exclude DNA contamination in the PCR mix. The double-stranded DNA was amplified in the Bio-Rad CFX96 Touch (Hercules, CA, USA) using the following PCR conditions: initial denaturing at 95 °C for 30 s, 40 cycles of denaturing at 95 °C for 15 s, various annealing temperatures according to Integrated DNA Technologies (IDT) for 30 s and elongating at 60 °C for 30 s.

The data on the expression levels of the genes were obtained as Cp values based on the "second derivative maximum" (automated method) as computed by Bio-Rad CFX Maestro 1.1 (Hercules, CA, USA). For each evaluated gene, the reactions were run in duplicate. All assays were quantified by including a standard curve in the real-time qPCR analysis. The next four points of the standard curve were prepared by a 1:10 dilution, in duplicate. A graph of Cp vs. log 10 concentrations was produced by the software and the efficiencies were calculated as 10[1/slope]. The specificity of the amplified real-time RT-PCR products was verified by melting curve analysis (60–95 °C) after 40 cycles, which should result in several different specific products, each with a specific melting temperature.

2.7. Intestinal Bacterial Population Assessment

2.7.1. Intestinal Sample Collection and DNA Extraction

The cecum samples (n = 5) were weighed (0.2 ± 0.02 g) under aseptic conditions. Details of the experiment can be found as previously described [31]. The tissues were sheared using beads and a vortex. EDTA and lysozyme were used for DNA extraction in addition to the Wizard Genomic DNA purification kit (Promega Corp., Madison, WI, USA).

2.7.2. Primer Design and PCR Amplification

Primers for genus *Clostridium*, *Bifidobacterium*, *Lactobacillus*, *Klebsiella*, and species *E. coli* and *L. plantarum* were designed as per previous literature [45–48]. The primers used in this study are detailed previously [49]. All known bacteria were identified using a Universal primer. The PCR amplification was carried out by adding the DNA extracted (as template) to a PCR premixture containing nuclease-free water, PCR buffer, Taq polymerase, dNTPs, and primer. PCR conditions were set as previously optimized [49]. The PCR products were quantified using gel electrophoresis with ethidium bromide. Visualized in Gel-Pro analyzer version 3.0 (Media Cybernetics LP, Rockville, MD, USA) [47].

2.8. Morphometric Examination of Duodenal Tissue

The duodenal morphological examination was conducted as previously described [31]. Briefly, the duodenal loop samples collected (n = 5) were fixed (with 4% (v/v) buffered formaldehyde) on a slide (four sections for each sample), deparaffinized in xylene, rehydrated in ethanol, and stained with a combination of Alcian blue/periodic acid-Schiff (PAS). Slides were examined under a light microscope (BX3M series, Olympus Waltham, MA, USA). To count and measure the CellSens Standard Software was used. A total of 40 villi per cross-section (4 sections per sample) were measured for villus surface area (1 length and an average of 3 widths). Villus surface area was calculated using the following equation:

$$Villus\ surface\ area = 2\pi\ x \frac{vw}{2} x\ VL$$

where vw is the average of three measurements of villus width, and VL is the villus length. Ten villi per sample were counted for number of goblet cells (blue–acidic, purple–mixed, and pink–neutral) and diameter. Ten circular crypts per sample were counted for Paneth cell number and diameter.

2.9. Statistical Analysis

The results are all (unless specified otherwise) expressed as means ± Standard Error Means (SEM) in tables and heatmaps. Heatmaps were created in Microsoft Excel (Microsoft

Corporation, Redmond, WA, USA) based on conditional formatting using color scales based on result means. Experiment groups were all assigned randomly ensuring even weight distribution to all groups. To assess distribution normality, the Shapiro-Wilk test was used. Normally distributed experimental group results were analyzed by one-way ANOVA. Analysis of Variance (ANOVA) was followed by a posthoc Duncan test using SPSS software version 27 (version 26.0, IBM, Armonk, NY, USA) with a significance value ($p < 0.05$).

3. Results

3.1. Polyphenol and Carbohydrate Analysis

The analysis revealed that the Grape Pomace (GP) extract had the most total polyphenols and highest (acid/neutral) detergent fiber content; whereas, the GPR corresponded to the highest amount of monomeric anthocyanins and non-fiber carbohydrates ($p < 0.05$, Table 2).

Table 2. Total polyphenol content (TPC), monomeric anthocyanins (MA), acid detergent fiber (ADF), neutral detergent fiber (NDF), and non-fiber carbohydrates (NFC) were estimated in the different grape fractions as appropriate.

Sample	TPC (mg/g GAE)	MA (CE/g)	ADF (%/DM)	NDF (%/DM)	NFC (%/DM)
GJ	2.4 ± 0.00^{c}	858 ± 256^{c}	NA	NA	NA
GP	11.6 ± 0.05^{a}	2353 ± 159^{b}	41.1	43.7	29.4
GPR	7.1 ± 0.30^{b}	2544 ± 91^{a}	5.7	6.3	82.8

Values are the means ± standard deviation, superscripts in the same column indicate a significant difference ($p < 0.05$). ADF cellulose, lignin, and insoluble minerals; NDF-cellulose, lignin, insoluble minerals, and hemicellulose; NFC-sugars, starches, organic acids, and pectin. GAE–gallic acid equivalence; CE-cyanidin-3-glucoside equivalents; DM–dry matter.

3.2. Hatchability and Body Weight

The hatchability with Grape Puree (GPR) and Grape Pomace (GP) was lower than expected. Whereas the more dilute treatment group-Grape Juice (GJ) resulted in 100% embryo survival as indicated in Table 3. No significant differences were observed in the body weight of the hatchlings.

Table 3. Hatchability, body weight, blood glucose, and pectoral glycogen values.

Treatment Group	Hatch/Injected	Body Weight (g)	Blood Glucose (mg/dL)	Glycogen (mg/g)
No Injection	9/10	40.8 ± 1.2^{a}	254 ± 24^{a}	0.40 ± 0.10^{a}
H_2O	9/10	38.3 ± 4.3^{a}	234 ± 11^{a}	0.30 ± 0.09^{a}
GJ	13/13	38.9 ± 1.7^{a}	226 ± 12^{a}	0.30 ± 0.06^{a}
GP	6/10	36.8 ± 1.2^{a}	314 ± 0.1^{a}	0.23 ± 0.11^{a}
GPR	9/12	39.6 ± 0.8^{a}	226 ± 12^{a}	0.32 ± 0.10^{a}

Values are the means ± SEM (n = 8). Treatment groups are all indicated by the same letter [a] hence statistically insignificant ($p < 0.05$).

3.3. Blood Glucose and Pectoral Glycogen Analysis

The different treatment groups did not result in significantly ($p < 0.05$) different values when compared to the controls and each other as seen in Table 3.

3.4. Duodenal Gene Expression

Figure 2 illustrates the differences in gene expression of proteins related to Zn, Fe, Vitamin A metabolism, inflammatory cytokines, and BBM functionality. The duodenum is the main site of Zn, Fe, and Vitamin A absorption in *Gallus gallus* [50,51]. The grape fractions did not result in any significant differences in Zn metabolism-related genes (ZnT1, ZnT7,

ZIP4, and ZIP1) between the different groups. Similarly, the expression of Fe metabolism-related genes (DcytB, DMT1, and Ferroportin), inflammation-related genes (*IL-6*, NF-κB, and TNF-α), and BBM functionality biomarkers (*OCLN*, *SI*, and *MUC6*) did not significantly change when comparing the H_2O Injection to the other treatment groups; however, the expression of *VDAC2* was upregulated in GP (6% grape pomace) compared to all other groups. Vitamin A metabolism-related gene *CRBP2* was upregulated in GP compared to all other groups, and *LRAT* was significantly different between GJ and GPR.

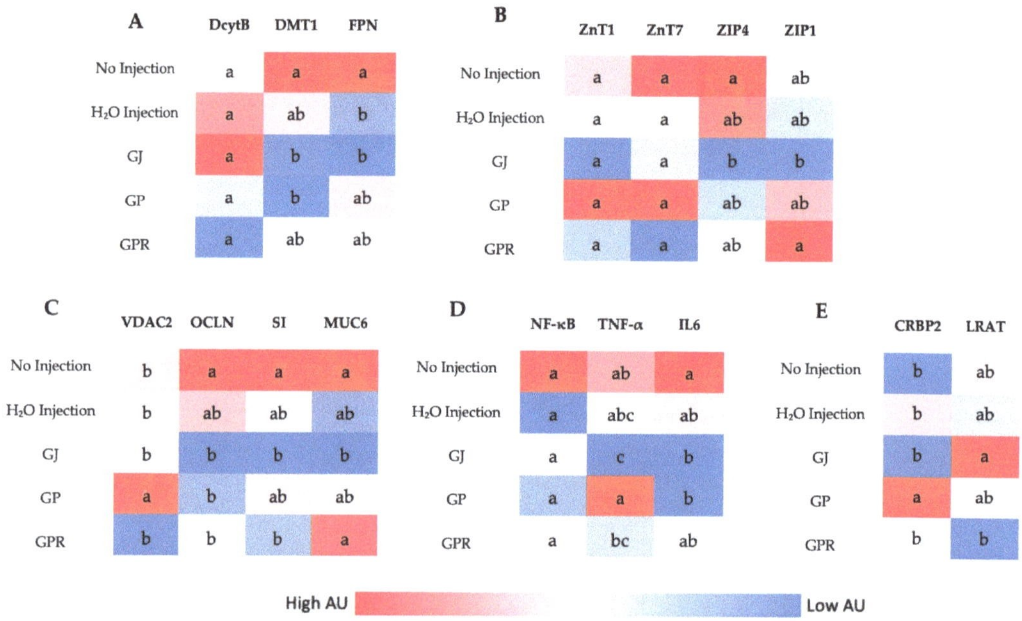

Figure 2. Heatmap showing the effect of intra-amniotic administration of grape juice (GJ), pomace (GP), and puree (GPR) on duodenal gene expression. (**A**)—iron metabolism-related genes, (**B**)—zinc metabolism-related genes, (**C**)—brush border membrane functionality genes, (**D**)—genes coding for inflammatory cytokines and (**E**)—Vitamin A metabolism-related genes. Values are in arbitrary units (AU). Genes not indicated by the same letter (a,b,c) are significantly different ($p < 0.05$). Key; Iron metabolism (DMT1: Divalent metal transporter 1; DcytB: Duodenal cytochrome b), Zinc metabolism (ZIP: Zrt-, Irt-like proteins; ZnT: zinc transporter), Vitamin A metabolism (CRBP2: Cellular retinol-binding protein; LRAT: Lecithin retinol acyltransferase), inflammatory cytokines (NF-κB: Nuclear factor-kappa beta; IL6: Interleukin 6; TNF-α: Tumor Necrosis Factor Alpha), Brush border membrane functionality (VDAC: Voltage-dependent anion channel; SI: Sucrose isomaltase; OCLN: Occludin; MUC6: Mucin).

3.5. Liver Gene Expression

Figure 3 depicts no significant differences in gene expression of Vitamin A metabolism-related genes *RBP4*, *STRA6*, and iron metabolism-related protein Hepcidin (when compared to H_2O Injection) assessed in the liver.

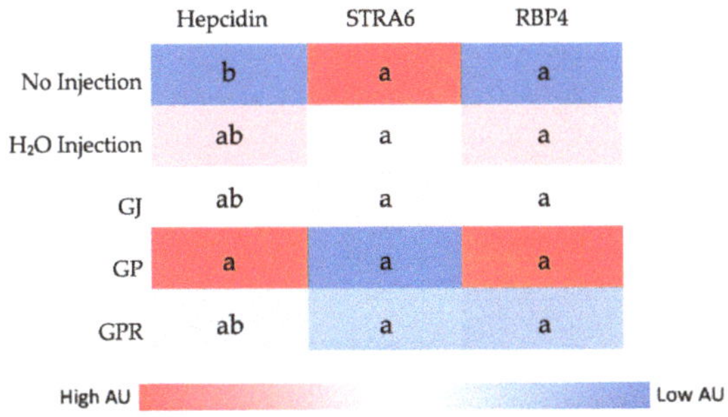

Figure 3. Heatmap showing the effect of intra-amniotic administration of grape juice (GJ), pomace (GP), and puree (GPR) on liver gene expression. Values are in arbitrary units (AU). Genes not indicated by the same letter (a,b) are significantly different ($p < 0.05$). Key; STRA6: Signaling receptor and transporter of retinol; RBP: Retinol-binding protein.

3.6. Duodenal Morphometric Parameters

The villi surface area was significantly reduced with the intra-amniotic administration of GP when compared to the controls (No injection and H_2O). Whereas no significant changes were observed in Paneth cell number or diameter between the treatment groups and the no injection control. A shorter crypt depth indicates increased villi change over rate. GP treatment corresponded to the smallest crypt depth whereas GPR was the largest ($p < 0.05$, Table 4).

Table 4. Effect of the intra-amniotic administration of grape juice (GJ), pomace (GP), and puree (GPR) on the duodenal villi surface area, crypt depth, Paneth cell number, and diameter.

Treatment Group	Villi Surface Area (μm²)	Crypt Depth (μm)	Paneth Cell Number	Paneth Cell Diameter (μm)
No Injection	164.6 ± 7.7^{a}	22.1 ± 0.8^{a}	1.2 ± 0.03^{a}	1.4 ± 0.02^{b}
H_2O	161.0 ± 3.8^{ab}	21.9 ± 0.7^{a}	1.0 ± 0.01^{b}	1.5 ± 0.02^{a}
GJ	153.3 ± 3.9^{abc}	16.2 ± 0.6^{b}	1.3 ± 0.03^{a}	1.4 ± 0.02^{b}
GP	145.4 ± 4.3^{c}	13.8 ± 0.5^{c}	1.3 ± 0.04^{a}	1.4 ± 0.02^{b}
GPR	148.8 ± 3.9^{bc}	15.4 ± 0.5^{bc}	1.3 ± 0.04^{a}	1.4 ± 0.02^{b}

Values are the means ± SEM (n = 5). [a–c] Treatment groups not indicated by the same letter are significantly different ($p < 0.05$).

The grape juice treated group showed the highest villi goblet cell number, significantly ($p < 0.05$) higher than the controls. Whereas the GP group was not significantly different from the (no injection) control but lower than GJ and GPR. Similarly, the goblet cell diameter was lower in GP when compared to GJ and GPR (Table 5).

Table 5. Effect of the intra-amniotic administration of grape juice (GJ), pomace (GP), and puree (GPR) on the duodenal villus goblet cell type, number, and diameter.

Treatment Group	Villi Goblet Cell Diameter (μM)	Total Villi Goblet Cell Number	Villus Goblet Cell Type-Number		
			Acidic	Neutral	Mixed
No Injection	3.5 ± 0.07^{a}	20.1 ± 0.60^{d}	18.3 ± 0.58^{d}	0.06 ± 0.02^{b}	1.8 ± 0.14^{c}
H_2O	3.2 ± 0.06^{bc}	33.2 ± 0.73^{c}	29.9 ± 0.69^{c}	0.22 ± 0.04^{b}	3.2 ± 0.20^{a}
GJ	3.3 ± 0.07^{ab}	41.7 ± 0.95^{a}	39.0 ± 0.87^{a}	0.14 ± 0.04^{b}	2.6 ± 0.23^{b}
GP	3.0 ± 0.06^{c}	21.6 ± 1.22^{d}	18.9 ± 1.20^{d}	0.62 ± 0.09^{a}	2.1 ± 0.14^{bc}
GPR	3.3 ± 0.07^{b}	38.4 ± 0.98^{b}	34.7 ± 0.87^{b}	0.21 ± 0.05^{b}	3.5 ± 0.26^{a}

Values are the means ± SEM (n = 5). [a–d] Treatment groups not indicated by the same letter are significantly different ($p < 0.05$).

Similar to the results of villi goblet cells, crypt goblet cells too were found to be lower ($p < 0.05$) in diameter in the GP when compared to GJ and GPR. Crypt goblet cell number was found to be the highest in the GJ group (Table 6).

Table 6. Effect of the intra-amniotic administration of grape juice (GJ), pomace (GP), and puree (GPR) on the duodenal crypt goblet cell type, number, and diameter.

Treatment Group	Crypt Goblet Cell Diameter (μM)	Total Crypt Goblet Cell Number	Crypt Goblet Cell Type-Number		
			Acidic	Neutral	Mixed
No Injection	3.0 ± 0.05 ^{b}	7.0 ± 0.24 ^{c}	5.8 ± 0.20 ^{c}	0.02 ± 0.02 ^{c}	1.2 ± 0.1 ^{c}
H_2O	2.9 ± 0.05 ^{b}	8.6 ± 0.32 ^{b}	6.9 ± 0.28 ^{b}	0.13 ± 0.03 ^{b}	1.5 ± 0.1 ^{bc}
GJ	2.9 ± 0.07 ^{b}	10.6 ± 0.36 ^{a}	8.0 ± 0.29 ^{a}	0.39 ± 0.05 ^{a}	2.2 ± 0.1 ^{a}
GP	2.7 ± 0.05 ^{c}	8.1 ± 0.27 ^{b}	6.4 ± 0.25 ^{bc}	0.0 ± 0.0 ^{c}	1.7 ± 0.1 ^{b}
GPR	3.3 ± 0.05 ^{a}	8.3 ± 0.28 ^{b}	6.7 ± 0.21 ^{b}	0.0 ± 0.0 ^{c}	1.6 ± 0.1 ^{b}

Values are the means ± SEM (n = 5). $^{a–c}$ Treatment groups not indicated by the same letter are significantly different ($p < 0.05$).

3.7. Analysis of the Gut Bacterial Populations

The 16s rDNA analysis of cecal bacterial populations showed no significant changes in the relative abundance of the *Lactobacillus*, *Klebsiella.*, *E. coli*, and *L. plantarum* when compared to the H_2O injection. However, GPR resulted in higher ($p < 0.05$) *Clostridium* numbers and lower *Bifidobacterium* numbers when compared to all other groups (Figure 4).

	Bifidobacterium	*Lactobacillus*	*E. coli*	*Clostridium*	*Klebsiella*	*L. plantarum*
No Injection	0.94 ± 0.02 ^{a}	0.80 ± 0.044 ^{a}	0.86 ± 0.03 ^{a}	0.23 ± 0.02 ^{b}	0.72 ± 0.05 ^{a}	0.76 ± 0.02 ^{a}
H_2O	0.93 ± 0.03 ^{a}	0.88 ± 0.048 ^{a}	0.87 ± 0.03 ^{a}	0.32 ± 0.1 ^{b}	0.80 ± 0.05 ^{a}	0.70 ± 0.01 ^{b}
GJ	0.92 ± 0.01 ^{a}	0.92 ± 0.029 ^{a}	0.92 ± 0.01 ^{a}	0.21 ± 0.01 ^{b}	0.74 ± 0.02 ^{a}	0.70 ± 0.01 ^{b}
GP	0.92 ± 0.01 ^{a}	0.91 ± 0.052 ^{a}	0.90 ± 0.01 ^{a}	0.21 ± 0.02 ^{b}	0.78 ± 0.04 ^{a}	0.72 ± 0.01 ^{ab}
GPR	0.73 ± 0.03 ^{b}	0.92 ± 0.025 ^{a}	0.84 ± 0.03 ^{a}	0.53 ± 0.06 ^{a}	0.80 ± 0.01 ^{a}	0.70 ± 0.03 ^{ab}

High Density (INT/mm^2) — Low Density (INT/mm^2)

Figure 4. Heatmap showing the difference in cecal bacterial population abundances following intra-amniotic administration of grape juice (GJ), pomace (GP), and puree (GPR). Values are the means ± standard error means (n = 5); Values followed by a different letter indicate statistically significant differences assessed by ANOVA followed by Duncan posthoc test.

4. Discussion

In the present study, we have shown that the intra-amniotic administration of grape pomace, puree, and juice can alter duodenal morphology, gene expression, and specific cecal bacterial populations in-vivo (*Gallus gallus*). These alterations are in-line with previous literature, suggesting that grape pomace and puree may be included in food products and feed provided their concentration is limited. Grape juice corresponds to the lowest fiber and polyphenol content of the three treatment groups and was found to produce the most desirable results in comparison.

A recent meta-analysis of randomized control trials showed that whole grapes and grape products increased fasting blood glucose levels [52], whereas another comprehensive review on grape wines concluded no effect on blood glucose [53]. In this study, we found no significant changes in blood glucose nor stored glucose (in the form of glycogen) in pectoral muscles. In the case of the second review, wine processing reduced the available sugars to alcohol, thereby preventing the rise in blood sugar. Additionally, in this study, the sugars were probably too diluted to cause any significant spikes in blood glucose (Table 3). In addition, no differences were observed in the body weight of the hatchlings. Other animal studies on pigs [54] and broilers [55,56] have similarly reported no significant changes in body weight following daily dietary supplementation of either grape seed, grape marc meal extract, or grape pomace for four weeks or more.

The morphology of a healthy intestine corresponds to large villi, the presence of many mucin-producing goblet cells, and numerous anti-microbial peptide-producing Paneth cells. These parameters often indicate high nutrient absorption efficiency and strong pathogen defenses [57,58]. In the present study, the morphometric assessment of the duodenum revealed a reduction ($p < 0.05$) in the villus absorptive surface area in the group administered with grape pomace (GP) when compared to the controls (Table 4). A general trend of reduced villi surface area is observed in the grape-treated groups. At the same time, no significant changes were observed in villi goblet cell number/diameter nor Paneth cell number/diameter when compared to both the controls (Tables 4 and 5). A similar reduction in duodenal villus height was observed in broilers following grape pomace dietary inclusion [55]. This suggests that the polyphenol "cocktail" found in grapes inhibits the proliferation of enterocytes [59,60]. Similarly, the *C. sativus* (saffron) floral bio-residue polyphenolic blend caused a dose-dependent decrease in villi surface area following intra-amniotic administration [31]. This may be due to the reduced digestibility of proteins. It has been shown that the reactive hydroxyl group of polyphenols interacts with the carbonyl group of proteins and forms undigestible complexes [55,61,62]. Proteins are essential for cell division and proliferation [63]. The interaction between polyphenols and proteins could explain the reduction in duodenal enterocyte proliferation and thereby, villus surface area. In this study, the total polyphenol concentration (Table 2) is inversely proportional to the surface area of villi (Table 4) and directly proportional to hatchability/survivability (Table 3).

This is the first study that investigated the effect of grapes on hepatic and brush border membrane micronutrient metabolism-related gene expression. For this investigation, we utilized a novel intra-amniotic approach using the embryonic phase of *Gallus gallus*. Previous studies have found the expression of pro-inflammatory, polyphenol absorption-related genes to be reduced and intestinal barrier integrity-related genes to be upregulated in mice and pigs when supplemented with grape polyphenols [54,64]. In the present study, we did not see any significant differences in iron (DMT1, FPN, and DcytB) and zinc (ZnT1, ZnT7, ZIP4, and ZIP1) metabolism-related genes in the treatment groups when compared to the controls. However, the expression of *CRBP2* (cellular retinol-binding protein 2) was significantly upregulated in the group treated with GP. *CRBP2* plays a crucial role in the intracellular transport of retinol and hence in retinoid signaling. It has been shown that resveratrol (a bioactive stilbene found in grapes and grape pomace) can upregulate *CRBP2* expression in the thyroid cancer cell line (THJ-11T) [65]. It is hypothesized that resveratrol (in GP) led to the increased expression of CRBP2. Similarly, the expression of *VDAC2* (Voltage-dependent anion-selective channel 2) was upregulated in the GP-administered group when compared to the others. *VDAC2* is abundant in the outer membrane of the mitochondria and has an anti-apoptotic role [66]. Previously, resveratrol has been shown to protect the mitochondria against age-related dysfunction [67]. Activation of *VDAC2* is perhaps one such mechanism. Surprisingly, no significant changes were observed in the expression of inflammatory cytokines as is expected with grape polyphenols treatment. A longer duration of consumption of grape polyphenols may be required to see the anti-inflammatory effects. Additionally, no significant differences were observed in genes assessed in the liver (Hepcidin coding gene, *STRA6*, and *RBP4*) as seen in Figure 3.

The relative abundance of bacterial populations assessed in the cecum remained largely unchanged with the intra-amniotic administration of grape puree, pomace, and juice compared to the controls. Similar results were reported in broilers fed 15 g/kg grape pomace dietary inclusion for 42 days. The study reported no significant differences ($p < 0.05$) in the gene copy number of *Lactobacillus* spp., *Enterococcus* spp., *E. coli*, *Campylobacter jejuni*,

S. aureus nor *C. perfringens* in cecum between the groups that were fed grape pomace and the control [56]. These results suggest that the dietary inclusion of grape pomace and grape juice maintains the gut microbiota without disturbing it. However, the GPR treatment resulted in a lower relative abundance of beneficial bacteria *Bifidobacterium* and a higher abundance of *Clostridium*. This finding could be explained by the presence of a high concentration of non-fibrous carbohydrates (NFC, Table 2) in the GPR group relative to the others. NFCs include sugars, starches, organic acids, and pectin. Dietary sugars are shown to fuel the proliferation of members of the genus *Clostridium* [68–70]. These changes in cecal bacterial populations, however, appear not to correlate with changes in duodenal morphology and gene expression. This suggests that overall, grape puree did not cause any severe detrimental effects to the gut based on the parameters assessed in this study.

5. Conclusions

This is the first study to comparatively assess grape pomace, juice, and puree on various aspects of intestinal health in-vivo. The inclusion of grape pomace (a low-cost by-product of the wine and grape juice industry) in feed and food could not only offset the negative environmental impact but also reduce their manufacturing cost without negatively affecting consumer intestinal health. The grape juice fraction seems to have the most desirable effect on gut health; however, a lower concentration of grape pomace and puree may generate similar effects. This is an exploratory interventional study; further long-term studies are now warranted to confirm the findings reported here.

Author Contributions: Sample collection–N.A., N.K., C.J., J.C. and E.T.; Sample analysis–N.K., N.A., C.J., O.I.P.-Z. and V.S.; Grape processing–V.S.; Writing–N.A.; Editing–N.A., O.I.P.-Z., E.T. and V.S.; N.A. is the primary investigator; E.T. is the principal investigator; N.A. and E.T. have the primary responsibility for the final content. All authors have read and agreed to the published version of the manuscript.

Funding: This research was partially funded by the New York Wine and Grape Foundation Research Grant (Tako); Cornell Atkinson Center for Sustainability Research Grant (Agarwal, Tako), and the New York State Department of Agriculture & Markets, Concord Grape Research Grant (Padilla-Zakour).

Institutional Review Board Statement: The animal protocol used in this study was conducted according to the guidelines of the Declaration of Helsinki and was approved by the Cornell University Institutional Animal Care and Use committee by the ethics approval code: 2020-0077.

Informed Consent Statement: Not applicable.

Data Availability Statement: Not applicable.

Conflicts of Interest: The authors declare no conflict of interest.

References

1. Li, Y.; Padilla-Zakour, O.I. High Pressure Processing vs. Thermal Pasteurization of Whole Concord Grape Puree: Effect on Nutritional Value, Quality Parameters and Refrigerated Shelf Life. *Foods* **2021**, *10*, 2608. [CrossRef] [PubMed]
2. Davis, T.J.; Gómez, M.I.; Martin, K. Cost of Establishment and Production of Concord Grapes in the Lake Erie Region of New York-2020. 2020. Available online: https://dyson.cornell.edu/wp-content/uploads/sites/5/2020/11/Cost_of_Establishment_and_Production_Concord_Grapes-VD.pdf (accessed on 6 July 2022).
3. FAOSTAT. Available online: https://www.fao.org/faostat/en/#data/QCL (accessed on 11 August 2022).
4. Beres, C.; Costa, G.N.S.; Cabezudo, I.; da Silva-James, N.K.; Teles, A.S.C.; Cruz, A.P.G.; Mellinger-Silva, C.; Tonon, R.v.; Cabral, L.M.C.; Freitas, S.P. Towards Integral Utilization of Grape Pomace from Winemaking Process: A Review. *Waste Manag.* **2017**, *68*, 581–594. [CrossRef] [PubMed]
5. Bordiga, M.; Travaglia, F.; Locatelli, M. Valorisation of Grape Pomace: An Approach That Is Increasingly Reaching Its Maturity—A Review. *Int. J. Food Sci. Technol.* **2019**, *54*, 933–942. [CrossRef]
6. Antonić, B.; Jančíková, S.; Dordević, D.; Tremlová, B. Grape Pomace Valorization: A Systematic Review and Meta-Analysis. *Foods* **2020**, *9*, 1627. [CrossRef]
7. Gupta, M.; Dey, S.; Marbaniang, D.; Pal, P.; Ray, S.; Mazumder, B. Grape Seed Extract: Having a Potential Health Benefits. *J. Food Sci. Technol.* **2020**, *57*, 1205–1215. [CrossRef]

8. Gomes, T.M.; Toaldo, I.M.; Haas, I.C.d.S.; Burin, V.M.; Caliari, V.; Luna, A.S.; de Gois, J.S.; Bordignon-Luiz, M.T. Differential Contribution of Grape Peel, Pulp, and Seed to Bioaccessibility of Micronutrients and Major Polyphenolic Compounds of Red and White Grapes through Simulated Human Digestion. *J. Funct. Foods* **2019**, *52*, 699–708. [CrossRef]
9. Abdrabba, S.; Hussein, S. Chemical Composition of Pulp, Seed and Peel of Red Grape from Lybia. *Glob. J. Sci. Res.* **2015**, *3*, 6–11.
10. Shinagawa, F.B.; de Santana, F.C.; Torres, L.R.O.; Mancini-Filho, J. Grape Seed Oil: A Potential Functional Food? *Food Sci. Technol.* **2015**, *35*, 399–406. [CrossRef]
11. Chowdhary, P.; Gupta, A.; Gnansounou, E.; Pandey, A.; Chaturvedi, P. Current Trends and Possibilities for Exploitation of Grape Pomace as a Potential Source for Value Addition. *Environ. Pollut.* **2021**, *278*, 116796. [CrossRef]
12. Pastrana-Bonilla, E.; Akoh, C.C.; Sellappan, S.; Krewer, G. Phenolic Content and Antioxidant Capacity of Muscadine Grapes. *J. Agric. Food Chem.* **2003**, *51*, 5497–5503. [CrossRef]
13. Poudel, P.R.; Tamura, H.; Kataoka, I.; Mochioka, R. Phenolic Compounds and Antioxidant Activities of Skins and Seeds of Five Wild Grapes and Two Hybrids Native to Japan. *J. Food Compos. Anal.* **2008**, *21*, 622–625. [CrossRef]
14. Falchi, M.; Bertelli, A.; lo Scalzo, R.; Morassut, M.; Morelli, R.; Das, S.; Cui, J.; Das, D.K. Comparison of Cardioprotective Abilities between the Flesh and Skin of Grapes. *J. Agric. Food Chem.* **2006**, *54*, 6613–6622. [CrossRef] [PubMed]
15. Mohansrinivasan, V.; Devi, C.S.; Deori, M.; Biswas, A.; Naine, S.J. Exploring the Anticancer Activity of Grape Seed Extract on Skin Cancer Cell Lines A431. *Braz. Arch. Biol. Technol.* **2015**, *58*, 540–546. [CrossRef]
16. Morré, D.M.; Morré, D.J. Anticancer Activity of Grape and Grape Skin Extracts Alone and Combined with Green Tea Infusions. *Cancer Lett.* **2006**, *238*, 202–209. [CrossRef] [PubMed]
17. Da Costa, G.F.; Santos, I.B.; de Bem, G.F.; Cordeiro, V.S.C.; da Costa, C.A.; de Carvalho, L.C.R.M.; Ognibene, D.T.; Resende, A.C.; de Moura, R.S. The Beneficial Effect of Anthocyanidin-Rich *Vitis vinifera* L. Grape Skin Extract on Metabolic Changes Induced by High-Fat Diet in Mice Involves Antiinflammatory and Antioxidant Actions. *Phytother. Res.* **2017**, *31*, 1621–1632. [CrossRef] [PubMed]
18. Baron, G.; Ferrario, G.; Marinello, C.; Carini, M.; Morazzoni, P.; Aldini, G. Effect of Extraction Solvent and Temperature on Polyphenol Profiles, Antioxidant and Anti-Inflammatory Effects of Red Grape Skin By-Product. *Molecules* **2021**, *26*, 5454. [CrossRef]
19. de Andrade, R.B.; Machado, B.A.S.; Barreto, G.d.A.; Nascimento, R.Q.; Corrêa, L.C.; Leal, I.L.; Tavares, P.P.L.G.; Ferreira, E.d.S.; Umsza-Guez, M.A. Syrah Grape Skin Residues Has Potential as Source of Antioxidant and Anti-Microbial Bioactive Compounds. *Biology* **2021**, *10*, 1262. [CrossRef]
20. Mattos, G.N.; Tonon, R.v.; Furtado, A.A.L.; Cabral, L.M.C. Grape By-Product Extracts against Microbial Proliferation and Lipid Oxidation: A Review. *J. Sci. Food Agric.* **2017**, *97*, 1055–1064. [CrossRef]
21. Ma, Z.F.; Zhang, H. Phytochemical Constituents, Health Benefits, and Industrial Applications of Grape Seeds: A Mini-Review. *Antioxidants* **2017**, *6*, 71. [CrossRef]
22. Unusan, N. Proanthocyanidins in Grape Seeds: An Updated Review of Their Health Benefits and Potential Uses in the Food Industry. *J. Funct. Foods* **2020**, *67*, 103861. [CrossRef]
23. Nowshehri, J.A.; Bhat, Z.A.; Shah, M.Y. Blessings in Disguise: Bio-Functional Benefits of Grape Seed Extracts. *Food Res. Int.* **2015**, *77*, 333–348. [CrossRef]
24. Shi, J.; Yu, J.; Pohorly, J.E.; Kakuda, Y. Polyphenolics in Grape Seeds—Biochemistry and Functionality. *J. Med. Food* **2004**, *6*, 291–299. [CrossRef] [PubMed]
25. Wilson, A.S.; Koller, K.R.; Ramaboli, M.C.; Nesengani, L.T.; Ocvirk, S.; Chen, C.; Flanagan, C.A.; Sapp, F.R.; Merritt, Z.T.; Bhatti, F.; et al. Diet and the Human Gut Microbiome: An International Review. *Dig. Dis. Sci.* **2020**, *65*, 723–740. [CrossRef]
26. Hasan, N.; Yang, H. Factors Affecting the Composition of the Gut Microbiota, and Its Modulation. *PeerJ* **2019**, *7*, e7502. [CrossRef] [PubMed]
27. Krajmalnik-Brown, R.; Ilhan, Z.E.; Kang, D.W.; DiBaise, J.K. Effects of Gut Microbes on Nutrient Absorption and Energy Regulation. *Nutr. Clin. Pract.* **2012**, *27*, 201. [CrossRef] [PubMed]
28. Brial, F.; le Lay, A.; Dumas, M.E.; Gauguier, D. Implication of Gut Microbiota Metabolites in Cardiovascular and Metabolic Diseases. *Cell. Mol. Life Sci.* **2018**, *75*, 3977–3990. [CrossRef]
29. Solis, A.G.; Klapholz, M.; Zhao, J.; Levy, M. The Bidirectional Nature of Microbiome-Epithelial Cell Interactions. *Curr. Opin. Microbiol.* **2020**, *56*, 45. [CrossRef]
30. Gassler, N. Paneth Cells in Intestinal Physiology and Pathophysiology. *World J. Gastrointest. Pathophysiol.* **2017**, *8*, 150. [CrossRef]
31. Agarwal, N.; Kolba, N.; Jung, Y.; Cheng, J.; Tako, E. Saffron (*Crocus sativus* L.) Flower Water Extract Disrupts the Cecal Microbiome, Brush Border Membrane Functionality, and Morphology In Vivo (Gallus Gallus). *Nutrients* **2022**, *14*, 220. [CrossRef]
32. Agarwal, N.; Kolba, N.; Khen, N.; Even, C.; Turjeman, S.; Koren, O.; Tako, E. Quinoa Soluble Fiber and Quercetin Alter the Composition of the Gut Microbiome and Improve Brush Border Membrane Morphology In Vivo (Gallus Gallus). *Nutrients* **2022**, *14*, 448. [CrossRef]
33. Gomes, M.J.C.; Kolba, N.; Agarwal, N.; Kim, D.; Eshel, A.; Koren, O.; Tako, E. Modifications in the Intestinal Functionality, Morphology and Microbiome Following Intra-Amniotic Administration (Gallus Gallus) of Grape (Vitis Vinifera) Stilbenes (Resveratrol and Pterostilbene). *Nutrients* **2021**, *13*, 3247. [CrossRef] [PubMed]

34. Beasley, J.T.; Johnson, A.A.T.; Kolba, N.; Bonneau, J.P.; Glahn, R.P.; Ozeri, L.; Koren, O.; Tako, E. Nicotianamine-Chelated Iron Positively Affects Iron Status, Intestinal Morphology and Microbial Populations in Vivo (Gallus Gallus). *Sci. Rep.* **2020**, *10*, 2297. [CrossRef] [PubMed]
35. Warkentin, T.; Kolba, N.; Tako, E. Low Phytate Peas (*Pisum sativum* L.) Improve Iron Status, Gut Microbiome, and Brush Border Membrane Functionality in Vivo (Gallus Gallus). *Nutrients* **2020**, *12*, 2563. [CrossRef]
36. Martino, H.S.D.; Kolba, N.; Tako, E. Yacon (Smallanthus Sonchifolius) Flour Soluble Extract Improve Intestinal Bacterial Populations, Brush Border Membrane Functionality and Morphology in Vivo (Gallus Gallus). *Food Res. Int.* **2020**, *137*, 109705. [CrossRef] [PubMed]
37. Reed, S.; Knez, M.; Uzan, A.; Stangoulis, J.C.R.; Glahn, R.P.; Koren, O.; Tako, E. Alterations in the Gut (Gallus Gallus) Microbiota Following the Consumption of Zinc Biofortified Wheat (*Triticum aestivum*)-Based Diet. *J. Agric. Food Chem.* **2018**, *66*, 6291–6299. [CrossRef]
38. Da Silva, B.P.; Kolba, N.; Martino, H.S.D.; Hart, J.; Tako, E. Soluble Extracts from Chia Seed (*Salvia hispanica* L.) Affect Brush Border Membrane Functionality, Morphology and Intestinal Bacterial Populations in Vivo (Gallus Gallus). *Nutrients* **2019**, *11*, 2457. [CrossRef]
39. Carboni, J.; Reed, S.; Kolba, N.; Eshel, A.; Koren, O.; Tako, E. Alterations in the Intestinal Morphology, Gut Microbiota, and Trace Mineral Status Following Intra-Amniotic Administration (Gallus Gallus) of Teff (Eragrostis Tef) Seed Extracts. *Nutrients* **2020**, *12*, 3020. [CrossRef]
40. Pacifici, S.; Song, J.; Zhang, C.; Wang, Q.; Glahn, R.P.; Kolba, N.; Tako, E. Intra Amniotic Administration of Raffinose and Stachyose Affects the Intestinal Brush Border Functionality and Alters Gut Microflora Populations. *Nutrients* **2017**, *9*, 304. [CrossRef]
41. Waterhouse, A.L. Determination of Total Phenolics. *Curr. Protoc. Food Anal. Chem.* **2002**, *6*, I1.1.1–I1.1.8. [CrossRef]
42. Lee, J.; Durst, R.W.; Wrolstad, R.E.; Barnes, K.W.; Eisele, T.; Giusti, M.M.; Haché, J.; Hofsommer, H.; Koswig, S.; Krueger, D.A.; et al. Determination of Total Monomeric Anthocyanin Pigment Content of Fruit Juices, Beverages, Natural Colorants, and Wines by the PH Differential Method: Collaborative Study. *J. Aoac Int.* **2005**, *88*, 1269–1278. [CrossRef]
43. Dreiling, C.E.; Brown, D.E.; Casale, L.; Kelly, L. Muscle Glycogen: Comparison of Iodine Binding and Enzyme Digestion Assays and Application to Meat Samples. *Meat Sci.* **1987**, *20*, 167–177. [CrossRef]
44. Dias, D.; Kolba, N.; Binyamin, D.; Ziv, O.; Regini Nutti, M.; Martino, H.; Glahn, R.; Koren, O.; Tako, E. Iron Biofortified Carioca Bean (*Phaseolus vulgaris* L.)—Based Brazilian Diet Delivers More Absorbable Iron and Affects the Gut Microbiota In Vivo (Gallus Gallus). *Nutrients* **2018**, *10*, 1970. [CrossRef] [PubMed]
45. Pisula, A. *Detecting a Probiotic Product within the Gut of Broiler Chickens*; California Polytechnic State University: San Luis Obispo, CA, USA, 2018.
46. Zhu, X.Y.; Zhong, T.; Pandya, Y.; Joerger, R.D. 16S RRNA-Based Analysis of Microbiota from the Cecum of Broiler Chickens. *Appl. Environ. Microbiol.* **2002**, *68*, 124. [CrossRef] [PubMed]
47. Amit-Romach, E.; Sklan, D.; Uni, Z. Microflora Ecology of the Chicken Intestine Using 16S Ribosomal DNA Primers. *Poult. Sci.* **2004**, *83*, 1093–1098. [CrossRef] [PubMed]
48. Langendijk, P.S.; Schut, F.; Jansen, G.J.; Raangs, G.C.; Kamphuis, G.R.; Wilkinson, M.H.F.; Welling, G.W. Quantitative Fluorescence in Situ Hybridization of Bifidobacterium Spp. with Genus-Specific 16S RRNA-Targeted Probes and Its Application in Fecal Samples. *Appl. Environ. Microbiol.* **1995**, *61*, 3069. [CrossRef]
49. Tako, E.; Glahn, R.P.; Welch, R.M.; Lei, X.; Yasuda, K.; Miller, D.D. Dietary Inulin Affects the Expression of Intestinal Enterocyte Iron Transporters, Receptors and Storage Protein and Alters the Microbiota in the Pig Intestine. *Br. J. Nutr.* **2008**, *99*, 472–480. [CrossRef]
50. Goff, J.P. Invited Review: Mineral Absorption Mechanisms, Mineral Interactions That Affect Acid–Base and Antioxidant Status, and Diet Considerations to Improve Mineral Status. *J. Dairy Sci.* **2018**, *101*, 2763–2813. [CrossRef]
51. Reboul, E. Absorption of Vitamin A and Carotenoids by the Enterocyte: Focus on Transport Proteins. *Nutrients* **2013**, *5*, 3563. [CrossRef]
52. Moodi, V.; Abedi, S.; Esmaeilpour, M.; Asbaghi, O.; Izadi, F.; Shirinbakhshmasoleh, M.; Behrouzian, M.; Shahriari, A.; Ghaedi, E.; Miraghajani, M. The Effect of Grapes/Grape Products on Glycemic Response: A Systematic Review and Meta-Analysis of Randomized Controlled Trials. *Phytother. Res.* **2021**, *35*, 5053–5067. [CrossRef]
53. Fragopoulou, E.; Choleva, M.; Antonopoulou, S.; Demopoulos, C.A. Wine and Its Metabolic Effects. A Comprehensive Review of Clinical Trials. *Metabolism* **2018**, *83*, 102–119. [CrossRef]
54. Fiesel, A.; Gessner, D.K.; Most, E.; Eder, K. Effects of Dietary Polyphenol-Rich Plant Products from Grape or Hop on pro-Inflammatory Gene Expression in the Intestine, Nutrient Digestibility and Faecal Microbiota of Weaned Pigs. *BMC Vet. Res.* **2014**, *10*, 196. [CrossRef] [PubMed]
55. Ebrahimzadeh, S.K.; Navidshad, B.; Farhoomand, P.; Farhoomand, P.; Aghjeh-Gheshlagh, F.M. Effects of Grape Pomace and Vitamin E on Performance, Antioxidant Status, Immune Response, Gut Morphology and Histopathological Responses in Broiler Chickens. *S. Afr. J. Anim. Sci.* **2018**, *48*, 324. [CrossRef]
56. Gungor, E.; Altop, A.; Erener, G. Effect of Raw and Fermented Grape Pomace on the Growth Performance, Antioxidant Status, Intestinal Morphology, and Selected Bacterial Species in Broiler Chicks. *Animals* **2021**, *11*, 364. [CrossRef]

57. Chee, S.H.; Iji, P.A.; Choct, M.; Mikkelsen, L.L.; Kocher, A. Characterisation and Response of Intestinal Microflora and Mucins to Manno-Oligosaccharide and Antibiotic Supplementation in Broiler Chickens. *Br. Poult. Sci.* **2010**, *51*, 368–380. [CrossRef] [PubMed]
58. Xu, Z.R.; Hu, C.H.; Xia, M.S.; Zhan, X.A.; Wang, M.Q. Effects of Dietary Fructooligosaccharide on Digestive Enzyme Activities, Intestinal Microflora and Morphology of Male Broilers. *Poult. Sci.* **2003**, *82*, 1030–1036. [CrossRef]
59. Hou, Y.; Wei, W.; Guan, X.; Liu, Y.; Bian, G.; He, D.; Fan, Q.; Cai, X.; Zhang, Y.; Wang, G.; et al. A Diet-Microbial Metabolism Feedforward Loop Modulates Intestinal Stem Cell Renewal in the Stressed Gut. *Nat. Commun.* **2021**, *12*, 1–15. [CrossRef] [PubMed]
60. Tandon, A.; Singh, S.J.; Chaturvedi, R.K. Stem Cells as Potential Targets of Polyphenols in Multiple Sclerosis and Alzheimer's Disease. *BioMed Res. Int.* **2018**, *2018*, 1–30. [CrossRef] [PubMed]
61. Ortiz, L.T.; Centeno, C.; Treviño, J. Tannins in Faba Bean Seeds: Effects on the Digestion of Protein and Amino Acids in Growing Chicks. *Anim. Feed. Sci. Technol.* **1993**, *41*, 271–278. [CrossRef]
62. Academic Publishers, W. Recent Advances of Research in Antinutritional Factors in Legume Seeds and Oilseeds. In Proceedings of the Fourth International Workshop on Antinutritional Factors in Legume Seeds and Oilseeds, Toledo, Spain, March 2004; p. 110. [CrossRef]
63. Otto, T.; Sicinski, P. Cell Cycle Proteins as Promising Targets in Cancer Therapy. *Nat. Rev. Cancer* **2017**, *17*, 93. [CrossRef]
64. Lu, F.; Li, Y.; Zhou, B.; Guo, Q.; Zhang, Y. Early-Life Supplementation of Grape Polyphenol Extract Promotes Polyphenol Absorption and Modulates the Intestinal Microbiota in Association with the Increase in MRNA Expression of the Key Intestinal Barrier Genes. *Food Funct.* **2021**, *12*, 602–613. [CrossRef]
65. Liu, X.; Li, H.; Wu, M.L.; Wu, J.; Sun, Y.; Zhang, K.L.; Liu, J. Resveratrol Reverses Retinoic Acid Resistance of Anaplastic Thyroid Cancer Cells via Demethylating CRABP2 Gene. *Front. Endocrinol.* **2019**, *10*, 734. [CrossRef] [PubMed]
66. Maurya, S.R.; Mahalakshmi, R. Mitochondrial VDAC2 and Cell Homeostasis: Highlighting Hidden Structural Features and Unique Functionalities. *Biol. Rev. Camb. Philos. Soc.* **2017**, *92*, 1843–1858. [CrossRef]
67. Ungvari, Z.; Sonntag, W.E.; de Cabo, R.; Baur, J.A.; Csiszar, A. Mitochondrial Protection by Resveratrol. *Exerc. Sport Sci. Rev.* **2011**, *39*, 128. [CrossRef] [PubMed]
68. Noble, E.E.; Hsu, T.M.; Jones, R.B.; Fodor, A.A.; Goran, M.I.; Kanoski, S.E. Early-Life Sugar Consumption Affects the Rat Microbiome Independently of Obesity. *J. Nutr.* **2017**, *147*, 20–28. [CrossRef] [PubMed]
69. Singh, R.K.; Chang, H.W.; Yan, D.; Lee, K.M.; Ucmak, D.; Wong, K.; Abrouk, M.; Farahnik, B.; Nakamura, M.; Zhu, T.H.; et al. Influence of Diet on the Gut Microbiome and Implications for Human Health. *J. Transl. Med.* **2017**, *15*, 1–17. [CrossRef]
70. Magnusson, K.R.; Hauck, L.; Jeffrey, B.M.; Elias, V.; Humphrey, A.; Nath, R.; Perrone, A.; Bermudez, L.E. Relationships between Diet-Related Changes in the Gut Microbiome and Cognitive Flexibility. *Neuroscience* **2015**, *300*, 128–140. [CrossRef]

nutrients

MDPI

Article

Intraamniotic Administration (*Gallus gallus*) of Genistein Alters Mineral Transport, Intestinal Morphology, and Gut Microbiota

Jacquelyn Cheng [1], Nikolai Kolba [1], Philip Sisser [2], Sondra Turjeman [3], Carmel Even [3], Omry Koren [3] and Elad Tako [1,*]

1 Department of Food Science, Cornell University, Stocking Hall, Ithaca, NY 14853, USA
2 Department of Environment & Sustainability, Cornell University, Kennedy Hall, Ithaca, NY 14853, USA
3 Azrieli Faculty of Medicine, Bar-Ilan University, Safed 1311502, Israel
* Correspondence: et79@cornell.edu; Tel.: +1-607-255-0884

Abstract: Genistein is an isoflavone naturally present in numerous staple food crops, such as soybeans and chickpeas. This study utilized the *Gallus gallus* intraamniotic administration procedure to assess genistein administration effects on trace mineral status, brush border membrane (BBM) functionality, intestinal morphology, and intestinal microbiome in vivo. Eggs were divided into five groups with 1 mL injection of the following treatments: no-injection, DI H_2O, 5% inulin, and 1.25% and 2.5% genistein (n = 8 per group). Upon hatch, blood, cecum, small intestine, and liver were collected for assessment of hemoglobin, intestinal microflora alterations, intestinal morphometric assessment, and mRNA gene expression of relevant iron and zinc transporter proteins, respectively. This study demonstrated that intraamniotic administration of 2.5% genistein increased villus surface area, number of acidic goblet cells, and hemoglobin. Additionally, genistein exposure downregulated duodenal cytochrome B (DcytB) and upregulated hepcidin expression. Further, genistein exposure positively altered the composition and function of the intestinal microbiota. Our results suggest a physiological role for genistein administration in improving mineral status, favorably altering BBM functionality and development, positively modulating the intestinal microbiome, as well as improving physiological status.

Keywords: genistein; brush border membrane; iron; microbiome; biomarkers

Citation: Cheng, J.; Kolba, N.; Sisser, P.; Turjeman, S.; Even, C.; Koren, O.; Tako, E. Intraamniotic Administration (*Gallus gallus*) of Genistein Alters Mineral Transport, Intestinal Morphology, and Gut Microbiota. *Nutrients* **2022**, *14*, 3473. https://doi.org/10.3390/nu14173473

Academic Editor: Anna Gramza-Michałowska

Received: 29 July 2022
Accepted: 21 August 2022
Published: 24 August 2022

1. Introduction

Genistein is a polyphenolic isoflavone naturally found in numerous staple crops, including soybeans and chickpeas. Many studies have reported genistein to possess various beneficial and protective physiological properties, with effects observed in metabolic syndrome, diabetes, and breast and prostate cancers in vivo [1,2]. The biological effects of isoflavone consumption have been attributed to structural similarity and function with human and animal estrogens. Specifically, due to structural similarity to 17b-estradiol, genistein has been observed to possess weak estrogenic activity and exhibit preferential binding to estrogen receptor ß [2,3].

The characterization of genistein metabolism and absorption is still ongoing, despite the well-studied physiological effects of genistein and other isoflavones. Dietary isoflavones exist as isoflavone-glycosides and are transformed by intestinal microbiota via bacterial enzymatic action to more potent metabolites, such as equol and O-desmethylan- lensin [4]. Thus, individual differences in gut microbiota will consequently be expected to influence the potential for physiological effects associated with isoflavone ingestion [5]. Current research has shown genistein administration in mice fed a high-fat diet ameliorated harmful effects associated with a high-fat diet through increasing populations of bacteria associated with reduced pro-inflammatory lipopolysaccharide and lower serum triglyceride levels [1].

Another recent study has shown that isoflavone administration in vitro promoted short-chain fatty acid (SCFA) production due to increased proliferation of SCFA-producing bacteria species from *Clostridium* cluster XIVa, *Roseburia* and *E. hallii* [4]. Additionally, maternal genistein intake perinatally and throughout pregnancy in mice mitigated harmful effects of a high-fat fed diet in dams and offspring and was associated with an increase in butyrate-producing gut bacteria [6]. Increased SCFA production has been associated with inhibiting harmful pathogen growth, decreased intestinal pH, and upregulated brush border membrane (BBM) gene expression [7,8]. Taken together, these effects enhance micronutrient bioavailability.

Emerging evidence suggests that genistein exposure could be implicated in the altered expression of proteins involved in iron (Fe) transport. Genistein significantly increased Fe export through estrogen receptor ß-dependent p38 MAPK up-regulation through ceruloplasmin and ferroportin-1 in glial cells [9]. However, another study found that genistein treatment of human hepatocytes increased both hepcidin transcription levels and promoter activity (hepcidin decreases intestinal Fe absorption by inhibiting ferroportin) [10].

Despite the investigation of specific health benefits attributed to dietary genistein administration and subsequent knowledge of genistein ingestion on gut microbiota modulation and Fe transport, there is a paucity of knowledge regarding how genistein affects the brush border membrane (BBM) of the small intestine. As BBM functional capacity (i.e., digestive enzyme production) dictates the extent of food digestion and absorption, it is key to investigate the interactions between bioactive compounds in the diet and the BBM. There is also a lack of studies that specifically utilize the embryonic stage of the *Gallus gallus* for elucidating the effects of genistein consumption on BBM development and functionality. Due to similarities in intestinal morphology, microbiota, and gene homology of duodenal mineral transporters between humans and *Gallus gallus*, the *Gallus gallus* has been used as a novel and cost-effective animal model to elucidate the physiological effects of plant bioactives and nutritional solutions relevant to human nutrition [11–15]. To study the impact of bioactive on the embryonic stage, the intraamniotic administration approach can be utilized for testing the effects of the solution administered into the amniotic fluid on the different systems of interest in a closed system, where the amniotic fluid is naturally and orally consumed by the embryo starting at day 17 and is entirely consumed by hatch [7,11,16–18].

In our present study, the effects of genistein intraamniotic administration on brush border membrane (BBM) functionality, intestinal morphology, and intestinal microbiome were studied in vivo using the embryonic stage of the *Gallus gallus*. It was previously demonstrated that daidzein, another major isoflavone found in soybeans with estrogenic effects, altered BBM Fe transport proteins and cecal bacterial populations in the embryonic stage of the *Gallus gallus* [19]. Therefore, the first objective of this study was to evaluate genistein administration effects on BBM functionality through evaluating duodenal gene expression of biomarkers of mineral status, BBM digestive and absorptive ability, and inflammation. To accomplish this objective, we assessed the expression of duodenal cytochrome B (DcytB, a Fe-specific cytochrome reductase on the luminal side of the enterocyte) and divalent metal transporter 1 (DMT1, the primary transporter of Fe^{2+} from the luminal side of the enterocyte), ferroportin (a basolateral exporter of dietary Fe^{2+}), liver hepcidin (decreases intestinal Fe absorption by inhibiting ferroportin), as well as duodenal ZnT7 (zinc transporter protein 7) and ZIP6 (zinc transporter) [10]. BBM digestive and absorptive ability were evaluated by assessing duodenum morphology and gene expression of biomarkers of BBM digestive and absorptive ability (AP—aminopeptidase, SI—sucrase-isomaltase, and NaK/ATPase—sodium-potassium adenosine triphosphatase). In addition, systemic inflammatory status was evaluated using the expression of immunoregulatory cytokines (TNF-α, tumor necrosis factor-alpha; and NF-κB, nuclear factor kappa B subunit 1). The second objective was to utilize PCR quantification to analyze duodenal microbial populations and next-generation sequencing to analyze the cecal microbiome to elucidate potential alterations in gut microbiota composition and function resulting from genistein administration. We hypothesize that when administered intraamniotically, genistein will alter mineral transport, cause favorable alterations in BBM functionality and development, and positively modulate the gut microbiota.

2. Materials and Methods

2.1. Animals and Experimental Design

Fertile Cornish-cross broiler eggs (*Gallus gallus*) were acquired (Moyer's chicks, Quakertown, PA, USA) and incubated utilizing optimum conditions at the Cornell University Animal Science Poultry Farm Incubator [20]. The protocol was approved by the Cornell University Institutional Animal Care and Use Committee (IACUC #2020-0077). On incubation day 17, viable embryos were weighed, and eggs were randomly distributed by weight into five groups (n = 8 per group, each group contained eggs of similar weight frequency distribution). Treatments in powder form were prepared in DI H_2O. The experimental groups were as follows: two treatment groups (1.25, 2.5% genistein), two controls (H_2O injection and no-injection), and a positive control (5% inulin). After identification of the injection site via candling, 1 mL of experimental solution was injected into the amniotic fluid of each egg using a 21-gauge needle. After injection, the injection holes were sterilized with 70% ethanol and sealed. Eggs were returned to the incubator with equal representation at each incubator location to reduce allocation bias. Immediately upon hatch (day 21), blood was collected, and all chicks were euthanized by CO_2 exposure. The small intestine, cecum, pectoral muscle, and liver were collected, placed in liquid nitrogen for immediate freezing, and stored at −80 °C until analysis.

2.2. Blood Hemoglobin (Hb) Measurements

Blood was collected in sodium heparin tubes (ThermoFisher Scientific, Waltham, MA, USA). The QuantiChrom™ Hemoglobin Assay (BioAssay Systems, Hayward, CA, USA) was utilized to quantify hemoglobin (Hb) concentrations spectrophotometrically following the manufacturer's instructions.

2.3. Total RNA Isolation from Duodenum and Liver Tissue Samples

Total RNA was extracted from 30 mg of duodenal (n = 6) or liver tissues (n = 6) according to the manufacturer's instructions under RNase-free conditions using the Qiagen RNeasy Mini Kit (Qiagen Inc., Valencia, CA, USA). RNA was quantified by the ratio of absorbance (260/280 nm) using a NanoDrop 2000 (ThermoFisher Scientific, Waltham, MA, USA). RNA samples were stored at −80 °C until use.

2.4. Real-Time Polymerase Chain Reaction (RT-PCR)

As was previously described [12,16,17,21], cDNA was made using a 20uL reverse transcriptase (RT) reaction in a BioRad C1000 Touch Thermal Cycler using the Improm-II Reverse Transcriptase Kit (Promega, Madison, WI, USA). The reverse transcriptase reaction consisted of the following: 1 µL total RNA template, 10 µM random hexanucleotide primers, and 2 mM of oligo(dT) primers. Reactions were completed in conditions as indicated: 94 °C for 5 min, 60 min at 42 °C, 70 °C for 15 min, and hold at 4 °C. cDNA concentration was determined using a NanoDrop 2000 (ThermoFisher Scientific, Waltham, MA, USA) by measuring the ratio of absorbance (260/280 nm).

2.4.1. Primer Design

As was previously described [12,16,17,21], primers were designed using the PrimerQuest Tool (IDT DNA, Coralvilla, IA, USA) based on 13 genetic sequences publicly available on the GenBank database. DNA sequences of primers utilized in this study are summarized in Table 1.

Table 1. Sequences of primers used in this study.

Analyte	Forward Primer (5′–3′)	Reverse Primer (3′–5′)	Base Pair	GI Identifier
Iron Metabolism				
DcytB	CATGTGCATTCTCTTCCAAAGTC	CTCCTTGGTGACCGCATTAT	103	20380692
DMT1	TTGATTCAGAGCCTCCCATTAG	GCGAGGAGTAGGCTTGTATTT	101	206597489
Ferroportin	CTCAGCAATCACTGGCATCA	ACTGGGCAACTCCAGAAATAAG	98	423984
Hepcidin	AGACGACAATGCAGACTAACC	CTGCAGCAATCCCACATTTC	132	SAMN08056490
Zinc Metabolism				
Δ-6-desaturase	GGCGAAAGTCAGCCTATTGA	AGGTGGGAAGATGAGGAAGA	93	261865208
ZIP6	GCTACTGGGTAATGGTGAAGAA	GCTGTGCCAGAACTGTAGAA	380	66735072
ZnT7	GGAAGATGTCAGGATGGTTCA	CGAAGGACAAATTGAGGCAAAG	87	56555152
BBM Functionality				
AP	CGTCAGCCAGTTTGACTATGTA	CTCTCAAAGAAGCTGAGGATGG	138	45382360
SI	CCAGCAATGCCAGCATATTG	CGGTTTCTCCTTACCACTTCTT	95	2246388
NaK/ATPase	CCTTGGAGGTTTCTTCACCTATT	GGTCATCCCACTGAAGTCTAATC	92	14330321
Inflammatory Response				
NF-κβ	CACAGCTGGAGGGAAGTAAAT	TTGAGTAAGGAAGTGAGGTTGAG	100	2130627
TNF-α	GACAGCCTATGCCAACAAGTA	TTACAGGAAGGGCAACTCATC	109	53854909
18S	GCAAGACGAACTAAAGCGAAAG	TCGGAACTACGACGGTATCT	100	7262899

DcytB, duodenal cytochrome b; DMT1, divalent metal transporter 1; ZIP6, zinc transport protein 6; ZnT7, Zinc transporter 7; AP, amino peptidase; SI, Sucrose isomaltase; NaK/ATPase, Sodium, Potassium and adenosine triphosphate; NF-κβ, nuclear factor kappa β subunit 1; TNF-α, tumor necrosis factor-α.

2.4.2. Real-Time qPCR Design

RT-qPCR was performed as was previously described [12,16,17,21]. Briefly, 10 μL RT-qPCR reactions comprised cDNA, SYBR Green Supermix (2X BioRad SSO Advanced Universal, Cat #1725274, Hercules, CA, USA), forward and reverse primers (as shown in Table 1), and nuclease-free H_2O. DNA amplification was performed under the following conditions: first denaturation at 95 °C for 30 s, 40 cycles of denaturation at 95 °C for 15 s, various annealing temperatures based on the primers utilized (PrimerQuest Tool, IDT DNA, Coralvilla, IA, USA) for 30 s and elongation at 60 °C for 30 s using a Bio-Rad CFX96 Touch (Hercules, CA, USA). Cp values were calculated using the automated "second derivative maximum" method (Bio-Rad CFX Maestro Software Version 4.1.2433.1219, Hercules, CA, USA). Gene expression was normalized to 18S gene expression [22]. RT-qPCR efficiency values for the 13 genes were as follows: DcytB, 1.046; DMT 1, 0.998; Ferroportin, 1.109; Hepcidin, 0.976; Δ-6-Desaturase, 0.925; ZIP6, 0.961; ZnT7, 0.916; NK-κβ, 1.113; TNF-α, 1.046; AP, 1.015; SI, 1.032; NaK/ATPase, 1.024; and 18S rRNA, 0.994.

2.5. Collection of Microbial Samples and Intestinal Contents DNA Isolation

As was previously described [16,21], intestinal contents were placed into a sterile 15 mL tube (Corning, Corning, NY, USA), 9 mL 1X phosphate buffered saline (PBS) was added, and the contents were vortexed with silicone beads (3 mm) for 3 min and centrifuged at 1000× *g* for 5 min. The supernatant was collected and centrifuged at 4000× *g* for 20 min, and the resulting pellet was washed twice with PBS. The pellet was dissolved in 50 mM EDTA and incubated with 10 mg/mL lysozyme (Sigma Aldrich CO., St. Louis, MO, USA) for 45 min at 37 °C. A Wizard Genomic DNA purification kit (Promega Corp., Madison, WI, USA) was used to isolate bacterial genomic DNA according to the manufacturer's instructions.

2.6. PCR Amplification of Bacterial 16S rDNA

Bifidobacterium, *Clostridium*, *Lactobacillus*, *E. coli*, and *L. plantarum* primers were designed as previously described [23,24]. Universal primers for the invariant region of bacterial 16S rRNA were utilized for results normalization. PCR products were separated using electrophoresis on 2% agarose gel, stained with ethidium bromide, and quantified with Quantity One 1D software (BioRad, Hercules, CA, USA).

2.7. 16S rRNA Gene Amplification, Sequencing and Analysis

Performed as previously described [25]. Briefly, cecal bacterial DNA was extracted as defined by the manufacturer (PowerSoil DNA isolation kit, MoBio Laboratories Ltd., Carlsbad, CA, USA). Bacterial 16S rRNA gene sequences were PCR-amplified using the 515F-806R primers for the V4 hypervariable region of the 16S rRNA gene [7,25–34]. Detailed methodology is provided in the supplementary materials.

2.8. Glycogen Analysis

Glycogen content quantification in the pectoralis muscle and liver was performed as previously described [7,35]. Briefly, the frozen pectoralis muscle or liver samples were homogenized for 1 min in perchloric acid (8% v/v) on ice, centrifuged at 12,000× g for 15 min at room temperature, and the resulting supernatant was discarded. A measurement of 1 mL of petroleum ether was added, the petroleum ether fraction was discarded, and the lower layer of each sample was transferred to a 96-well plate containing iodine reagent (300 μL). Samples were read at 450 nm in a plate reader (Epoch, BioTek, VT, USA). The glycogen content was calculated using a standard curve.

2.9. Tissue Morphology Examination

Intestinal tissue morphometric assessment was performed as was previously described on duodenal sections [7,17,21]. Duodenum sections were fixed in 4% (v/v) buffered formaldehyde, dehydrated, cleared, and embedded in paraffin. Sections were cut (5 μm thickness) and positioned on glass slides, deparaffinized in xylene, rehydrated in ethanol, and stained with Alcian Blue/Periodic acid-Schiff. Villus height, villus width, crypt depth, Paneth cell number per crypt, Paneth cell width, goblet cell number, goblet cell diameter, goblet cell type within the villi, and goblet cell type within the crypts were assessed using a light microscope (CellSens Standard software, Olympus, Waltham, MA, USA). Five biological samples per treatment group (n = 5) and four segments for each biological sample were analyzed. Ten randomly selected villi and crypts were analyzed per segment and cell size measurements and counts were counted in ten randomly selected villi and/or crypts per segment (40 replicates per biological sample). Villus surface area was calculated using the following equation:

$$Villus\ surface\ area = 2\pi \times \frac{VW}{2} \times VL \quad (1)$$

where VW is the average of three measurements of villus width, and VL is the villus length.

2.10. Statistical Analysis

Results are shown as mean ± standard error, n = 6–12, in tables and heatmaps. Heatmaps were created in Microsoft Excel (Microsoft Corporation, Redmond, WA, USA) based on conditional formatting using color scales based on result means. Gene expression was normalized to 18S gene expression [22] and presented in arbitrary units (AU). To assess distribution normality, the Shapiro–Wilk test was used. Normally distributed results were analyzed by one-way ANOVA and Duncan post-hoc test. The Kruskal–Wallis test was utilized for non-parametric data. Differences were considered significant at $p < 0.05$. Statistical analyses were carried out using SPSS software (version 20.0, IBM, Armonk, NY, USA).

3. Results

3.1. Body Weight and Cecum Weight

The body weight of the 2.5% genistein group is significantly higher than the no-injection group ($p < 0.05$, Table 2). For cecum weights, the no-injection and H_2O groups demonstrate significantly greater values when compared to the 5% inulin and 2.5% genistein groups ($p < 0.05$).

Table 2. Effect of genistein exposure on body weight and cecum weight [1].

Treatment Group	Average Body Weight (g)	Average Cecum Weight (g)
No Injection	43.23 ± 1.44^{b}	0.60 ± 0.05^{a}
H_2O	44.62 ± 1.43^{ab}	0.59 ± 0.05^{a}
5% Inulin	46.04 ± 1.18^{ab}	0.43 ± 0.05^{b}
1.25% Genistein	45.83 ± 0.99^{ab}	0.50 ± 0.04^{ab}
2.5% Genistein	47.69 ± 1.30^{a}	0.44 ± 0.03^{b}

[1] Values are means ± SEM, $n = 6$. ^{a,b} Treatment groups not indicated by the same letter in the same column are significantly different ($p < 0.05$) according to one-way ANOVA with post-hoc Duncan test.

3.2. *Hemoglobin and Glycogen Concentrations*

Blood hemoglobin (Hb) levels in the 1.25% genistein group are significantly elevated compared to the no-injection, H_2O, and 5% inulin groups ($p < 0.05$, Table 3). The blood hemoglobin of the 2.5% genistein group is higher than the no-injection group and significantly higher versus the H_2O and 5% inulin groups. Among average glycogen, there were no significant differences between the genistein-treated and no-injection groups ($p > 0.05$).

Table 3. Blood hemoglobin (Hb) concentrations (g/dL) and pectoral muscle glycogen concentrations (mg/g) following genistein exposure [1].

Treatment Group	Average Hb (g/dL)	Average Glycogen (mg/g)
No Injection	10.10 ± 2.40^{bc}	0.019 ± 0.005^{a}
H_2O	9.68 ± 2.50^{c}	0.014 ± 0.003^{a}
5% Inulin	9.56 ± 0.92^{c}	0.002 ± 0.001^{b}
1.25% Genistein	14.98 ± 0.45^{a}	0.008 ± 0.003^{ab}
2.5% Genistein	14.23 ± 0.79^{ab}	0.015 ± 0.004^{a}

[1] Values are the means ± SEM, $n = 6$–12. $^{a–c}$ Treatment groups not indicated by the same letter in the same column are significantly different ($p < 0.05$) according to one-way ANOVA with post-hoc Duncan test.

3.3. *Gene Expression of Fe, Zn, BBM Functionality, and Inflammation Related Proteins*

3.3.1. Fe-Related Proteins

As depicted in Figure 1, gene expression of DMT1 is downregulated in the 2.5% genistein when compared to all other experimental groups ($p < 0.05$). DcytB was significantly downregulated ($p < 0.05$) in the genistein treatment groups compared to the no-injection, H_2O, and inulin groups. Hepcidin was significantly upregulated ($p < 0.05$) with genistein exposure compared to the no-injection group. There were no significant differences in ferroportin expression between groups.

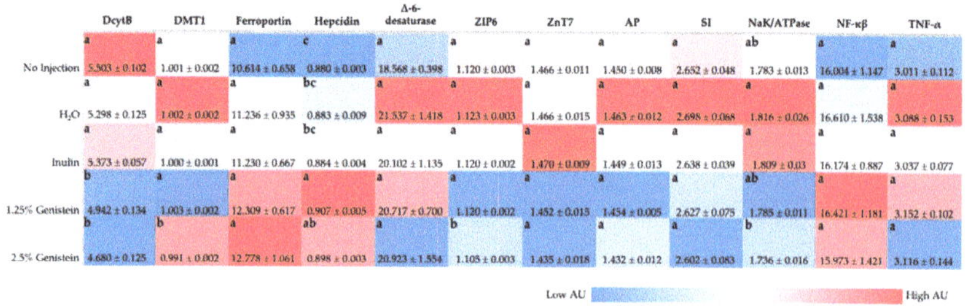

Figure 1. Effect of intraamniotic administration of genistein and controls on duodenal and liver (hepcidin) mRNA gene expression. Gene expression has been normalized to the 18S housekeeping gene and is in arbitrary units (AU). Values are presented as mean ± SEM, $n = 6$. $^{a–c}$ Per gene (in the same column), treatments groups not indicated by the same letter are significantly different ($p < 0.05$) according to one-way ANOVA with post-hoc Duncan test. DcytB, duodenal cytochrome b; DMT1, divalent metal transporter 1; ZIP6, zinc transport protein 6; ZnT7, zinc transporter 7; AP, amino peptidase; SI, sucrose isomaltase; NaK/ATPase, sodium, potassium and adenosine triphosphate; NF-κβ, nuclear factor kappa β subunit 1; TNF-α, tumor necrosis factor-α.

3.3.2. Zn-Related Proteins

ZIP6 was significantly downregulated ($p < 0.05$) in the 2.5% genistein group compared to all other treatment groups (Figure 1). There were no significant differences in ZnT7 or Δ-6-desaturase expression between groups.

3.3.3. Inflammatory Cytokines and BBM Functionality

No significant differences in gene expression of aminopeptidase (AP), sucrose isomaltase (SI), sodium, potassium, and adenosine triphosphate (NaK/ATPase) were found when comparing the treatment groups to the no-injection group (Figure 1). No significant differences in gene expression of nuclear transcription factor (NF-κβ) and tumor necrosis factor-α (TNF-α) between groups were found.

3.4. Morphometric Analysis of Duodenal Villi, Depth of Crypts, Goblet Cells, and Paneth Cells

The villus height, width, and surface area of the 2.5% genistein were significantly increased ($p < 0.05$) compared to the no-injection and H_2O groups (Table 4). The 1.25% genistein group had significantly ($p < 0.05$) greater villus width than the no-injection and H_2O groups. The 2.5% genistein group had significantly higher ($p < 0.05$) villus height, width and surface area compared to the 1.25% genistein.

Table 4. Effects of genistein intraamniotic administration on duodenal small intestinal villus [1].

Treatment Group	Villus Height (μm)	Villus Width (μm)	Villus Surface Area (μm^2)
No Injection	201.18 ± 4.94 ^{b}	33.73 ± 0.67 ^{e}	112.51 ± 4.28 ^{d}
H_2O	204.74 ± 4.52 ^{b}	41.92 ± 1.01 ^{d}	143.33 ± 5.27 ^{c}
5% Inulin	246.64 ± 5.14 ^{a}	50.98 ± 1.03 ^{a}	206.92 ± 6.37 ^{a}
1.25% Genistein	204.18 ± 3.73 ^{b}	44.51 ± 0.86 ^{c}	146.97 ± 4.55 ^{c}
2.5% Genistein	238.22 ± 3.17 ^{a}	48.27 ± 0.87 ^{b}	184.13 ± 4.66 ^{b}

[1] Values are presented as mean ± SEM, n = 5. $^{a–e}$ Treatment groups not indicated by the same letter in the same column are significantly different ($p < 0.05$) according to one-way ANOVA with post-hoc Duncan test.

The villi goblet cell diameter and total goblet cell number were significantly higher ($p < 0.05$) in the genistein-exposed groups than in the no-injection, H_2O, and inulin groups (Table 5). More specifically, the acidic villi goblet cell count was significantly increased ($p < 0.05$) in the 1.25% genistein and 2.5% genistein groups relative to the 5% inulin, no-injection, and H_2O control groups. The neutral villi goblet cell count of 1.25% genistein, 2.5% genistein, and 5% inulin groups were significantly higher ($p < 0.05$) compared with the no-injection and H_2O injection controls, and the mixture villi goblet cells were significantly reduced ($p < 0.05$) with genistein exposure when compared with no-injection, H_2O, and 5% inulin control groups.

Table 5. Effects of genistein intraamniotic administration on villi goblet cells [1].

Treatment Group	Villi Goblet Cell Diameter (μm)	Villi Goblet Cell Number (Unit)			
		Acidic	Neutral	Mixture	Total
No Injection	2.86 ± 0.02 ^{d}	13.59 ± 0.39 ^{d}	0.01 ± 0.01 ^{c*}	3.50 ± 0.23 ^{c}	17.09 ± 0.49 ^{d}
H_2O	3.11 ± 0.03 ^{c}	15.03 ± 0.39 ^{c}	0.01 ± 0.01 ^{c*}	5.76 ± 0.30 ^{b}	20.80 ± 0.47 ^{c}
5% Inulin	2.74 ± 0.03 ^{e}	16.39 ± 0.54 ^{c}	0.10 ± 0.02 ^{b*}	6.53 ± 0.30 ^{a}	23.02 ± 0.60 ^{b}
1.25% Genistein	3.41 ± 0.03 ^{b}	23.49 ± 0.67 ^{a}	0.09 ± 0.03 ^{b*}	1.69 ± 0.13 ^{e}	25.26 ± 0.67 ^{a}
2.5% Genistein	3.50 ± 0.03 ^{a}	22.01 ± 0.51 ^{b}	0.19 ± 0.04 ^{a*}	2.70 ± 0.16 ^{d}	24.89 ± 0.54 ^{a}

[1] Values are presented as mean ± SEM, n = 5. $^{a–e}$ Treatment groups not indicated by the same letter in the same column are significantly different ($p < 0.05$) according to one-way ANOVA with post-hoc Duncan test. $^{a*–c*}$ Treatment groups indicated are significantly different ($p < 0.05$) based on Kruskal–Wallis.

As shown in Table 6, the crypt goblet cell diameter of the 1.25% genistein group was significantly larger ($p < 0.05$) than all control groups. The 2.5% genistein group had a significantly higher diameter than the 5% inulin group. Genistein exposure resulted in a significantly higher ($p < 0.05$) total crypt goblet cell count when compared with the no-injection and H_2O groups. More specifically, the acidic crypt goblet cell count of both genistein treatment groups was significantly higher ($p < 0.05$) when compared with the no-injection, H_2O, and inulin groups. Genistein exposure significantly reduced mixed crypt goblet cells ($p < 0.05$) compared with the no-injection, H_2O, and inulin groups.

Table 6. Effects of genistein intraamniotic administration on crypt goblet cells [1].

Treatment Group	Crypt Goblet Cell Diameter (μm)	Crypt Goblet Cell Number (Unit)			
		Acidic	Neutral	Mixture	Total
No Injection	2.68 ± 0.02 [b]	5.46 ± 0.18 [d]	0.00 ± 0.00 [a]	1.49 ± 0.09 [b]	6.95 ± 0.21 [d]
H_2O	2.65 ± 0.02 [b]	6.07 ± 0.18 [c]	0.00 ± 0.00 [a]	1.76 ± 0.08 [a]	7.83 ± 0.19 [c]
5% Inulin	2.51 ± 0.02 [c]	7.97 ± 0.16 [b]	0.00 ± 0.00 [a]	1.19 ± 0.08 [c]	9.15 ± 0.16 [b]
1.25% Genistein	2.89 ± 0.02 [a]	8.51 ± 0.14 [a]	0.00 ± 0.00 [a]	0.74 ± 0.06 [d]	9.25 ± 0.14 [ab]
2.5% Genistein	2.63 ± 0.02 [b]	8.81 ± 0.19 [a]	0.00 ± 0.00 [a]	0.88 ± 0.06 [d]	9.68 ± 0.19 [a]

[1] Values are presented as mean ± SEM, $n = 5$. [a–d] treatment groups not indicated by the same letter in the same column are significantly different ($p < 0.05$) according to one-way ANOVA with post-hoc Duncan test.

The number of crypt Paneth cells was significantly greater ($p < 0.05$) for the genistein treatment groups compared to the no-injection, H_2O, and inulin groups (Table 7). The crypt depth for the genistein treatment groups was significantly lower ($p < 0.05$) compared to the H_2O-injection group. The 1.25% genistein group had a significantly ($p < 0.05$) higher crypt Paneth cell diameter than the no-injection and 5% inulin groups.

Table 7. Effects of genistein intraamniotic administration on crypt depth and Paneth cells [1].

Treatment Group	Crypt Depth (μm)	# Crypt Paneth Cells	Crypt Paneth cell Diameter (μm)
No Injection	22.45 ± 0.39 [d]	1.48 ± 0.05 [d]	1.67 ± 0.03 [b]
H_2O	39.07 ± 0.80 [a]	2.46 ± 0.11 [c]	1.82 ± 0.04 [a]
5% Inulin	35.00 ± 0.43 [b]	2.56 ± 0.09 [c]	1.68 ± 0.03 [b]
1.25% Genistein	26.36 ± 0.46 [c]	2.92 ± 0.10 [b]	1.78 ± 0.04 [a]
2.5% Genistein	23.65 ± 0.46 [d]	3.24 ± 0.11 [a]	1.65 ± 0.03 [b]

[1] Values are the means ± SEM, $n = 5$. [a–d] treatment groups not indicated by the same letter in the same column are significantly different ($p < 0.05$) according to one-way ANOVA with post-hoc Duncan test. The # symbol refers to the number of Paneth cells.

3.5. Intestinal Content Bacterial Expression

Figure 2 shows the duodenal genera and species-level bacterial populations. The relative abundance of *Bifidobacterium* spp., considered a probiotic bacteria, was significantly increased ($p < 0.05$) with 2.5% genistein exposure compared with all other treatment groups. *Lactobacillus* spp. relative abundance was significantly increased ($p < 0.05$) with genistein exposure compared to the no-injection control. *L. plantarum*, a probiotic bacteria associated with increased Fe absorption, was significantly increased ($p < 0.05$) in the genistein-exposed groups and 5% inulin control compared with the H_2O-injected control. Genistein exposure significantly decreased ($p < 0.05$) the relative abundance of E. coli compared with all other experimental groups. *Clostridium* spp. relative abundance was significantly increased ($p < 0.05$) in the genistein-treated groups and 5% inulin control compared to the no-injection and H_2O injection controls.

	Bifidobacterium	*Lactobacillus*	*L. plantarum*	*E. coli*	*Clostridium*
No Injection	b 0.87 ± 0.02	b 0.93 ± 0.08	ab 0.95 ± 0.12	a 0.89 ± 0.03	b 0.82 ± 0.02
H_2O	b 0.84 ± 0.05	a 1.14 ± 0.02	b 0.84 ± 0.06	a 0.84 ± 0.03	b 0.77 ± 0.06
5% Inulin	b 0.94 ± 0.09	a 1.19 ± 0.04	a 1.13 ± 0.08	a 0.92 ± 0.03	a 2.37 ± 0.15
1.25% Genistein	b 1.05 ± 0.04	a 1.16 ± 0.02	a 1.11 ± 0.02	b 0.67 ± 0.02	a 1.80 ± 0.08
2.5% Genistein	a 1.46 ± 0.18	a 1.34 ± 0.10	a 1.18 ± 0.07	b 0.57 ± 0.03	a 2.13 ± 0.17

Low Density (INT/mm^2) — High Density (INT/mm^2)

Figure 2. Effects of intraamniotic injections of genistein and the controls on duodenal genera and species-level bacterial populations. Values are presented as mean ± SEM, $n = 5$, as relative intensity of bands per mm^2 of gel. ^{a,b} per bacterial category (in the same column), treatment groups that do not share any letters are significantly different ($p < 0.05$) according to one-way ANOVA with post-hoc Duncan test.

4. Discussion

In the current study, we have evaluated the effect of intraamniotic genistein administration on mineral transport, duodenal brush border membrane development and functionality, and intestinal microbiota. Although the ingestion of genistein has been associated with marked physiological changes associated with cancer and metabolic syndrome, further understanding of tissue-level effects associated with genistein exposure is needed [6,36,37]. Presently, there is a paucity of studies in the literature that directly measure the effects of genistein on the combination of mineral transport, BBM morphology or functionality, and intestinal microbiota.

The intraamniotic administration of genistein positively affected intestinal development, as demonstrated by increased enterocyte proliferation. The duodenal morphometric analysis demonstrated a significant ($p < 0.05$) dose-responsive effect of genistein treatment on increasing villus surface area versus the no-injection control (Table 4), indicative of improved digestive enzyme and absorptive capacity [7]. A significantly ($p < 0.05$) reduced crypt depth was observed with genistein administration when compared to the H_2O injection control group (Table 7), which has been shown to be a marker of efficient tissue turnover and good condition of the gut [38]. The increase in villus surface area and reduction in crypt depth are in accordance with other genistein administration trials using the in vivo *Gallus gallus* model [39,40]. Additionally, increased proliferation in total villi and crypt goblet cells and an increase in the proportion of villi acidic and crypt acidic ($p < 0.05$) goblet cells were observed with genistein exposure compared to the no-injection and H_2O injection controls (Tables 5 and 6). This indicates increased synthesis and secretion of acidic luminal mucin by duodenal goblet cells [11,12]. The major goblet cell mucins in the small intestine are mucin 2 proteins, gel-forming secretory mucins that facilitate hydrolysis and absorption of nutrients [18,41–43]. In addition to serving as a protective intestinal epithelial barrier, this mucin (mucin 2) also functions as a habitat that supports probiotic populations and promotes epithelial cell function [44,45]. Taken as a whole, this demonstrates that the intraamniotic administration of genistein can positively modulate BBM development and functionality.

The intestinal microbiota of the *Gallus gallus* model is significantly and directly influenced by host genetics, environment, and diet [23,46]. At the phylum level, there is a significant resemblance between the gut microbiota of *Gallus gallus* and humans, with *Bacteroidetes*, *Firmicutes*, *Proteobacteria*, and *Actinobacteria* representing the dominant bacterial phyla [47]. Soy isoflavone treatment has been shown to alter intestinal bacterial populations in vivo, including increases in populations of SCFA-producing bacteria [1,36,48]. In the duodenum, the relative abundance of *Bifidobacterium* spp. considered a probiotic bacteria species, significantly increased with 2.5% genistein exposure compared with all other treatment groups (Figure 2). *Lactobacillus* spp. relative abundance was significantly increased with genistein exposure compared to the no-injection control. Further, linear discriminant analysis effect size (LefSe) analysis found that genistein treatment enriched bacterial pathways associated with *de novo* synthesis of vitamin B_{12} (Figure S1), where bacteria from the *Lactobacillus* genus represent a small number of bacteria known to encode the complete *de novo* biosynthetic pathway of vitamin B_{12} [49,50]. *L. plantarum*, a probiotic bacteria species associated with increased Fe absorption, was significantly increased in the genistein exposed-groups and 5% inulin control compared with the H_2O-injected control [51]. *L. plantarum* produces glucosidases that can hydrolyze isoflavones (glycosides) into metabolites (aglycones) with increased antioxidant activity [52]. Increased populations of health-promoting bacteria, *Bifidobacterium* spp., *Lactobacillus* spp., and *L. plantarum*, resulting from genistein exposure, can be attributed to increased acidic mucin production [45,53,54]. Increased acidic mucin synthesis provides an environment conducive to the proliferation of these probiotic bacterial populations, which can be associated with an increased Paneth cell number per crypt and number of villi and crypt acidic goblet cells associated with genistein administration [45,53]. *Clostridium* spp. was significantly increased in the genistein-treated groups, and butyrate-producing (SCFA) bacteria, such as *Roseburia* spp. and *E. hallii* from *Clostridium* cluster XIVa, have previously been observed to be increased with genistein exposure in vitro [4]. The increase in *Lactobacillus* spp., *Bifidobacterium* spp., and *Clostridium* spp. abundance may further contribute to increased mineral bioavailability as these genera house SCFA-producing species, where SCFAs reduce the intestinal pH and thus may increase mineral (Fe and Zn) solubility and absorption [7,18,55].

Our previous research suggested soy isoflavone (daidzein) intraamniotic administration has the potential to improve dietary Fe bioavailability [19]. In our current study, BBM gene expression analysis (Figure 1) demonstrated that genistein downregulated DMT1 (transports Fe^{2+} into duodenal enterocyte) and DcytB (reduces Fe^{3+} to Fe^{2+}) and upregulated ferroportin (transports Fe^{2+} into blood) and hepcidin (binds to ferroportin, causes ferroportin internalization and degradation), relative to the control group, though these results were not necessarily dose-dependent or significant [56–59]. Based on protein functionalities in Fe sufficient or excess scenarios, it is expected that DcytB, DMT1, and ferroportin would be downregulated, whereas hepcidin would be upregulated [57,58,60–63]. Though upregulation of ferroportin has previously been associated with Fe deficiency, genistein treatment was found to upregulate ferroportin expression in glial cells through estrogen receptor ß-dependent p38 MAPK activation, independent of Fe status [9,63]. Genistein administration has been shown to upregulate hepcidin expression, directly influencing ferroportin expression in in vivo and in vitro liver cell models [10]. Blood Hb levels were increased with genistein administration compared with the controls, which, taken together with Fe gene expression analysis, may indicate Fe status was improved by genistein administration. Genistein exposure resulted in ZIP6 (imports zinc across cell membrane) downregulation in comparison with the no-injection control, potentially indicative of improved zinc status with genistein administration [64,65], or could be associated with estrogenic effects of soy isoflavones, where ZIP6 expression was found to be modulated with anti-estrogen treatment in breast cells [66,67]. Although Zn absorption occurs in the duodenum, it has been suggested that the ileum is the leading site of Zn absorption in *Gallus gallus* [68], where future studies should focus on the Zn-transporter gene expression in the ileum to further understand the effects of genistein administration on Zn transport and absorption. Overall, alterations in mineral transport and hemoglobin concentration associated with improvements in mineral status can potentially be attributed to the combination of increased bacterial production of SCFA and increased proportion of acidic goblet cells associated with genistein exposure, resulting in a lowered intestinal pH and increased mineral solubility, thus improving mineral absorption [12,18,69].

Increases in body weight were observed in a dose-dependent manner compared with the controls, with the 2.5% genistein treatment group being significantly higher ($p < 0.05$) than the no-injection control (Table 2). Given the short exposure time, a significant increase in body weight is unexpected, but when taken with improved Fe status and BBM development, and given that the in vivo *Gallus gallus* model is sensitive to dietary Fe and Zn deficiencies [55,70], a significant increase in body weight confirms the positive developmental effects related to genistein exposure [71]. Additional studies are warranted to assess shifts in mineral status, intestinal functionality and development, and intestinal microbiota post-hatch and during a long-term feeding trial associated with genistein consumption.

5. Conclusions

This present study demonstrates intraamniotic administration of genistein improved brush border membrane functionality through improvements in villus architecture, goblet cell expansion, and related mucin production. Additionally, increases in the relative abundance of bacterial populations associated with SCFA production were found. Consequently, the combination of these factors contributed to alterations in the relative expression of various duodenal and hepatic proteins responsible for mineral absorption and transport associated with improved Fe status. Given these findings, genistein represents a promising plant bioactive and should be further evaluated in long-term animal and controlled human efficacy trials.

Supplementary Materials: The following supporting information can be downloaded at: https://www.mdpi.com/article/10.3390/nu14173473/s1, Figure S1: Bacterial pathways identified by the LEfSe method with the greatest differences in the control, 5% inulin, and genistein treated groups.

Author Contributions: Conceptualization, P.S. and E.T.; methodology, N.K., J.C. and E.T.; formal analysis, N.K., J.C., S.T., C.E., O.K. and E.T.; investigation, J.C., P.S., N.K. and E.T.; resources, O.K. and E.T.; data curation, N.K., J.C., S.T., C.E., O.K. and E.T.; writing—original draft preparation, J.C. and P.S.; writing—review and editing, J.C., N.K., S.T., C.E., O.K. and E.T.; supervision, O.K. and E.T.; funding acquisition, E.T. All authors have read and agreed to the published version of the manuscript.

Funding: This research received no external funding.

Institutional Review Board Statement: The animal protocol used in this study was conducted according to the guidelines of the Declaration of Helsinki and approved by the Cornell University Institutional Animal Care and Use Committee by ethic approval code 2020-0077.

Informed Consent Statement: Not applicable.

Data Availability Statement: Data available upon reasonable request.

Conflicts of Interest: The authors declare no conflict of interest.

References

1. Lopez, P.; Sanchez, M.; Perez-Cruz, C.; Velazquez-Villegas, L.A.; Syeda, T.; Aguilar-Lopez, M.; Rocha-Viggiano, A.K.; Del Carmen Silva-Lucero, M.; Torre-Villalvazo, I.; Noriega, L.G.; et al. Long-Term Genistein Consumption Modifies Gut Microbiota, Improving Glucose Metabolism, Metabolic Endotoxemia, and Cognitive Function in Mice Fed a High-Fat Diet. *Mol. Nutr. Food Res.* **2018**, *62*, e1800313. [CrossRef]
2. Spagnuolo, C.; Russo, G.L.; Orhan, I.E.; Habtemariam, S.; Daglia, M.; Sureda, A.; Nabavi, S.F.; Devi, K.P.; Loizzo, M.R.; Tundis, R.; et al. Genistein and cancer: Current status, challenges, and future directions. *Adv. Nutr.* **2015**, *6*, 408–419. [CrossRef]
3. Setchell, K. Phytoestrogens: The biochemistry, physiology, and implications for human health of soy isoflavones. *Am. J. Clin. Nutr.* **1998**, *68*, 1333S–1346S. [CrossRef]
4. Guadamuro, L.; Dohrmann, A.B.; Tebbe, C.C.; Mayo, B.; Delgado, S. Bacterial communities and metabolic activity of faecal cultures from equol producer and non-producer menopausal women under treatment with soy isoflavones. *BMC Microbiol.* **2017**, *17*, 93. [CrossRef]
5. Matthies, A.; Blaut, M.; Braune, A. Isolation of a human intestinal bacterium capable of daidzein and genistein conversion. *Appl. Environ. Microbiol.* **2009**, *75*, 1740–1744. [CrossRef]
6. Zhou, L.; Xiao, X.; Zhang, Q.; Zheng, J.; Li, M.; Wang, X.; Deng, M.; Zhai, X.; Liu, J. Gut microbiota might be a crucial factor in deciphering the metabolic benefits of perinatal genistein consumption in dams and adult female offspring. *Food Funct.* **2019**, *10*, 4505–4521. [CrossRef]

7. Carboni, J.; Reed, S.; Kolba, N.; Eshel, A.; Koren, O.; Tako, E. Alterations in the Intestinal Morphology, Gut Microbiota, and Trace Mineral Status Following Intra-Amniotic Administration (*Gallus gallus*) of Teff (*Eragrostis tef*) Seed Extracts. *Nutrients* **2020**, *12*, 3020. [CrossRef]
8. Dias, D.M.; Costa, N.M.B.; Nutti, M.R.; Tako, E.; Martino, H.S.D. Advantages and limitations of in vitro and in vivo methods of iron and zinc bioavailability evaluation in the assessment of biofortification program effectiveness. *Crit. Rev. Food Sci. Nutr.* **2018**, *58*, 2136–2146. [CrossRef]
9. Persichini, T.; Maio, N.; di Patti, M.C.; Rizzo, G.; Colasanti, M.; Musci, G. Genistein up-regulates the iron efflux system in glial cells. *Neurosci. Lett.* **2010**, *470*, 145–149. [CrossRef] [PubMed]
10. Zhen, A.W.; Nguyen, N.H.; Gibert, Y.; Motola, S.; Buckett, P.; Wessling-Resnick, M.; Fraenkel, E.; Fraenkel, P.G. The small molecule, genistein, increases hepcidin expression in human hepatocytes. *Hepatology* **2013**, *58*, 1315–1325. [CrossRef]
11. Hou, T.; Tako, E. The In Ovo Feeding Administration (*Gallus gallus*)—An Emerging In Vivo Approach to Assess Bioactive Compounds with Potential Nutritional Benefits. *Nutrients* **2018**, *10*, 418. [CrossRef] [PubMed]
12. Dias, D.M.; Kolba, N.; Hart, J.J.; Ma, M.; Sha, S.T.; Lakshmanan, N.; Nutti, M.R.; Martino, H.S.D.; Glahn, R.P.; Tako, E. Soluble extracts from carioca beans (*Phaseolus vulgaris* L.) affect the gut microbiota and iron related brush border membrane protein expression in vivo (*Gallus gallus*). *Food Res. Int.* **2019**, *123*, 172–180. [CrossRef] [PubMed]
13. Mahler, G.J.; Esch, M.B.; Tako, E.; Southard, T.L.; Archer, S.D.; Glahn, R.P.; Shuler, M.L. Oral exposure to polystyrene nanoparticles affects iron absorption. *Nat. Nanotechnol.* **2012**, *7*, 264–271. [CrossRef] [PubMed]
14. Consortium, I.C.G.S. Sequence and comparative analysis of the chicken genome provide unique perspectives on vertebrate evolution. *Nature* **2004**, *432*, 695–777.
15. Warkentin, T.; Kolba, N.; Tako, E. Low Phytate Peas (*Pisum sativum* L.) Improve Iron Status, Gut Microbiome, and Brush Border Membrane Functionality In Vivo (*Gallus gallus*). *Nutrients* **2020**, *12*, 2563. [CrossRef]
16. Kolba, N.; Guo, Z.; Olivas, F.M.; Mahler, G.J.; Tako, E. Intra-amniotic administration (*Gallus gallus*) of TiO2, SiO2, and ZnO nanoparticles affect brush border membrane functionality and alters gut microflora populations. *Food Chem. Toxicol.* **2020**, *135*, 110896. [CrossRef]
17. Pereira da Silva, B.; Kolba, N.; Stampini Duarte Martino, H.; Hart, J.; Tako, E. Soluble Extracts from Chia Seed (*Salvia hispanica* L.) Affect Brush Border Membrane Functionality, Morphology and Intestinal Bacterial Populations In Vivo (*Gallus gallus*). *Nutrients* **2019**, *11*, 2457. [CrossRef]
18. Pacifici, S.; Song, J.; Zhang, C.; Wang, Q.; Glahn, R.P.; Kolba, N.; Tako, E. Intra Amniotic Administration of Raffinose and Stachyose Affects the Intestinal Brush Border Functionality and Alters Gut Microflora Populations. *Nutrients* **2017**, *9*, 304. [CrossRef]
19. Hartono, K.; Reed, S.; Ankrah, N.A.; Glahn, R.P.; Tako, E. Alterations in gut microflora populations and brush border functionality following intra-amniotic daidzein administration. *RSC Adv.* **2015**, *5*, 6407–6412. [CrossRef]
20. Decuypere, E.; Michels, H. Incubation temperature as a management tool: A review. *World's Poult. Sci. Assoc.* **1992**, *48*, 28–38. [CrossRef]
21. Martino, H.S.D.; Kolba, N.; Tako, E. Yacon (*Smallanthus sonchifolius*) flour soluble extract improve intestinal bacterial populations, brush border membrane functionality and morphology in vivo (*Gallus gallus*). *Food Res. Int.* **2020**, *137*, 109705. [CrossRef] [PubMed]
22. Kuchipudi, S.V.; Tellabati, M.; Nelli, R.K.; White, G.A.; Perez, B.B.; Sebastian, S.; Slomka, M.J.; Brookes, S.M.; Brown, I.H.; Dunham, S.P.; et al. 18S rRNAis a reliable normalisation gene for real time PCR based on influenza virus infected cells. *Virology J.* **2012**, *9*, 230. [CrossRef] [PubMed]
23. Zhu, X.Y.; Zhong, T.; Pandya, Y.; Joerger, R.D. 16S rRNA-based analysis of microbiota from the cecum of broiler chickens. *Appl. Environ. Microbiol.* **2002**, *68*, 124–137. [CrossRef] [PubMed]
24. Gorsuch, J.; LeSaint, D.; VanderKelen, J.; Buckman, D.; Kitts, C.L. A comparison of methods for enumerating bacteria in direct fed microbials for animal feed. *J. Microbiol. Methods* **2019**, *160*, 124–129. [CrossRef] [PubMed]
25. Reed, S.; Neuman, H.; Glahn, R.P.; Koren, O.; Tako, E. Characterizing the gut (*Gallus gallus*) microbiota following the consumption of an iron biofortified Rwandan cream seeded carioca (*Phaseolus vulgaris* L.) bean-based diet. *PLoS ONE* **2017**, *12*, e0182431. [CrossRef] [PubMed]
26. Bolyen, E.; Rideout, J.R.; Dillon, M.R.; Bokulich, N.A.; Abnet, C.C.; Al-Ghalith, G.A.; Alexander, H.; Alm, E.J.; Arumugam, M.; Asnicar, F.; et al. Reproducible, interactive, scalable and extensible microbiome data science using QIIME 2. *Nat. Biotechnol.* **2019**, *37*, 852–857. [CrossRef]
27. McDonald, D.; Price, M.N.; Goodrich, J.; Nawrocki, E.P.; DeSantis, T.Z.; Probst, A.; Andersen, G.L.; Knight, R.; Hugenholtz, P. An improved Greengenes taxonomy with explicit ranks for ecological and evolutionary analyses of bacteria and archaea. *ISME J.* **2012**, *6*, 610–618. [CrossRef]
28. Langille, M.G.; Zaneveld, J.; Caporaso, J.G.; McDonald, D.; Knights, D.; Reyes, J.A.; Clemente, J.C.; Burkepile, D.E.; Vega Thurber, R.L.; Knight, R.; et al. Predictive functional profiling of microbial communities using 16S rRNA marker gene sequences. *Nat. Biotechnol.* **2013**, *31*, 814–821. [CrossRef]
29. Segata, N.; Izard, J.; Waldron, L.; Gevers, D.; Miropolsky, L.; Garrett, W.S.; Huttenhower, C. Metagenomic biomarker discovery and explanation. *Genome Biol.* **2011**, *12*, R60. [CrossRef]

30. Caporaso, J.G.; Kuczynski, J.; Stombaugh, J.; Bittinger, K.; Bushman, F.D.; Costello, E.K.; Fierer, N.; Pena, A.G.; Goodrich, J.K.; Gordon, J.I.; et al. QIIME allows analysis of high-throughput community sequencing data. *Nat. Methods* **2010**, *7*, 335–336. [CrossRef]
31. Price, M.N.; Dehal, P.S.; Arkin, A.P. FastTree: Computing large minimum evolution trees with profiles instead of a distance matrix. *Mol. Biol. Evol.* **2009**, *26*, 1641–1650. [CrossRef] [PubMed]
32. Faith, D. Conservation evaluation and phylogenetic diversity. *Biol. Conserv.* **1992**, *61*, 1–10. [CrossRef]
33. Lozupone, C.; Knight, R. UniFrac: A new phylogenetic method for comparing microbial communities. *Appl. Environ. Microbiol.* **2005**, *71*, 8228–8235. [CrossRef] [PubMed]
34. Caporaso, J.G.; Bittinger, K.; Bushman, F.D.; DeSantis, T.Z.; Andersen, G.L.; Knight, R. PyNAST: A flexible tool for aligning sequences to a template alignment. *Bioinformatics* **2010**, *26*, 266–267. [CrossRef]
35. Dreiling, C.; Brown, D.; Casale, L.; Kelly, L. Muscle Glycogen: Comparison of Iodine Binding and Enzyme Digestion Assays and Application to Meat Samples. *Meat Sci.* **1987**, *20*, 167–177. [CrossRef]
36. Huang, G.; Xu, J.; Lefever, D.E.; Glenn, T.C.; Nagy, T.; Guo, T.L. Genistein prevention of hyperglycemia and improvement of glucose tolerance in adult non-obese diabetic mice are associated with alterations of gut microbiome and immune homeostasis. *Toxicol. Appl. Pharmacol.* **2017**, *332*, 138–148. [CrossRef]
37. Guevara-Cruz, M.; Godinez-Salas, E.T.; Sanchez-Tapia, M.; Torres-Villalobos, G.; Pichardo-Ontiveros, E.; Guizar-Heredia, R.; Arteaga-Sanchez, L.; Gamba, G.; Mojica-Espinosa, R.; Schcolnik-Cabrera, A.; et al. Genistein stimulates insulin sensitivity through gut microbiota reshaping and skeletal muscle AMPK activation in obese subjects. *BMJ Open Diabetes Res. Care* **2020**, *8*, e000948. [CrossRef]
38. Sobolewska, A.; Bogucka, J.; Dankowiakowska, A.; Elminowska-Wenda, G.; Stadnicka, K.; Bednarczyk, M. The impact of synbiotic administration through in ovo technology on the microstructure of a broiler chicken small intestine tissue on the 1st and 42nd day of rearing. *J. Anim. Sci. Biotechnol.* **2017**, *8*, 61. [CrossRef]
39. Kamboh, A.A.; Zhu, W.Y. Individual and combined effects of genistein and hesperidin on immunity and intestinal morphometry in lipopolysacharide-challenged broiler chickens. *Poult. Sci.* **2014**, *93*, 2175–2183. [CrossRef]
40. Glisic, M.; Boskovic, M.; Baltic, M.Z.; Sefer, D.; Radovanovic, A.; Djordjevic, V.; Raseta, M.; Markovic, R. Performance, intestinal histomorphology and bone composition of broiler chickens fed diets supplemented with genistein. *S. Afr. J. Anim. Sci.* **2020**, *50*, 241–252. [CrossRef]
41. Limage, R.; Tako, E.; Kolba, N.; Guo, Z.; Garcia-Rodriguez, A.; Marques, C.N.H.; Mahler, G.J. TiO_2 Nanoparticles and Commensal Bacteria Alter Mucus Layer Thickness and Composition in a Gastrointestinal Tract Model. *Small* **2020**, *16*, e2000601. [CrossRef] [PubMed]
42. Nikolenko, V.N.; Oganesyan, M.V.; Sankova, M.V.; Bulygin, K.V.; Vovkogon, A.D.; Rizaeva, N.A.; Sinelnikov, M.Y. Paneth cells: Maintaining dynamic microbiome-host homeostasis, protecting against inflammation and cancer. *Bioessays* **2021**, *43*, e2000180. [CrossRef] [PubMed]
43. Kim, Y.S.; Ho, S.B. Intestinal goblet cells and mucins in health and disease: Recent insights and progress. *Curr. Gastroenterol. Rep.* **2010**, *12*, 319–330. [CrossRef] [PubMed]
44. Engevik, M.A.; Luk, B.; Chang-Graham, A.L.; Hall, A.; Herrmann, B.; Ruan, W.; Endres, B.T.; Shi, Z.; Garey, K.W.; Hyser, J.M.; et al. Bifidobacterium dentium Fortifies the Intestinal Mucus Layer via Autophagy and Calcium Signaling Pathways. *mBio* **2019**, *10*, 3. [CrossRef]
45. O'Callaghan, A.; van Sinderen, D. Bifidobacteria and Their Role as Members of the Human Gut Microbiota. *Front. Microbiol.* **2016**, *7*, 925. [CrossRef]
46. Yegani, M.; Korver, D.R. Factors affecting intestinal health in poultry. *Poult. Sci.* **2008**, *87*, 2052–2063. [CrossRef]
47. Qin, J.; Li, R.; Raes, J.; Arumugam, M.; Burgdorf, K.S.; Manichanh, C.; Nielsen, T.; Pons, N.; Levenez, F.; Yamada, T.; et al. A human gut microbial gene catalogue established by metagenomic sequencing. *Nature* **2010**, *464*, 59–65. [CrossRef]
48. Huang, G.; Xu, J.; Cai, D.; Chen, S.Y.; Nagy, T.; Guo, T.L. Exacerbation of Type 1 Diabetes in Perinatally Genistein Exposed Female Non-Obese Diabetic (NOD) Mouse Is Associated With Alterations of Gut Microbiota and Immune Homeostasis. *Toxicol. Sci.* **2018**, *165*, 291–301. [CrossRef]
49. Torres, A.C.; Vannini, V.; Bonacina, J.; Font, G.; Saavedra, L.; Taranto, M.P. Cobalamin production by Lactobacillus coryniformis: Biochemical identification of the synthetized corrinoid and genomic analysis of the biosynthetic cluster. *BMC Microbiol.* **2016**, *16*, 240. [CrossRef]
50. De Angelis, M.; Bottacini, F.; Fosso, B.; Kelleher, P.; Calasso, M.; Di Cagno, R.; Ventura, M.; Picardi, E.; van Sinderen, D.; Gobbetti, M. Lactobacillus rossiae, a vitamin B12 producer, represents a metabolically versatile species within the Genus Lactobacillus. *PLoS ONE* **2014**, *9*, e107232. [CrossRef]
51. Axling, U.; Onning, G.; Martinsson Niskanen, T.; Larsson, N.; Hansson, S.R.; Hulthen, L. The effect of Lactiplantibacillus plantarum 299v together with a low dose of iron on iron status in healthy pregnant women: A randomized clinical trial. *Acta Obstet. Gynecol. Scand.* **2021**, *100*, 1602–1610. [CrossRef] [PubMed]
52. Landete, J.M.; Curiel, J.A.; Rodríguez, H.; de las Rivas, B.; Muñoz, R. Aryl glycosidases from Lactobacillus plantarum increase antioxidant activity of phenolic compounds. *J. Funct. Foods* **2014**, *7*, 322–329. [CrossRef]

53. Sánchez, B.; Champomier-Vergès, M.-C.; Collado, M.d.C.; Anglade, P.; Baraige, F.; Sanz, Y.; de los Reyes-Gavilán, C.G.; Margolles, A.; Zagorec, M. Low-pH adaptation and the acid tolerance response of Bifidobacterium longum biotype longum. *Appl. Environ. Microbiol.* **2007**, *73*, 6450–6459. [CrossRef] [PubMed]
54. Sánchez, B.; Schmitter, J.M.; Urdaci, M.C. Identification of novel proteins secreted by Lactobacillus plantarum that bind to mucin and fibronectin. *J. Mol. Microbiol. Biotechnol.* **2009**, *17*, 158–162. [CrossRef]
55. Tako, E.; Glahn, R.P.; Knez, M.; Stangoulis, J.C.R. The effect of wheat prebiotics on the gut bacterial population and iron status of iron deficient broiler chickens. *Nutr. J.* **2014**, *13*, 58. [PubMed]
56. Mackenzie, B.; Garrick, M.D. Iron Imports. II. Iron uptake at the apical membrane in the intestine. *Am. J. Physiol. Gastrointest. Liver Physiol.* **2005**, *289*, G981–G986. [CrossRef]
57. Morgan, E.H.; Oates, P.S. Mechanisms and regulation of intestinal iron absorption. *Blood Cells Mol. Dis.* **2002**, *29*, 384–399. [CrossRef]
58. Fuqua, B.K.; Vulpe, C.D.; Anderson, G.J. Intestinal iron absorption. *J. Trace Elem. Med. Biol.* **2012**, *26*, 115–119. [CrossRef]
59. Patchen, B.; Koppe, T.; Cheng, A.; Seo, Y.A.; Wessling-Resnick, M.; Fraenkel, P.G. Dietary supplementation with ipriflavone decreases hepatic iron stores in wild type mice. *Blood Cells Mol. Dis.* **2016**, *60*, 36–43. [CrossRef]
60. Imam, M.U.; Zhang, S.; Ma, J.; Wang, H.; Wang, F. Antioxidants Mediate Both Iron Homeostasis and Oxidative Stress. *Nutrients* **2017**, *9*, 671. [CrossRef]
61. Lane, D.J.; Bae, D.H.; Merlot, A.M.; Sahni, S.; Richardson, D.R. Duodenal cytochrome b (DCYTB) in iron metabolism: An update on function and regulation. *Nutrients* **2015**, *7*, 2274–2296. [CrossRef] [PubMed]
62. Gulec, S.; Anderson, G.J.; Collins, J.F. Mechanistic and regulatory aspects of intestinal iron absorption. *Am. J. Physiol. Gastrointest. Liver Physiol.* **2014**, *307*, G397–G409. [CrossRef] [PubMed]
63. Zimmermann, M.B.; Hurrell, R.F. Nutritional iron deficiency. *Lancet* **2007**, *370*, 511–520. [CrossRef]
64. Maares, M.; Haase, H. A Guide to Human Zinc Absorption: General Overview and Recent Advances of In Vitro Intestinal Models. *Nutrients* **2020**, *12*, 762. [CrossRef] [PubMed]
65. Takagishi, T.; Hara, T.; Fukada, T. Recent Advances in the Role of SLC39A/ZIP Zinc Transporters In Vivo. *Int. J. Mol. Sci.* **2017**, *18*, 2708. [CrossRef]
66. Takatani-Nakase, T. Zinc Transporters and the Progression of Breast Cancers. *Biol. Pharm. Bull.* **2018**, *41*, 1517–1522. [CrossRef]
67. Lopez, V.; Kelleher, S.L. Zip6-attenuation promotes epithelial-to-mesenchymal transition in ductal breast tumor (T47D) cells. *Exp. Cell Res.* **2010**, *316*, 366–375. [CrossRef]
68. Yu, Y.; Lu, L.; Luo, X.G.; Liu, B. Kinetics of Zinc Absorption by In Situ Ligated Intestinal Loops of Broilers Involved in Zinc Transporters1. *Poult. Sci.* **2008**, *87*, 1146–1155. [CrossRef]
69. Wang, X.; Kolba, N.; Liang, J.; Tako, E. Alterations in gut microflora populations and brush border functionality following intra-amniotic administration (*Gallus gallus*) of wheat bran prebiotic extracts. *Food Funct.* **2019**, *10*, 4834–4843. [CrossRef]
70. Reed, S.; Qin, X.; Ran-Ressler, R.; Brenna, J.T.; Glahn, R.P.; Tako, E. Dietary zinc deficiency affects blood linoleic acid: Dihomo-gamma-linolenic acid (LA:DGLA) ratio; a sensitive physiological marker of zinc status in vivo (*Gallus gallus*). *Nutrients* **2014**, *6*, 1164–1180. [CrossRef]
71. Tako, E.; Glahn, R.P. Iron Status of the Late Term Broiler (*Gallus gallus*) Embryo and Hatchling. *Int. J. Poult. Sci.* **2011**, *10*, 42–48. [CrossRef]

nutrients

Article

Alterations in Intestinal Brush Border Membrane Functionality and Bacterial Populations Following Intra-Amniotic Administration (*Gallus gallus*) of Nicotinamide Riboside and Its Derivatives

Nikolai Kolba †, Amin Zarei †, Jacquelyn Cheng, Nikita Agarwal, Younas Dadmohammadi, Leila Khazdooz, Alireza Abbaspourrad * and Elad Tako *

Department of Food Science, Cornell University, Ithaca, NY 14853, USA; nk598@cornell.edu (N.K.); az534@cornell.edu (A.Z.); jyc53@cornell.edu (J.C.); na494@cornell.edu (N.A.); younas@cornell.edu (Y.D.); lk532@cornell.edu (L.K.)

* Correspondence: alireza@cornell.edu (A.A.); et79@cornell.edu (E.T.); Tel.: +1-607-255-2923 (A.A.); +1-607-255-0884 (E.T.)

† These authors contributed equally to this work.

Abstract: Nicotinamide riboside (NR) acts as a nicotinamide adenine dinucleotide (NAD+) precursor where NR supplementation has previously been shown to be beneficial. Thus, we synthesized and characterized nicotinamide riboside tributyrate chloride (NRTBCl, water-soluble) and nicotinamide riboside trioleate chloride (NRTOCl, oil-soluble) as two new ester derivatives of nicotinamide riboside chloride (NRCl). NRCl and its derivatives were assessed in vivo, via intra-amniotic administration (*Gallus gallus*), with the following treatment groups: (1) non-injected (control); and injection of (2) deionized H_2O (control); (3) NRCl (30 mg/mL dose); (4) NRTBCl (30 mg/mL dose); and (5) NRTOCl (30 mg/mL dose). Post-intervention, the effects on physiological markers associated with brush border membrane morphology, intestinal bacterial populations, and duodenal gene expression of key proteins were investigated. Although no significant changes were observed in average body weights, NRTBCl exposure increased average cecum weight. NR treatment significantly increased *Clostridium* and NRCl treatment resulted in increased populations of *Bifidobacterium*, *Lactobacillus*, and *E. coli*. Duodenal gene expression analysis revealed that NRCl, NRTBCl, and NRTOCl treatments upregulated the expression of ZnT1, MUC2, and IL6 compared to the controls, suggesting alterations in brush border membrane functionality. The administration of NRCl and its derivatives appears to trigger increased expression of brush border membrane digestive proteins, with added effects on the composition and function of cecal microbial populations. Additional research is now warranted to further elucidate the effects on inflammatory biomarkers and observe changes in the specific intestinal bacterial populations post introduction of NR and its derivatives.

Keywords: intra-amniotic administration; brush border membrane; nicotinamide riboside derivatives; microbiome

Citation: Kolba, N.; Zarei, A.; Cheng, J.; Agarwal, N.; Dadmohammadi, Y.; Khazdooz, L.; Abbaspourrad, A.; Tako, E. Alterations in Intestinal Brush Border Membrane Functionality and Bacterial Populations Following Intra-Amniotic Administration (*Gallus gallus*) of Nicotinamide Riboside and Its Derivatives. *Nutrients* **2022**, *14*, 3130. https://doi.org/10.3390/nu14153130

Academic Editors: Michael Conlon and Maria Traka

Received: 26 April 2022
Accepted: 27 July 2022
Published: 29 July 2022

1. Introduction

The duodenal brush border membrane (BBM), the digestive and absorptive surface of the small intestine, is an essential component of the digestive tract where elucidation of gut health can be ascertained by BBM functionality and morphology [1]. A critical factor in intestine health maintenance is its resident microbiota's composition and function [1]. The intestinal microbiota comprises trillions of microorganisms that live symbiotically with their host. These microbes play vital roles in the digestive and immunological functions of the gastrointestinal tract by preventing colonization of potentially pathogenic organisms regulating the mucosal immune system, and maintaining intestinal homeostasis [2]. Further, this commensal microbial community contributes to the digestion of dietary fibers and minerals [3,4]; and interacts with epithelial cells to maintain an effective gut barrier [5,6]. Prebiotics, foods fortified with micronutrients such as vitamins and minerals, and polyphenolic compounds such as anthocyanins have been shown to exert positive effects on the growth and activity of bacteria that improve host health [7–11]. These health-promoting food ingredients are fermented to short-chain fatty acids (SCFAs) by specific bacteria in the colon [12–15]. The presence of SCFAs has been associated with increasing populations of bacteria (i.e., *Lactobacillus* and *Bifidobacterium*), acidifying the intestinal luminal pH and subsequently inhibiting the growth of potentially pathogenic bacteria.

Butyrate, a well-studied SCFA, plays a critical role in gastrointestinal tract health, where butyrate is utilized as an energy source for intestinal epithelial cells and can assist in maintaining intestinal epithelium integrity. Moreover, butyrate can protect the host from potential immune and inflammatory diseases associated with the translocation of antigens and pathogens [16]. Increased SCFA production has been associated with increased enterocyte proliferation, demonstrated through increases in villus surface area and enhancements in duodenal BBM functionality [17]. BBM functional capacity dictates the extent of food hydrolysis and micronutrient uptake. Thus, it is essential to examine the interactions between dietary bioactives and the BBM functionality and morphology.

It was previously demonstrated that the *Gallus gallus* is physiologically relevant in vivo models for evaluating the absorption and bioavailability of biofortified foods and bioactive compounds [18–24]. Specifically, the intra-amniotic administration approach, where the amniotic fluid is naturally and orally consumed by the embryo, allows for an assessment of the effects of the solution administered into the amniotic fluid on the different systems of interest. Hence, and as was previously demonstrated, this in vivo model has been utilized to study the impact of various biofortified foods, prebiotics, amino acids, carbohydrates, food additives, and antioxidants on BBM morphology and functionality, the intestinal bacterial populations, and mineral status [6,16,17,25–30].

Nicotinamide riboside (NR) is a bioavailable form of vitamin B_3 naturally present in the diet and acts as a cellular nicotinamide adenine dinucleotide (NAD+) precursor [31,32]. NAD+ is an essential cofactor and substrate for numerous critical cellular processes, where NAD+ precursors were demonstrated to have protective roles in several disease states [33]. NR is reportedly more effective than other NAD+ precursors, such as niacin and nicotinamide [32]. As NR-enriched foods have not yet been identified and well-characterized [34], supplementation of NR as a prebiotic can be a promising option to boost NAD+ levels [31,32,35]. NR supplementation was previously studied, in vivo and in humans, and was found to have beneficial effects, including the reduction of DNA and mitochondria damage [36] and providing therapeutic benefits for Alzheimer's disease [37], obesity [38,39], diabetes [39,40], muscle degeneration [35], and aging [41].

Moreover, NR has been shown to improve the ability to fight pathogenic infections by increasing the innate immune response [41,42]. Currently, NR is an FDA-approved nutritional supplement and is commercially available as a chloride salt of NR (NRCl) in capsule form under the brand name Niagen^{TM} [43]. Recently, Gonzalez et al. 2020 studied the intra-amniotic administration of NR on myogenesis, where significant beneficial effects were observed with pectoralis muscle development in vivo [34,44]. However, the impact of NR and its derivatives on intestinal health, including BBM functionality and bacterial populations, have not yet been studied.

In this work, for the first time, we studied the functionalization of currently commercially available NRCl to nicotinamide riboside tributyrate chloride (NRTBCl) as a water-soluble derivative and nicotinamide riboside trioleate chloride (NRTOCl) as an oil-soluble derivative (Figure 1). The primary objective of this study was to assess the effects of intra-amniotic administration of NRCl and two new derivatives (NRTBCl and NRTOCl) on BBM functionality via the evaluation of duodenal gene expression of BBM biomarkers, specifically key digestive and absorptive proteins, immune function proteins, and inflammation biomarkers, in vivo. Further, we assessed the effects of NRCl and its new derivatives on cecal bacterial populations. We hypothesized that intra-amniotic administration of NRCl and its derivatives causes favorable alterations in brush border functionality and gut microbiota.

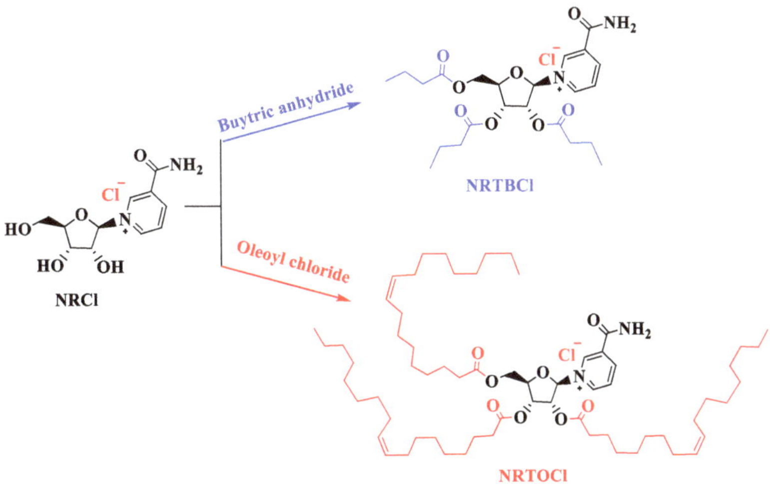

Figure 1. Functionalization of NRCl to NRTBCl and NRTOCl.

2. Materials and Methods

2.1. Animals

For the following study, a commercial hatchery (Moyer's chicks, Quakertown, PA, USA) provided Cornish cross-fertile broiler chicken eggs (*n* = 50). The eggs were incubated under controlled conditions at the Cornell University Animal Science poultry farm incubator. All animal protocols were approved by the Cornell University Institutional Animal Care and Use Committee (IACUC #2020-0077).

2.2. Materials

Nicotinamide riboside chloride (NRCl, beta form) was received as a donation from ChromaDex Company (Los Angeles, CA, USA). Oleoyl chloride, butyric anhydride, and 4-dimethylamino pyridine were purchased from Sigma Aldrich with 89, 98, and 99% purity, respectively. Pyridine was bought from Fluka with 99.9% purity. Silica gel (P60, 40–63 µm, 60 Å) was purchased from SiliCycle (Québec, QC, Canada), and Silica Gel 60 F254 Coated Aluminum-Backed TLC (thin layer chromatography) sheets were obtained from EMD Millipore (Billerica, MA, USA).

2.2.1. Synthesis of Nicotinamide Riboside Tributyrate Chloride (NRTBCl)

Using a round-bottom flask in an ice bath, 300 mg (1.035 mmol) of NRCl, 25 mg of 4-dimethylamino pyridine (0.205 mmol), 1.5 mL of butyric anhydride (9.170 mmol), and 9 mL of acetonitrile (CH_3CN) were added and stirred for 5 h under nitrogen blanket. A thin-layer chromatography (TLC) test was employed to track the progress of the reaction. Next, the solvent was evaporated by a rotary evaporator under reduced pressure, and the excess amount of butyric anhydride was washed out with *n*-hexane. Finally, the crude product was purified by column chromatography on SiO_2. The eluent was a mixture of CH_3OH (35%) and ethyl acetate (EtOAc) (65%). The purified NR-tributyrate chloride was obtained in 71% (367 mg) as a pale-yellow-colored greasy product (λ_{max} in water was 266 nm).

2.2.2. Synthesis of Nicotinamide Riboside Trioleate Chloride (NRTOCl)

Using a round-bottom flask in an ice bath, 200 mg (0.690 mmol) of NRCl, 0.55 mL (6.81 mmol) of pyridine, and 4.75 mL of dimethylformamide (DMF) were added. Then, 2.0 mL (5.38 mmol) of oleoyl chloride was dropwise added, and the reaction mixture was stirred for 3 h under a nitrogen blanket. A TLC test was used to track the progress of the reaction. After 3 h, 5 mL of methanol was added to the reaction mixture to neutralize the extra amount of oleoyl chloride. After that, the solvent was evaporated using a rotary evaporator under reduced pressure. The crude product was extracted in *n*-hexane and purified using column chromatography on SiO_2. The eluent was a mixture of CH_3OH (12%) and EtOAc (88%). The purified NRTOCl was obtained in 64.3% (479.2 mg) as a pale-creamy-colored greasy product (λ_{max} in methanol was 267 nm).

2.3. Characterization of NRTBCl and NRTOCl

2.3.1. Nuclear Magnetic Resonance (NMR) Spectroscopy

A 500 MHz NMR (Bruker AVANCE) spectrometer was used for ^1H NMR (500 MHz) and ^{13}C NMR (125 MHz) spectra in deuterated chloroform ($CDCl_3$). The chemical shifts were expressed in δ (ppm) relative to tetramethylsilane (TMS) as the internal standard and coupling constants (J) were measured in Hz. Spin multiplicities were described as singlet (s), doublet (d), triplet (t), quartet (q), and multiplet (m).

2.3.2. Attenuated Total Reflectance—Fourier-Transform Infrared (ATR-FTIR) Spectroscopy

The ATR-FTIR spectra were recorded on a Shimadzu IRAffinity-1S spectrophotometer in transmittance mode in the range of 400–4000 cm^{-1} wave number.

2.3.3. UV-Vis Spectroscopy

UV-Vis was recorded on a Shimadzu UV-2600 spectrophotometer in the range of 200–800 nm.

2.3.4. Liquid Chromatography-Mass Spectrometry (LC-MS) Analysis

Liquid Chromatograph (Agilent 1100 series) was coupled with a mass spectrometer for LC-MS analysis. Before injection, all samples were passed through a 13 mm nylon syringe filter with a 0.22 μm pore size. Reverse-phase chromatography was used with a Phenomenex Luna Omega (Phenomenex) LC column with the following specifications: 100 × 4.6 mm, 3 μm, polar C18, 100 Å pore size with a flow rate of 0.3 mL min^{-1}. LC eluents include MilliQ-water (solvent A) and acetonitrile (solvent B) using gradient elution (solution A: B composition change with time: 0 min: 95:5, 3 min: 95:5, 15 min: 85:15, 17 min: 90:10, and 20 min 95:5). The mass spectrometer (Finnigan LTQ mass spectrometer) was equipped with an electrospray interface (ESI) set in positive electrospray ionization mode for analyzing the NRTOCl and NRTBCl. The optimized parameters were sheath gas flow rate at 20 arbitrary units, spray voltage set at 4.00 kV, capillary temperature at 350 °C, capillary voltage at 41.0 V, and tube lens set at 125.0 V.

2.3.5. Particle Characterization

The particle size distribution, mean particle diameter (average zeta size), and zeta-potential of NRTOCl in 1% ethanol in DI water were measured using a commercial dynamic light-scattering device (Nano-ZS, Malvern Instruments, Worcestershire, UK).

2.4. Intra-Amniotic Administration Solution Preparation

After the synthesis and characterization of NRTBCl and NRTOCl, these compounds and NRCl were used for intra-amniotic administration. NRCl and NRTBCl were dissolved in DI H_2O at 30 mg/mL. NRTOCl was insoluble in water; thus, it was dispersed using 1% ethanol as a cosolvent.

2.5. Intra-Amniotic Administration Procedure and Study Design

The intra-amniotic administration procedure was previously described by Tako et al. [25,28,30,45–48]. On Day 17 of embryonic incubation, eggs with viable embryos were weighed and allocated into treatment groups (n = 10) with equal weight distribution. The intra-amniotic injection solution (1 mL) was injected with a 21-gauge needle into the amniotic fluid, recognized by candling. Following injection, the injection sites were sterilized with 70% ethanol and sealed with cellophane tape. Eggs were then placed in hatching baskets, with each treatment equally represented at each incubator location. The treatment groups are as follows: (1) non-injected (control); and injection of (2) DI H_2O (control); (3) NR (30 mg/mL dose); (4) NRTBCl (30 mg/mL dose); and (5) NRTOCl (30 mg/mL dose).

2.6. Tissue Collection

Immediately post-hatch (Day 21), birds were weighed and euthanized with CO_2 exposure. The duodenum, ceca, and pectoral muscles were immediately collected and frozen in liquid nitrogen. Samples were stored at −80 °C until analysis [49].

2.7. Isolation of Total RNA from Chicken Duodenum

Total RNA was extracted from 30 mg of the proximal duodenal tissue using a Qiagen RNeasy Mini Kit (Qiagen Inc., Germantown, MD, USA). Total RNA was eluted in 50 µL of RNase-free water. All steps were carried out under RNase-free conditions. RNA was quantified with a NanoDrop 2000 (ThermoFisher Scientific, Waltham, MA, USA) at A 260/280. RNA was stored at −80 °C until use.

2.8. Real-Time Polymerase Chain Reaction

All procedures were conducted as previously described [30,47,50]. Briefly, the primers used in the real-time polymerase chain reactions (RT-PCR) were designed using Real-time Primer Design Tool software (IDT DNA, Coralville, IA, USA) based on 11 gene sequences from the GenBank database (Table 1). cDNA was generated using a C1000 Touch thermocycler (Biorad, Hercules, CA, USA) and a Promega-Improm-II Reverse Transcriptase Kit (Catalog #A1250) 20 µL reverse transcriptase reaction following the manufacturer's protocols. The concentration of cDNA was determined with a NanoDrop 2000 at A 260/280 with an extinction coefficient of 33 for single-stranded DNA.

Table 1. Real-time polymerase chain reaction (RT-PCR) primer sequences.

Gene	Forward Primer (5′→3′)	Reverse Primer (5′→3′)	Base Pair	GI Identifier
		Iron Metabolism		
DMT1	TTGATTCAGAGCCTCCCATTAG	GCGAGGAGTAGGCTTGTATTT	101	206597489
		Zinc Metabolism		
ZIP1	TGCCTCAGTTTCCCTCAC	GGCTCTTAAGGGCACTTCT	144	107055139
ZnT1	GGTAACAGAGCTGCCTTAACT	GGTAACAGAGCTGCCTTAACT	105	54109718
		Inflammatory Response		
TNF-α	GACAGCCTATGCCAACAAGTA	TTACAGGAAGGGCAACTCATC	109	53854909
IL-8	TCATCCATCCCAAGTTCATTCA	GACACACTTCTCTGCCATCTT	105	395872
IL-6	ACCTCATCCTCCGAGACTTTA	GCACTGAAACTCCTGGTCTT	105	302315692
IL-1β	CTCACAGTCCTTCGACATCTTC	TGTTGAGCCTCACTTTCTGG	119	88702685
		BBM functionality		
SGLT-1	GCATCCTTACTCTGTGGTACTG	TATCCGCACATCACACATCC	106	8346783
SI	CCAGCAATGCCAGCATATTG	CGGTTTCTCCTTACCACTTCTT	95	2246388
MUC2	CCTGCTGCAAGGAAGTAGAA	GGAAGATCAGAGTGGTGCATAG	155	423101
18S	GCAAGACGAACTAAAGCGAAAG	TCGGAACTACGACGGTATCT	100	7262899

DMT-1, Divalent metal transporter; ZIP1, Zinc transporter 1; ZnT1, zinc transporter 1; TNF-α, tumor necrosis factor; IL-8, interleukin 8; IL-6, interleukin 6; IL-1β, interleukin 1 beta; SGLT-1, sodium-glucose transporter 1; SI, sucrose isomaltase; Muc2, Mucin 2; 18S rRNA, 18S ribosomal subunit.

RT-PCR procedure was conducted with a Bio-RadCFX96 Touch (Hercules, CA, USA). Ten µL RT-PCR mixtures consisted of cDNA (2 µg), 2X BioRad SSO Advanced Universal SYBR Green Supermix (Catalog #1725274, Hercules, CA, USA), forward and reversed primers, and nuclease-free water (no template control). The no-template control of nuclease-free water was included to eliminate DNA contamination in the PCR mix. Reactions were performed in duplicates and under the following reaction conditions: initial denaturing (95 °C, 30 s), followed by 40 cycles of denaturing (95 °C, 15 s), several annealing temperatures (according to IDT for 30 s), and elongating (60 °C, 30 s). After the cycling process was accomplished, melting curves were determined from 65.0 °C to 95.0 °C with increments of 0.5 °C for 5 s to confirm the amplification of a single product. RT-PCR efficiency values for the eleven genes were Muc2, 1.022; 18s rRNA, 0.934. Gene expression levels were determined from Ct values based on the 'second derivative maximum' calculated by the Bio-Rad CFX Maestro Software (Bio-Rad, Hercules, CA, USA). Gene expression was standardized to the expression of 18S.

2.9. Cecal Microbial DNA Isolation and Analysis

All procedures were conducted as previously described [18,22,24,30,47]. Briefly, Ceca contents were inserted into a sterile 50 mL tube (Corning, NY, USA) with 9 mL of sterile 1X phosphate saline (PBS) and then vortexed with glass beads (3 mm size) for 3 min. Particles and remains were removed by centrifugation at 700× *g* for 1 min, and the supernatant was collected and centrifuged at 12,000× *g* for 5 min. The pellet was rinsed twice with 1X PBS and kept at −20 °C for DNA extraction.

For DNA extraction, the pellet was mixed with 50 mM EDTA and treated with 10 mg/mL lysozyme (Sigma Aldrich Co., St. Louis, MO, USA) for 45 min at 37 °C. The bacterial genomic DNA was recovered using the Wizard Genomic DNA purification kit (Promega Corp., Madison, WI, USA), following the manufacturer's instructions.

2.10. Cecal Short-Chain Fatty Acids (SCFA) Analysis and Cecal Content pH

As was previously described [28], cecal samples were homogenized in HCl (2 mL, 3%, 1 M), centrifuged and combined with ethyl acetate (100 µL) and acetic acid-d4 (1 µg/mL) before collecting the organic phase to determine short-chain fatty acid (SCFA) composition. Samples were quantified via GC-MS using a TRACE™ 1310 gas chromatograph (Thermo Fisher Scientific, Waltham, MA, USA) and a TraceGOLD™ TG-WaxMS A column (Thermo Fisher Scientific, Waltham, MA, USA). The pH of cecum content was determined using an Oakton® model 700 digital pH meter (Oakton Instruments, Vermon Hills, IL, USA). Before testing, the potentiometer was calibrated with pH buffers at 1.68, 4.01, 7.00, 10.01, and 12.45 according to the manufacturer's recommendations.

2.11. PCR Amplification of Bacterial 16s rDNA

Primers for *Lactobacillus*, *Bifidobacterium*, *E. coli*, and *Clostridium* were designed as previously described [18,22,24,30,47,51]. The universal primers were prepared with the invariant sequence regions in the 16S rRNA of bacteria and used as an internal standard to normalize data. PCR reaction products were isolated by electrophoresis (2% agarose gel), stained with ethidium bromide, and quantified using the Quantity One 1-D analysis software (Bio-Rad, Hercules, CA, USA).

2.12. Statistical Analysis

Experimental treatments for the intra-amniotic administration assay were arranged in a completely randomized design and checked for normality of data utilizing the Shapiro–Wilk test before analyzing data further. Once the Gaussian distribution was confirmed, the one-way multiple analysis of variance (ANOVA) was conducted. Differences between treatment groups were compared with a post hoc Duncan test, with results considered statistically different at $p < 0.05$. Statistical analyses were carried out using SPSS version 27.0 software (IBM, Armonk, NY, USA). Results are expressed as mean ± standard error, $n \geq 8$.

3. Results

3.1. Fourier Transform Infrared (FTIR) of NRTBCl

The FTIR of NRTBCl shows two bands at 3340 and 3130 cm^{-1}, which are asymmetric and symmetric stretching bonds of NH_2 in the amide functional group. The exitance of two bands at 2964 and 2877 cm^{-1} is attributed to asymmetric and symmetric stretching vibrations of aliphatic C-H. A strong band at 1738 cm^{-1} confirms the carbonyl of ester groups in this compound. The carbonyl of the amide functional group appears at 1678 cm^{-1}. The band at 1620 cm^{-1} is evidence of the C=C bond in the pyridinium ring. Two bands at 1462 and 1384 cm^{-1} show out-of-plane C-H bending vibrations of the methylene and methyl groups, respectively. The stretching vibrations of the C-O bonds in the ester groups and ribose ring appear at 1163 and 1097 cm^{-1} (Figure 2). The obtained FTIR results confirm the functional groups in the NRTBCl structure.

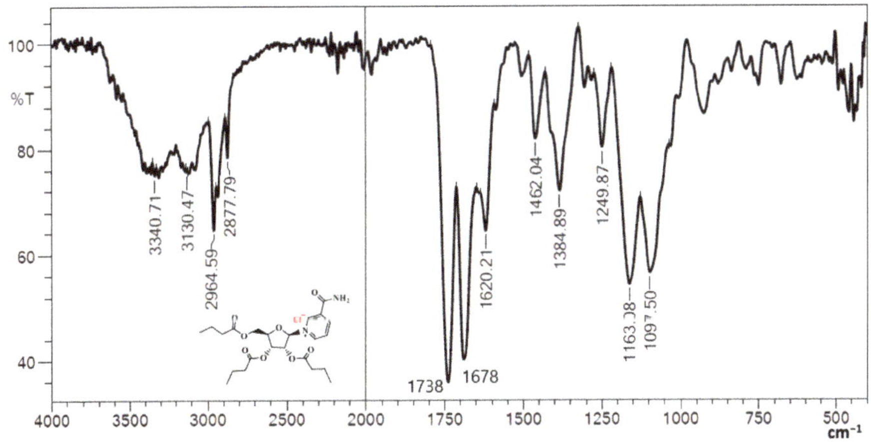

Figure 2. FTIR of NRTBCl.

3.2. ^{1}H NMR of NRTBCl

The ^{1}H NMR (500 MHz) of NRTBCl was performed in $CDCl_3$ at room temperature (Figure 3). The expanded ^{1}H NMR of this compound displays that the most deshielded proton (H1) at 9.93 ppm is attributed to the hydrogen located on the pyridinium ring between the positive nitrogen and amide group (Figure 4). A doublet (J = 6 Hz) at 9.60 ppm is attributed to H5 located on the pyridinium ring in a position ortho to the positive nitrogen. The chemical shift of H3 in the para position with respect to the positive nitrogen appears at 9.41 ppm as a doublet peak (J = 7.5 Hz). Because of the interaction between nitrogen lone pair and carbonyl of the amide group, the chemical shifts of NH_2 protons are not equivalent in NRTBCl. In this compound, one of the NH_2 protons appears at 9.37 ppm and another at 7.25 ppm. The final hydrogen on the pyridinium ring is H4 which appears as a triplet peak (J = 7 Hz) at 8.43 ppm. In the structure of NRTBCl, there are four hydrogens on the ribose ring. The anomeric hydrogen (H1′) is impacted more by the oxygen atom of the ribose ring and the positive nitrogen of the pyridinium ring so that this hydrogen appears at 6.88 ppm as a doublet peak (J = 3.5 Hz). H2′ and H3′ are neighbors and appear as two triplet peaks (J = 5 Hz) and (J = 6 Hz) with chemical shifts of 5.74 and 5.48 ppm, respectively. Since H2′ is closer to the anomeric center than H3′, its chemical shift is more deshielded than H3′. H4′ in the ribose ring and one of the hydrogens of the methylene group (H5′) bonded to the single oxygen of the ester group overlap and appeared as a multiplet at 4.67 ppm with integral 2. Another hydrogen of this methylene group appears at 4.54 ppm as a doublet (J = 11 Hz). In the three chain ester groups of NRTBCl, there are three CH_2 groups near the ester carbonyl groups, which appear as multiplets between 2.33–2.52 ppm. The multiplet at 1.62 ppm can be attributed to the other three methylene groups near the CH_2 groups bonded to the carbonyl groups. Finally, a multiplet peak between 0.90–0.96 ppm with an integral of 9 confirms the existence of three methyl groups at the end of the butyrate esters arms. The ^{1}H NMR results verified the structure of NRTBCl.

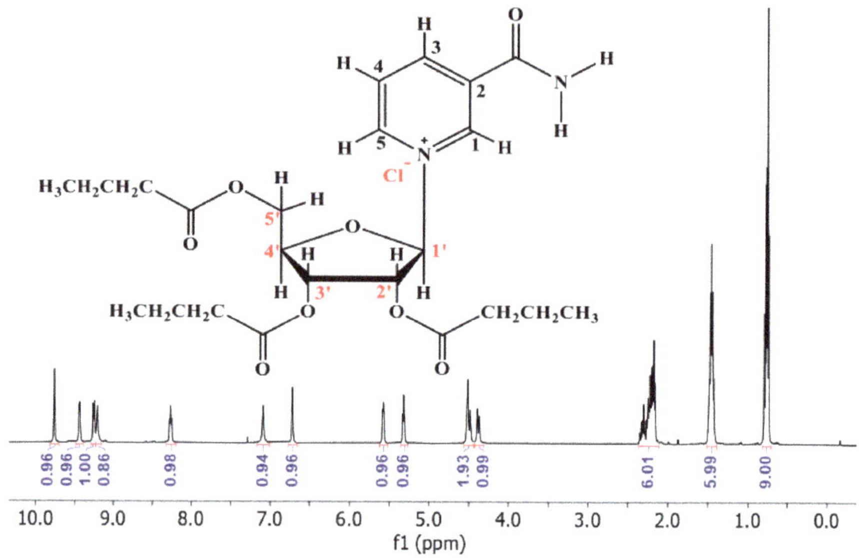

Figure 3. ^{1}H NMR of NRTBCl in $CDCl_3$.

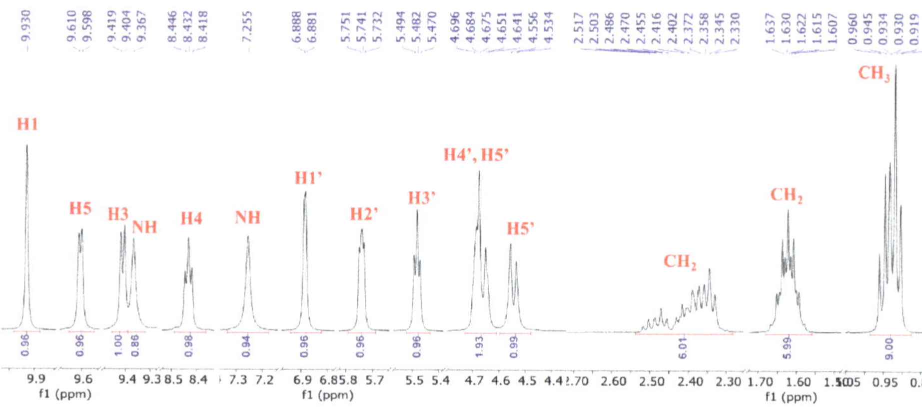

Figure 4. Expanded ^{1}H NMR of NRTBCl.

3.3. ^{13}C NMR of NRTBCl

The ^{13}C NMR (125 MHz) of NRTBCl in $CDCl_3$ was also studied at room temperature (Figure 5). The ^{13}C NMR of this compound exhibits three peaks at 173.2, 172.7, and 172.2 ppm, attributed to the three different carbonyl carbons of the ester groups in the structure of NRTBCl. A peak at 163.1 ppm confirms the carbonyl of the amide group in this compound. There are five distinct peaks at 147.3, 143.4, 141.6, 134.3, and 128.8 ppm for the carbons in the pyridinium ring. Four peaks at 98.1, 82.4, 75.6, and 69.1 ppm confirmed the existence of a ribose ring in the structure of NRTBCl, and the chemical shift of the methylene carbon bonded to the single oxygen of the ester group appears at 62.4 ppm. In the three short ester chains of NRTBCl, three distinct peaks at 35.73, 35.66, and 35.5 ppm are attributed to the three CH_2 groups near the carbonyl carbons of the ester groups (Figure 5). The methylene groups in these chains appear at 18.25, 18.2, and 18.1 ppm. Because the chemical shifts of the methyl groups are very close to each other, one of the methyl groups overlaps with the other one, and these three methyl groups appear as two peaks at 13.62 and 13.60 ppm. The obtained results of ^{13}C NMR corroborated well with the ^{1}H NMR results to verify the NRTBCl structure.

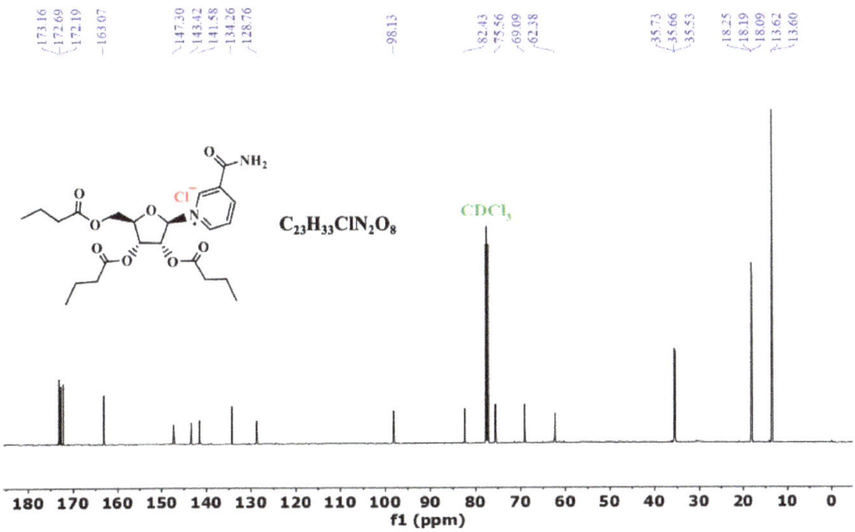

Figure 5. ^{13}C NMR of NRTBCl in $CDCl_3$.

3.4. LC-MS Analysis of NRTBCl

To confirm the presence of three butyrate groups, LC-MS was performed to find the molecular weight of NRTBCl (Figure 6). The selected reaction monitoring (SRM) results show a single peak with 465.05 m/z (M-Cl) that agrees with the structure of the NRTB cation. Interestingly, a fragment with 343.22 m/z is attributed to the ribose-trioleate molecule formed by eliminating the nicotinamide molecule from NRTBCl.

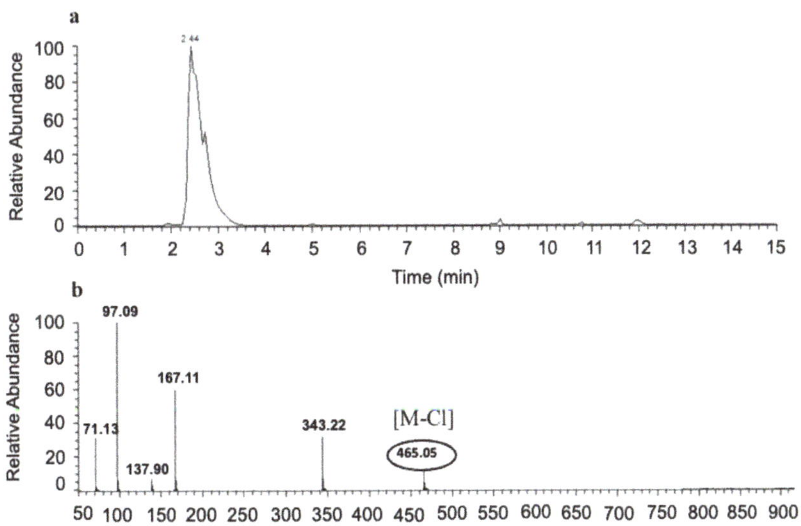

Figure 6. SRM LC-MS of NRTBCl. (**a**) SRM LC of NRTBCl. (**b**) Mass spectrum of NRTBCl.

3.5. FTIR of NRTOCl

FTIR (cm^{-1}): 3288 (asymmetric stretching vibration of N-H), 3122 (symmetric stretching vibration of N-H), 3005 (stretching vibration of vinyl and aromatic C-H), 2922 (asymmetric stretching vibration of aliphatic C-H), 2852 (symmetric stretching vibration of aliphatic C-H), 1743 (stretching vibration of C=O in ester groups), 1689 (stretching vibration of C=O in the amide group), 1622 (stretching vibration of C=C), 1458 (out-of-plane C-H bending vibrations of CH_2), 1379 (out-of-plane C-H bending vibrations of CH_3), 1161 (stretching vibration of C-O), 1116 (stretching vibration of C-O), 914 (out-of-plane C-H bending vibration of aromatic ring), 721 (out-of-plane C-H bending vibration of vinyl groups), 677 and 632 (out-of-plane C-H bending vibration of aromatic ring). (Supporting information, Figure S1 FTIR of NRTOCl). The obtained FTIR results verified the functional groups of NRTOCl.

3.6. ^1H NMR of NRTOCl

^1H NMR (500 MHz, $CDCl_3$) δ (ppm): 10.34 (s, 1 H, pyridinium ring), 9.86 (s, 1 H, NH), 9.44 (d, J = 10 Hz, 1 H, pyridinium ring), 9.34 (d, J = 10 Hz, 1 H, pyridinium ring), 8.20 (t, J = 10 Hz, 1 H, pyridinium ring), 6.75 (d, J = 5 Hz, 1 H, ribose ring), 6.28 (s, 1 H, NH), 5.57 (t, J = 5 Hz, 1 H, ribose ring), 5.43 (t, J = 5 Hz, 1 H, ribose ring), 5.35 (m, 6 H, H-C=C-H groups), 4.70 (m, 2 H, ribose ring and one of the diastereotopic methylene group), 4.50 (dd, J_1 = 14, J_2 = 4 Hz, 1 H, diastereotopic methylene group), 2.37–2.55 (m, 6 H, three CH_2 groups), 2.02 (m, 12 H, six CH_2 groups), 1.63 (m, 3 H, three CH_2 groups), 1.30 (m, 60 H, thirty CH_2 groups), 0.89 (t, 9 H, three CH_3 groups) (Supporting information, Figure S2 ^1H NMR of NRTOCl in $CDCl_3$, and Figure S3 Expanded ^1H NMR of NRTOCl).

3.7. ^{13}C NMR of NRTOCl

^{13}C NMR (125 MHz, $CDCl_3$) δ (ppm): 173.1, 172.9 and 172.3 (carbonyl of ester groups), 162.5 (carbonyl of amide group), 146.7, 142.5, 141.8 and 134.6 (pyridinium ring), 130.07, 130.06, 130.04, 129.67 and 129.62 (C=C), 127.9 (pyridinium ring), 98.0, 82.9, 75.8 and 69.1 (ribose ring), 62.2 (diastereotopic methylene), 33.9, 33.8, 33.7, 31.9, 29.8, 29.74, 29.73, 29.72, 29.5, 29.34, 29.32, 29.23, 29.22, 29.16, 29.15, 29.11, 29.10, 29.09, 27.24, 27.18, 24.76, 24.73, 24.6, 22.7 and 14.1 (aliphatic carbons in oleate chains) (Supporting information, Figure S4 ^{13}C NMR of Netcool in $CDCl_3$, and Figure S5 Expanded ^{13}C NMR of NRTOCl). The results of ^1H NMR and ^{13}C NMR confirm the structure of NRTOCl.

3.8. LC-MS of NRTOCl

A single peak with 1047.52 m/z (M-Cl) agrees with the structure of the NRTO cation. A fragment with 925.70 m/z is attributed to the ribose-trioleate molecule formed by removing the nicotinamide molecule from NRTOCl (Supporting information, Figure S6 SRM LC-MS of NRTOCl. (a) SRM LC of NRTOCl. (b) Mass spectrum of NRTOCl).

3.9. Particle Size and Zeta Potential of NRTOCl

NRTOCl was dispersed in DI water using 1% (v/v) ethanol as a cosolvent, and the average size and zeta potential of the NRTOCl particles were 192 nm and +65 mV, respectively (Supporting information, Figure S7 Particle size of NRTOCl in DI water containing 1% ethanol).

3.10. Gross Physiological Parameters

There were no significant differences in body weight between treatment groups. Compared with the non-injected control, the average cecum weight was significantly increased ($p < 0.05$) with NRCl exposure. When compared with NRCl exposure, NRTBCl exposure resulted in significantly increased ($p < 0.05$) cecum weight. Compared with the non-injected and H_2O control, no significant differences were found with NRCl and NRTBCl and NRTOCl exposure between cecum: bodyweight ratios (Table 2).

Table 2. Gross physiological parameters measured on the day of hatch (Day 21) [1].

Group Name	Average Body Weight (g)	Average Cecum Weight (g)	CW: BW
NI	42.06 ± 1.35 [a]	0.45 ± 0.05 [a]	0.011 ± 0.002 [a]
H_2O	42.16 ± 1.28 [a]	0.29 ± 0.04 [bc]	0.007 ± 0.001 [ab]
NRCl	42.74 ± 1.30 [a]	0.23 ± 0.05 [c]	0.005 ± 0.001 [b]
NRTBCl	43.80 ± 0.74 [a]	0.38 ± 0.06 [ab]	0.009 ± 0.001 [ab]
NRTOCl	43.17 ± 0.85 [a]	0.35 ± 0.04 [abc]	0.008 ± 0.001 [ab]

[1] Values are the means ± SEM, n = 10. [a–c] Treatment groups not indicated by the same letter within the same column are significantly different ($p < 0.05$). NI = non-injected, CW = cecum weight, BW = body weight.

3.11. Ceca Bacterial Analysis

Cecal genera bacterial populations are shown in Figure 7. NRCl exposure resulted in a significant increase ($p < 0.05$) in the relative abundance of *Bifidobacterium* spp. when compared with all other treatment groups. NRCl derivative exposure did not significantly alter *Bifidobacterium* spp. relative abundance when compared with the controls. *Bifidobacterium* spp. and *Lactobacillus* spp. relative abundance was significantly elevated with NRCl exposure, while NRTOCl exposure decreased relative abundance compared with the H_2O control ($p < 0.05$). Compared with the non-injected control, NRTOCl treatment resulted in a significant decrease in the relative abundance of *E. coli*. NRCl, NRTBCl, and NTROCl exposure significantly elevated *Clostridium* populations compared with the controls ($p < 0.05$).

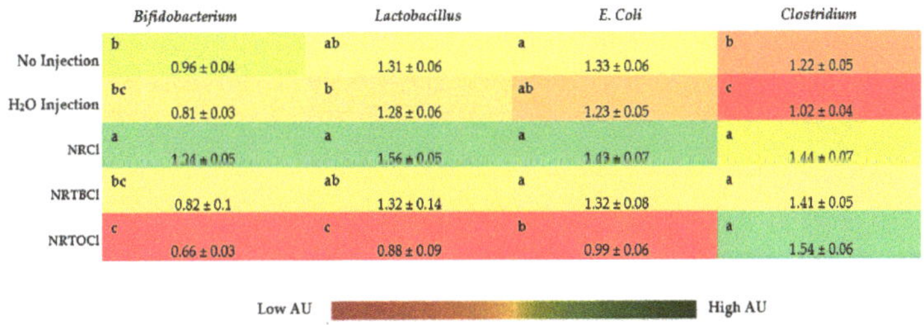

Figure 7. Effects of intra-amniotic administration of experimental solutions on cecal genera and species-level bacterial populations (day of the hatch). Values are means ± SEM, n = 10. [a–c] Per bacterial category, treatments groups that do not share any letters within the same column are significantly different according to a one-way ANOVA with post hoc Duncan test ($p < 0.05$). AU = arbitrary units.

3.12. Short-Chain Fatty Acids and pH Concentrations in Cecal Contents

Short-chain fatty acid (SCFA) production significantly increased for butyrate, and as a result, cecal chyme pH significantly decreased in NRCl, NRTBCl, and NRTOCl groups compared to the non-injected and water-injected groups (Figure 8).

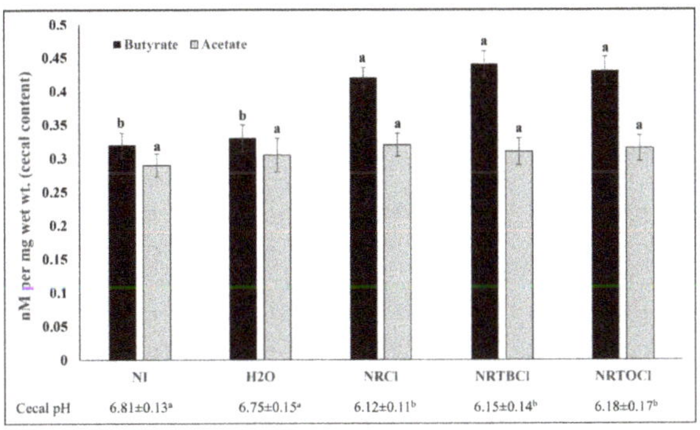

Figure 8. Cecal short-chain fatty acid (SCFA) composition and cecal pH. Values are the means ± SEM, n = 5. [a,b] Per SCFA (butyrate or acetate) and pH, treatments groups not indicated by the same letter within the same column are significantly different ($p < 0.05$).

3.13. Duodenal Brush Border Membrane Gene Expression

For iron-related protein gene expression, divalent metal transporter 1 (DMT1), there was no significant difference between any treatment groups; however, there was a general trend of increasing expression of the experimental groups (NRCl, NRTBCl, and NTROCl) compared to the controls (non-injected and H_2O injected groups) (Figure 9). For zinc transporters gene expression, while there were no significant differences in ZIP1 expression, there was a significant increase ($p < 0.05$) in zinc transporter 1 (ZnT1) with NR and NR derivative exposure (NRCl, NRTBCl, and NTROCl) when compared to the non-injected and H_2O controls.

	DMT1	ZIP1	ZnT1	TNF-α	IL-8	IL-6	IL-1β	SGLT-1	SI	MUC2
No Injection	^{a} 1.016 ± 0.001	^{a} 0.227 ± 0.004	^{b} 5.819 ± 0.046	^{a} 2.88 ± 0.022	^{a} 1.02 ± 0.001	^{ab} $2.33 \times 10^5 \pm 5.87 \times 10^4$	^{a} 4.423 ± 0.100	^{a} 0.942 ± 0.001	^{a} 2.431 ± 0.02	^{c} 1.613 ± 0.006
H_2O Injection	^{a} 1.018 ± 0.001	^{a} 0.217 ± 0.003	^{b} 5.843 ± 0.092	^{a} 2.96 ± 0.038	^{a} 1.02 ± 0.001	^{a} $3.98 \times 10^5 \pm 1.09 \times 10^5$	^{a} 4.385 ± 0.079	^{a} 0.942 ± 0.002	^{a} 2.439 ± 0.007	^{c} 1.611 ± 0.006
NRCl	^{a} 1.024 ± 0.003	^{a} 0.187 ± 0.006	^{a} 8.137 ± 0.301	^{a} 2.827 ± 0.044	^{a} 1.027 ± 0.003	^{b} $9.52 \times 10^4 \pm 1.63 \times 10^4$	^{b} 3.54 ± 0.069	^{a} 0.951 ± 0.003	^{a} 2.504 ± 0.025	^{b} 1.793 ± 0.015
NRTBCl	^{a} 1.028 ± 0.008	^{a} 0.188 ± 0.002	^{a} 7.795 ± 0.232	^{a} 2.775 ± 0.053	^{a} 1.028 ± 0.008	^{b} $4.95 \times 10^4 \pm 4.54 \times 10^3$	^{b} 3.403 ± 0.065	^{a} 0.951 ± 0.007	^{a} 2.497 ± 0.031	^{b} 1.76 ± 0.019
NRTOCl	^{a} 1.023 ± 0.004	^{a} 0.343 ± 0.156	^{a} 8.688 ± 0.915	^{a} 2.799 ± 0.149	^{a} 1.056 ± 0.032	^{b} $1.19 \times 10^5 \pm 7.92 \times 10^4$	^{b} 3.511 ± 0.219	^{a} 0.945 ± 0.005	^{a} 2.443 ± 0.089	^{a} 1.841 ± 0.016

Low AU — High AU

Figure 9. Effects of the intraamniotic administration of experimental solutions on duodenal gene expression. Values are the means ± SEM, n = 10. $^{a-c}$ Per gene, treatments groups not indicated by the same letter within the same column are significantly different ($p < 0.05$). DMT1, Divalent metal transporter 1; ZIP1, Zinc Transport Protein 1; ZnT1, Zinc transporter 1; Sodium/Glucose cotransporter 1; SI, Sucrose isomaltase; MUC2, Mucin 2; TNF-α, Tumor necrosis factor-alpha; IL-8, Interleukin 8; IL-6, Interleukin 6; IL-1β, Interleukin 1 Beta. AU = arbitrary units.

As for inflammatory gene expression, NR exposure (NRCl, NRTBCl, and NTROCl) did not alter ($p > 0.05$) the expression of TNF-α and IL-8 relative to the non-injected and H_2O injection groups. However, there was a significant ($p < 0.05$) down-regulation in the expression of IL-6 and IL-1β in the NR experimental groups (NRCl, NRTBCl, and NRTOCl) compared to the H_2O injected group.

NRCl and NRTBCl exposure increased gene expression of brush border membrane absorptive proteins, sodium-glucose transporter 1 (SGLT-1), and sucrose isomaltase (SI) relative to the non-injected and H_2O injected groups. Additionally, the digestive viscoelastic gels formed by mucin (MUC2) (protecting the intestinal cells) were altered significantly ($p < 0.05$) in gene expression within NRCl, NRTBCl, and NRTOCl groups compared to the non-injected and water injected groups (Figure 9).

4. Discussion

NR was discovered as a form of vitamin B_3 that can increase NAD(P) levels, as NAD^+ is utilized as a coenzyme for oxidoreductases and a source of ADP ribosyl group for adding one or more ADP-ribose moieties to a protein [52–54]. This study focuses on NR derivatives, NRTBCl and NRTOCl, synthesized using butyric anhydride and oleoyl chloride, respectively. The reactions were executed under mild conditions and resulted in acceptable yields. The final purified products were characterized using FTIR, NMR, and LC-MS to determine their structures.

The effects of synthesized and characterized NRCl, NRTBCl, and NRTOCl as prebiotic supplementation on duodenal brush border molecular and cecal microbial populations were investigated in vivo. These compounds did not show any significant effects on average body weight (Table 2). However, NRCl treatment substantially reduced the average cecum weight compared to the non-injection control. A potential explanation for this observation may be the three butyrate groups that are integral structural parts of NRTBCl, which can be metabolized to butyric acid by the gut microbiota [55–57]. Previously, Aghazadeh & TahaYazdi (2012) found that butyric acid dietary supplementation increased the weight of the liver and intestines compared to butyric acid-free diets [56]. Additionally, Panda et al. (2009) demonstrated that 0.4% dietary butyrate supplementation significantly increased

($p < 0.05$) body weight, intestinal tract health, and villi development in vivo (*Gallus gallus*) in comparison to antibiotic-treated and control groups [57].

Further, we studied the effects of the investigated NR compounds on cecum bacterial populations (Figure 7). Significant ($p < 0.05$) increases in *Clostridium* populations were observed with NR exposure compared to the controls, where the *Clostridium* genus houses a well-known butyrate producer, Clostridial cluster XIVa; a microbial cluster that may affect intestinal butyrate levels [58–61]. As was previously demonstrated, butyrate, a short-chain fatty acid, plays a key role as a significant energy source for gut bacteria and the induction of epithelial cell proliferation, supporting intestinal development and health [28,55–57,62–64]. In the current study, butyrate cecal contents concentrations were higher in treatment groups relative to controls ($p < 0.05$, Figure 8). However, the observed increased abundance in *Clostridium* class is not a direct indicator of butyrate production. Previously, Lozada-Fernandez et al. (2022) found similar results, demonstrating that NR-treated mice had an increase in fecal propionate, butyrate, valerate and isobutyrate concentrations compared to controls while having an increased population of *Firmicutes* (oxidizing butyrate for growth) [65]. This result was hypothesized to be caused by *Firmicutes* metagenome-assembled genomes (MAGs) utilizing acetyl coenzyme A (acetyl- CoA) butyrate synthesis pathway, thus indicating the NR supplementation enriches butyrate-producing *Firmicutes* (e.g., *Clostridium* sp.) [65]. However, current observations indicate that further and detailed investigation, especially the specific quantification of butyrate-producing bacteria is essential to better understand the microbial basis of an increase in butyrate, as a result of NR consumption.

The NRCl treatment group demonstrated a significant ($p < 0.05$) increase in *Bifidobacterium*, *Lactobacillus*, and *E. coli* populations compared to the other experimental groups. This may be attributed to the potential and indirect targeting of *Clostridium* due to the immunomodulatory role that nicotinamide holds, as was similarly found under *M. tuberculosis* and HIV infections circumstances [66–71]. Interestingly, the impact of NRCl on the increase of the populations of *Bifidobacterium* and *Lactobacillus* was more than that of *E. coli* in comparison to H_2O, and non-injected controls. Elevated levels of commensal bacterial populations (*Bifidobacterium* and *Lactobacillus*) may be due to NRCl, and its derivatives consumed by beneficial bacteria within the gut. These commensal bacteria are preferred by the host, and this preference decreases the populations of potentially invasive bacteria through the production of antimicrobial defensins and cathelicidins (i.e., muramidase, α-defensins, β-defensins) by the host's innate immune system in the small intestine [72–75].

Contrary to NRCl, the NRTBCl treatment did not alter the populations of *Bifidobacterium*, *Lactobacillus*, or *E. coli*. Further, relative to NRCl and NRTBCl, the NRTOCl treatment significantly ($p < 0.05$) reduced the populations of *Bifidobacterium*, *Lactobacillus*, and *E. coli*. Structurally, NRTOCl comprises a quaternary ammonium group with a positive charge and three hydrophobic oleate branches. This structure can act as a cationic surfactant and negatively affect these bacterial populations. Reuerio et al. (2016) demonstrated a significant increase in Bacteroidetes and Firmicutes (i.e., *Lactobacillus*) in subjects with elevated ammonia levels, leading to a decreased population of Actinobacteria (i.e., *Bifidobacterium*), which supports the microbial findings presented here [76]. Therefore, the increase in ammonium associated with supplementation of NRTOCl led to a bacterial profile shift towards Bacteroidetes, possibly lowering the population of *Bifidobacterium*, *Lactobacillus*, and *E. coli*.

Previous studies have demonstrated that the intra-amniotic administration of polyphenols and other dietary substances (as soluble fiber extracts) has affected BBM functionality through alterations in gene expression of key BBM digestive and absorptive proteins [29,45,47,50,77,78]. In the current study, we investigated the effects of NRCl and its derivatives on BBM gene expression (Figure 9), and the results demonstrated that NRCl, NRTBCl, and NRTOCl increased the expression of ZnT1 and MUC2 and decreased the expression of IL-6. In comparison with the other investigated compounds and controls, NRTOCl significantly increased MUC2 gene expression level, which is a valuable factor in maintaining intestinal health, as MUC2 that encodes mucin 2 protein that is secreted onto mucosal surfaces and protects the intestinal epithelium cells where its disruption causes several pathologies [79–81]. A consistently high level of MUC2 expression can be associated with an essential protective barrier against external pathogens (as indicated by microbial findings, Figure 7) due to MUC2's diverse functions in intestinal homeostasis. It was previously demonstrated that there was embryonic development of MUC2 at nine weeks of gestation, making it an important marker for the differentiation of secretory cell lineages [47].

The significant decrease in the gene expression of IL-6, which acts as a pleiotropic pro-inflammatory cytokine, is indirectly activated by the supplementation of NRCl, NRTBCl, and NRTOCl, to protect the host against invasive bacteria, which further explains the reduction in the abundance of specific invasive bacterial populations (*E. coli*). This finding agrees with previous observations by Elhassan et al. (2019), where oral NR supplementation resulted in significantly decreased IL-6 expression, in vivo [82].

Moreover, gene expression of the ZnT1 transporter was significantly increased in NRCl, NRTBCl, and NRTOCl, compared to the control groups; this observation can be linked to the potential antioxidative effect NRCl, which may lead to increased cellular zinc [52,83,84]. Specifically, ZnT1 acts as a rescue agent under excess zinc conditions to export zinc from cellular organelles to the cytosol [85,86]. This finding may indicate increased zinc content within the cellular organelles of the enterocytes due to NR supplementation, resulting in excess zinc being transported via the basolateral membrane [87]. Previously, it was demonstrated that the intra-amniotic administration of zinc-methionine increased ZnT1 expression due to the introduction of additional zinc [87]. Additionally, ZnT1 expression is increased during times of inflammation as a shuttle of zinc content into the plasma for circulation, which coincides with the increase in IL-6 gene expression [88–92].

Therefore, the administration of NRCl and its derivatives triggers altered expression of key BBM genes involved in digestion, and absorption, with additional effects on intestinal microbiota composition and function. Further and as was previously suggested, these functional changes, as demonstrated via gene expression of key BBM proteins (zinc transporter, inflammatory cytokines, absorptive proteins, and mucin), were previously associated with BBM tissue physiological and morphometric alterations, as increased villi size [24,50,93–97]. These alterations may potentially also be associated with increased proliferation of cellular populations that hold essential roles in BBM function, including, enterocytes, and therefore, increased villus surface area (the intestinal digestive and absorptive surface), and goblet cells (produce and secrete mucus), both number and diameter in intestinal villi and crypt [18,22,24,47,50,93,98–105]. In this context, it was previously demonstrated that colonocytes and enterocytes oxidation pathways utilize butyrate as fuel for cell metabolism [106–108], via SL16 monocarboxylate transporter 1 (MCT1, SCL16A1) and Sodium-coupled monocarboxylate transporter 1 (SMCT1, SLC5A8) that transport butyrate through epithelial cells in the small intestine [109,110]. This allows butyrate and its metabolites to enter the bloodstream, and to potentially affect anti-inflammatory cytokines [111,112], by mediating the binding of free fatty acid receptors (FFARs) [113–115], and by that to indirectly support rapid glycolytic energy extraction for undifferentiated stem cells [116–118]. Further, Kien et al. demonstrated in piglets that cecal infu-sion of butyrate significantly ($p = 0.007$) supported small intestinal enterocytes proliferation (ileum and jejunum) [119]. Similarly, Zhang et al. suggested that propionate's cecal fermentation plays a significant role in jejunum development and gut health [120]. In addition, De Vadder et al. demonstrated that FOS and SCFAs (propionate and butyrate) increased cell proliferation, via FFAR3 receptors in rats [101]. However, it is important to emphasize that further assessments are necessary to confirm the potential effects of NR and its derivatives on intestinal morphology and functionality due to SCFAs production (specifically butyrate) by the cecal and/or small intestinal microbiome.

Overall, current results introduce an innovative approach to evaluating the impact of NR and its derivatives on BBM functional biomarkers, and intestinal microbial populations, in vivo.

5. Conclusions

This study is the first to demonstrate the effects of nicotinamide riboside and its derivatives on duodenal BBM gene expression and cecal microbial profiles, in-vivo. We have synthesized and characterized two derivatives of vitamin B_3–NRTOCl and NRTBCl. Through the in vivo study, we found that NRTOCl has the potential to improve BBM functionality by increasing MUC2, and ZnT1 and reducing the expression of inflammatory cytokine IL-1β. Additionally, we detected significant differences in cecal bacterial populations, which suggests that NR and its derivatives positively modulate the intestinal microbial profile, composition, and function. Further studies are warranted to validate the findings of the current research and establish the safety of the synthesized compounds.

Supplementary Materials: The following are available online at https://www.mdpi.com/article/10.3390/nu14153130/s1, Figure S1: FT-IR of NRTOCl, Figure S2: ^{1}H NMR of NRTOCl in $CDCl_3$, Figure S3: Expanded ^{1}H NMR of NRTOCl, Figure S4: ^{13}C NMR of NRTOCl in $CDCl_3$, Figure S5: Expanded ^{13}C NMR of NRTOCl, Figure S6: SRM LC-MS of NRTOCl. (a) SRM LC of NRTOCl. (b) Mass spectrum of NRTOCl, Figure S7: Particle size of NRTOCl in DI water containing 1% ethanol.

Author Contributions: Chemical synthesis and analysis; A.Z., L.K., Y.D. and A.A.; In vivo experiment, data analysis, collection, and methodology, N.K., J.C., N.A. and E.T.; A.Z., L.K., N.K., N.A. and J.C. wrote the manuscript draft; E.T. and A.A. finalized the manuscript and were the primary investigators who led the research conducted and presented in this manuscript. All authors have read and agreed to the published version of the manuscript.

Funding: This research received no external funding.

Institutional Review Board Statement: The animal protocol used in this study was conducted according to the guidelines of the Declaration of Helsinki and approved by the Cornell University Institutional Animal Care and Use Committee by ethic approval code 2020-0077.

Informed Consent Statement: Not applicable.

Data Availability Statement: The data presented in this study are available on request from the corresponding author.

Acknowledgments: Nicotinamide riboside chloride was a gift from ChromaDex Company.

Conflicts of Interest: The authors declare no conflict of interest.

References

1. Da Silva, B.P.; Martino, H.S.D.; Tako, E. Plant Origin Prebiotics Affect Duodenal Brush Border Membrane Functionality and Morphology, in Vivo (Gallus Gallus). *Food Funct.* **2021**, *12*, 6157–6166. [CrossRef] [PubMed]
2. Kataoka, K. The Intestinal Microbiota and Its Role in Human Health and Disease. *J. Med. Investig.* **2016**, *63*, 27–37. [CrossRef] [PubMed]
3. Tanes, C.; Bittinger, K.; Gao, Y.; Friedman, E.S.; Nessel, L.; Paladhi, U.R.; Chau, L.; Panfen, E.; Fischbach, M.A.; Braun, J.; et al. Role of Dietary Fiber in the Recovery of the Human Gut Microbiome and Its Metabolome. *Cell Host Microbe* **2021**, *29*, 394–407. [CrossRef] [PubMed]
4. Brisbin, J.T.; Gong, J.; Sharif, S. Interactions between Commensal Bacteria and the Gut-Associated Immune System of the Chicken. *Anim. Health Res. Rev.* **2008**, *9*, 101–110. [CrossRef] [PubMed]
5. Zhang, Y.-J.; Li, S.; Gan, R.-Y.; Zhou, T.; Xu, D.-P.; Li, H.-B. Impacts of Gut Bacteria on Human Health and Diseases. *Int. J. Mol. Sci.* **2015**, *16*, 7493–7519. [CrossRef] [PubMed]
6. Reed, S.; Neuman, H.; Glahn, R.P.; Koren, O.; Tako, E. Characterizing the Gut (Gallus Gallus) Microbiota Following the Consumption of an Iron Biofortified Rwandan Cream Seeded Carioca (Phaseolus Vulgaris L.) Bean-Based Diet. *PLoS ONE* **2017**, *12*, e0182431. [CrossRef]
7. Bouis, H.E.; Hotz, C.; McClafferty, B.; Meenakshi, J.V.; Pfeiffer, W.H. Biofortification: A New Tool to Reduce Micronutrient Malnutrition. *Food Nutr. Bull.* **2011**, *32*, S31–S40. [CrossRef]
8. Welch, R.M. Biotechnology, Biofortification, and Global Health. *Food Nutr. Bull.* **2005**, *26*, S304–S306. [CrossRef]
9. Mayer, J.E.; Pfeiffer, W.H.; Beyer, P. Biofortified Crops to Alleviate Micronutrient Malnutrition. *Curr. Opin. Plant Biol.* **2008**, *11*, 166–170. [CrossRef]
10. Juste Contin Gomes, M.; Stampini Duarte Martino, H.; Tako, E. Effects of Iron and Zinc Biofortified Foods on Gut Microbiota in Vivo (*Gallus gallus*): A Systematic Review. *Nutrients* **2021**, *13*, 189. [CrossRef]

11. Tian, L.; Tan, Y.; Chen, G.; Wang, G.; Sun, J.; Ou, S.; Chen, W.; Bai, W. Metabolism of Anthocyanins and Consequent Effects on the Gut Microbiota. *Crit. Rev. Food Sci. Nutr.* **2019**, *59*, 982–991. [CrossRef] [PubMed]
12. Gibson, G.R.; Scott, K.P.; Rastall, R.A.; Tuohy, K.M.; Hotchkiss, A.; Dubert-Ferrandon, A.; Gareau, M.; Murphy, E.F.; Saulnier, D.; Loh, G. Dietary Prebiotics: Current Status and New Definition. *Food Sci. Technol. Bull. Funct. Foods* **2010**, *7*, 1–19. [CrossRef]
13. van der Beek, C.M.; Canfora, E.E.; Kip, A.M.; Gorissen, S.H.; Damink, S.W.O.; van Eijk, H.M.; Holst, J.J.; Blaak, E.E.; Dejong, C.H.; Lenaerts, K. The Prebiotic Inulin Improves Substrate Metabolism and Promotes Short-Chain Fatty Acid Production in Overweight to Obese Men. *Metabolism* **2018**, *87*, 25–35. [CrossRef] [PubMed]
14. McLoughlin, R.F.; Berthon, B.S.; Jensen, M.E.; Baines, K.J.; Wood, L.G. Short-Chain Fatty Acids, Prebiotics, Synbiotics, and Systemic Inflammation: A Systematic Review and Meta-Analysis. *Am. J. Clin. Nutr.* **2017**, *106*, 930–945. [CrossRef]
15. Preidis, G.A.; Versalovic, J. Targeting the Human Microbiome with Antibiotics, Probiotics, and Prebiotics: Gastroenterology Enters the Metagenomics Era. *Gastroenterology* **2009**, *136*, 2015–2031. [CrossRef]
16. Verediano, T.A.; Stampini Duarte Martino, H.; Dias Paes, M.C.; Tako, E. Effects of Anthocyanin on Intestinal Health: A Systematic Review. *Nutrients* **2021**, *13*, 1331. [CrossRef]
17. Hou, T.; Tako, E. The in Ovo Feeding Administration (Gallus Gallus)—An Emerging in Vivo Approach to Assess Bioactive Compounds with Potential Nutritional Benefits. *Nutrients* **2018**, *10*, 418. [CrossRef]
18. Wiesinger, J.A.; Glahn, R.P.; Cichy, K.A.; Kolba, N.; Hart, J.J.; Tako, E. An in Vivo (*Gallus gallus*) Feeding Trial Demonstrating the Enhanced Iron Bioavailability Properties of the Fast Cooking Manteca Yellow Bean (*Phaseolus vulgaris* L.). *Nutrients* **2019**, *11*, 1768. [CrossRef]
19. Knez, M.; Tako, E.; Glahn, R.P.; Kolba, N.; de Courcy-Ireland, E.; Stangoulis, J.C. Linoleic Acid: Dihomo-γ-Linolenic Acid Ratio Predicts the Efficacy of Zn-Biofortified Wheat in Chicken (*Gallus gallus*). *J. Agric. Food Chem.* **2018**, *66*, 1394–1400. [CrossRef]
20. Dias, D.M.; Kolba, N.; Binyamin, D.; Ziv, O.; Regini Nutti, M.; Martino, H.S.D.; Glahn, R.P.; Koren, O.; Tako, E. Iron Biofortified Carioca Bean (*Phaseolus vulgaris* L.)—Based Brazilian Diet Delivers More Absorbable Iron and Affects the Gut Microbiota in Vivo (*Gallus gallus*). *Nutrients* **2018**, *10*, 1970. [CrossRef]
21. Tako, E.; Bar, H.; Glahn, R.P. The Combined Application of the Caco-2 Cell Bioassay Coupled with in Vivo (Gallus Gallus) Feeding Trial Represents an Effective Approach to Predicting Fe Bioavailability in Humans. *Nutrients* **2016**, *8*, 732. [CrossRef]
22. Agrizzi Verediano, T.; Stampini Duarte Martino, H.; Kolba, N.; Fu, Y.; Cristina Dias Paes, M.; Tako, E. Black Corn (*Zea mays* L.) Soluble Extract Showed Anti-Inflammatory Effects and Improved the Intestinal Barrier Integrity in Vivo (*Gallus gallus*). *Food Res. Int.* **2022**, *157*, 111227. [CrossRef]
23. Agrizzi Verediano, T.; Agarwal, N.; Gomes, M.J.C.; Duarte Martino, H.S.; Tako, E. Effects of Dietary Fiber on Intestinal Iron Absorption, and Physiological Status: A Systematic Review of in Vivo and Clinical Studies. *Crit. Rev. Food Sci. Nutr.* **2022**, *1*, 1–16. [CrossRef]
24. Gomes, M.J.C.; Martino, H.S.D.; Kolba, N.; Cheng, J.; Agarwal, N.; De Moura Rocha, M.; Tako, E. Zinc Biofortified Cowpea (*Vigna unguiculata* L. Walp.) Soluble Extracts Modulate Assessed Cecal Bacterial Populations and Gut Morphology in Vivo (*Gallus gallus*). *Front. BioScience Landmark* **2022**, *27*, 140–153. [CrossRef]
25. Uni, Z.; Ferket, P.R.; Tako, E.; Kedar, O. In Ovo Feeding Improves Energy Status of Late-Term Chicken Embryos. *Poult. Sci.* **2005**, *84*, 764–770. [CrossRef]
26. Liu, H.H.; Wang, J.W.; Chen, X.; Zhang, R.P.; Yu, H.Y.; Jin, H.B.; Li, L.; Han, C.C. In Ovo Administration of RhIGF-1 to Duck Eggs Affects the Expression of Myogenic Transcription Factors and Muscle Mass during Late Embryo Development. *J. Appl. Physiol.* **2011**, *111*, 1789–1797. [CrossRef]
27. Selim, S.A.; Gaafar, K.M.; El-ballal, S.S. Influence of In-Ovo Administration with Vitamin E and Ascorbic Acid on Theperformance of Muscovy Ducks. *Emir. J. Food Agric.* **2012**, *24*, 264–271.
28. Beasley, J.T.; Johnson, A.A.; Kolba, N.; Bonneau, J.P.; Glahn, R.P.; Ozeri, L.; Koren, O.; Tako, E. Nicotianamine-Chelated Iron Positively Affects Iron Status, Intestinal Morphology and Microbial Populations in Vivo (*Gallus gallus*). *Sci. Rep.* **2020**, *10*, 1–14. [CrossRef]
29. Carboni, J.; Reed, S.; Kolba, N.; Eshel, A.; Koren, O.; Tako, E. Alterations in the Intestinal Morphology, Gut Microbiota, and Trace Mineral Status Following Intra-Amniotic Administration (*Gallus gallus*) of Teff (*Eragrostis tef*) Seed Extracts. *Nutrients* **2020**, *12*, 3020. [CrossRef]
30. Pereira da Silva, B.; Kolba, N.; Duarte Martino, H.S.; Hart, J.J.; Tako, E. Soluble Extracts from Chia Seed (*Salvia hispanica* L.) Affect Brush Border Membrane Functionality, Morphology and Intestinal Bacterial Populations In Vivo (*Gallus gallus*). *Nutrients* **2019**, *11*, 2457. [CrossRef]
31. Martens, C.R.; Denman, B.A.; Mazzo, M.R.; Armstrong, M.L.; Reisdorph, N.; McQueen, M.B.; Chonchol, M.; Seals, D.R. Chronic Nicotinamide Riboside Supplementation Is Well-Tolerated and Elevates NAD+ in Healthy Middle-Aged and Older Adults. *Nat. Commun.* **2018**, *9*, 1286. [CrossRef]
32. Yang, T.; Chan, N.Y.-K.; Sauve, A.A. Syntheses of Nicotinamide Riboside and Derivatives: Effective Agents for Increasing Nicotinamide Adenine Dinucleotide Concentrations in Mammalian Cells. *J. Med. Chem.* **2007**, *50*, 6458–6461. [CrossRef] [PubMed]
33. Braidy, N.; Berg, J.; Clement, J.; Khorshidi, F.; Poljak, A.; Jayasena, T.; Grant, R.; Sachdev, P. Role of Nicotinamide Adenine Dinucleotide and Related Precursors as Therapeutic Targets for Age-Related Degenerative Diseases: Rationale, Biochemistry, Pharmacokinetics, and Outcomes. *Antioxid. Redox Signal.* **2019**, *30*, 251–294. [CrossRef]
34. Gonzalez, J.M.; Jackson, A.R. In Ovo Feeding of Nicotinamide Riboside Affects Broiler Pectoralis Major Muscle Development. *Transl. Anim. Sci.* **2020**, *4*, txaa126. [CrossRef]

35. Zhang, H.; Ryu, D.; Wu, Y.; Gariani, K.; Wang, X.; Luan, P.; D'Amico, D.; Ropelle, E.R.; Lutolf, M.P.; Aebersold, R. NAD+ Repletion Improves Mitochondrial and Stem Cell Function and Enhances Life Span in Mice. *Science* **2016**, *352*, 1436–1443. [CrossRef]
36. Cerutti, R.; Pirinen, E.; Lamperti, C.; Marchet, S.; Sauve, A.A.; Li, W.; Leoni, V.; Schon, E.A.; Dantzer, F.; Auwerx, J. NAD+-Dependent Activation of Sirt1 Corrects the Phenotype in a Mouse Model of Mitochondrial Disease. *Cell Metab.* **2014**, *19*, 1042–1049. [CrossRef]
37. Gong, B.; Pan, Y.; Vempati, P.; Zhao, W.; Knable, L.; Ho, L.; Wang, J.; Sastre, M.; Ono, K.; Sauve, A.A. Nicotinamide Riboside Restores Cognition through an Upregulation of Proliferator-Activated Receptor-γ Coactivator 1α Regulated β-Secretase 1 Degradation and Mitochondrial Gene Expression in Alzheimer's Mouse Models. *Neurobiol. Aging* **2013**, *34*, 1581–1588. [CrossRef]
38. Cantó, C.; Houtkooper, R.H.; Pirinen, E.; Youn, D.Y.; Oosterveer, M.H.; Cen, Y.; Fernandez-Marcos, P.J.; Yamamoto, H.; Andreux, P.A.; Cettour-Rose, P. The NAD+ Precursor Nicotinamide Riboside Enhances Oxidative Metabolism and Protects against High-Fat Diet-Induced Obesity. *Cell Metab.* **2012**, *15*, 838–847. [CrossRef]
39. Dollerup, O.L.; Christensen, B.; Svart, M.; Schmidt, M.S.; Sulek, K.; Ringgaard, S.; Stødkilde-Jørgensen, H.; Møller, N.; Brenner, C.; Treebak, J.T. A Randomized Placebo-Controlled Clinical Trial of Nicotinamide Riboside in Obese Men: Safety, Insulin-Sensitivity, and Lipid-Mobilizing Effects. *Am. J. Clin. Nutr.* **2018**, *108*, 343–353. [CrossRef]
40. Lee, H.J.; Hong, Y.-S.; Jun, W.; Yang, S.J. Nicotinamide Riboside Ameliorates Hepatic Metaflammation by Modulating NLRP3 Inflammasome in a Rodent Model of Type 2 Diabetes. *J. Med. Food* **2015**, *18*, 1207–1213. [CrossRef]
41. Mehmel, M.; Jovanović, N.; Spitz, U. Nicotinamide Riboside—the Current State of Research and Therapeutic Uses. *Nutrients* **2020**, *12*, 1616. [CrossRef] [PubMed]
42. Zhou, Y.; Fu, B.; Zheng, X.; Wang, D.; Zhao, C.; Qi, Y.; Sun, R.; Tian, Z.; Xu, X.; Wei, H. Pathogenic T-Cells and Inflammatory Monocytes Incite Inflammatory Storms in Severe COVID-19 Patients. *Natl. Sci. Rev.* **2020**, *7*, 998–1002. [CrossRef] [PubMed]
43. Campbell, M.T.; Jones, D.S.; Andrews, G.P.; Li, S. Understanding the Physicochemical Properties and Degradation Kinetics of Nicotinamide Riboside, a Promising Vitamin B3nutritional Supplement. *Food Nutr. Res.* **2019**, *63*. [CrossRef] [PubMed]
44. Xu, X.; Jackson, A.R.; Gonzalez, J.M. The Effects of in Ovo Nicotinamide Riboside Dose on Broiler Myogenesis. *Poult. Sci.* **2021**, *100*, 100926. [CrossRef]
45. Pacifici, S.; Song, J.; Zhang, C.; Wang, Q.; Glahn, R.; Kolba, N.; Tako, E. Intra Amniotic Administration of Raffinose and Stachyose Affects the Intestinal Brush Border Functionality and Alters Gut Microflora Populations. *Nutrients* **2017**, *9*, 304. [CrossRef]
46. Tako, E. Dietary Plant-Origin Bio-Active Compounds, Intestinal Functionality, and Microbiome. *Nutrients* **2020**, *12*, 3223. [CrossRef]
47. Martino, H.S.D.; Kolba, N.; Tako, E. Yacon (Smallanthus Sonchifolius) Flour Soluble Extract Improve Intestinal Bacterial Populations, Brush Border Membrane Functionality and Morphology in Vivo (*Gallus gallus*). *Food Res. Int.* **2020**, *137*, 109705. [CrossRef]
48. Tako, E.; Ferket, P.R.; Uni, Z. Effects of in Ovo Feeding of Carbohydrates and Beta-Hydroxy-Beta-Methylbutyrate on the Development of Chicken Intestine. *Poult. Sci.* **2004**, *83*, 2023–2028. [CrossRef]
49. Tako, E.; Glahn, R.P.; Knez, M.; Stangoulis, J.C. The Effect of Wheat Prebiotics on the Gut Bacterial Population and Iron Status of Iron Deficient Broiler Chickens. *Nutr. J.* **2014**, *13*, 1. [CrossRef]
50. Gomes, M.J.C.; Kolba, N.; Agarwal, N.; Kim, D.; Eshel, A.; Koren, O.; Tako, E. Modifications in the Intestinal Functionality, Morphology and Microbiome Following Intra-Amniotic Administration (*Gallus gallus*) of Grape (*Vitis vinifera*) Stilbenes (Resveratrol and Pterostilbene). *Nutrients* **2021**, *13*, 3247. [CrossRef]
51. Zhu, X.Y.; Zhong, T.; Pandya, Y.; Joerger, R.D. 16S RRNA-Based Analysis of Microbiota from the Cecum of Broiler Chickens. *Appl. Environ. Microbiol.* **2002**, *68*, 124–137. [CrossRef]
52. Kourtzidis, I.A.; Dolopikou, C.F.; Tsiftsis, A.N.; Margaritelis, N.V.; Theodorou, A.A.; Zervos, I.A.; Tsantarliotou, M.P.; Veskoukis, A.S.; Vrabas, I.S.; Paschalis, V.; et al. Nicotinamide Riboside Supplementation Dysregulates Redox and Energy Metabolism in Rats: Implications for Exercise Performance. *Exp. Physiol.* **2018**, *103*, 1357–1366. [CrossRef]
53. Bieganowski, P.; Brenner, C. Discoveries of Nicotinamide Riboside as a Nutrient and Conserved NRK Genes Establish a Preiss-Handler Independent Route to NAD+ in Fungi and Humans. *Cell* **2004**, *117*, 495–502. [CrossRef]
54. Bürkle, A. Physiology and Pathophysiology of Poly(ADP-Ribosyl)Ation *: Review Articles. *Bioessays* **2001**, *23*, 795–806. [CrossRef]
55. Chen, R.; Xu, Y.; Wu, P.; Zhou, H.; Lasanajak, Y.; Fang, Y.; Tang, L.; Ye, L.; Li, X.; Cai, Z.; et al. Transplantation of Fecal Microbiota Rich in Short Chain Fatty Acids and Butyric Acid Treat Cerebral Ischemic Stroke by Regulating Gut Microbiota. *Pharmacol. Res.* **2019**, *148*, 104403. [CrossRef]
56. Aghazadeh, A.; TahaYazdi, M. Effect of Butyric Acid Supplementation and Whole Wheat Inclusion on the Performance and Carcass Traits of Broilers. *SA J. An. Sci.* **2012**, *42*, 241–248. [CrossRef]
57. Panda, A.K.; Rao, S.V.R.; Raju, M.V.L.N.; Sunder, G.S. Effect of Butyric Acid on Performance, Gastrointestinal Tract Health and Carcass Characteristics in Broiler Chickens. *Asian Australas. J. Anim. Sci* **2009**, *22*, 1026–1031. [CrossRef]
58. Guo, P.; Zhang, K.; Ma, X.; He, P. Clostridium Species as Probiotics: Potentials and Challenges. *J. Anim. Sci. Biotechnol.* **2020**, *11*, 24. [CrossRef]
59. Van den Abbeele, P.; Belzer, C.; Goossens, M.; Kleerebezem, M.; De Vos, W.M.; Thas, O.; De Weirdt, R.; Kerckhof, F.-M.; Van de Wiele, T. Butyrate-Producing Clostridium Cluster XIVa Species Specifically Colonize Mucins in an in Vitro Gut Model. *ISME J* **2013**, *7*, 949–961. [CrossRef]

60. Lopetuso, L.R.; Scaldaferri, F.; Petito, V.; Gasbarrini, A. Commensal Clostridia: Leading Players in the Maintenance of Gut Homeostasis. *Gut Pathog.* **2013**, *5*, 23. [CrossRef]
61. Wu, C.-S.; Muthyala, S.D.V.; Klemashevich, C.; Ufondu, A.U.; Menon, R.; Chen, Z.; Devaraj, S.; Jayaraman, A.; Sun, Y. Age-Dependent Remodeling of Gut Microbiome and Host Serum Metabolome in Mice. *Aging* **2021**, *13*, 6330–6345. [CrossRef] [PubMed]
62. Kaczmarek, S.A.; Barri, A.; Hejdysz, M.; Rutkowski, A. Effect of Different Doses of Coated Butyric Acid on Growth Performance and Energy Utilization in Broilers. *Poult. Sci.* **2016**, *95*, 851–859. [CrossRef] [PubMed]
63. Zou, X.; Ji, J.; Qu, H.; Wang, J.; Shu, D.M.; Wang, Y.; Liu, T.F.; Li, Y.; Luo, C.L. Effects of Sodium Butyrate on Intestinal Health and Gut Microbiota Composition during Intestinal Inflammation Progression in Broilers. *Poult. Sci.* **2019**, *98*, 4449–4456. [CrossRef] [PubMed]
64. Aalamifar, H.; Soltanian, S.; Vazirzadeh, A.; Akhlaghi, M.; Morshedi, V.; Gholamhosseini, A.; Torfi Mozanzadeh, M. Dietary Butyric Acid Improved Growth, Digestive Enzyme Activities and Humoral Immune Parameters in Barramundi (*Lates Calcarifer*). *Aquacult. Nutr.* **2020**, *26*, 156–164. [CrossRef]
65. Lozada-Fernández, V.V.; de Leon, O.; Kellogg, S.L.; Saravia, F.L.; Hadiono, M.A.; Atkinson, S.N.; Grobe, J.L.; Kirby, J.R. Nicotinamide Riboside-Conditioned Microbiota Deflects High-Fat Diet-Induced Weight Gain in Mice. *mSystems* **2022**, *7*, e00230-21. [CrossRef]
66. Schein, P.S.; Loftus, S. Streptozotocin: Depression of Mouse Liver Pyridine Nucleotides. *Cancer Res.* **1968**, *28*, 1501–1506.
67. Nagai, A.; Matsumiya, H.; Hayashi, M.; Yasui, S.; Okamoto, H.; Konno, K. Effects of Nicotinamide and Niacin on Bleomycin-Induced Acute Injury and Subsequent Fibrosis in Hamster Lungs. *Exp. Lung Res.* **1994**, *20*, 263–281. [CrossRef]
68. LeClaire, R.D.; Kell, W.; Bavari, S.; Smith, T.J.; Hunt, R.E. Protective Effects of Niacinamide in Staphylococcal Enterotoxin-B-Induced Toxicity. *Toxicology* **1996**, *107*, 69–81. [CrossRef]
69. Hiromatsu, Y.; Sato, M.; Yamada, K.; Nonaka, K. Inhibitory Effects of Nicotinamide on Recombinant Human Interferon-Gamma-Induced Intercellular Adhesion Molecule-1 (ICAM-1) and HLA-DR Antigen Expression on Cultured Human Endothelial Cells. *Immunol. Lett.* **1992**, *31*, 35–39. [CrossRef]
70. Otsuka, A.; Hanafusa, T.; Miyagawa, J.-I.; Kono, N.; Tarui, S. Nicotinamide and 3-Aminobenzamide Reduce Interferon-γ -Induced Class II MHC (HLA-DR and -DP) Molecule Expression on Cultured Human Endothelial Cells and Fibroblasts. *Immunopharmacol. Immunotoxicol.* **1991**, *13*, 263–280. [CrossRef]
71. Murray, M.F. Nicotinamide: An Oral Antimicrobial Agent with Activity against Both Mycobacterium Tuberculosis and Human Immunodeficiency Virus. *Clin. Infect. Dis.* **2003**, *36*, 453–460. [CrossRef]
72. Yang, D.; Chertov, O.; Oppenheim, J.J. Participation of Mammalian Defensins and Cathelicidins in Anti-Microbial Immunity: Receptors and Activities of Humandefensins and Cathelicidin (LL-37). *J. Leukoc. Biol.* **2001**, *69*, 1–7. [CrossRef]
73. Baquero, F.; Nombela, C. The Microbiome as a Human Organ. *Clin. Microbiol. Infect.* **2012**, *18*, 2–4. [CrossRef]
74. Biesalski, H.K. Nutrition Meets the Microbiome: Micronutrients and the Microbiota: Nutrition Meets the Microbiome. *Ann. N. Y. Acad. Sci.* **2016**, *1372*, 53–64. [CrossRef]
75. Takakuwa, A.; Nakamura, K.; Kikuchi, M.; Sugimoto, R.; Ohira, S.; Yokoi, Y.; Ayabe, T. Butyric Acid and Leucine Induce α-Defensin Secretion from Small Intestinal Paneth Cells. *Nutrients* **2019**, *11*, 2817. [CrossRef]
76. Regueiro, L.; Carballa, M.; Lema, J.M. Microbiome Response to Controlled Shifts in Ammonium and LCFA Levels in Co-Digestion Systems. *J. Biotechnol.* **2016**, *220*, 35–44. [CrossRef]
77. Limage, R.; Tako, E.; Kolba, N.; Guo, Z.; García-Rodríguez, A.; Marques, C.N.H.; Mahler, G.J. TiO_2 Nanoparticles and Commensal Bacteria Alter Mucus Layer Thickness and Composition in a Gastrointestinal Tract Model. *Small* **2020**, 2000601. [CrossRef]
78. Kolba, N.; Guo, Z.; Olivas, F.M.; Mahler, G.J.; Tako, E. Intra-Amniotic Administration (*Gallus gallus*) of TiO_2, SiO_2, and ZnO Nanoparticles Affect Brush Border Membrane Functionality and Alters Gut Microflora Populations. *Food Chem. Toxicol.* **2019**, 110896. [CrossRef]
79. Allen, A.; Hutton, D.A.; Pearson, J.P. The MUC2 Gene Product: A Human Intestinal Mucin. *Int. J. Biochem. Cell Biol.* **1998**, *30*, 797–801. [CrossRef]
80. Bergstrom, K.S.B.; Kissoon-Singh, V.; Gibson, D.L.; Ma, C.; Montero, M.; Sham, H.P.; Ryz, N.; Huang, T.; Velcich, A.; Finlay, B.B.; et al. Muc2 Protects against Lethal Infectious Colitis by Disassociating Pathogenic and Commensal Bacteria from the Colonic Mucosa. *PLoS Pathog* **2010**, *6*, e1000902. [CrossRef]
81. Zarepour, M.; Bhullar, K.; Montero, M.; Ma, C.; Huang, T.; Velcich, A.; Xia, L.; Vallance, B.A. The Mucin Muc2 Limits Pathogen Burdens and Epithelial Barrier Dysfunction during Salmonella Enterica Serovar Typhimurium Colitis. *Infect. Immun.* **2013**, *81*, 3672–3683. [CrossRef] [PubMed]
82. Elhassan, Y.S.; Kluckova, K.; Fletcher, R.S.; Schmidt, M.S.; Garten, A.; Doig, C.L.; Cartwright, D.M.; Oakey, L.; Burley, C.V.; Jenkinson, N.; et al. Nicotinamide Riboside Augments the Aged Human Skeletal Muscle NAD+ Metabolome and Induces Transcriptomic and Anti-Inflammatory Signatures. *Cell Rep.* **2019**, *28*, 1717–1728. [CrossRef] [PubMed]
83. Dolopikou, C.F.; Kourtzidis, I.A.; Margaritelis, N.V.; Vrabas, I.S.; Koidou, I.; Kyparos, A.; Theodorou, A.A.; Paschalis, V.; Nikolaidis, M.G. Acute Nicotinamide Riboside Supplementation Improves Redox Homeostasis and Exercise Performance in Old Individuals: A Double-Blind Cross-over Study. *Eur. J. Nutr.* **2020**, *11*, 505–515. [CrossRef] [PubMed]
84. Carrera-Juliá, S.; Moreno, M.L.; Barrios, C.; de la Rubia Ortí, J.E.; Drehmer, E. Antioxidant Alternatives in the Treatment of Amyotrophic Lateral Sclerosis: A Comprehensive Review. *Front. Physiol.* **2020**, *11*, 63. [CrossRef]
85. Andrews, G.K.; Wang, H.; Dey, S.K.; Palmiter, R.D. Mouse Zinc Transporter 1 Gene Provides an Essential Function during Early Embryonic Development. *genesis* **2004**, *40*, 74–81. [CrossRef]

86. Segal, D.; Ohana, E.; Besser, L.; Hershfinkel, M.; Moran, A.; Sekler, I. A Role for ZnT-1 in Regulating Cellular Cation Influx. *Biochem. Biophys. Res. Commun.* **2004**, *323*, 1145–1150. [CrossRef]
87. Tako, E.; Ferket, P.; Uni, Z. Changes in Chicken Intestinal Zinc Exporter MRNA Expression and Small Intestinal Functionality Following Intra-Amniotic Zinc-Methionine Administration. *J. Nutr. Biochem.* **2005**, *16*, 339–346. [CrossRef]
88. Ghashut, R.A.; McMillan, D.C.; Kinsella, J.; Vasilaki, A.T.; Talwar, D.; Duncan, A. The Effect of the Systemic Inflammatory Response on Plasma Zinc and Selenium Adjusted for Albumin. *Clin. Nutr.* **2016**, *35*, 381–387. [CrossRef]
89. Vasto, S.; Mocchegiani, E.; Candore, G.; Listì, F.; Colonna-Romano, G.; Lio, D.; Malavolta, M.; Giacconi, R.; Cipriano, C.; Caruso, C. Inflammation, Genes and Zinc in Ageing and Age-Related Diseases. *Biogerontology* **2006**, *7*, 315–327. [CrossRef]
90. Vasto, S.; Mocchegiani, E.; Malavolta, M.; Cuppari, I.; Listi, F.; Nuzzo, D.; Ditta, V.; Candore, G.; Caruso, C. Zinc and Inflammatory/Immune Response in Aging. *Ann. N. Y. Acad. Sci.* **2007**, *1100*, 111–122. [CrossRef]
91. Mburu, A.S.W.; Thurnham, D.I.; Mwaniki, D.L.; Muniu, E.M.; Alumasa, F.M. The Influence of Inflammation on Plasma Zinc Concentration in Apparently Healthy, HIV+ Kenyan Adults and Zinc Responses after a Multi-Micronutrient Supplement. *Eur. J. Clin. Nutr.* **2010**, *64*, 510–517. [CrossRef]
92. McDonald, C.M.; Suchdev, P.S.; Krebs, N.F.; Hess, S.Y.; Wessells, K.R.; Ismaily, S.; Rahman, S.; Wieringa, F.T.; Williams, A.M.; Brown, K.H.; et al. Adjusting Plasma or Serum Zinc Concentrations for Inflammation: Biomarkers Reflecting Inflammation and Nutritional Determinants of Anemia (BRINDA) Project. *Am. J. Clin. Nutr.* **2020**, *111*, 927–937. [CrossRef]
93. Agarwal, N.; Kolba, N.; Jung, Y.; Cheng, J.; Tako, E. Saffron (*Crocus sativus* L.) Flower Water Extract Disrupts the Cecal Microbiome, Brush Border Membrane Functionality, and Morphology In Vivo (*Gallus gallus*). *Nutrients* **2022**, *14*, 220. [CrossRef]
94. Agarwal, N.; Kolba, N.; Khen, N.; Even, C.; Turjeman, S.; Koren, O.; Tako, E. Quinoa Soluble Fiber and Quercetin Alter the Composition of the Gut Microbiome and Improve Brush Border Membrane Morphology In Vivo (*Gallus gallus*). *Nutrients* **2022**, *14*, 448. [CrossRef]
95. Salam, S.; Iqbal, Z.; Khan, A.A.; Mahmood, R. Oral Administration of Thiram Inhibits Brush Border Membrane Enzymes, Oxidizes Proteins and Thiols, Impairs Redox System and Causes Histological Changes in Rat Intestine: A Dose Dependent Study. *Pestic. Biochem. Physiol.* **2021**, *178*, 104915. [CrossRef]
96. Lucea, S.; Guillén, N.; Sosa, C.; Sorribas, V. Inhibition of Phosphate Transport by NAD+/NADH in Brush Border Membrane Vesicles. *Am. J. Physiol.-Cell Physiol.* **2022**, *322*, C803–C813. [CrossRef]
97. Shahid, F.; Farooqui, Z.; Khan, A.A.; Khan, F. Oral Nigella Sativa Oil and Thymoquinone Administration Ameliorates the Effect of Long-Term Cisplatin Treatment on the Enzymes of Carbohydrate Metabolism, Brush Border Membrane, and Antioxidant Defense in Rat Intestine. *Naunyn-Schmiedeberg's Arch Pharm.* **2018**, *391*, 145–157. [CrossRef]
98. Li, H.; Cheng, J.; Yuan, Y.; Luo, R.; Zhu, Z. Age-related Intestinal Monosaccharides Transporters Expression and Villus Surface Area Increase in Broiler and Layer Chickens. *J. Anim. Physiol. Anim. Nutr.* **2020**, *104*, 144–155. [CrossRef]
99. Coméra, C.; Cartier, C.; Gaultier, E.; Catrice, O.; Panouille, Q.; El Hamdi, S.; Tirez, K.; Nelissen, I.; Théodorou, V.; Houdeau, E. Jejunal Villus Absorption and Paracellular Tight Junction Permeability Are Major Routes for Early Intestinal Uptake of Food-Grade TiO2 Particles: An in Vivo and Ex Vivo Study in Mice. *Part Fibre Toxicol.* **2020**, *17*, 26. [CrossRef]
100. Tysoe, O. Dietary Fructose Acts on Gut to Increase Nutrient Uptake. *Nat. Rev. Endocrinol.* **2021**, *17*, 639. [CrossRef]
101. De Vadder, F.; Kovatcheva-Datchary, P.; Goncalves, D.; Vinera, J.; Zitoun, C.; Duchampt, A.; Bäckhed, F.; Mithieux, G. Microbiota-Generated Metabolites Promote Metabolic Benefits via Gut-Brain Neural Circuits. *Cell* **2014**, *156*, 84–96. [CrossRef] [PubMed]
102. Kelly, D. Regulation of Gut Function and Immunity. In *Formula for the Future: Nutrition or Pathology? Evaluating Performance and Health in Pigs and Poultry*; Nottingham University Press: Nottingham, UK, 2008; Volume 151, p. 131. ISBN 978-90-8686-088-3.
103. Snel, J.; Harmsen, H.; van der Wielen, P.; Williams, B. Dietary Strategies to Influence the Gastrointestinal Microflora of Young Animals, and Its Potential to Improve Intestinal Health. In *Nutrition and Health in the Gastrointestinal Tract*; Wageningen Academic Publishers: Wageningen, The Netherlands, 2002; pp. 37–69.
104. Lewis, A.J.; Southern, L.L. Intestinal Bacteria and Their Influence on Swine Growth. In *Swine Nutrition*; CRC Press: Boca Raton, FL, USA, 2000; pp. 585–611. ISBN 978-0-429-11507-3.
105. Apajalahti, J.; Kettunen, A.; Graham, H. Characteristics of the Gastrointestinal Microbial Communities, with Special Reference to the Chicken. *World's Poult. Sci. J.* **2004**, *60*, 223–232. [CrossRef]
106. He, W.; Wu, G. Oxidation of Amino Acids, Glucose, and Fatty Acids as Metabolic Fuels in Enterocytes of Developing Pigs. *Amino Acids* **2022**, *54*, 1025–1039. [CrossRef] [PubMed]
107. Litvak, Y.; Byndloss, M.X.; Bäumler, A.J. Colonocyte Metabolism Shapes the Gut Microbiota. *Science* **2018**, *362*, eaat9076. [CrossRef]
108. Salvi, P.S.; Cowles, R.A. Butyrate and the Intestinal Epithelium: Modulation of Proliferation and Inflammation in Homeostasis and Disease. *Cells* **2021**, *10*, 1775. [CrossRef]
109. Borthakur, A.; Saksena, S.; Gill, R.K.; Alrefai, W.A.; Ramaswamy, K.; Dudeja, P.K. Regulation of Monocarboxylate Transporter 1 (MCT1) Promoter by Butyrate in Human Intestinal Epithelial Cells: Involvement of NF-KB Pathway. *J. Cell. Biochem.* **2008**, *103*, 1452–1463. [CrossRef]
110. Gonçalves, P.; Araújo, J.R.; Martel, F. Characterization of Butyrate Uptake by Nontransformed Intestinal Epithelial Cell Lines. *J Membr. Biol* **2011**, *240*, 35–46. [CrossRef]
111. Kim, M.H.; Kang, S.G.; Park, J.H.; Yanagisawa, M.; Kim, C.H. Short-Chain Fatty Acids Activate GPR41 and GPR43 on Intestinal Epithelial Cells to Promote Inflammatory Responses in Mice. *Gastroenterology* **2013**, *145*, 396–406.e10. [CrossRef]
112. Tazoe, H.; Otomo, Y.; Kaji, I.; Tanaka, R.; Karaki, S.-I.; Kuwahara, A. Roles of Short-Chain Fatty Acids Receptors, GPR41 and GPR43 on Colonic Functions. *J. Physiol. Pharm.* **2008**, *59* (Suppl. 2), 251–262.

113. Zhao, Y.; Chen, F.; Wu, W.; Sun, M.; Bilotta, A.J.; Yao, S.; Xiao, Y.; Huang, X.; Eaves-Pyles, T.D.; Golovko, G.; et al. GPR43 Mediates Microbiota Metabolite SCFA Regulation of Antimicrobial Peptide Expression in Intestinal Epithelial Cells via Activation of MTOR and STAT3. *Mucosal. Immunol.* **2018**, *11*, 752–762. [CrossRef]
114. Fujiwara, H.; Docampo, M.D.; Riwes, M.; Peltier, D.; Toubai, T.; Henig, I.; Wu, S.J.; Kim, S.; Taylor, A.; Brabbs, S.; et al. Microbial Metabolite Sensor GPR43 Controls Severity of Experimental GVHD. *Nat. Commun.* **2018**, *9*, 3674. [CrossRef]
115. Brown, A.J.; Goldsworthy, S.M.; Barnes, A.A.; Eilert, M.M.; Tcheang, L.; Daniels, D.; Muir, A.I.; Wigglesworth, M.J.; Kinghorn, I.; Fraser, N.J.; et al. The Orphan G Protein-Coupled Receptors GPR41 and GPR43 Are Activated by Propionate and Other Short Chain Carboxylic Acids. *J. Biol. Chem.* **2003**, *278*, 11312–11319. [CrossRef]
116. Barker, N.; van Es, J.H.; Kuipers, J.; Kujala, P.; van den Born, M.; Cozijnsen, M.; Haegebarth, A.; Korving, J.; Begthel, H.; Peters, P.J.; et al. Identification of Stem Cells in Small Intestine and Colon by Marker Gene Lgr5. *Nature* **2007**, *449*, 1003–1007. [CrossRef]
117. Kaiko, G.E.; Ryu, S.H.; Koues, O.I.; Collins, P.L.; Solnica-Krezel, L.; Pearce, E.J.; Pearce, E.L.; Oltz, E.M.; Stappenbeck, T.S. The Colonic Crypt Protects Stem Cells from Microbiota-Derived Metabolites. *Cell* **2016**, *165*, 1708–1720. [CrossRef]
118. Neelis, E.; Koning, B.; Rings, E.; Wijnen, R.; Nichols, B.; Hulst, J.; Gerasimidis, K. The Gut Microbiome in Patients with Intestinal Failure: Current Evidence and Implications for Clinical Practice. *J. Parenter. Enter. Nutr.* **2019**, *43*, 194–205. [CrossRef]
119. Kien, C.L.; Blauwiekel, R.; Bunn, J.Y.; Jetton, T.L.; Frankel, W.L.; Holst, J.J. Cecal Infusion of Butyrate Increases Intestinal Cell Proliferation in Piglets. *J. Nutr.* **2007**, *137*, 916–922. [CrossRef]
120. Zhang, Y.; Chen, H.; Zhu, W.; Yu, K. Cecal Infusion of Sodium Propionate Promotes Intestinal Development and Jejunal Barrier Function in Growing Pigs. *Animals* **2019**, *9*, 284. [CrossRef]

nutrients

Review

Dietary Trehalose as a Bioactive Nutrient

Anqi Chen [1,2] and Patrick A. Gibney [1,*]

[1] Department of Food Science, Cornell University, Ithaca, NY 14853, USA
[2] Science Center for Future Foods, Jiangnan University, Wuxi 214122, China
* Correspondence: pag235@cornell.edu

Abstract: Trehalose is a naturally occurring, non-reducing disaccharide comprising two covalently-linked glucose molecules. It possesses unique physiochemical properties, which account for multiple biological roles in a variety of prokaryotic and eukaryotic organisms. In the past few decades, intensive research on trehalose has uncovered its functions, and extended its uses as a sweetener and stabilizer in the food, medical, pharmaceutical, and cosmetic industries. Further, increased dietary trehalose consumption has sparked research on how trehalose affects the gut microbiome. In addition to its role as a dietary sugar, trehalose has gained attention for its ability to modulate glucose homeostasis, and potentially as a therapeutic agent for diabetes. This review discusses the bioactive effects of dietary trehalose, highlighting its promise in future industrial and scientific contributions.

Keywords: trehalose; *Clostridioides difficile*; diabetes; gut microbiome; sweeteners

Citation: Chen, A.; Gibney, P.A. Dietary Trehalose as a Bioactive Nutrient. *Nutrients* **2023**, *15*, 1393. https://doi.org/10.3390/nu15061393

Academic Editors: Elad Tako and Adam Matkowski

Received: 20 February 2023
Revised: 6 March 2023
Accepted: 11 March 2023
Published: 14 March 2023

1. Introduction

Trehalose, a naturally occurring, non-reducing disaccharide, comprises two glucopyranosyl units connected via an α, α−1, 1-glycosidic bond, as shown in Figure 1 [1]. It is widely distributed in many organisms, including bacteria, fungi, insects, plants, and some invertebrate animals [2]. Trehalose contains a non-reducing end hydroxyl group, rendering it a stable molecule not subject to glycation reactions [3]. The structure of trehalose also makes it a molecule highly resistant to acid hydrolysis, as well as a dehydration or cryo-protectant in a variety of microorganisms, plants, and animals [4–6].

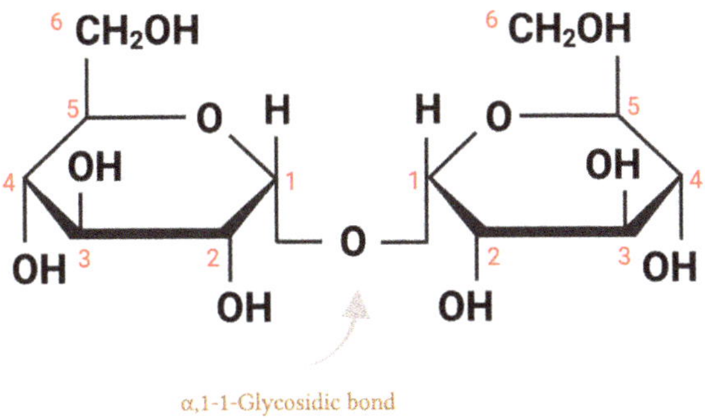

Figure 1. Structural illustration of trehalose, a disaccharide formed by combining two glucose molecules with an α-1,1 glycosidic linkage (α-D-glucopyranosyl-1,1-α-D-glucopyranoside).

In recent decades, trehalose has sparked growing interest in a variety of industrial applications [3,7]. This can be attributed to the growing body of research demonstrating the bioprotective properties of trehalose, as well as its absence of toxicity [8]. The introduction of an efficient biocatalytic technology for its synthesis in 1995, enzymatic conversion of starches into trehalose, has also contributed to its remarkable expansion in industrial use over the last two decades [9]. Its distinctive properties that bring about stabilization and sweetness have allowed its various applications as a packaging material, desiccant, drug excipient, and preservative in the food industry [10]. Furthermore, the ability of trehalose to protect a broad array of biological materials, such as DNA, proteins, cell lines, and tissues, has opened up possibilities in the pharmaceutical and biotechnology industries [11]. Trehalose is also gaining attention as a potential human therapeutic, as it has been proposed to mitigate the disease burden in models of neurodegenerative and cardiometabolic diseases [12]. It has also been reported to have various biological effects, such as suppression of bone resorption and inflammation, induction of autophagy, and alleviation of Huntington's disease [13,14]. However, the precise mechanism or mechanisms underlying these observations remain unclear, and may be complicated by the enzymatic conversion of trehalose into glucose in the human digestive system by either human-encoded or microbially-encoded trehalase enzymes. This leads to open questions regarding how much free trehalose is accessible to diseased tissues, if and how trehalose enters human cells to elicit a physiological response such as induction of autophagy, and whether these or other trehalose-related effects on human physiology are indirect results of the microbial metabolism of trehalose by human microflora [14]. Further research into the role of trehalose and its effect on these disease states is warranted, especially with the appropriate control experiments and validation in a clinical setting.

Taken together, as a dietary sugar, trehalose has been increasingly consumed over recent years, especially since technological progress has enabled a reduction in its cost. As a bioactive nutrient, studies have proposed that trehalose is a potential tool for regulating blood sugar levels and alleviating diabetes symptoms. Additionally, recent work has focused on the effects of trehalose in the human gut microbiome modulation. The goal of this mini review is to discuss trehalose as a bioactive nutrient, including how trehalose affects the human metabolism and gut microflora populations (Figure 2).

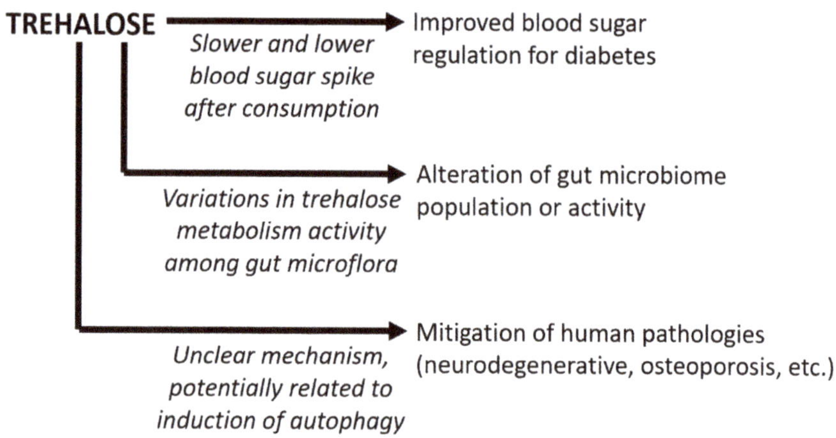

Figure 2. Bioactive roles of trehalose as a dietary nutrient. The three main bioactive roles of trehalose discussed in this manuscript are indicated, along with arrows to indicate the proposed mechanisms of action (text in italics).

2. Trehalose Chemical Structure and Biochemical Properties

Trehalose is composed of two glucose molecules connected by an α, $\alpha-1$,1-glycosidic bond (Figure 1). It is believed that α-,α-trehalose is the only form of trehalose that can be isolated from, and biosynthesized in, nature, due to its low conformation energy compared to the other two isomers, α, β-trehalose, and β, β-trehalose [2]. No living organism has been found to contain either α, β-trehalose or β, β-trehalose; however, both can be created through the Koenigs–Knorr reaction [15].

As one of the least chemically reactive sugars in nature, trehalose is non-reducing because the two glucose molecules are linked together by an O-glycosidic bond at the anomeric carbons of the glucopyranose rings [16]. Trehalose does not possess any free aldehyde groups that could take part in reduction processes, such as the Maillard reaction, since both of its anomeric carbons are involved in the disaccharide bond. Its glycosidic bond energy is less than 1 kcal/mol, making it exceptionally stable [17]. In comparison, sucrose, another commonly found nonreducing disaccharide, has a high glycosidic bond energy of 27 kcal/mol [18]. Due to its unique structure, trehalose only breaks down into two reducing monosaccharides under extreme hydrolysis conditions, or in the presence of enzymes, while sucrose quickly decomposes when exposed to reactive amino groups.

Trehalose stands out from other disaccharides due to its non-reducing nature and low glycosidic bond energy, as well as its exceptionally high glass transition temperature [7]. At its glass transition temperature (T_g), 110–120 °C, trehalose shifts from a solid to a viscous or rubbery state, which can be used as a reference point to explain its stabilizing properties across the solid, glass, and liquid phases [19]. The widespread use of both sucrose and trehalose as a stabilizing agent for a variety of storage purposes is well-known, with disaccharides in general offering stabilizing effects for both cryo-storage and anhydrobiotic preservation (e.g., freeze-drying) [20]. However, trehalose proves to be more advantageous than sucrose in a variety of biological preservation applications, such as protection from heat shock and lyophilization [21]. The high glass transition temperature of trehalose makes it an attractive option for the high-sucrose foods industry to consider for amorphous sucrose–trehalose mixtures [22]. If trehalose is able to effectively reduce the rate of sucrose crystallization in typical storage environments that surpass T_g, then product shelf life can be prolonged.

3. Trehalose as an Increasingly Consumed Dietary Sugar

Before 1995 in Japan, and before the turn of the century in the U.S. and Europe, the vast majority of consumed trehalose was derived from natural sources [23]. It is found in mushrooms, yeast, honey, beans, seaweeds, and shellfish, though the dietary contribution of trehalose from the consumption of all these foods combined is low, compared to added sugars in processed foods [2,24]. The total trehalose consumption in the U.S. was estimated to be at approximately 21 g (per person per year), with natural sources contributing less than 0.3 g [23]. In the mid-1990s, a revolutionary enzymatic technology, based on the liberation of trehalose from starch, was developed in Japan for the industrial scale manufacture of trehalose, resulting in a steep decline in its unit price from USD 700/kg to USD 5–6/kg [7]. Trehalose has been incorporated into various food items, including baked goods, rice and pasta, breakfast cereals, dried fruits, processed vegetables, dairy products, chewing gum, seafood, and ice cream [25]. Confectionery items make up a substantial portion of the total use of trehalose, which has been incorporated into more than 8000 different types of food [3]. The growing use of trehalose in confectionary products can be attributed to its less intense sweetness and longer persistence of sweetness [26]. Due to its growing applications, the average daily intake of trehalose for consumers of all ages in 2019, which was reported by a Generally Recognized As Safe (GRAS) notice, had increased to 42.37 g (per person per day), doubling the amount from two decades ago [27].

While human cells do not encode enzymes to synthesize trehalose, trehalose ingested from food can be hydrolyzed into two glucose molecules in the small intestine by the trehalase enzyme, an enzyme that specifically acts on trehalose [28]. Trehalase activity has also been discovered in kidney, liver, and peripheral lymphocytes, with the proximal and mid-small intestine having the highest concentration, and the distal ileum having the lowest [25]. The glucose molecules released from trehalose can then be actively absorbed and metabolized by intestinal mucosal cells through the glucose transporter 1 (*SGLT1*) [29].

In the human digestive tract, apart from host-produced trehalase, trehalose is also metabolized by microbial-produced trehalases [30]. Microbes can use trehalase enzymes to break down trehalose into glucose, which can then be used in glycolysis. Many intestinal bacteria produce trehalase enzymes, including *E. coli*, *C. difficile*, *Blautia* spp., and *Bacillus* spp. [31,32]. *Bacillus* species use trehalose as a carbon and energy source during exponential growth, while *E. coli* utilizes trehalose as a carbon source through an osmolarity-dependent process [33,34].

Although human consumption and use of trehalose have risen significantly, there have been no reported toxic side effects, aside from rare cases of malabsorption due to trehalose deficiency [25]. Trehalose deficiency is a metabolic condition in which the body lacks functional trehalase enzymes and is not able to convert trehalose into glucose [35]. Affected individuals suffer from abdominal discomfort, vomiting, and diarrhea after eating foods containing trehalose. It is not a common occurrence, yet, at least 8% of Greenland's population is known to have trehalase deficiency as a result of autosomal dominant inheritance [36]. However, since it is a rare disease with no official means of estimating incidence or prevalence, it is probably underreported.

4. Trehalose as a Low Glycemic-Index Sugar for Diabetes Mitigation

The over-consumption of added sugars has been associated with a heightened probability of many chronic diseases, including, diabetes, obesity, cardiovascular and liver diseases, cancer, and cognitive impairments [37]. Among these, diabetes is one of the most prevalent metabolic disorders, resulting in serious health complications [38]. Various treatments and medications have been formulated to modulate glucose homeostasis in diabetic individuals [39].

One of the proposed biological roles of trehalose is that it can potentially stabilize blood sugar levels through slower blood glucose release and a milder insulin response compared to other monosaccharides and disaccharides [40,41]. Oku et al. reported that in comparison to glucose, trehalose does not result in a rapid rise in blood sugar levels or a reduction in insulin secretion in female college students [42]. Yoshizane et al. found that 25 g trehalose (in 100 mL water) stimulated insulin and active incretin (a gut hormone released into the blood after eating that promotes insulin production) secretion less than glucose [41]. In contrast, when sucrose (composed of glucose and fructose) is consumed, it triggers a prompt increase in both blood glucose and insulin levels [43]. The findings of these studies indicate that, when compared to glucose ingestion, trehalose intake resulted in a diminished peak in blood glucose levels, not only suppressing the initial spike, but also producing a considerably reduced cumulative effect [44].

However, there are open questions remaining about the role of trehalose and its effects on blood glucose concentration. It is noteworthy that trehalose and glucose are metabolized differently in the human body, as the glucose released from trehalose is taken up by different parts of the gastrointestinal tract than that of trehalose itself. It was suggested that trehalose may impede glucose transport via the solute carrier 2A (SLC2A) transporter, which may explain the decrease in blood glucose levels [45]. However, once the brush border trehalase transforms trehalose into glucose, these glucose molecules can then be taken up by the uninhibited SLC2A transporters in the small intestine. Thus, it is possible that not every trehalose molecule gets broken down into glucose and taken up, and an unknown quantity of undecomposed trehalose can be found throughout the bloodstream after ingestion. These intact trehalose molecules may have physiological activities independent of providing two glucose molecules upon hydrolysis.

The molecular effects associated with trehalose modulating glucose homeostasis have been discussed and summarized in detail in a recent review [40]. The evidence points to trehalose consumption having a beneficial effect on hyperglycemic conditions by alleviating underlying pathological processes, through (1) enhancing insulin sensitivity by modulating the glucose signaling pathway, (2) reducing insulin secretion by reducing fat cell buildup, which is triggered by consuming carbohydrates and requires insulin to be processed, (3) normalizing glucose metabolism by modulating postprandial glucose levels, (4) modulating lipid metabolism by regulating postprandial insulin secretion, (5) enhancing pancreatic islet function by improving pancreatic beta cell function and preventing apoptotic processes associated with beta cell malfunction, (6) attenuating oxidative stress, which reduces free radical overload and improves insulin resistance, and (7) inhibiting inflammatory responses by ameliorating inflammatory mediators [40].

Taken together, these studies point to trehalose as a promising non-pharmaceutical agent for the control of glycemia in diabetic patients. More clinical trials, however, are needed to better characterize the effect of trehalose in mitigating diabetes symptoms. Further work to characterize the associated mechanism can be used to corroborate clinical effects, and potentially pave the way for the discovery of novel therapeutic avenues for other metabolic diseases.

5. Trehalose Effects on the Gut Microbiome

5.1. Alterations in Gut Microbiome in Response to Trehalose and Other Sugars

Changes in diet can have a variety of effects on the gut microbiome. Hundreds of species coexist in the human gastrointestinal tract, each thriving within a niche habitat determined by environmental factors, including nutrient availability, predators, and competitors [46]. It has been demonstrated that microbial composition can shift rapidly following a diet change [47]. A change in the amount of available nutrients in the environment can cause evolutionary pressure to be exerted on the whole gut microbial population [48]. New or existing mutations that improve nutrient sensing, transport, or metabolism and provide an advantage to their host are more favored. As a result, the microbiota is constantly adapting to changes in the availability of various sugars and other dietary components.

Trehalose and other dietary sugars or sweeteners impact the microbial populations that comprise the gut microbiome [49]. In the U.S., roughly half of all calories consumed come from carbohydrates, of which 13% derive from added sugars [50]. Simple sugars (e.g., glucose, fructose, high-fructose corn syrup, sucrose, and trehalose), sugar alcohols (e.g., sorbitol, erythritol, xylitol, and mannitol), and synthetic sugars (e.g., stevia, aspartame, saccharin, sucralose, and acesulfame potassium) are examples of commonly added sugars and sweeteners in modern Western diets [51]. In the past decades, synthetic sweeteners have been formulated and various natural sugars have been incorporated into food products [52]. The introduction of these sugars and sugar substitutes has prompted evaluations of how they affect the human body through changes in the composition and function of the gut microbiota. Indeed, multiple studies have demonstrated that microorganisms can swiftly alter their metabolism when exposed to a novel sugar [53,54]. A recent review highlighted three major strategies the gut microbiome can utilize to adjust to an increased consumption of sugar or sweeteners, including transcriptional changes, population compositional changes, and genetic changes [49].

First, microorganisms can adjust their transcription and metabolic activity in various ways, to adapt to different conditions in their environment, using regulatory signaling pathways, such as catabolite repression [49]. One example is that bacteria often modify the transcription and translation levels of key metabolic and transport proteins as the nutrient pool changes. The catabolite repression of glucose and fructose in *Bacteroides thetaiotaomicron* leads to the suppression of polysaccharide utilization genes, thus hindering its ability to colonize mouse gastrointestinal tracts [53]. Second, shifts in the microbiome composition can occur, allowing the microbes that are best suited to a given environment to become more prevalent. Research on trehalose as a prebiotic found that it significantly stimulated the growth of bacteriocin-producing lactic acid bacteria, particularly *Lactococcus lactis* spp. and *Lactococcus* sp. [55]. An investigation of the influence of a diet high in glucose or fructose on gut microbiota and intestinal permeability showed that these two highly consumed dietary sugars induce changes in the mouse microbiota, leading to a decrease in *Bacteroides* diversity and abundance, with a concurrent increase in *Proteobacteria* abundance [56]. Consumption of the artificial sweeteners sucralose and saccharin may also cause an imbalance in the microbiota composition [57]. The blood glucose regulation of a group of non-diabetic human participants was shown to be inferior when consuming saccharin, as indicated by their raised blood glucose levels [58]. Third, microorganism populations within the gut are able to modify their genetic makeup in response to changes in their environment, enabling them to make the most of new microhabitats, and allowing them to take advantage of new food sources. For example, the addition of galactitol, a sugar alcohol derived from galactose, caused the emergence and co-existence of a bacterial strain able to metabolize the galactitol alongside the original bacterial population [59]. It was found that these galactitol-positive strains were capable of using a rarely tapped galactitol ecosystem while coexisting with the galactitol-negative strains, which competed for alternative carbon sources with the collective microbiome.

There are many unexplored routes which have yet to be investigated regarding how microbes interact with sugars and sweeteners, and their effect on the host. Sugar alcohols, for example, were found to promote multiple beneficial microbes, although it is unclear whether this occurs through simple growth stimulation or more intricate connections [60]. It is critical to recognize that any change in the sugar and sweetener profile we consume redefines the nutrient environment available to our gut microflora. Further research into how trehalose, among other types of sugars and sweeteners, is able to affect the gut microflora population, activity, and effect on human physiology will continue to shed light on the interactions between diet, nutrition, and health.

5.2. Trehalose and Clostridioides difficile

A number of recent studies on the role of trehalose in microbiome modification has been focused on the pathogen *C. difficile*. It was first reported in 2018 that hypervirulent strains of *C. difficile*, a spore-forming Gram-positive bacterium, have acquired novel mechanisms for trehalose utilization, leading to an enhanced virulence and a more severe intestinal infection [61]. The two variant *C. difficile* strains, ribotypes RT027 and RT078, exhibited a noticeably higher efficiency in trehalose uptake, allowing for better growth in low trehalose concentrations. It was speculated that the rise in trehalose consumption, which was prompted by the approval of trehalose as Generally Recognized as Safe (GRAS) in 2000, may have contributed to the proliferation of these strains in humans [61].

All of the examined strains of *C. difficile* harbor trehalase TreA, a phosphotrehalase enzyme that is regulated by an upstream transcriptional regulator TreR [61]. TreA is important for *C. difficile* to grow on trehalose-containing media, as it catalyzes the transformation of trehalose-6-phosphate into glucose and glucose-6-phosphate. In RT027 isolates, sequence analysis implicated a single nucleotide polymorphism (SNP) within the *treR* gene causing de-repression of *treA*, allowing RT027 to activate the *treA* gene at 500-fold lower trehalose concentrations, explaining the increased trehalose utilization efficiency [23]. In the RT078 lineage, enhanced trehalose utilization is achieved through an alternative mechanism. In all RT078 strains that have been sequenced thus far, a four-gene addition was discovered, which is believed to encode a second copy of a phosphotrehalase (TreA2, with 55% amino acid identity to TreA), a potential trehalose-specific transport protein (PtsT), a putative glycan debranching enzyme (TreX) from the trehalase family, and a second copy of the TreR repressor (TreR2, with 44% amino acid identity to TreR) [62]. Mutational analysis of the transporter gene *ptsT* in RT078 was found to confer improved growth when trehalose levels were low [61]. Thus, increased trehalose consumption, combined with enhanced utilization of trehalose in these *C. difficile* epidemic strains, was suggested as potentially playing a causative role in the development of these hypervirulent ribotypes.

The potential of hypervirulent *C. difficile* strains to utilize dietary trehalose as a carbon source has led to the development of trehalase-resistant trehalose analogues as a more advanced metabolic treatment, which can retain the sweetener and potentially metabolic advantages of trehalose, and reduce the potential risks associated with its enzymatic breakdown. Two trehalose analogues, lactotrehalose and 5-thiotrehalose, have been discovered to resist enzymatic degradation, thus not able to serve as a carbon source for *C. difficile* [63].

Lactotrehalose inhibition was ribotype-dependent, as increasing lactotrehalose concentrations decreased growth of RT027, whereas RT078 was unaffected. The trehalose analog 5-thiotrehalose, on the other hand, can inhibit the growth of both RT027 and RT078 strains while remaining non-toxic to mammalian cells [63]. Thus, research into trehalose analogues that are resistant to degradation could be beneficial for their use as trehalase inhibitors, and as alternatives, or combined with trehalose in situations where controlling microbial growth and disease are important.

However, despite the suggestion that increasing trehalose consumption is related to increasing incidence of *C. difficile* infections, multiple subsequent studies have indicated this correlation is unlikely to be causative. In one study, a human gut model, which has been previously validated, was used to assess how trehalose influences the composition of human microbiota and the emergence of *C. difficile* infection (CDI) [64]. The researchers discovered that the human microbiome adapted swiftly to process all of the available trehalose, and when antibiotics altered the microbiota, trehalose supplementation did not cause CDI in comparison to glucose or saline supplementation [30]. A different study examining the impact of trehalose and lactotrehalose on the microbiome and severity of CDI reported that oral administration of trehalose reduced *C. difficile* abundance [65]. Eyre et al. investigated the prevalence of mutations influencing trehalose metabolism among CDI patients and the correlation of these mutations to the severity of the illness by analyzing the genetic diversity of *C. difficile* strains which had been previously sequenced [66]. The population was found to have a high prevalence of trehalose utilization variants, and the administration of trehalose did not generate greater concentrations of *C. difficile* or its spores than those that were seen with glucose or saline supplementation. On the contrary, trehalose administration lowered toxin detection to undetectable levels, though the cause was not clear [66]. Similarly, Saund et al. reported no correlation between trehalose utilization variants and severe cases of CDI among hospitalized patients, as demonstrated by the clinical data from 1144 CDI patient samples [62]. Both studies suggested that the presence of trehalose variants was far more extensive than previously anticipated, reflecting that the ability to utilize low concentrations of trehalose was acquired before the recent surge in trehalose usage. However, one limitation of these studies is that the majority of the examined isolates originated from the U.S. or Europe, despite the fact that trehalose consumption is higher in Asia. While further study could be useful to evaluate these geographic differences, a meta-analysis of 51 studies reported that CDI occurrence in Asia is comparable to that in Europe and North America [67]. Taken together, available evidence indicates trehalose consumption is not linked to increased *C. difficile* infections or the prevalence of strain variants more able to consume trehalose among *C. difficile* infections. However, this example highlights the utility of performing well-controlled experiments to evaluate the consequences of trehalose and other sugars or sweeteners on the gut microflora. Such studies have potential for the development of beneficial prebiotic supplements to promote healthy gut microflora.

6. Summary and Future Prospects

Trehalose is a naturally occurring, non-reducing sugar found in abundance in nature and has been used in many everyday products. For the past several decades, research has suggested that it may be a viable therapeutic agent to activate autophagy under several conditions in which autophagy is vital, such as neurodegenerative diseases, cancer, aging, cardiometabolic disorders, and infectious diseases [14,68–71]. It is speculated that the protective properties of trehalose are from autophagy induction and clearance of protein aggregates, however, these observations require further confirmation and validation in a clinical setting.

Trehalose has also been proposed as a non-pharmacological agent to regulate glucose levels in diabetic individuals. Studies have indicated that it could be effective in modulating glucose metabolism and maintaining glucose homeostasis by ameliorating inflammation and apoptosis, reducing postprandial insulin secretion, improving beta cell function, and normalizing the lipid profile [40]. This evidence suggests that trehalose could be a viable alternative to drug therapy for controlling blood sugar in diabetic individuals, however, further clinical trials are also required to validate this hypothesis.

In addition to potential therapeutic benefits, trehalose incorporated into diets can alter the composition of the human intestinal microbiota. Despite a lack of comprehensive research on the effects of trehalose on microbial populations, the introduction of novel sugars and sweeteners can drastically remodel the gut habitat, which in turn affects microbial metabolism and metabolite secretion [49]. One illustration is the recently proposed association between a *C. difficile* outbreak and increased trehalose dietary consumption, though more recent research presented contradictory results and indicated no direct correlation between trehalose consumption and *C. difficile* infections [30,72]. However, this raised the general concern that pathogens may use trehalose or other rare sugars to gain an advantage in the gut microflora, which has led to developing trehalase-resistant trehalose analogues. These molecules may help retain or enhance the metabolic benefits of trehalose, and mitigate the potential risks of microbial drug metabolism.

Taken together, the studies examining trehalose as a bioactive nutrient have yielded interesting insights into its potential to treat various human diseases, as well as its importance in the gut microbiota. It is likely that trehalose has multiple, distinct effects that contribute to its ability to reduce the severity of metabolic diseases. Further work remains to be performed to elucidate the complete scope of the biological impacts associated with trehalose.

Author Contributions: Conceptualization and writing: A.C. and P.A.G. All authors have read and agreed to the published version of the manuscript.

Funding: This research was supported by startup funds provided by E&J Gallo Winery and Cornell University.

Conflicts of Interest: The authors declare no conflict of interest.

References

1. Liu, Q.; Schmidt, R.K.; Teo, B.; Karplus, P.A.; Brady, J.W. Molecular dynamics studies of the hydration of α,α-trehalose. *J. Am. Chem. Soc.* **1997**, *119*, 7851–7862. [CrossRef]
2. Elbein, A.D.; Pan, Y.T.; Pastuszak, I.; Carroll, D. New insights on trehalose: A multifunctional molecule. *Glycobiology* **2003**, *13*, 17R–27R. [CrossRef] [PubMed]
3. Ohtake, S.; Wang, Y.J. Trehalose: Current Use and Future Applications. *J. Pharm. Sci.* **2011**, *100*, 2020–2053. [CrossRef] [PubMed]
4. Kahraman, H. The Importance of Trehalose Sugar. *Biomed. J. Sci. Tech. Res.* **2019**, *21*, 15917–15918. [CrossRef]
5. Tapia, H.; Young, L.; Fox, D.; Bertozzi, C.R.; Koshland, D. Increasing intracellular trehalose is sufficient to confer desiccation tolerance to *Saccharomyces cerevisiae*. *Proc. Natl. Acad. Sci. USA* **2015**, *112*, 6122–6127. [CrossRef]
6. Chen, A.; Gibney, P.A. Intracellular trehalose accumulation via the Agt1 transporter promotes freeze–thaw tolerance in *Saccharomyces cerevisiae*. *J. Appl. Microbiol.* **2022**, *133*, 2390–2402. [CrossRef]
7. Chen, A.; Tapia, H.; Goddard, J.M.; Gibney, P.A. Trehalose and its applications in the food industry. *Compr. Rev. Food Sci. Food Saf.* **2022**, *21*, 5004–5037. [CrossRef]
8. Laihia, J.; Kaarniranta, K. Trehalose for Ocular Surface Health. *Biomolecules* **2020**, *10*, 809. [CrossRef]
9. Cai, X.; Seitl, I.; Mu, W.; Zhang, T.; Stressler, T.; Fischer, L.; Jiang, B. Biotechnical production of trehalose through the trehalose synthase pathway: Current status and future prospects. *Appl. Microbiol. Biotechnol.* **2018**, *102*, 2965–2976. [CrossRef]
10. Bosch, S.; de Beaurepaire, L.; Allard, M.; Mosser, M.; Heichette, C.; Chrétien, D.; Jegou, D.; Bach, J.-M. Trehalose prevents aggregation of exosomes and cryodamage. *Sci. Rep.* **2016**, *6*, 36162. [CrossRef]
11. O'Neill, M.K.; Piligian, B.F.; Olson, C.D.; Woodruff, P.J.; Swarts, B.M. Tailoring trehalose for biomedical and biotechnological applications. *Pure Appl. Chem.* **2017**, *89*, 1223–1249. [CrossRef]
12. Sahebkar, A.; Khalifeh, M.; Barreto, G.E. Therapeutic potential of trehalose in neurodegenerative diseases: The knowns and unknowns. *Neural Regen. Res.* **2021**, *16*, 2026–2027. [CrossRef]
13. Liu, Z.; Chen, D.; Chen, X.; Bian, F.; Qin, W.; Gao, N.; Xiao, Y.; Li, J.; Pflugfelder, S.C.; Li, D.-Q. Trehalose Induces Autophagy Against Inflammation by Activating TFEB Signaling Pathway in Human Corneal Epithelial Cells Exposed to Hyperosmotic Stress. *Investig. Opthalmol. Vis. Sci.* **2020**, *61*, 26. [CrossRef]
14. Lee, H.-J.; Yoon, Y.-S.; Lee, S.-J. Mechanism of neuroprotection by trehalose: Controversy surrounding autophagy induction. *Cell Death Dis.* **2018**, *9*, 1–12. [CrossRef]

15. Helferich, B.; Weis, K. Zur Synthese von Glucosiden und von nicht-reduzierenden Disacchariden. *Eur. J. Inorg. Chem.* **1956**, *89*, 314–321. [CrossRef]
16. Stick, R.; Spencer, W. Disaccharides, oligosaccharides and polysaccharides. In *Carbohydrates: The Essential Molecules of Life*; Elsevier Science: Amsterdam, The Netherlands, 2009; pp. 321–341. [CrossRef]
17. Iturriaga, G.; Suárez, R.; Nova-Franco, B. Trehalose Metabolism: From Osmoprotection to Signaling. *Int. J. Mol. Sci.* **2009**, *10*, 3793–3810. [CrossRef] [PubMed]
18. Paiva, C.L.; Panek, A.D. Biotechnological Applications of the Disaccharide Trehalose. *Biotechnol. Annu. Rev.* **1996**, *2*, 293–314. [PubMed]
19. Patist, A.; Zoerb, H. Preservation mechanisms of trehalose in food and biosystems. *Colloids Surf. B Biointerfaces* **2005**, *40*, 107–113. [CrossRef] [PubMed]
20. Crowe, J.H.; Carpenter, J.F.; Crowe, L.M.; Hoekstra, F.A.; Hand, S.C.; Menze, M.A.; Toner, M.; Boswell, L.; Moore, D.; Berne, B.J.; et al. The role of vitrification in anhydrobiosis. *Annu. Rev. Physiol.* **1998**, *60*, 73–103. [CrossRef]
21. Olsson, C.; Swenson, J. Structural Comparison between Sucrose and Trehalose in Aqueous Solution. *J. Phys. Chem. B* **2020**, *124*, 3074–3082. [CrossRef]
22. Roe, K.D.; Labuza, T.P. Glass Transition and Crystallization of Amorphous Trehalose-sucrose Mixtures. *Int. J. Food Prop.* **2005**, *8*, 559–574. [CrossRef]
23. Collins, J.; Danhof, H.; Britton, R.A. The role of trehalose in the global spread of epidemic *Clostridium difficile*. *Gut Microbes* **2019**, *10*, 204–209. [CrossRef] [PubMed]
24. Colaço, C.; Sen, S.; Thangavelu, M.; Pinder, S.; Roser, B. Extraordinary Stability of Enzymes Dried in Trehalose: Simplified Molecular Biology. *Nat. Biotechnol.* **1992**, *10*, 1007–1011. [CrossRef]
25. Richards, A.; Krakowka, S.; Dexter, L.; Schmid, H.; Wolterbeek, A.; Waalkens-Berendsen, D.; Shigoyuki, A.; Kurimoto, M. Trehalose: A review of properties, history of use and human tolerance, and results of multiple safety studies. *Food Chem. Toxicol.* **2002**, *40*, 871–898. [CrossRef]
26. Portmann, M.-O.; Birch, G. Sweet taste and solution properties of α,α-trehalose. *J. Sci. Food Agric.* **1995**, *69*, 275–281. [CrossRef]
27. GRAS Notice No GRN 912. 2019. Available online: https://www.fda.gov/food/generally-recognized-safe-gras/gras-notice-inventory (accessed on 19 February 2023).
28. Elbein, A.D. The Metabolism of α,α-Trehalose. *Adv. Carbohydr. Chem. Biochem.* **1974**, *30*, 227–256. [CrossRef] [PubMed]
29. Koepsell, H. Glucose transporters in the small intestine in health and disease. *Pflug. Arch. Eur. J. Physiol.* **2020**, *472*, 1207–1248. [CrossRef]
30. Buckley, A.M.; Moura, I.B.; Arai, N.; Spittal, W.; Clark, E.; Nishida, Y.; Harris, H.C.; Bentley, K.; Davis, G.; Wang, D.; et al. Trehalose-Induced Remodelling of the Human Microbiota Affects Clostridioides difficile Infection Outcome in an In Vitro Colonic Model: A Pilot Study. *Front. Cell. Infect. Microbiol.* **2021**, *11*, 670935. [CrossRef]
31. The UniProt Consortium. UniProt: The universal protein knowledgebase in 2021. *Nucleic Acids Res.* **2021**, *49*, D480–D489. [CrossRef]
32. Argüelles, J.C. Physiological roles of trehalose in bacteria and yeasts: A comparative analysis. *Arch. Microbiol.* **2000**, *174*, 217–224.
33. Horlacher, R.; Boos, W. Characterization of TreR, the major regulator of the *Escherichia coli* trehalose system. *J. Biol. Chem.* **1997**, *272*, 13026–13032. [CrossRef] [PubMed]
34. Boos, W.; Ehmann, U.; Forkl, H.; Klein, W.; Rimmele, M.; Postma, P. Trehalose transport and metabolism in *Escherichia coli*. *J. Bacteriol.* **1990**, *172*, 3450–3461. [CrossRef] [PubMed]
35. Arola, H.; Koivula, T.; Karvonen, A.L.; Jokela, H.; Ahola, T.; Isokoski, M. Low trehalase activity is associated with abdominal symptoms caused by edible mushrooms. *Scand. J. Gastroenterol.* **1999**, *34*, 898–903. [PubMed]
36. Gudmand-Høsyer, E.; Fenger, H.J.; Skovbjerg, H.; Kern-Hansen, P.; Madsen, P.R. Trehalase deficiency in Greenland. *Scand. J. Gastroenterol.* **1988**, *23*, 775–778. [CrossRef] [PubMed]
37. Rippe, J.M.; Angelopoulos, T.J. Relationship between Added Sugars Consumption and Chronic Disease Risk Factors: Current Understanding. *Nutrients* **2016**, *8*, 697. [CrossRef]
38. Deshpande, A.D.; Harris-Hayes, M.; Schootman, M. Epidemiology of Diabetes and Diabetes-Related Complications. *Phys. Ther.* **2008**, *88*, 1254–1264. [CrossRef]
39. Kalra, S.; Choudhary, N.; Unnikrishnan, A.G.; Ajish, T. Preventive pharmacotherapy in type 2 diabetes mellitus. *Indian J. Endocrinol. Metab.* **2012**, *16*, 33–43. [CrossRef]
40. Yaribeygi, H.; Yaribeygi, A.; Sathyapalan, T.; Sahebkar, A. Molecular mechanisms of trehalose in modulating glucose homeostasis in diabetes. *Diabetes Metab. Syndr. Clin. Res. Rev.* **2019**, *13*, 2214–2218. [CrossRef]
41. Yoshizane, C.; Mizote, A.; Yamada, M.; Arai, N.; Arai, S.; Maruta, K.; Mitsuzumi, H.; Ariyasu, T.; Ushio, S.; Fukuda, S. Glycemic, insulinemic and incretin responses after oral trehalose ingestion in healthy subjects. *Nutr. J.* **2017**, *16*, 9. [CrossRef]
42. Oku, T.; Nakamura, S. Estimation of intestinal trehalase activity from a laxative threshold of trehalose and lactulose on healthy female subjects. *Eur. J. Clin. Nutr.* **2000**, *54*, 783–788. [CrossRef] [PubMed]
43. Keyhani-Nejad, F.; Kemper, M.; Schueler, R.; Pivovarova, O.; Rudovich, N.; Pfeiffer, A.F. Effects of Palatinose and Sucrose Intake on Glucose Metabolism and Incretin Secretion in Subjects with Type 2 Diabetes. *Diabetes Care* **2015**, *39*, e38–e39. [CrossRef]

44. Mizote, A.; Yamada, M.; Yoshizane, C.; Arai, N.; Maruta, K.; Arai, S.; Endo, S.; Ogawa, R.; Mitsuzumi, H.; Ariyasu, T.; et al. Daily Intake of Trehalose Is Effective in the Prevention of Lifestyle-Related Diseases in Individuals with Risk Factors for Metabolic Syndrome. *J. Nutr. Sci. Vitaminol.* **2016**, *62*, 380–387. [CrossRef] [PubMed]
45. DeBosch, B.J.; Heitmeier, M.R.; Mayer, A.L.; Higgins, C.B.; Crowley, J.R.; Kraft, T.E.; Chi, M.; Newberry, E.P.; Chen, Z.; Finck, B.N.; et al. Trehalose inhibits solute carrier 2A (SLC2A) proteins to induce autophagy and prevent hepatic steatosis. *Sci. Signal.* **2016**, *9*, ra21. [CrossRef]
46. Thursby, E.; Juge, N. Introduction to the human gut microbiota. *Biochem. J.* **2017**, *474*, 1823–1836. [CrossRef]
47. Leeming, E.R.; Johnson, A.J.; Spector, T.D.; Le Roy, C.I. Effect of Diet on the Gut Microbiota: Rethinking Intervention Duration. *Nutrients* **2019**, *11*, 2862. [CrossRef] [PubMed]
48. Scanlan, P.D. Microbial evolution and ecological opportunity in the gut environment. *Proc. R. Soc. B Boil. Sci.* **2019**, *286*, 20191964. [CrossRef]
49. Di Rienzi, S.C.; Britton, R.A. Adaptation of the Gut Microbiota to Modern Dietary Sugars and Sweeteners. *Adv. Nutr. Int. Rev. J.* **2019**, *11*, 616–629. [CrossRef]
50. Ervin, R.B.; Ogden, C.L. *Consumption of Added Sugars among U.S. Adults, 2005–2010*; NCHS Data Brief: Hyattsville, MD, USA, 2013; pp. 1–8.
51. Chattopadhyay, S.; Raychaudhuri, U.; Chakraborty, R. Artificial sweeteners—A review. *J. Food Sci. Technol.* **2014**, *51*, 611–621. [CrossRef]
52. Saraiva, A.; Carrascosa, C.; Raheem, D.; Ramos, F.; Raposo, A. Natural sweeteners: The relevance of food naturalness for consumers, food security aspects, sustainability and health impacts. *Int. J. Environ. Res. Public Health* **2020**, *17*, 6285. [CrossRef]
53. Townsend, G.E.; Han, W.; Schwalm, N.D., III; Raghavan, V.; Barry, N.A.; Goodman, A.L.; Groisman, E.A. Dietary sugar silences a colonization factor in a mammalian gut symbiont. *Proc. Natl. Acad. Sci. USA* **2019**, *116*, 233–238. [CrossRef] [PubMed]
54. Silva, J.C.P.; Mota, M.; Martins, F.O.; Nogueira, C.; Gonçalves, T.; Carneiro, T.; Pinto, J.; Duarte, D.; Barros, A.S.; Jones, J.G.; et al. Intestinal Microbial and Metabolic Profiling of Mice Fed with High-Glucose and High-Fructose Diets. *J. Proteome Res.* **2018**, *17*, 2880–2891. [CrossRef]
55. Chen, Y.-S.; Srionnual, S.; Onda, T.; Yanagida, F. Effects of prebiotic oligosaccharides and trehalose on growth and production of bacteriocins by lactic acid bacteria. *Lett. Appl. Microbiol.* **2007**, *45*, 190–193. [CrossRef]
56. Do, M.H.; Lee, E.; Oh, M.-J.; Kim, Y.; Park, H.-Y. High-Glucose or -Fructose Diet Cause Changes of the Gut Microbiota and Metabolic Disorders in Mice without Body Weight Change. *Nutrients* **2018**, *10*, 761. [CrossRef]
57. Ramne, S.; Brunkwall, L.; Ericson, U.; Gray, N.; Kuhnle, G.G.; Nilsson, P.M.; Orho-Melander, M.; Sonestedt, E. Gut microbiota composition in relation to intake of added sugar, sugar-sweetened beverages and artificially sweetened beverages in the Malmö Offspring Study. *Eur. J. Nutr.* **2021**, *60*, 2087–2097. [CrossRef] [PubMed]
58. Azeez, O.H.; Alkass, S.Y.; Persike, D.S. Long-Term Saccharin Consumption and Increased Risk of Obesity, Diabetes, Hepatic Dysfunction, and Renal Impairment in Rats. *Medicina* **2019**, *55*, 681. [CrossRef] [PubMed]
59. Sousa, A.; Ramiro, R.S.; Barroso-Batista, J.; Güleresi, D.; Lourenço, M.; Gordo, I. Recurrent Reverse Evolution Maintains Polymorphism after Strong Bottlenecks in Commensal Gut Bacteria. *Mol. Biol. Evol.* **2017**, *34*, 2879–2892. [CrossRef]
60. Peuranen, S.; Tiihonen, K.; Apajalahti, J.; Kettunen, A.; Saarinen, M.; Rautonen, N. Combination of polydextrose and lactitol affects microbial ecosystem and immune responses in rat gastrointestinal tract. *Br. J. Nutr.* **2004**, *91*, 905–914. [CrossRef] [PubMed]
61. Collins, J.; Robinson, C.; Danhof, H.; Knetsch, C.W.; Van Leeuwen, H.C.; Lawley, T.D.; Auchtung, J.; Britton, R.A. Dietary trehalose enhances virulence of epidemic Clostridium difficile. *Nature* **2018**, *553*, 291–294. [CrossRef]
62. Saund, K.; Rao, K.; Young, V.; Snitkin, E.S. Genetic Determinants of Trehalose Utilization Are Not Associated With Severe Clostridium difficile Infection Outcome. *Open Forum Infect. Dis.* **2020**, *7*, ofz548. [CrossRef]
63. Danielson, N.D.; Collins, J.; Stothard, A.I.; Dong, Q.Q.; Kalera, K.; Woodruff, P.J.; DeBosch, B.J.; Britton, R.A.; Swarts, B.M. Degradation-resistant trehalose analogues block utilization of trehalose by hypervirulent *Clostridioides difficile. Chem. Commun.* **2019**, *55*, 5009–5012. [CrossRef]
64. Freeman, J.; O'Neill, F.J.; Wilcox, M.H. Effects of cefotaxime and desacetylcefotaxime upon Clostridium difficile proliferation and toxin production in a triple-stage chemostat model of the human gut. *J. Antimicrob. Chemother.* **2003**, *52*, 96–102. [CrossRef]
65. Zhang, Y.; Shaikh, N.; Ferey, J.L.; Wankhade, U.D.; Chintapalli, S.V.; Higgins, C.B.; Crowley, J.R.; Heitmeier, M.R.; Stothard, A.I.; Mihi, B.; et al. Lactotrehalose, an Analog of Trehalose, Increases Energy Metabolism Without Promoting Clostridioides difficile Infection in Mice. *Gastroenterology* **2020**, *158*, 1402–1416.e2. [CrossRef]
66. Eyre, D.W.; Didelot, X.; Buckley, A.M.; Freeman, J.; Moura, I.B.; Crook, D.W.; Peto, T.E.; Walker, A.S.; Wilcox, M.H.; Dingle, K.E. Clostridium difficile trehalose metabolism variants are common and not associated with adverse patient outcomes when variably present in the same lineage. *Ebiomedicine* **2019**, *43*, 347–355. [CrossRef]
67. Borren, N.; Ghadermarzi, S.; Hutfless, S.; Ananthakrishnan, A.N. The emergence of Clostridium difficile infection in Asia: A systematic review and meta-analysis of incidence and impact. *PLoS ONE* **2017**, *12*, e0176797. [CrossRef]
68. Chaitanya, N.S.N.; Devi, A.; Sahu, S.; Alugoju, P. Molecular mechanisms of action of Trehalose in cancer: A comprehensive review. *Life Sci.* **2021**, *269*, 118968. [CrossRef]
69. Khalifeh, M.; Barreto, G.E.; Sahebkar, A. Trehalose as a promising therapeutic candidate for the treatment of Parkinson's disease. *Br. J. Pharmacol.* **2019**, *176*, 1173–1189. [CrossRef]

70. Portbury, S.D.; Hare, D.J.; Finkelstein, D.I.; Adlard, P.A. Trehalose improves traumatic brain injury-induced cognitive impairment. *PLoS ONE* **2017**, *12*, e0183683. [CrossRef] [PubMed]
71. Martinon, D.; Borges, V.F.; Gomez, A.C.; Shimada, K. Potential Fast COVID-19 Containment With Trehalose. *Front. Immunol.* **2020**, *11*, 1623. [CrossRef] [PubMed]
72. Buckley, A.M.; Moura, I.B.; Wilcox, M.H. Is there a causal relationship between trehalose consumption and Clostridioides difficile infection? *Curr. Opin. Gastroenterol.* **2021**, *37*, 9–14. [CrossRef] [PubMed]

nutrients

MDPI

Brief Report

Resistant Maltodextrin Consumption in a Double-Blind, Randomized, Crossover Clinical Trial Induces Specific Changes in Potentially Beneficial Gut Bacteria

Volker Mai [1,*], Alyssa M. Burns [2], Rebecca J. Solch [2], Jennifer C. Dennis-Wall [1], Maria Ukhanova [1] and Bobbi Langkamp-Henken [1]

1 Department of Epidemiology, Emerging Pathogens Institute, University of Florida, Gainesville, FL 32610, USA; jenniswall@gmail.com (J.C.D.-W.); mukhanova@peds.ufl.edu (M.U.); henken@ufl.edu (B.L.-H.)

2 Food Science and Human Nutrition Department, University of Florida, 572 Newell Drive, Gainesville, FL 32611, USA; aburns22@ufl.edu (A.M.B.); rsolch@tulane.edu (R.J.S.)

* Correspondence: vmai@ufl.edu; Tel.: +1-352-273-9398; Fax: +1-352-273-9399

Abstract: Background: We have previously reported that the addition of resistant maltodextrin (RMD), a fermentable functional fiber, to the diet increases fecal weight as well as the amount of fecal bifidobacteria. Here, we report on the targeted analysis of changes in potentially beneficial gut bacteria associated with the intervention. Objective: The primary objective of this study was to determine the effect of adding 0, 15 and 25 g RMD to the diets of healthy free-living adults on potentially beneficial gut bacteria. Methods: We expanded on our previously reported microbiota analysis in a double-blind, placebo-controlled feeding study (NCT02733263) by performing additional qPCR analyses targeting fecal lactic acid bacteria (LAB), *Akkermansia muciniphila*, *Faecalibacterium prausnitzii* and *Fusicatenibacter saccharivorans* in samples from 49 participants. Results: RMD resulted in an approximately two-fold increase in fecal *Fusicatenibacter saccharivorans* ($p = 0.024$ for 15 g/day RMD and $p = 0.017$ for 25 g/day RMD). For *Akkermansia muciniphila* and *Faecalibacterium prausnitzii*, we obtained borderline evidence that showed increased amounts in participants that had low baseline levels of these bacteria ($p < 0.1$ for 25 g/day RMD). We did not detect any effects of RMD on LAB. Conclusions: RMD supplementation in healthy individuals increases *Fusicatenibacter saccharivorans*. Albeit to a lesser extent, RMD at the higher intake level may also increase *Akkermansia muciniphila* and *Faecalibacterium prausnitzii* in individuals with low baseline levels of those two species. Potential benefits associated with these microbiota changes remain to be established in studies with quantifiable health-related endpoints.

Keywords: resistant maltodextrin; bifidobacteria; healthy adults; gastrointestinal function

Citation: Mai, V.; Burns, A.M.; Solch, R.J.; Dennis-Wall, J.C.; Ukhanova, M.; Langkamp-Henken, B. Resistant Maltodextrin Consumption in a Double-Blind, Randomized, Crossover Clinical Trial Induces Specific Changes in Potentially Beneficial Gut Bacteria. *Nutrients* **2022**, *14*, 2192. https://doi.org/10.3390/nu14112192

Academic Editor: Elad Tako

Received: 28 April 2022
Accepted: 20 May 2022
Published: 25 May 2022

1. Introduction

Health benefits of consuming dietary fiber (DF) have been well established and include a decreased risk of obesity, cardiovascular disease, type 2 diabetes and some cancers, especially those of the gastrointestinal (GI) tract. However, questions remain about the generalizability across various DF in mechanisms mediating these health benefits. Some of the previously investigated mechanisms include laxation through fecal bulking, binding and excretion of toxins and support of a balanced gut microbiota with beneficial metabolic capabilities. A better understanding of the contributions of these mechanisms to mediate DF-associated health benefits will facilitate a more personalized approach for adding specific DF that can target health benefits in susceptible individuals. Functional fibers, purified DF that can be added to food or drinks, offer an effective means for increasing DF consumption to reach recommended intake guidelines and confer health benefits. While a wide array of distinct functional fibers have been developed, each functional fiber has distinct characteristics and thus requires data to support its unique health benefits, specific mechanisms involved and acceptability and tolerability by the consumer. Of particular interest, changes in the microbiota during aging might contribute to reduced immune surveillance as well as compromised gut barrier function that can result in increased levels of chronic inflammation. Thus, functional fiber, which targets specific gut microbes might be able to improve symptoms in individuals with chronic inflammatory diseases of the GI tract, including ulcerative colitis (UC).

Resistant maltodextrin (RMD), a water-soluble, fermentable functional fiber derived from heat treatment of cornstarch has been shown to be well tolerated [1]. We have previously shown that supplementing 25 g RMD/day increased DF intake, stool wet weight (WW) and fecal bifidobacteria counts in adults with DF intake below recommendations [2]. To expand upon previous microbiota findings from this intervention study, we now report additional observations of changes in potentially beneficial gut bacteria, particularly, lactic acid bacteria (LAB), *Fusicatenibacter saccharivorans (F. saccharivorans)*, *Faecalibacterium prausnitzii* (*F. prausnitzii*) and *Akkermansia muciniphila* (*A. muciniphila*) associated with RMD intake. We selected these taxa due to the recognized or emerging signature in disease states and their potential beneficial effect on the host, as reported in the literature [3–9]. In these additional analyses, we also stratified participants by pre-intervention baseline levels of the targeted bacteria. We hypothesized that RMD increases levels of targeted bacteria, especially in individuals with lower levels at baseline. The rationale for this approach is that individuals with low or undetectable levels of a potentially beneficial microbial taxon might be more likely to benefit from an effort to increase those numbers than those that already have high levels present.

2. Methods

Details of the study have been published previously [2]. In brief, 49 volunteers between the ages of 18 to 50 years old from the southeastern United States completed a randomized, placebo-controlled, double-blind, three-intervention (0 g, 15 g or 25 g) crossover study with a 2-week baseline before each 3-week intervention. This study was conducted from August 2016 to November 2016, was registered at Clinicaltrials.gov #NCT02733263 and conducted according to the ethical guidelines in the Declaration of Helsinki and approved by the University of Florida Institutional Review Board (IRB201501168). Written informed consent was obtained from all participants by trained study coordinators prior to beginning the study. Participants were instructed to mix powdered study supplement in water or a beverage of choice and consume it throughout the day of each intervention. A 2-day total stool collection was obtained during the second week of each baseline and during the last week of each of the three interventions.

2.1. Microbiota Analyses

For 16S rDNA-based microbiota analysis, the first stool sample from each day was combined into a single sample per 2-day collection. Total DNA was extracted from 0.1–0.2 g of feces using a modified protocol (QIAamp DNA Mini Kits, Qiagen) that included a bead-beating step [10].

2.2. qPCR to Quantify Counts of Targeted Bacteria

Reactions containing 10 ng of DNA and 0.2 µmol/L of each primer [11,12] were carried out in duplicate with an initial melting step at 95 °C for 10 min followed by 40 cycles of 95 °C for 30 s, 58 °C for 60 s and then 72 °C for 60 s.

2.3. Statistical Analyses

All analyses were performed on $\log_{10}$ transformed counts of targeted bacteria per 10 ng of DNA. For intervention effects on fecal bacteria counts, the difference between the $\log_{10}$ counts at the beginning of the first intervention period (baseline) and $\log_{10}$ counts at the end of the third week of each intervention was calculated for each participant (Table 1).

Comparisons of $\log_{10}$ counts at the end of each period among RMD dosage levels (0 g, 15 g or 25 g) were also performed. In exploratory analyses, we stratified participants by baseline levels of each taxon into "low", "medium" and "high" visually based on the distribution of counts for each targeted probe and compared the effects of RMD in the "low" group by comparing differences of $\log_{10}$ counts at baseline with those at the end of the intervention period for RMD dosage levels (15 g or 25 g).

Table 1. qPCR-detected changes in $\log_{10}$ counts of targeted fecal bacteria.

Target	Baseline	Placebo	15 g RMD	25 g RMD
F. saccharivorans	3.44×10^5	3.35×10^5	6.98×10^5	6.97×10^5
A. muciniphila	2.40×10^4	2.35×10^4	3.57×10^4	1.79×10^4
F. prausnitzii	2.56×10^5	2.43×10^5	3.14×10^5	2.80×10^5
LAB	2.13×10^4	1.48×10^4	2.43×10^4	1.59×10^4

Mean $\log_{10}$ counts of targeted bacterial taxa by qPCR during the intervention trial. *F. saccharivorans*, *Fusicatenibacter saccharivorans*; *F. prausnitzii*, *Faecalibacterium prausnitzii*; *A. muciniphila*, *Akkermansia muciniphila*; LAB, lactic acid bacteria ($n = 49$). Values in bold are significantly different ($p < 0.05$) from both baseline and end of placebo period (0 g).

3. Results

Using samples collected in a previously reported intervention trial, we successfully quantified RMD-associated changes in three of the four taxa targeted in this study, all of which are commonly found in the human gut microbiota. While the targeted taxa were detected above the minimal detection level in more than 80% of all samples, for some of the targets we failed to detect their presence in one or more samples from a few individuals. While this had minor effects on statistical power to detect potential effects of RMD for some of the targets, overall, these effects were minor as we detected a signal in a vast majority of samples for all four targets.

In the primary analysis, compared to both the baseline and the placebo (i.e., 0 g RMD) period, the $\log_{10}$ counts of *F. saccharivorans* increased during the intervention period by more than two-fold upon RMD consumption at either intake level ($p < 0.05$) (Table 1). We did not detect an effect of RMD on any of the other three microbiota targets in the primary analysis.

To further explore our data, we then stratified participants by baseline levels of targeted taxa. An example of the stratification into "high" and "low" baseline levels based on the distribution is shown for LAB (see Figure 1). While for LAB there appeared to be more of an increase during the two RMD periods compared to the placebo and this difference was not statistically significant (Figure 2). In individuals with low baseline levels of the respective other bacterial targets, we detected minor effects for *F. prausnitzii* ($p = 0.03$) and borderline effects for *A. muciniphila* ($p = 0.09$) at 25 g RMD (Table 2). While these effects were borderline statistically significant, they did not reach a 2-fold threshold.

Table 2. p-values for exploratory analysis of effects of RMD on *A. muciniphila*, *F. prausnitzii* and LAB in individuals with low baseline $\log_{10}$ counts of respective targets.

Target	15 g RMD	25 g RMD
A. muciniphila	>0.1	0.09
F. prausnitzii	>0.1	0.03
LAB	>0.1	>0.1

p-values for differences between baseline and end of intervention in individuals with low baseline $\log_{10}$ counts of targeted bacterial taxa by qPCR. *F. prausnitzii*, *Faecalibacterium prausnitzii*; *A. muciniphila*, *Akkermansia muciniphila*; LAB, lactic acid bacteria ($n = 49$).

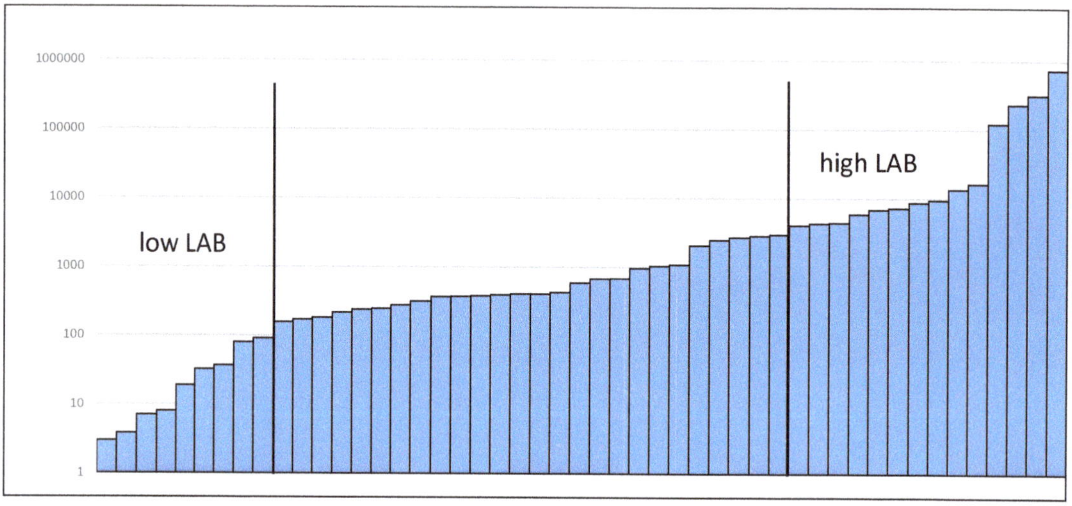

Figure 1. Distribution of LAB at baseline.

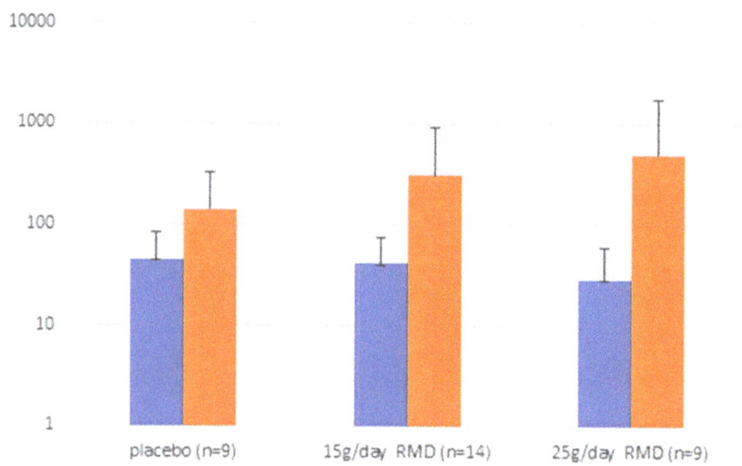

Figure 2. LAB genome copies/10 ng DNA in samples with low LAB content at beginning of each period. Blue: baseline of intervention period; Orange: end of intervention period.

4. Discussion

We previously reported that the consumption of 25 g/day of RMD increased total fecal bacterial output and number of bifidobacteria but decreased the Firmicutes to Bacteroidetes ratio [2]. Here, we provide further support to the previous observation that adding up to 25 g/day of RMD to the diet is a well-tolerated approach for modifying fecal microbiota towards a potentially more beneficial composition.

Specifically, we now show that dietary supplementation with both 15 g and 25 g/day of RMD increased the amount of fecal *F. saccharivorans* by more than two-fold regardless of the baseline levels ($p < 0.05$). This observation is of interest as such a consistent and significant effect across dietary intake levels is rare when targeting specific fecal bacterial taxa in an intervention study. This finding suggests that RMD, even at a lower intake, can increase the amount of *F. saccharivorans*, which is of interest due to potential associated health benefits. Previous observations were not as strong and occurred at higher intake levels (up to 50 g/day RMD) [13]. A recent study of microbiota differences in patients with UC has shown significantly lower numbers of *F. saccharivorans* compared to healthy controls [8]. It has been suggested that an increase in *F. saccharivorans* may reduce intestinal inflammation as it stimulates the production of anti-inflammatory interleukin-10 in intestinal lamina propria cells isolated from patients with UC [7]. Additionally, in a study where patients with UC received fecal microbiota transplantation, higher *F. saccharivorans* counts in donors were significantly correlated with clinical remission of UC [9].

We also obtained evidence that, in individuals with low baseline levels of *A. muciniphila* and *F. prausnitzii*, RMD supplementation may have increased said taxa compared to the baseline albeit mostly at the higher intake level of 25 g/day. In contrast, we did not detect an effect of RMD in individuals with high baseline levels of the respective bacterial targets. We targeted these species due to their potential to exhibit beneficial effects on the host [4,14]. LAB are bacteria that produce lactic acid as a major end product of carbohydrate fermentation and are frequently used in probiotic formulations due to their beneficial effects (strains from genera *Lactobacillus* and *Bifidobacterium* are most commonly used). While in vitro and rodent in vivo evidence demonstrates anti-inflammatory and immunomodulatory potential of certain LAB species, human trials result in inconsistent findings regarding the association of LAB species with metabolic diseases [5]. *A. muciniphila* is a commensal bacterium for which a low abundance in the gut is recognized as a signature of obesity-related metabolic disorders [14]. As a mucin degrader, *A. muciniphila* is necessary for maintaining optimal thickness of the mucosal layer and the integrity of the intestinal epithelial cells [15], which are components of intestinal health. In rodent studies, prebiotic intake increased *A. muciniphila* abundance in obese and diabetic mouse models [16], while *A. muciniphila* administration reduced endotoxin concentrations in obese mice fed a high-fat diet [17]. In a human trial, supplementation with *A. muciniphila* improved several metabolic parameters related to insulin and lipid metabolism in overweight and obese insulin-resistant participants [4]. In a recent systematic review and meta-analysis of 16 studies, butyrate producer *F. prausnitzii* has been shown to be present in lower abundance in patients with inflammatory bowel disease (IBD) compared to healthy controls, as well as in patients with active IBD compared to patients with IBD in remission [6]. Both *A. muciniphila* and *F. prausnitzii* have been extensively researched in human dietary interventions. Supplementation with ingredients such as inulin-type fructans, fructooligosacharides, polydextrose and soluble corn fiber can increase these species; however, the findings in inulin-oligofructose, resistant starch and wheat bran foods are inconsistent [3]. This inconsistency likely is at least partially due to studies being performed in populations with underlying differences in dietary habits, gut microbiota, health status, etc.

The observation of increases in potentially beneficial bacteria in participants with low baseline levels needs to be interpreted carefully. It might make biological sense that individuals with low baseline levels are more likely to benefit from dietary supplements that can increase fecal numbers of targeted bacteria. However, one also needs to consider that over time, counts of bacteria tend to vary around a mean. Thus, in individuals with a single low baseline measure, the expected trend would be an increase towards the mean, which is what we observed for LAB (see Figure 2). Similarly, in individuals with a high baseline count, the trend in the absence of any intervention effect would be a decrease towards the mean. While observations of changes during the placebo period in comparison to intervention periods can help address this concern, potential remaining carryover effects of intervention periods might have interfered in this analysis. We also note that "high" and "low" baseline levels were set arbitrarily visually based on the observed distribution. However, to demonstrate a meaningful health benefit, cutoffs in future studies should be based on biologically relevant levels that are associated with the likelihood of an increase in a particular microbial taxon rather than statistically determined cutoffs. Further studies that link microbiota composition with well-defined health endpoints are required to establish such biologically relevant levels for each microbial taxon of interest. This issue is further complicated as multiple taxa might mediate similar health benefits and interactions between taxa might also contribute.

As our study provides strong evidence that RMD, even at lower intake levels, increases *F. saccharivorans* regardless of baseline levels, it seems reasonable to explore if RMD can reduce some UC-associated symptoms in future studies. Prebiotics have largely been studied in healthy individuals, often with rather diffuse endpoints that are rarely objectively quantifiable. Various barriers, often regulatory in nature, exist for developing probiotics and prebiotics to target specific disease symptoms, rather than supporting more general and often ambiguous health maintenance claims. It might be timely to invest additional resources into more effectively utilizing the large potential of interventions targeting specific microbiota, including RMD supplements, to improve distinct disease outcomes, especially those associated with the GI tract.

5. Conclusions

This secondary analysis of a previously published clinical trial demonstrated an objective, quantifiable effect of RMD on a specific microbial taxon that is potentially beneficial to human health, *F. saccharivorans*. Studies assessing the effects of dietary fibers such as RMD on specific microbial taxa and their related health benefits continue to be warranted.

Author Contributions: The author responsibilities were as follows—B.L.-H. and V.M. designed the research; R.J.S., J.C.D.-W., M.U., A.M.B., V.M. and B.L.-H., conducted the research; M.U., V.M., J.C.D.-W., A.M.B. and B.L.-H., analyzed data; and V.M., J.C.D.-W., A.M.B. and B.L.-H., wrote the paper. V.M. had primary responsibility for final content. All authors have read and agreed to the published version of the manuscript.

Funding: ADM/Matsutani: LLC; USDA National Institute of Food and Agriculture, Hatch project FLA-FOS-005636.

Institutional Review Board Statement: The study was approved by the University of Florida Institutional Review Board (IRB201501168).

Informed Consent Statement: Informed consent was obtained from all subjects involved in the study.

Data Availability Statement: De-identified qPCR data will be shared upon request.

Acknowledgments: The authors would like to thank the participants and the undergraduate and graduate volunteers for their assistance with this project.

Conflicts of Interest: The authors declare no conflict of interest.

Abbreviations

RMD	Resistant maltodextrin
DF	Dietary fiber
BSS	Bristol stool form scale
GI	Gastrointestinal
UC	Ulcerative colitis
OUT	Operational taxonomic units
WW	Wet weight
LAB	Lactic acid bacteria
F. saccharivorans	Fusicatenibacter saccharivorans
F. prausnitzii	Faecalibacterium prausnitzii
A. muciniphila	Akkermansia muciniphila

References

1. Fastinger, N.D.; Karr-Lilienthal, L.K.; Spears, J.K.; Swanson, K.S.; Zinn, K.E.; Nava, G.M.; Ohkuma, K.; Kanahori, S.; Gordon, D.T.; Fahey, G.C., Jr. A novel resistant maltodextrin alters gastrointestinal tolerance factors, fecal characteristics, and fecal microbiota in healthy adult humans. *J. Am. Coll. Nutr.* **2008**, *27*, 356–366. [CrossRef] [PubMed]
2. Burns, A.M.; Solch, R.J.; Dennis-Wall, J.C.; Ukhanova, M.; Nieves, C., Jr.; Mai, V.; Christman, M.C.; Gordon, D.T.; Langkamp-Henken, B. In healthy adults, resistant maltodextrin produces a greater change in fecal bifidobacteria counts and increases stool wet weight: A double-blind, randomized, controlled crossover study. *Nutr. Res.* **2018**, *60*, 33–42. [CrossRef]
3. Verhoog, S.; Taneri, P.E.; Roa Díaz, Z.M.; Marques-Vidal, P.; Troup, J.P.; Bally, L.; Franco, O.H.; Glisic, M.; Muka, T. Dietary Factors and Modulation of Bacteria Strains of *Akkermansia muciniphila* and *Faecalibacterium prausnitzii*: A Systematic Review. *Nutrients* **2019**, *11*, 1565. [CrossRef] [PubMed]
4. Depommier, C.; Everard, A.; Druart, C.; Plovier, H.; Van Hul, M.; Vieira-Silva, S.; Falony, G.; Raes, J.; Maiter, D.; Delzenne, N.M.; et al. Supplementation with *Akkermansia muciniphila* in overweight and obese human volunteers: A proof-of-concept exploratory study. *Nat. Med.* **2019**, *25*, 1096–1103. [CrossRef] [PubMed]
5. De Filippis, F.; Pasolli, E.; Ercolini, D. The food-gut axis: Lactic acid bacteria and their link to food, the gut microbiome and human health. *FEMS Microbiol. Rev.* **2020**, *44*, 454–489. [CrossRef] [PubMed]
6. Zhao, H.; Xu, H.; Chen, S.; He, J.; Zhou, Y.; Nie, Y. Systematic review and meta-analysis of the role of *Faecalibacterium prausnitzii* alteration in inflammatory bowel disease. *J. Gastroenterol. Hepatol.* **2021**, *36*, 320–328. [CrossRef] [PubMed]
7. Takeshita, K.; Mizuno, S.; Mikami, Y.; Sujino, T.; Saigusa, K.; Matsuoka, K.; Naganuma, M.; Sato, T.; Takada, T.; Tsuji, H.; et al. A Single Species of Clostridium Subcluster XIVa Decreased in Ulcerative Colitis Patients. *Inflamm. Bowel Dis.* **2016**, *22*, 2802–2810. [CrossRef] [PubMed]
8. Gryaznova, M.V.; Solodskikh, S.A.; Panevina, A.V.; Syromyatnikov, M.Y.; Dvoretskaya, Y.D.; Sviridova, T.N.; Popov, E.S.; Popov, V.N. Study of microbiome changes in patients with ulcerative colitis in the Central European part of Russia. *Heliyon* **2021**, *7*, e06432. [CrossRef] [PubMed]
9. Osaki, H.; Jodai, Y.; Koyama, K.; Omori, T.; Horiguchi, N.; Kamano, T.; Funasaka, K.; Nagasaka, M.; Nakagawa, Y.; Shibata, T.; et al. Clinical response and changes in the fecal microbiota and metabolite levels after fecal microbiota transplantation in patients with inflammatory bowel disease and recurrent *Clostridioides difficile* infection. *Fujita Med. J.* **2021**, *7*, 87–98. [PubMed]
10. Miller, D.N.; Bryant, J.E.; Madsen, E.L.; Ghiorse, W.C. Evaluation and optimization of DNA extraction and purification procedures for soil and sediment samples. *Appl. Environ. Microbiol.* **1999**, *65*, 4715–4724. [CrossRef] [PubMed]
11. Nabizadeh, E.; Jazani, N.; Bagheri, M.; Shahabi, S. Association of altered gut microbiota composition with chronic urticaria. *Ann. Allergy Asthma Immunol.* **2017**, *119*, 48–53. [CrossRef] [PubMed]
12. Dubernet, S.; Desmasures, N.; Guéguen, M. A PCR-based method for identification of lactobacilli at the genus level. *FEMS Microbiol. Lett.* **2002**, *214*, 271–275. [CrossRef] [PubMed]
13. Ukhanova, M.; Culpepper, T.; Baer, D.; Gordon, D.; Kanahori, S.; Valentine, J.; Neu, J.; Sun, Y.; Wang, X.; Mai, V. Gut microbiota correlates with energy gain from dietary fibre and appears to be associated with acute and chronic intestinal diseases. *Clin. Microbiol. Infect* **2012**, *18* (Suppl. S4), 62–66. [CrossRef] [PubMed]
14. Cani, P.D. Human gut microbiome: Hopes, threats and promises. *Gut* **2018**, *67*, 1716–1725. [CrossRef] [PubMed]
15. Reunanen, J.; Kainulainen, V.; Huuskonen, L.; Ottman, N.; Belzer, C.; Huhtinen, H.; de Vos, W.M.; Satokari, R. *Akkermansia muciniphila* Adheres to Enterocytes and Strengthens the Integrity of the Epithelial Cell Layer. *Appl. Environ. Microbiol.* **2015**, *81*, 3655–3662. [CrossRef] [PubMed]
16. Everard, A.; Lazarevic, V.; Derrien, M.; Girard, M.; Muccioli, G.G.; Neyrinck, A.M.; Possemiers, S.; Van Holle, A.; François, P.; de Vos, W.M.; et al. Responses of gut microbiota and glucose and lipid metabolism to prebiotics in genetic obese and diet-induced leptin-resistant mice. *Diabetes* **2011**, *60*, 2775–2786. [CrossRef] [PubMed]
17. Everard, A.; Belzer, C.; Geurts, L.; Ouwerkerk, J.P.; Druart, C.; Bindels, L.B.; Guiot, Y.; Derrien, M.; Muccioli, G.G.; Delzenne, N.M.; et al. Cross-talk between *Akkermansia muciniphila* and intestinal epithelium controls diet-induced obesity. *Proc. Natl. Acad. Sci. USA* **2013**, *110*, 9066–9071. [CrossRef] [PubMed]

nutrients MDPI

Article

Effects of Intra-Amniotic Administration of the Hydrolyzed Protein of Chia (*Salvia hispanica* L.) and *Lacticaseibacillus paracasei* on Intestinal Functionality, Morphology, and Bacterial Populations, In Vivo (*Gallus gallus*)

Marcella Duarte Villas Mishima [1], Hércia Stampini Duarte Martino [2], Nikolai Kolba [1], Drashti Dhirenkumar Shah [1], Mariana Grancieri [2], Karina Maria Olbrich Dos Santos [3], Janine Passos Lima [3], Bárbara Pereira Da Silva [2], Elvira Gonzalez de Mejia [4] and Elad Tako [1,*]

1 Department of Food Science, Stocking Hall, Cornell University, Ithaca, NY 14853, USA
2 Department of Nutrition and Health, Federal University of Viçosa, Av. Purdue, s/n, Campus Universitário, Viçosa 36570-900, MG, Brazil
3 Embrapa Agroindústria de Alimentos, Av. das Américas 29.501, Rio de Janeiro 23020-470, RJ, Brazil
4 Department of Food Science & Human Nutrition, University of Illinois at Urbana-Champaign, Urbana, IL 61801, USA
* Correspondence: et79@cornell.edu

Abstract: As a protein source, chia contains high concentrations of bioactive peptides. Probiotics support a healthy digestive tract and immune system. Our study evaluated the effects of the intra-amniotic administration of the hydrolyzed chia protein and the probiotic *Lacticaseibacillus paracasei* on intestinal bacterial populations, the intestinal barrier, the inflammatory response, and brush border membrane functionality *in ovo* (*Gallus gallus*). Fertile broiler (*Gallus gallus*) eggs ($n = 9$/group) were divided into 5 groups: (NI) non-injected; (H_2O) 18 MΩ H_2O; (CP) 10 mg/mL hydrolyzed chia protein; (CPP) 10 mg/mL hydrolyzed chia protein + 10^6 colony-forming unit (CFU) *L. paracasei*; (P) 10^6 CFU *L. paracasei*. The intra-amniotic administration was performed on day 17 of incubation. At hatching (day 21), the animals were euthanized, and the duodenum and cecum content were collected. The probiotic downregulated the gene expression of NF-κβ, increased *Lactobacillus* and *E. coli*, and reduced *Clostridium* populations. The hydrolyzed chia protein downregulated the gene expression of TNF-α, increased OCLN, MUC2, and aminopeptidase, reduced *Bifidobacterium*, and increased *Lactobacillus*. The three experimental groups improved in terms of intestinal morphology. The current results suggest that the intra-amniotic administration of the hydrolyzed chia protein or a probiotic promoted positive changes in terms of the intestinal inflammation, barrier, and morphology, improving intestinal health.

Keywords: gut health; microbiota; intestinal barrier; inflammation; chia seed; bioactive peptides; in vivo; probiotic

Citation: Mishima, M.D.V.; Martino, H.S.D.; Kolba, N.; Shah, D.D.; Grancieri, M.; Dos Santos, K.M.O.; Lima, J.P.; Da Silva, B.P.; Gonzalez de Mejia, E.; Tako, E. Effects of Intra-Amniotic Administration of the Hydrolyzed Protein of Chia (*Salvia hispanica* L.) and *Lacticaseibacillus paracasei* on Intestinal Functionality, Morphology, and Bacterial Populations, In Vivo (*Gallus gallus*). *Nutrients* **2023**, *15*, 1831. https://doi.org/10.3390/nu15081831

Academic Editor: Adam Matkowski

Received: 21 February 2023
Revised: 20 March 2023
Accepted: 7 April 2023
Published: 11 April 2023

1. Introduction

Chia (*Salvia hispanica* L.) is considered a potentially bioactive food source since it has demonstrated multiple health benefits, such as a decrease in adiposity, modulation of the glucose metabolism [1], a hypoglycemic effect, a reduction in hepatic fat deposition [2], an improved lipid profile [2–5], reductions in inflammatory processes [4–6], antioxidant protection [4,7] and reduced fat content of the heart [7]. Regarding intestinal health, chia seeds have been shown to improve the intestinal brush border membrane and favored its functionality [2,8–10], increasing the production of short-chain fatty acids and increasing the richness of the microbiota [9,10].

Chia seeds have high concentrations of lipids (30.17 g.100 g^{-1}), proteins (19.72 g.100 g^{-1}), total dietary fiber (37.18 g.100 g^{-1}), and bioactive compounds such as phenolic compounds, tocopherols, and tocotrienols [11]. As a protein source, the hydrolyzed chia protein seeds contains high concentrations of bioactive peptides with promising compositions and sequences. The amino acid sequences of chia seed proteins have been identified; these proteins are associated with hypoglycemic and hypotensive activity, an antioxidant effect, and glucose uptake stimulating peptides [12]. Furthermore, the hydrolyzed chia protein demonstrated these antioxidant, antihypertensive, and hypoglycemic properties in vitro [13]; meanwhile, in vivo, it improved the biochemical profile, reduced TNF-α expression, and reduced the production of activated NF-$\kappa\beta$ and affected adipogenesis, with antilipidemic and antiadipogenic actions [14]. Bioactive peptides have diverse functions, including modulating intestinal homeostasis and affecting barrier function, the villus surface area, mucosal immune responses, inflammation, and the gut microbiota [15–18]. Studies are still needed to understand the effects of the intestinal environment on the bioavailability of bioactive peptides, their interactions with other compounds, and fundamental questions regarding the effectiveness of hydrolyzed protein as a source of bioactive peptides on intestinal health, as well as the interactions between intestinal barrier function, microbiota, and the immune system [19].

Probiotics are live microorganisms that, when administered in adequate amounts, confer benefits to the health of the host. Probiotic use supports a healthy digestive tract and a healthy immune system, so the overall benefit of probiotics on the gut microbiota derives from creating a favorable gut environment [20–22]. Probiotic food products represent a market trend, being associated with the increasing consumer awareness regarding the link between diet and well-being, and especially gut health. Strains pertaining to several *Lactobacilli* were shown to exhibit probiotic properties, particularly within the *Lacticaseibacillus* species [23–25]. *Lactobacilli* are considered autochthonous residents in the gastrointestinal tracts of animals such as chickens, rodents, and humans [24,25]. They may promote the host's intestinal health and immune function in different ways, such as by strengthening the epithelial barrier, competitively rejecting pathogenic microorganisms, producing antimicrobial substances, and interacting with immune cells by stimulating pattern recognition receptors [24]. Studies have shown that *Lacticaseibacillus paracasei* upregulated the expression of tight junction proteins, downregulated the production of pro-inflammatory cytokines, and altered the structure of the intestinal microbiota [26].

The *Gallus gallus* is an animal model that presents a complex gut microbiota with gene sequences that are, at the phylum level, highly homologous to humans [27]. The intra-amniotic feeding model is widely used as an in vivo method to assess bioactive compounds with potential nutritional effects, specifically prebiotic effects. The intra-amniotic administration is conducted on day 17 of embryonic development, prior to the embryo's oral consumption of the amniotic fluid [28]. The intra-amniotic administration of proteins can stimulate the maturation and functionality of the small intestine by promoting cell proliferation and differentiation and the expansion of the absorptive surface area [29]. The intra-amniotic administration of duck egg white peptides was able to increase the villus surface area and diameter of goblet cells by promoting the proliferation of enterocytes, promoting beneficial bacterial populations, and limiting potentially pathogenic bacterial populations, in addition to promoting and contributing to calcium uptake [15]. The use of probiotics in poultry nutrition has shown several benefits, such as improvements in the microbiological homeostasis of the intestine, and in the immune response and growth. These benefits vary due to differences differences in, for example, the strains and the doses of the probiotics [30]. The intra-amniotic administration of a probiotic can upregulate the mRNA expression of intestinal-function-related genes and nutrient-transporter-related genes [31].

To date, no studies have explored the combined effects of the intra-amniotic application of hydrolyzed chia protein and the probiotic *L. paracasei*. As such, the objective of this study was to conduct an in vivo assessment of the effects of the intra-amniotic administration of hydrolyzed chia protein and the probiotic *L. paracasei* on intestinal bacterial populations, the intestinal barrier, the inflammatory response, and the brush border membrane's (BBM) functionality. Bioactive peptides might have functions that affect intestinal homeostasis, in addition to probiotics, which have been demonstrated to play a supporting role in the digestive tract's health and function; we therefore hypothesized that, combined with the probiotic, the hydrolyzed chia protein would increase the abundance of beneficial bacterial populations. This effect will further improve the intestinal BBM's functionality via regulation of the gene expression of key proteins that are required for tight junction development and inflammatory responses.

2. Materials and Methods

2.1. Sample Material

2.1.1. Hydrolyzed Chia protein

Chia seeds grown in Rio Grande do Sul/Brazil were prepared according to Orona-Tamayo (2015) [32]. The total digested protein was produced and analyzed in a Laboratory at the University of Illinois, Urbana-Champaign, IL, USA, as detailed by Grancieri (2019) [13]. Briefly, for mucilage formation, the seeds were immersed for 1 h in distilled water (1 g/20 mL), frozen overnight (−80 °C), and freeze-dried (Labconco Freeze Dryer 4.5; Kansas, MO, USA). The mucilage was removed from the seeds using a sieve (500 µm per 35 mesh). Using a coffee grinder (Mr. Coffee®, Cleveland, OH, USA), the mucilage-free seeds were ground, sieved (500 µm per 35 mesh), and degreased using hexane (1 g/10 mL) at 60 °C for 2 h under constant stirring. The mixture was centrifuged (6000× *g*, 15 min, 4 °C) and the resulting flour was left overnight under a hood and stored at 4 °C until use. Then, deionized water was added to the mucilage and fat-free chia flour (1 g/20 mL); the pH was adjusted to 8.0 and it was kept under constant stirring (35 °C per 1 h). The mixture was centrifuged (5000× *g*; 15 min; 25 °C) and the supernatant was collected, freeze-dried, and stored at −20 °C. This fraction is referred to as "total proteins".

For the digestion of the extracted protein, we used a previously described technique [33] in which gastrointestinal digestion is simulated. Briefly, the extracted protein was suspended in deionized water (1 g/20 mL), the pH was adjusted to 2.0, pepsin was added at a concentration of 1:20 (enzyme:protein), and it was stirred for 2 h at 37 °C. Afterwards, the pH was adjusted to 7.5, pancreatin 1:20 (enzyme:protein) was added, and then the digestion was carried out as above. Finally, the simulated digestion was stopped by placing the mixture in a water bath (75 °C, 20 min); it was then centrifuged twice at 20,000× *g* (4 °C, 15 min). The supernatant was collected (total protein digested) and dialyzed using a 100–500 Da molecular weight exclusion membrane (Spectra/Por®, Biotech CE Membrane); it was then lyophilized and stored at −20 °C until use.

2.1.2. Probiotic

Freeze-dried *L. paracasei* (strain TRA038563) was provided by the Brazilian Agriculture Research Corporation (EMBRAPA, Rio de Janeiro, Brazil).

2.2. The Intra-Amniotic Administration

Cornish Cross fertile broiler eggs ($n = 50$), obtained from a commercial hatchery (Moyer's chicks, Quakertown, PA, USA), were incubated under optimal conditions (37 ± 2 °C and 89.6 ± 2% humidity) using Cornell University's Animal Science poultry farm incubator. All animal protocols were approved by Cornell University's Institutional Animal Care and Use Committee (protocol code: 2020–0077).

The hydrolyzed chia protein in powdered form was diluted in 18 MΩ H_2O. The osmolarity of the solution was tested to determine the concentration necessary to maintain an osmolarity value (Osm) of less than 320 Osm; this also ensured that the chicken embryos would not be dehydrated upon the injection of the solution. The intra-amniotic administration followed the methodology previously described [34–36].

At 17 days of embryonic incubation, eggs were candled to discard those that presented as infertile, cracked, contaminated, or early dead embryos. Eggs containing viable embryos were weighed and randomly allocated into 5 groups with a similar weight frequency distribution ($n = 10$/group). Briefly, all eggs were disinfected with 70% ethanol. Each group was then injected with the specified solution (1 mL per egg) with a 19 mm gauge needle that was vertically inserted into the amniotic fluid, which was identified by candling.

The 5 groups were assigned as follows: (NI) non-injected; (H_2O) 18 MΩ H_2O; (CP) 10 mg/mL (1%) hydrolyzed chia protein; (CPP) 10 mg/mL (1%) hydrolyzed chia protein + 10^6 colony-forming unit (CFU) *Lacticaseibacillus paracasei* (800 μL of hydrolyzed chia protein + 200 μL of probiotic/egg); (P) 10^6 CFU *Lacticaseibacillus paracasei.*

After the injections, the injection holes were sealed with cellophane tape and the eggs placed in hatching baskets. Immediately after hatching (21 days), chicks from each treatment group were weighed and then euthanized by CO_2 exposure. Their duodenum, cecum, and cecum content were collected.

2.3. Extraction of the Total RNA from the Duodenum Tissue Samples

According to the manufacture's protocol (Rneasy Mini Kit, Qiagen Inc., Valencia, CA, USA), the total RNA was extracted from 30 mg of the duodenum (n = 5 animals/group) [35–37]. The total RNA was eluted in 50 μL of RNase-free water. All steps were carried out under RNase-free conditions. The RNA was quantified by absorbance at 260/280, and the integrity of the 18S ribosomal RNAs was verified using 1.5% agarose gel electrophoresis, followed by ethidium bromide staining. The samples were stored at −80 °C until the analysis.

2.4. Real-Time Polymerase Chain Reaction (RT-PCR) and Prime Design

The cDNA was created with a 20 μL reverse transcriptase (RT) reaction in a BioRad C1000 touch thermocycler using the Improm-II Reverse Transcriptase Kit (Catalog #A1250; Promega, Madison, WI, USA). The concentration of cDNA obtained was determined by measuring the absorbance at 260 nm and 280 nm using an extinction coefficient of 33 (for single-stranded DNA). The gene expression of the duodenum was determined using real-time polymerase chain reaction (RT-PCR).

The primers used in the real-time qPCR (Table 1) were designed based on gene sequences from the GenBank database using Real-Time Primer Design Tool software (IDT DNA, Coralvilla, IA, USA) [36,38]. The specificity of the primers was tested by performing a BLAST search against the genomic National Center for Biotechnology Information (NCBI) database. The *Gallus gallus* primer 18S rRNA was designed as a reference gene.

Table 1. Sequence of experimental primers used in this study.

Gene	Oligonucleotides (5′-3′)	
	Forward Primer (5′-3′)	Reverse Primer (5′-3′)
	BBM functionality	
AP	CGTCAGCCAGTTTGACTATGTA	CTCTCAAAGAAGCTGAGGATGG
SI	CCAGCAATGCCAGCATATTG	CGGTTTCTCCTTACCACTTCTT
18S rRNA	GCAAGACGAACTAAAGCGAAAG	TCGGAACTACGACGGTATCT
	Inflammation	
TNF-α	GACAGCCTATGCCAACAAGTA	TTACAGGAAGGGCAACTCATC
NF-κβ1	CACAGCTGGAGGGAAGTAAAT	TTGAGTAAGGAAGTGAGGTTGAG
	Intestinal Barrier	
MUC2	CCTGCTGCAAGGAAGTAGAA	GGAAGATCAGAGTGGTGCATAG
OCLN	GTCTGTGGGTTCCTCATCGT	GTTCTTCACCCACTCCTCCA

BBM: brush border membrane; AP: aminopeptidase; SI: sucrose isomaltase; TNF-α: tumor necrosis factor-alpha; NF-κβ1: nuclear factor kappa beta; MUC2: mucin 2; OCLN: occludin.

2.5. Real-Time qPCR Design

All procedures were performed as previously described [8,34,36,39]. Briefly, cDNA was used for each 10 μL reaction containing 2 × BioRad SSO Advanced Universal SYBR Green Supermix (Hercules, CA, USA). For each reaction, 8 μL of the master mix and 2 μL of cDNA were pipetted into a 96-well plate; meanwhile, for the standard curve, 7 points were evaluated in duplicate. A "no template" control of nuclease-free water was included to exclude DNA contamination in the PCR mix. Table 1 shows the primers used in this study. The double-stranded DNA was amplified using Bio-Rad CFX96 Touch (Hercules, CA, USA) under the following PCR conditions: initial denaturing at 95 °C for 30 s, 40 cycles of denaturing at 95 °C for 15 s, various annealing temperatures according to Integrated DNA Technologies (IDT) for 30 s, and elongating at 60 °C for 30 s.

The data on the expression levels of the genes were obtained as Cp values based on the "second derivative maximum" (automated method) as computed by Bio-Rad CFX Maestro 1.1 (Version 4.1.2433.1219, Hercules, CA, USA). The assays were quantified by including a standard curve in the real-time qPCR analysis, and a standard curve with 4 points was prepared with a 1:10 dilution (duplicates). The software produced a Cp vs. log 10 concentrations graph, and the efficiencies were calculated as 10 (1/slope). The specificity of the amplified real-time RT-PCR products was verified using melting curve analysis (60–95 °C) after 40 cycles, resulting in several different specific products with specific melting temperatures.

2.6. Collection of Microbial Samples and DNA Extraction of Intestinal Content

The cecum was removed in a sterile manner and treated as described previously [15,34]. Briefly, the cecum contents were homogenized using a vortex and glass beads. Debris was removed and the supernatant was collected and centrifuged. The pellet was washed and stored at −20 °C until DNA extraction. Then, the pellet was resuspended in ethylenediaminetetraacetic acid (EDTA) and treated with lysozyme (Sigma Aldrich CO., St. Louis, MO, USA; final concentration 10 mg/mL). A Wizard Genomic DNA purification kit (Promega Corp., Madison, WI, USA) was used to isolate the bacterial genomic DNA.

2.7. Primer Design and PCR Amplification of Bacterial 16S rDNA

Primers for *Bifidobacterium*, *Lactobacillus*, *Escherichia coli*, and *Clostridium* were used [10]. The universal primers were designed with the invariant region in the 16S rRNA of the bacteria and were used as internal standards. The relative abundance of each examined bacterium was evaluated as previously described [38,41]. Briefly, the PCR products were quantified using 2% agarose gel and stained with ethidium bromide. All products are expressed relative to the content of the universal 16S rRNA primer product and the proportions of each examined bacterial product.

2.8. Morphological Examination of Duodenal Tissue

Intestinal morphology was performed as previously described [8,36]. Briefly, the duodenum samples were fixed in fresh 4% (v/v) buffered formaldehyde; numerous sections were cut and placed on glass slides. The sections were deparaffinized in xylene and rehydrated in ethanol. Afterwards, the slides were stained with Alcian Blue/Periodic acid-Schiff and examined using light microscopy (CellSens Standard software, Olympus, Waltham, MA, USA). The following morphometric measurements were evaluated: villus height (μM), villus surface area (μM), depth of crypts (μM), Paneth cell number and diameter (μM), goblet cell number and goblet cell diameter (μM) in the villi and crypts, and goblet cell type (acidic, neutral, and mixed). In total, 4 segments for each biological sample (n = 3/treatment group) were performed and 10 randomly selected villi and crypts were analyzed per segment (40 replicates per biological sample). For the Alcian Blue and Periodic acid-Schiff stain, the segments were only counted for the type of goblet cells (acid, neutral, or mixed) in the villi epithelium and in the crypts. The goblet cells were enumerated at 10 villi or crypts/sample, and the means were calculated for statistical analysis. The *villus surface area* was obtained using the following equation:

$$Villus\ surface\ area = 2\frac{VW}{2} \times VL \quad (1)$$

where VW = the villus width average of three measurements, and VL = the villus length.

A representative image of the histological cross-section of the duodenum from each experimental group indicates the villus surface area measurements.

2.9. Statistical Analysis

Experimental treatments for the intra-amniotic administration procedure were arranged in a completely randomized design. All the results were expressed as means ± standard error deviation (SED) from 7 to 9 biological samples per treatment group (according to hatching). Differences were considered significant when $p < 0.05$.

The Shapiro–Wilk normality test was used to evaluate values for normal distribution and variance homogeneity. Normally distributed results were analyzed using a one-way analysis of variance (ANOVA). For a significant "p-value," the post hoc Duncan test was used for those with a normal distribution. The means without normal distribution were analyzed using Kruskal–Wallis and a post hoc Dunn's test. The statistical analyses were performed using the statistical software IBM SPSS Statistics®, version 25.

The correlation between the biomarkers of intestinal health, the bacterial population, and histological parameters was analyzed using Spearman's rank correlation coefficient. GraphPad Prism® version 9.0 software packages (GraphPad Software Inc., San Diego, CA, USA) were used for graphics.

3. Results

3.1. Body Weight

The body weight was similar among all the groups compared according to one way ANOVA followed by the post hoc Duncan test: non-injected (35.00 ± 0.82), 18 MΩ H_2O (35.13 ± 1.09), hydrolyzed chia protein (36.56 ± 0.65), hydrolyzed chia protein + *Lacticaseibacillus paracasei* (34.38 ± 1.28), and probiotic (*Lacticaseibacillus paracasei*) (34.78 ± 0.95).

3.2. Effect of the Chia Protein and/or Probiotic on the Gene Expression of Intestinal Inflammation, Intestinal Barrier Proteins, and Brush Border Membrane Functional Proteins

In the duodenum, the expression of tumor necrosis factor alpha (TNF-α) decreased in the groups that received the hydrolyzed chia protein and the hydrolyzed chia protein + a probiotic compared to the non-injected group, but was similar to the group injected with H_2O. Furthermore, the probiotic group was not able to reduce the expression of TNF-α when compared to both control groups (NI and H_2O); however, it decreased expression of nuclear factor kappa beta (NF-κβ1) relative to all other groups. Regarding intestinal barrier proteins, the groups that received the hydrolyzed chia protein or the hydrolyzed chia protein + the probiotic showed a higher expression of occluding (OCLN) compared to both control groups (NI and H_2O); the probiotic group did not show an increase in this parameter. The same occurred for mucin-2 (MUC2); the hydrolyzed chia protein and hydrolyzed chia protein + probiotic groups showed an increase in MUC2 expression, but only in relation to the group injected with H_2O, and the probiotic group showed no change for this parameter. When we evaluated the expression of intestinal functionality proteins, we observed that the hydrolyzed chia protein group had higher AP expression compared to the control groups (NI and H_2O), and the hydrolyzed chia protein + probiotic group had higher AP expression in relation to the control group injected with H_2O; meanwhile, the probiotic group presented AP expression similar to the group injected with H_2O and lower than that of the other groups. In relation to SI expression, there was no difference between any of the experimental groups (Figure 1).

	NF-κβ1	TNF-α	OCLN	MUC2	AP	SI
NI	15.48 ± 0.54 [a]	4.09 ± 0.09 [a]	1.06 ± 0.00 [b]	13.27 ± 0.58 [a]	1.53 ± 0.01 [bc]	3.25 ± 0.02 [a]
H_2O	15.26 ± 0.81 [a]	3.54 ± 0.10 [c]	1.06 ± 0.00 [b]	10.69 ± 0.97 [b]	1.49 ± 0.01 [cd]	3.25 ± 0.01 [a]
CP	16.79 ± 1.07 [a]	3.56 ± 0.10 [c]	1.08 ± 0.00 [a]	12.89 ± 0.66 [a]	1.58 ± 0.02 [a]	3.30 ± 0.03 [a]
CPP	14.67 ± 0.56 [a]	3.64 ± 0.08 [bc]	1.08 ± 0.00 [a]	12.90 ± 0.45 [a]	1.57 ± 0.01 [ab]	3.31 ± 0.02 [a]
P	12.33 ± 0.61 [b]	3.88 ± 0.07 [ab]	1.06 ± 0.00 [b]	9.25 ± 0.33 [b]	1.46 ± 0.01 [d]	3.25 ± 0.02 [a]

High AU

Low AU

Figure 1. Effect of the intra-amniotic administration of chia protein and/or a probiotic on intestinal (duodenum) gene expression. NI: non-injected; H_2O: 18 MΩ H_2O; CP: hydrolyzed chia protein; CPP: hydrolyzed chia protein + *Lacticaseibacillus paracasei*; P: probiotic (*Lacticaseibacillus paracasei*); NF-κβ1: nuclear factor kappa beta; TNF-α: tumor necrosis factor-alpha; OCLN: occludin; MUC2: mucin 2; AP: aminopeptidase; SI: sucrose isomaltase; AU: arbitrary unit. Values are expressed as means ± SEM, $n = 5$/group. Per gene (in the same column), red depicts lower gene expression levels and green depicts higher gene expression levels. $^{a–d}$ Per gene (in the same column), treatment group means not indicated by the same letter are significantly different ($p < 0.05$) according to one way ANOVA followed by the post hoc Duncan test.

3.3. *Effect of Chia Protein and/or a Probiotic on the Bacterial Population in Cecum Contents*

The intra-amniotic administration of hydrolyzed chia protein and hydrolyzed chia protein + a probiotic reduced the abundance of *Bifidobacterium* compared to both control groups (NI and H_2O) and the P group. The group probiotic group showed a similar abundance of *Bifidobacterium* to the NI and H_2O groups. The hydrolyzed chia protein group showed a similar abundance of *Lactobacillus* compared to the H_2O injection control group and a higher abundance than the NI control group. The abundance of *Lactobacillus* in the hydrolyzed chia protein + probiotic group was similar to the CP and higher than both control groups (NI and H_2O); additionally, the probiotic group showed a higher abundance of *Lactobacillus* compared to all other groups. Furthermore, the probiotic group showed a higher abundance of *E. coli* compared to all other groups, and all other groups presented a similar abundance of *E. coli*. The abundance of *Clostridium* was similar among the NI, H_2O, hydrolyzed chia protein, and hydrolyzed chia protein + probiotic groups; it was reduced in the probiotic group (Figure 2).

	Bifidobacterium	*Lactobacillus*	*E. coli*	*Clostridium*
NI	0.87 ± 0.01 [a]	0.71 ± 0.04 [d]	0.88 ± 0.03 [b]	0.77 ± 0.11 [a]
H_2O	0.89 ± 0.02 [a]	0.73 ± 0.02 [cd]	0.94 ± 0.01 [b]	0.59 ± 0.01 [a]
CP	0.58 ± 0.02 [b]	0.82 ± 0.02 [bc]	0.80 ± 0.01 [b]	0.60 ± 0.03 [a]
CPP	0.55 ± 0.02 [b]	0.84 ± 0.01 [b]	0.82 ± 0.01 [b]	0.52 ± 0.17 [a]
P	0.89 ± 0.04 [a]	1.18 ± 0.05 [a]	1.29 ± 0.10 [a]	0.00 ± 0.00 [b]

High density (AU)

Low density (AU)

Figure 2. Effect of the intra-amniotic administration of chia protein and/or a probiotic on genera- and species-level bacterial populations from cecal content. NI: non-injected; H_2O: 18 MΩ H_2O; CP: hydrolyzed chia protein; CPP: hydrolyzed chia protein + *Lacticaseibacillus paracasei*; P: probiotic (*Lacticaseibacillus paracasei*); AU: arbitrary unit. Values are expressed as means ± SEM, $n = 5$/group. Per bacterial category (in the same column), red depicts lower gene expression levels and green depicts higher gene expression levels. $^{a–d}$ Per bacterial category (in the same column), treatment group means not indicated by the same letter are significantly different ($p < 0.05$) according to one way ANOVA followed by a post hoc Duncan test.

3.4. Effect of Chia Protein and/or a Probiotic on Duodenal Morphological Parameters

The villi height and villus surface area were increased in all treatment groups (CP, CPP, and P). The hydrolyzed chia protein (CP) group showed villi height and villus surface area higher than all other groups. The hydrolyzed chia protein + probiotic (CPP) group showed villi height and villus surface area higher than both of control groups (NI and H_2O), and the probiotic (P) group presented villi height greater than both control groups (NI and H_2O) and CPP group; the villus surface area was higher than both control groups (NI and H_2O) and similar to the CPP and CP groups (Figures 3 and 4).

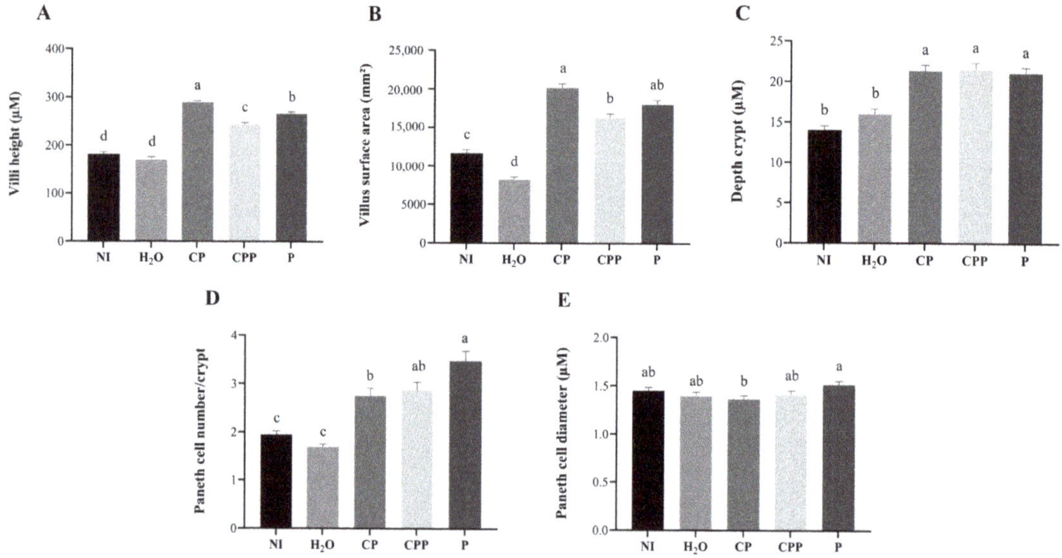

Figure 3. Effect of the intra-amniotic administration of chia protein and/or a probiotic on duodenal morphological parameters. (**A**) Villi height (μM); (**B**) villus surface area (mm^2); (**C**) crypt depth (μM); (**D**) Paneth cell number/crypt; (**E**) Paneth cell diameter (μM); NI: non-injected; H_2O: 18 MΩ H_2O; CP: hydrolyzed chia protein; CPP: hydrolyzed chia protein + *Lacticaseibacillus paracasei*; P: probiotic (*Lacticaseibacillus paracasei*). Values are expressed as means ± SEM, n = 3 animals/group, 4 sections, 10 measurements. Treatment group means for specific variables followed by the same letter are not significantly different ($p > 0.05$) according to Kruskal–Wallis and a post hoc Dunn's test. [a–d] Treatment group means not indicated by the same letter are significantly different ($p < 0.05$) according to Kruskal–Wallis and a post hoc Dunn's test.

The depth of the crypts was increased in all treatment groups (CP, CPP, and P) compared to both of the control groups (NI and H_2O) and was similar among the experimental groups (Figure 3).

The treatment groups (CP, CPP, and P) had a higher number of Paneth cells compared to the control groups (NI and H_2O), and the probiotic group had a higher number of Paneth cells compared to the CP group. The Paneth diameter was similar among all the groups, except when comparing the CP and P groups; the administration of the probiotic increased the diameter of Paneth cells compared to those in the group injected with hydrolyzed chia protein (Figure 3).

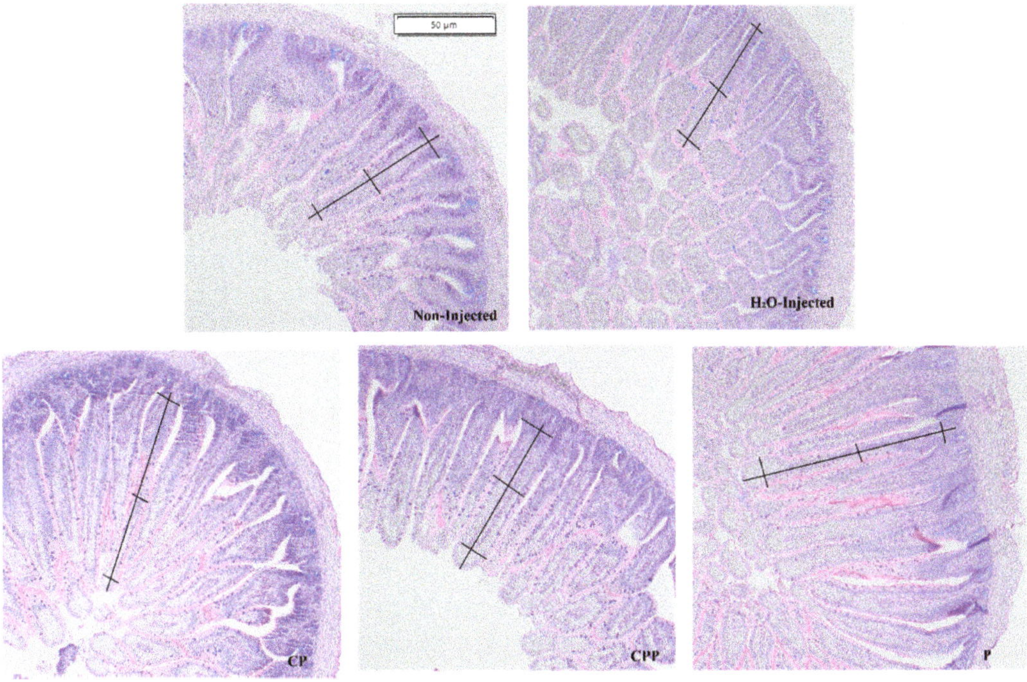

Figure 4. Representation of duodenal cross section morphology for each treatment group. The *villi surface area*, depicted in black, was obtained using the following equation: $Villus\ surface\ area = 2\frac{VW}{2} \times VL$, where *VW* = average villus width of three measurements, and *VL* = villus length. NI: non-injected; H_2O: 18 MΩ H_2O; CP: hydrolyzed chia protein; CPP: hydrolyzed chia protein + *Lacticaseibacillus paracasei*; P: probiotic (*Lacticaseibacillus paracasei*).

In the villi, the goblet cell numbers were not different among the treatment groups (CP, CPP, and P); these groups showed numbers of goblet cells similar to the NI group, and lower than the H_2O group. However, the intra-amniotic administration of hydrolyzed chia protein, CPP, and P increased the diameter of the goblet cells. The CPP and P groups were similar and presented goblet cells with larger diameters compared to both of the control groups (NI and H_2O). CPP was similar to CP, and P was higher than CP. Similar results were found regarding the diameter of the goblet cells of the crypts. Regarding the goblet cell numbers in the crypt, the H_2O, CP, CPP, and P groups showed fewer goblet cells than the NI group, and the CPP group had lower numbers than CP and P (Table 2).

In the villi, the treatment groups (CP, CPP, and P) exhibited reduced numbers of acidic and neutral goblet cells and increased numbers of mixed goblet cells. In the crypt, the treatment groups (CP, CPP, and P) reduced neutral goblet cells; for acidic goblet cells, the CP and P groups were similar to both of the controls, and CPP was similar to the H_2O group. Regarding the mixed goblet cells, the P group was similar to both of the controls and the CP and CPP groups were similar to the H_2O group (Table 2).

3.5. Correlation Analysis

Spearman correlation analysis was used to assess the relationships between the intestinal parameters investigated in this study, and significant correlations were observed. Positive correlations were observed between OCLN and MUC2, as well as between AP and the villi surface area. Furthermore, negative correlations were observed between *E. coli* and OCLN, MUC2, AP, and the villi surface area (Figure 5).

Table 2. Effect of chia protein and/or a probiotic on goblet cells.

	NI	H_2O	CP	CPP	P
Villi Goblet Cell Number	24.68 ± 0.74^{b}	38.38 ± 0.91^{a}	26.40 ± 0.72^{b}	27.46 ± 0.71^{b}	25.79 ± 0.83^{b}
Villi Goblet Cell Diameter (μM)	2.45 ± 0.06^{c}	2.20 ± 0.05^{d}	2.60 ± 0.06^{bc}	2.94 ± 0.10^{ab}	3.13 ± 0.08^{a}
Villi Goblet Cell Type Number					
Acidic	15.28 ± 0.71^{b}	26.71 ± 1.12^{a}	10.29 ± 0.59^{c}	8.49 ± 0.43^{c}	8.82 ± 0.52^{c}
Neutral	0.79 ± 0.13^{a}	0.10 ± 0.04^{b}	0.28 ± 0.09^{b}	0.43 ± 0.12^{b}	0.47 ± 0.12^{b}
Mixed	8.68 ± 0.57^{c}	11.57 ± 0.66^{a}	15.83 ± 0.66^{a}	18.54 ± 0.65^{a}	16.51 ± 0.73^{a}
Crypt Goblet Cell Number	12.67 ± 0.55^{a}	10.95 ± 0.62^{b}	10.03 ± 0.36^{b}	8.04 ± 0.38^{c}	10.18 ± 0.40^{b}
Crypt Goblet Cell Diameter (μM)	2.92 ± 0.05^{d}	3.13 ± 0.05^{cd}	3.20 ± 0.06^{bc}	3.40 ± 0.06^{ab}	3.62 ± 0.07^{a}
Crypt Goblet Cell Type Number					
Acidic	8.53 ± 0.43^{a}	7.88 ± 0.51^{ab}	7.63 ± 0.30^{a}	5.98 ± 0.30^{b}	7.14 ± 0.33^{ab}
Neutral	0.41 ± 0.06^{a}	0.50 ± 0.07^{a}	0.09 ± 0.03^{b}	0.08 ± 0.03^{b}	0.04 ± 0.02^{b}
Mixed	3.73 ± 0.27^{a}	2.58 ± 0.21^{bc}	2.33 ± 0.15^{bc}	1.98 ± 0.17^{c}	3.00 ± 0.17^{ab}

NI: non-injected; H_2O: 18 MΩ H_2O; CP: hydrolyzed chia protein; CPP: hydrolyzed chia protein + *Lacticaseibacillus paracasei*; P: probiotic (*Lacticaseibacillus paracasei*). Values are expressed as means ± SED, n = 3 animals/group, 4 sections, 10 measurements. $^{a\text{–}d}$ Treatment group means not indicated by the same letter are significantly different ($p < 0.05$) according to Kruskal–Wallis and a post hoc Dunn's test.

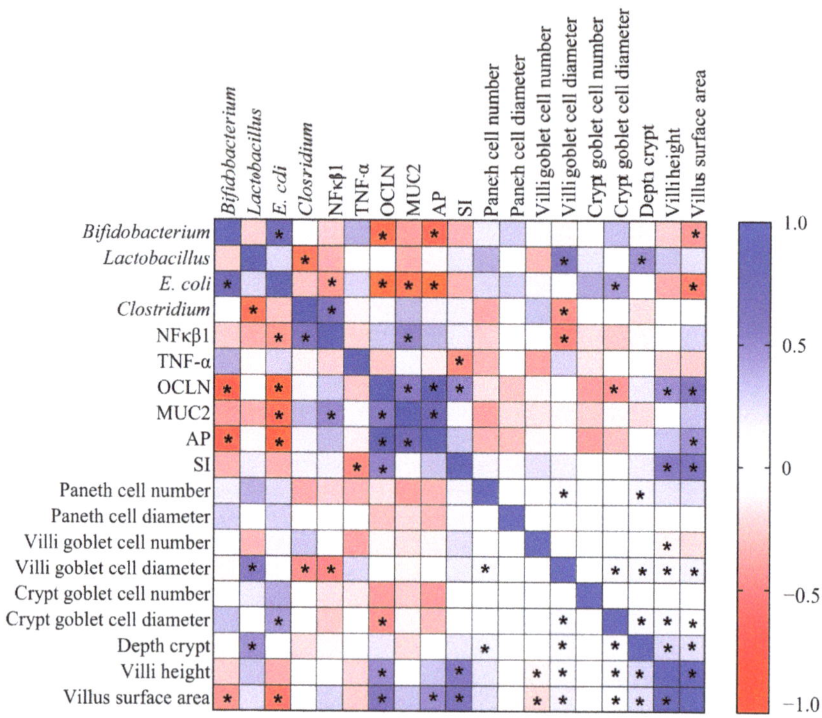

Figure 5. Heatmap of Spearman correlation analysis. NF-κβ1: nuclear factor kappa beta; TNF-α: tumor necrosis factor alpha; OCLN: occludin; MUC2: mucin 2; AP: aminopeptidase; SI: sucrose isomaltase. * Indicates a statistically significant difference ($p < 0.05$).

4. Discussion

Chia is a seed with a rich nutritional composition; it is a good source of bioactive peptides [11,12]. However, the potential effects of chia seeds' hydrolyzed proteins in combination with the probiotic *Lacticaseibacillus paracasei* on intestinal bacterial populations, the intestinal barrier, inflammatory responses, and BBM functionality in vivo have not been investigated. Our results demonstrated that the intra-amniotic administration of the probiotic downregulated NF-κβ1 gene expression, increased the cecal *Lactobacillus* and *E. coli* populations, decreased *Clostridium*, and increased the diameter of goblet cells (Figures 1 and 2 and Table 2). Furthermore, hydrolyzed chia protein downregulated TNF-α expression; moreover, the gene expression of OCLN, MUC2, and AP were improved in the presence of the hydrolyzed protein, either alone or with the probiotic. Although the hydrolyzed chia protein decreased the number of goblet cells, their diameter was increased. The villi height, villi surface area, crypt depth, and number of Paneth cells were increased in all treatment groups (CP, CPP, and P) compared to the control groups (NI and H_2O).

In the present study, administration of the probiotic downregulated NF-κβ1 gene expression compared to both of the control groups (NI and H_2O) and to the other treatment groups (CP and CPP). Moreover, TNF-α expression decreased in the groups administered the hydrolyzed chia protein (Figure 1). NF-κβ1 is responsible for inducing cytokine gene expression to control inflammatory and immune responses [42]. NF-κβ is translocated to the cytosol and into the nucleus to initiate the pathway to drive the expression of target genes such as TNF-α, IL1β, IL6, and IL10 [43]. The hydrolyzed chia protein also demonstrated effects in reducing the secretion of TNF-α in an in vitro study and showed effects in the adipose tissue in an in vivo study [14,44]. These observations suggest that the probiotic has an anti-inflammatory effect, as was previously reported [20,45]. Likewise, the hydrolyzed chia protein might have similar effects, as demonstrated in the current study. Probiotic bacteria play a role in the regulation of the immune response by stimulating immune cells through the modulation of intestinal microbiota and downregulating inflammation [21,22]. Some of these bacteria might be useful in mitigating intestinal inflammation; specifically, *Lactobacillus* and *Bifidobacteria* demonstrate antimicrobial effects by affecting both local and systemic immune responses [20,45]. Peptides with hydrophobic amino acids have more ionizable groups that block free radicals and then increase antioxidant activity [46]. This activity was also observed in peptides with fewer than 20 amino acid residues per molecule, since small peptides are better able to cross the intestinal barrier and exert their biological effects [47]. Furthermore, peptides with hydrophobicity $\leq$20 kcal mol^{-1} are more effective at penetrating the cell membrane and exercising effects on the molecule [48]. Most of the peptides found in our hydrolyzed chia protein showed these characteristics, which may explain their beneficial effects against inflammation [12,13]. The main storage protein fractions found in chia seeds are albumin, globulin, glutelin, and prolamin [12,13].

Intestinal inflammation is associated with impaired barrier function, leading to increased intestinal permeability and bacterial translocation [43]. In our study, both groups administered hydrolyzed chia protein (CP and CPP) showed increased gene expression of OCLN and MUC2 compared to the H_2O-injected control group. TNF-α is a pro-inflammatory cytokine that plays a key role in the inflammatory cascade that causes increased intestinal epithelial permeability and then leads to chronic intestinal inflammation [26,49,50]. Thus, as this study found, the bioactive peptides present in hydrolyzed chia protein can reduce the disruption of the intestinal barrier by reducing inflammatory factors such as TNF-α [51].

The mucus layer acts as the first barrier on the surface of the gastrointestinal tract; it is an innate host defense mechanism [52]. Mucus is essential for regulating the homeostasis of the intestinal microbiota and preventing disease, which it does by protecting the gastrointestinal barrier from pathogenic microorganisms, toxins, and other irritants [53–55]. MUC2 is the main mucin specifically produced in goblet cells [52]. Under intestinal inflammatory conditions, with a reduction in goblet cells, the expression of MUC2 is reduced, and the mucus loses its barrier function and exposes the mucous membrane to inflammatory substances, such as pathogenic microorganisms, toxins, and lipopolysaccharides [52,55]. We observed that hydrolyzed chia protein, either alone or when combined with a probiotic, increased the gene expression of MUC2; this is probably due to the increased diameter of the goblet cells, which improves the protection of the intestinal barrier. As the tight junction is interdependent with the mucus barrier, the loss of one can reduce the other; this interdependence may result from the regulation of signals that regulate both the mucus and the tight junction [52]. Based on the correlation analysis (Figure 5), the gene expression of OCLN was positively correlated with the gene expression of MUC2. Moreover, with the increase in OCLN and MUC2, the intestinal barrier can protect the host against the permeability of pathogenic components.

Aminopeptidase (AP) is an exopeptidase that cleaves amino acids from the N-terminus of peptides; it is located on the enterocyte's BBM [28]. The upregulation of the BBM and functional genes' expression reflects intestinal development and digestive capabilities. Thus, it also affects the potential increased absorption of nutrients [28,56]. In the current study, AP gene expression was upregulated, probably due to the increase in the villi height and villi surface area, which is accordance with our correlation analysis. This increase in the villi surface area leads to more enterocytes and, consequently, higher functionality and absorption capacity [56]. This improvement in absorption capacity is important for the absorption of peptides that play an anti-inflammatory role. In addition, the administration of CP, CPP, and P maintain the gene expression of sucrose isomaltase (SI), compared to the non-injected and H_2O-injected groups (Figure 1).

The intra-amniotic administration of hydrolyzed chia protein, hydrolyzed chia protein + probiotic or just the probiotic (*L. paracasei*) were able to increase villi height, villi surface area and crypt depth, but the increases were higher when these compounds were administered separately (Figure 3). The intestinal villi play an essential role in the digestion and absorption of nutrients, as the villi increase the internal surface area and, in turn, the digestive and absorptive capacities of the BBM [8,28,56]. The crypts are comprised of continuously proliferating stem cells; these are responsible for the formation of enterocytes, which play a key role due to their nutrients' absorptive ability [57]. Deeper crypts lead to an increase in the secretion of digestive enzymes [57]. Thus, the surface area of the villi and the crypts' depth are indictors of intestinal developmental and functional status. Therefore, an increase in these morphometric parameters due to CP, CPP, or P can improve the digestive and absorptive capabilities of the BBM [28], as demonstrated by the correlation analysis (Figure 5).

Paneth cells are found in the crypts; they are secretory cells that produce antimicrobial peptides, proteins, and other important components of host defense and immunity [58]. Dysfunction of Paneth cells can disrupt these functions, leading to an imbalance in the gastrointestinal tract. Decreased numbers of Paneth cells, followed by the leakage of bacteria, are often seen in various infectious diseases [58]. Here, we found that, in the treatment groups (CP, CPP, and P), the number of Paneth cells, but not their diameter, was increased; therefore, we suggest that the intra-amniotic administration of hydrolyzed chia protein, hydrolyzed chia protein + a probiotic, or just the probiotic (*L. paracasei*) improved Paneth cell development without affecting the size of the cells. This indicates that the antimicrobial peptides secreted by Paneth cells were not produced, since there were no stimuli such as inflammation or pathogenic bacteria. Similar results were demonstrated after the intra-amniotic administration of black corn soluble extract, a source of phenolics [34].

Goblet cells produce mucin, the most important substance in the mucus layer, which forms a gel barrier against pathogenic bacteria [54]. Our results showed that the investigated probiotic increased the diameter of goblet cells; meanwhile, the intra-amniotic administration of hydrolyzed chia protein + a probiotic decreased the number of goblet cells, but their diameter was increased. It is possible that the observed reduction in the inflammatory biomarker (TNF-α) may have led to an increased goblet cell diameter and, therefore, increased MUC2 gene expression. Regarding the types of goblet cells, the treatment groups (CP, CPP, and P) reduced the number of acidic goblet cells in the villi. An acidic pH in the intestine facilitates the growth of beneficial bacteria over detrimental bacteria [59]. Therefore, the reduction in acidic goblet cells might be associated with the increase in the *E. coli* population, as seen in the probiotic group [60].

The microbial analysis revealed that the probiotic *L. paracasei* increased the abundance of *Lactobacillus* and *E. coli* and reduced *Clostridium* populations in the cecal contents, in comparison to all the other groups (Figure 2). Furthermore, the intra-amniotic administration of hydrolyzed chia protein and hydrolyzed chia protein + a probiotic (*L. paracasei*) reduced the abundance of *Bifidobacterium* but increased *Lactobacillus*. *Lactobacillus* species are considered probiotics due to their immunomodulatory and anti-inflammatory actions, their competition with pathogens, and their stimulation of the release of antimicrobial substances, especially mucin, which activates the MUC2 gene and prevents pathogens from adhering to the epithelial barrier [45]. In contrast, the E. coli genus may impair the epithelial barrier by disrupting tight junction proteins [53], as indicated by the negative correlation between *E. coli* and OCLN (Figure 5). The intra-amniotic administration of the probiotic (*L. paracasei*) did not improve the intestinal barrier, as we found in the group administered the hydrolyzed chia protein; this finding might be related to the increase in *E. coli*. Mucus is an important barrier against potentially pathogenic bacteria [52,55]. In the probiotic group, the *E. coli* population was increased and MUC2 gene expression remained constant; however, when the hydrolyzed chia protein was added to this probiotic, MUC2 expression was upregulated, and *E. coli* abundance remained similar ($p > 0.05$) to that seen in the control groups. Therefore, it is likely that the presence of the hydrolyzed chia protein can increase MUC2 and prevent the increase in *E. coli*.

It is important to highlight that, in this study, we investigated the administration of only one strain of a probiotic, but the literature highlights the importance of administering different strains of probiotics, to obtain greater benefits [20]. The *Lacticaseibacillus paracasei* used in this study is an innovative culture to be tested on plant products without fermentation, which has a promising probiotic effect.

The administration of hydrolyzed chia protein and the administration of the probiotic (*L. paracasei*) were able to improve intestinal health, and we found more pronounced effects when they were administered separately. When the probiotic is administered with the protein, it can protect the probiotic during digestion, and it can arrive in the intestine in a more intact state or in different quantities. This may be the reason for the differences in effects when the compounds were administered alone or together. The probiotic had effects on inflammation and microbiota composition, the hydrolyzed chia protein had effects on inflammation, barrier function, and functionality, and both had an impact on intestinal morphology. Thus, the administration of both can bring about benefits for intestinal health (Figure 6). The current study is the first to investigate the intestinal effects of hydrolyzed chia protein and the probiotic *Lacticaseibacillus paracasei* in vivo; thus, future studies that aim to assess the effects in a long-term feeding trial should be conducted to better understand these effects, since the hydrolyzed chia protein and the probiotic *L. paracasei* presented good results in relation to the improvement of intestinal health.

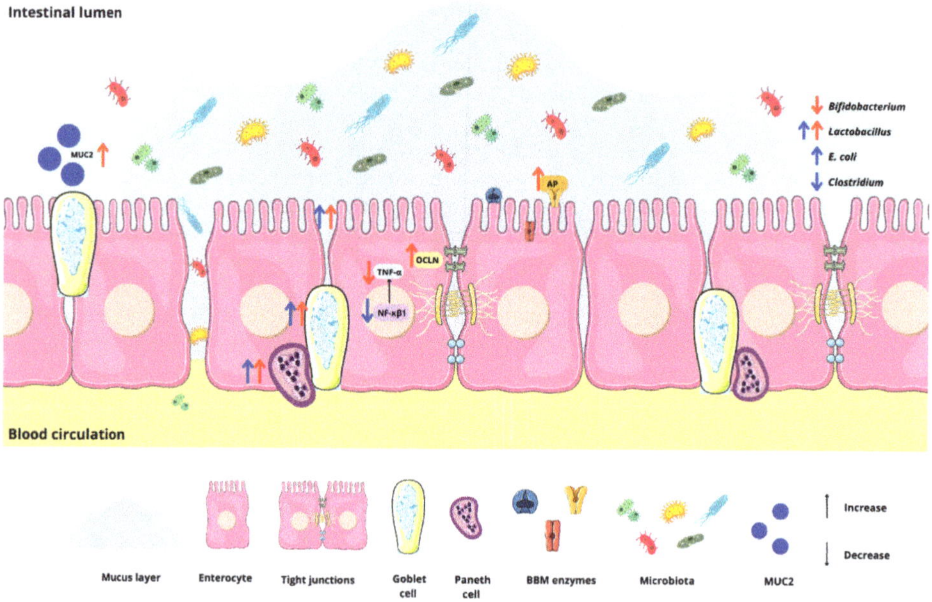

Figure 6. Graphical representation of the results. The intra-amniotic administration of hydrolyzed chia protein reduced TNF-α and the *Bifidobacterium* population and increased OCLN, MUC2, AP, the *Lactobacillus* population, villi height, villi surface area and crypt depth, the Paneth cell number, and the diameter of goblet cells. The administration of the probiotic reduced NF-κβ1 and *Clostridium* populations, and increased *Lactobacillus* and *E. coli*, the villi height, villi surface area, and crypt depth, the Paneth cell number, and the diameter of the goblet cells. NF-κβ1: nuclear factor kappa beta; TNF-α: tumor necrosis factor alpha; OCLN: occludin; MUC2: mucin 2; AP: aminopeptidase; red arrow: effects of hydrolyzed chia protein; blue arrow: effects of *L. paracasei*.

5. Conclusions

The intra-amniotic administration of hydrolyzed chia protein decreased TNF-α, increased the *Lactobacillus* population, OCLN, MUC2, and AP, increased the villi height, villi surface area, and crypt depth, and increased the Paneth cell number and goblet cell diameter. The administration of the probiotic (*L. paracasei*) promoted NF-κβ1 reduction, increased *Lactobacillus* and reduced *Clostridium*, and increased the villi height, villi surface area, crypt depth, Paneth cell number, and goblet cell diameter. Hydrolyzed chia protein and the probiotic (*L. paracasei*) modulate some aspects of intestinal health, and we found more pronounced effects when they were administered separately. These findings suggest that the intra-amniotic administration of hydrolyzed chia protein or a probiotic (*L. paracasei*) improved intestinal health. Therefore, the current study indicates that additional long-term in vivo feeding trials are now warranted to further investigate the observed effects of dietary hydrolyzed chia protein and the experimental probiotic.

Author Contributions: Conceptualization, M.D.V.M., H.S.D.M., and E.T.; methodology, M.D.V.M., N.K., M.G., D.D.S. and B.P.D.S.; formal analysis, M.D.V.M., N.K. and D.D.S.; investigation, M.D.V.M., H.S.D.M., N.K., D.D.S., B.P.D.S. and E.T.; resources, H.S.D.M., K.M.O.D.S., J.P.L., E.G.d.M. and E.T.; data curation, M.D.V.M. and N.K.; writing—original draft preparation, M.D.V.M.; writing—review and editing, M.D.V.M., H.S.D.M., M.G., K.M.O.D.S., E.G.d.M. and E.T.; supervision, H.S.D.M. and E.T.; project administration, H.S.D.M. and E.T.; funding acquisition, E.T. All authors have read and agreed to the published version of the manuscript.

Funding: This research received no external funding.

Institutional Review Board Statement: The animal protocol used in this study was conducted according to the guidelines of the Declaration of Helsinki and approved by the Cornell University Institutional Animal Care and Use Committee by ethic approval code 2020-0077.

Informed Consent Statement: Not applicable.

Data Availability Statement: Not applicable.

Acknowledgments: The authors thank the Coordination for the Improvement of Higher Educational Personnel (CAPES), Brazil, for providing the scholarship support in the Capes-Print Program (process number 88887.694292/2022-00), and the National Council of Technological and Scientific Development (CNPq, Brazil) for the Research Productivity fellowships [PQ2—grant number 310910/2020-0]. Additionally, they thank EMBRAPA (Brazilian Agricultural Research Corporation, SEG 20.20.03.054.00.00), Rio de Janeiro, Brazil. All authors have consented to all the acknowledgements included here.

Conflicts of Interest: The authors declare no conflict of interest.

References

1. Enes, B.N.; Moreira, L.D.P.D.; Toledo, R.C.L.; Moraes, É.A.; de Castro Moreira, M.E.; Hermsdorff, H.H.M.; Noratto, G.; Mertens-Talcott, S.U.; Talcott, S.; Martino, H.S.D. Effect of Different Fractions of Chia (*Salvia hispanica* L.) on Glucose Metabolism, in Vivo and in Vitro. *J. Funct. Foods* **2020**, *71*, 157–164. [CrossRef]
2. Da Silva, B.P.; Dias, D.M.; de Castro Moreira, M.E.; Toledo, R.C.L.; da Matta, S.L.P.; Della Lucia, C.M.; Martino, H.S.D.; Pinheiro-Sant'Ana, H.M. Chia Seed Shows Good Protein Quality, Hypoglycemic Effect and Improves the Lipid Profile and Liver and Intestinal Morphology of Wistar Rats. *Plant Foods Hum. Nutr.* **2016**, *71*, 225–230. [CrossRef] [PubMed]
3. Da Silva, B.P.; Toledo, R.C.L.; Grancieri, M.; de Castro Moreira, M.E.; Medina, N.R.; Silva, R.R.; Costa, N.M.B.; Martino, H.S.D. Effects of Chia (*Salvia hispanica* L.) on Calcium Bioavailability and Inflammation in Wistar Rats. *Food Res. Int.* **2019**, *116*, 592–599. [CrossRef] [PubMed]
4. Da Silva, B.P.; Toledo, R.C.L.; Mishima, M.D.V.; de Castro Moreira, M.E.; Vasconcelos, C.M.; Pereira, C.E.R.; Favarato, L.S.C.; Costa, N.M.B.; Martino, H.S.D. Effects of Chia (*Salvia hispanica* L.) on Oxidative Stress and Inflammation in Ovariectomized Adult Female Wistar Rats. *Food Funct.* **2019**, *10*, 4036–4045. [CrossRef] [PubMed]
5. Silva, L.D.A.; Verneque, B.J.F.; Mota, A.P.L.; Duarte, C.K. Chia Seed (*Salvia hispanica* L.) Consumption and Lipid Profile: A Systematic Review and Meta-Analysis. *Food Funct.* **2021**, *12*, 8835–8849. [CrossRef]
6. Vega Joubert, M.B.; Degrave, V.; Ingaramo, P.; Oliva, M.E.; D'Alessandro, M.E. *Salvia hispanica* L. (Chia) Seed Improves Liver Inflammation and Endothelial Dysfunction in an Experimental Model of Metabolic Syndrome. *Food Funct.* **2022**, *13*, 11249–11261. [CrossRef] [PubMed]
7. Mishima, M.D.V.; Ladeira, L.C.M.; da Silva, B.P.; Toledo, R.C.L.; de Oliveira, T.V.; Costa, N.M.B.; Martino, H.S.D. Cardioprotective Action of Chia (*Salvia hispanica* L.) in Ovariectomized Rats Fed a High Fat Diet. *Food Funct.* **2021**, *12*, 3069–3082. [CrossRef] [PubMed]
8. Da Silva, B.P.; Kolba, N.; Martino, H.S.D.; Hart, J.; Tako, E. Soluble Extracts from Chia Seed (*Salvia hispanica* L.) Affect Brush Border Membrane Functionality, Morphology and Intestinal Bacterial Populations in Vivo (*Gallus gallus*). *Nutrients* **2019**, *11*, 2457. [CrossRef]
9. Mishima, M.D.V.; Da Silva, B.P.; Gomes, M.J.C.; Toledo, R.C.L.; Mantovani, H.C.; de São José, V.P.B.; Costa, N.M.B.; Tako, E.; Martino, H.S.D. Effect of Chia (*Salvia hispanica* L.) Associated with High-Fat Diet on the Intestinal Health of Wistar Rats. *Nutrients* **2022**, *14*, 4924. [CrossRef]
10. Mishima, M.D.V.; Da Silva, B.P.; Gomes, M.J.C.; Toledo, R.C.L.; Pereira, C.E.R.; Costa, N.M.B.; Martino, H.S.D. Effect of Chia Flour Associated with High Fat Diet on Intestinal Health in Female Ovariectomized Wistar Rats. *Eur. J. Nutr.* **2022**, *62*, 905–919. [CrossRef]
11. Da Silva, B.P.; Anunciação, P.C.; da Silva Matyelka, J.C.; Della Lucia, C.M.; Martino, H.S.D.; Pinheiro-Sant'Ana, H.M. Chemical Composition of Brazilian Chia Seeds Grown in Different Places. *Food Chem.* **2017**, *221*, 1709–1716. [CrossRef] [PubMed]
12. Grancieri, M.; Stampini, H.; Martino, D.; Mejia, E.G. De Chia Seed (*Salvia hispanica* L.) as a Source of Proteins and Bioactive Peptides with Health Benefits: A Review. *Compr. Rev. Food Sci. Food Saf.* **2019**, *18*, 480–499. [CrossRef] [PubMed]
13. Grancieri, M.; Stampini, H.; Martino, D.; Gonzalez, E.; Mejia, D. Digested Total Protein and Protein Fractions from Chia Seed (*Salvia hispanica* L.) Had High Scavenging Capacity and Inhibited 5-LOX, COX-1-2, and INOS Enzymes. *Food Chem.* **2019**, *289*, 204–214. [CrossRef] [PubMed]
14. Grancieri, M.; Verediano, T.A.; Sant'Ana, C.T.; de Assis, A.; Toledo, R.L.; de Mejia, E.G.; Martino, H.S.D. Digested Protein from Chia Seed (*Salvia hispanica* L) Prevents Obesity and Associated Inflammation of Adipose Tissue in Mice Fed a High-Fat Diet. *PharmaNutrition* **2022**, *21*, 100298. [CrossRef]
15. Hou, T.; Kolba, N.; Glahn, R.P.; Tako, E. Intra-Amniotic Administration (*Gallus gallus*) of Cicer Arietinum and Lens Culinaris Prebiotics Extracts and Duck Egg White Peptides Affects Calcium Status and Intestinal Functionality. *Nutrients* **2017**, *9*, 785. [CrossRef] [PubMed]

16. Requena, T.; Miguel, M.; Garcés-Rimón, M.; Martínez-Cuesta, M.C.; López-Fandiño, R.; Peláez, C. Pepsin Egg White Hydrolysate Modulates Gut Microbiota in Zucker Obese Rats. *Food Funct.* **2017**, *8*, 437–443. [CrossRef]
17. Ma, Y.; Ding, S.; Liu, G.; Fang, J.; Yan, W.; Duraipandiyan, V.; Al-Dhabi, N.A.; Ali Esmail, G.; Jiang, H. Egg Protein Transferrin-Derived Peptides IRW and IQW Regulate Citrobacter Rodentium-Induced, Inflammation-Related Microbial and Metabolomic Profiles. *Front. Microbiol.* **2019**, *10*, 643. [CrossRef]
18. Jiao, H.; Zhang, Q.; Lin, Y.; Gao, Y.; Zhang, P. The Ovotransferrin-Derived Peptide IRW Attenuates Lipopolysaccharide-Induced Inflammatory Responses. *Biomed Res. Int.* **2019**, *2019*, 8676410. [CrossRef]
19. Bao, X.; Wu, J. Impact of Food-Derived Bioactive Peptides on Gut Function and Health. *Food Res. Int.* **2021**, *147*, 110485. [CrossRef]
20. Hill, C.; Guarner, F.; Reid, G.; Gibson, G.R.; Merenstein, D.J.; Pot, B.; Morelli, L.; Canani, R.B.; Flint, H.J.; Salminen, S.; et al. Expert Consensus Document: The International Scientific Association for Probiotics and Prebiotics Consensus Statement on the Scope and Appropriate Use of the Term Probiotic. *Nat. Rev. Gastroenterol. Hepatol.* **2014**, *11*, 506–514. [CrossRef]
21. Ballan, R.; Battistini, C.; Xavier-santos, D.; Marta, S.; Saad, I. *Interactions of Probiotics and Prebiotics with the Gut Microbiota*, 1st ed.; Elsevier Inc.: Amsterdam, The Netherlands, 2020; Volume 171.
22. Araújo, M.M.; Botelho, P.B. Probiotics, Prebiotics, and Synbiotics in Chronic Constipation: Outstanding Aspects to Be Considered for the Current Evidence. *Front. Nutr.* **2022**, *9*, 2941. [CrossRef] [PubMed]
23. Zheng, J.; Wittouck, S.; Salvetti, E.; Franz, C.M.A.P.; Harris, H.M.B.; Mattarelli, P.; O'toole, P.W.; Pot, B.; Vandamme, P.; Walter, J.; et al. A Taxonomic Note on the Genus Lactobacillus: Description of 23 Novel Genera, Emended Description of the Genus Lactobacillus Beijerinck 1901, and Union of Lactobacillaceae and Leuconostocaceae. *Int. J. Syst. Evol. Microbiol.* **2020**, *70*, 2782–2858. [CrossRef] [PubMed]
24. Duan, A.Y.; Ju, A.Q.; Zhang, Y.N.; Qin, Y.J.; Xue, L.G.; Ma, X.; Luan, W.M.; Yang, S.B. The Effects of *In Ovo* Injection of Synbiotics on the Early Growth Performance and Intestinal Health of Chicks. *Front. Vet. Sci.* **2021**, *8*, 658301. [CrossRef]
25. Duar, R.M.; Lin, X.B.; Zheng, J.; Martino, M.E.; Grenier, T.; Pérez-Muñoz, M.E.; Leulier, F.; Gänzle, M.; Walter, J. Lifestyles in Transition: Evolution and Natural History of the Genus Lactobacillus. *FEMS Microbiol. Rev.* **2017**, *41*, S27–S48. [CrossRef]
26. Ren, S.; Wang, C.; Chen, A.; Lv, W.; Gao, R. The Probiotic Lactobacillus Paracasei Ameliorates Diarrhea Cause by *Escherichia coli* O_8 via Gut Microbiota Modulation1. *Front. Nutr.* **2022**, *9*, 814. [CrossRef] [PubMed]
27. Yegani, M.; Korver, D.R. Factors Affecting Intestinal Health in Poultry. *Poult. Sci.* **2008**, *87*, 2052–2063. [CrossRef]
28. Hou, T.; Tako, E. The *in Ovo* Feeding Administration (*Gallus gallus*)—An Emerging in Vivo Approach to Assess Bioactive Compounds with Potential Nutritional Benefits. *Nutrients* **2018**, *10*, 418. [CrossRef]
29. Reicher, N.; Melkman-Zehavi, T.; Dayan, J.; Wong, E.A.; Uni, Z. Nutritional Stimulation by In-Ovo Feeding Modulates Cellular Proliferation and Differentiation in the Small Intestinal Epithelium of Chicks. *Anim. Nutr.* **2022**, *8*, 91–101. [CrossRef]
30. Jha, R.; Das, R.; Oak, S.; Mishra, P. Probiotics (Direct-fed Microbials) in Poultry Nutrition and Their Effects on Nutrient Utilization, Growth and Laying Performance, and Gut Health: A Systematic Review. *Animals* **2020**, *10*, 1863. [CrossRef]
31. Shehata, A.M.; Paswan, V.K.; Attia, Y.A.; Abougabal, M.S.; Khamis, T.; Alqosaibi, A.I.; Alnamshan, M.M.; Elmazoudy, R.; Abaza, M.A.; Salama, E.A.A.; et al. *In Ovo* Inoculation of Bacillus Subtilis and Raffinose Affects Growth Performance, Cecal Microbiota, Volatile Fatty Acid, Ileal Morphology and Gene Expression, and Sustainability of Broiler Chickens (*Gallus gallus*). *Front. Nutr.* **2022**, *9*, 3847. [CrossRef]
32. Orona-Tamayo, D.; Valverde, M.E.; Nieto-Rendón, B.; Paredes-López, O. Inhibitory Activity of Chia (*Salvia hispanica* L.) Protein Fractions against Angiotensin I-Converting Enzyme and Antioxidant Capacity. *LWT-Food Sci. Technol.* **2015**, *64*, 236–242. [CrossRef]
33. Megías, C.; Del Mar Yust, M.; Pedroche, J.; Lquari, H.; Girón-Calle, J.; Alaiz, M.; Millán, F.; Vioque, J. Purification of an ACE Inhibitory Peptide after Hydrolysis of Sunflower (*Helianthus annuus* L.) Protein Isolates. *J. Agric. Food Chem.* **2004**, *52*, 1928–1932. [CrossRef] [PubMed]
34. Verediano, T.A.; Stampini Duarte Martino, H.; Kolba, N.; Fu, Y.; Cristina Dias Paes, M.; Tako, E. Black Corn (*Zea mays* L.) Soluble Extract Showed Anti-Inflammatory Effects and Improved the Intestinal Barrier Integrity in Vivo (*Gallus gallus*). *Food Res. Int.* **2022**, *157*, 111227. [CrossRef] [PubMed]
35. Gomes, M.J.C.; Martino, H.S.D.; Tako, E. Effects of Iron and Zinc Biofortified Foods on Gut Microbiota in Vivo (*Gallus gallus*): A Systematic Review. *Nutrients* **2021**, *13*, 189. [CrossRef] [PubMed]
36. Martino, H.S.D.; Kolba, N.; Tako, E. Yacon (*Smallanthus sonchifolius*) Flour Soluble Extract Improve Intestinal Bacterial Populations, Brush Border Membrane Functionality and Morphology in Vivo (*Gallus gallus*). *Food Res. Int.* **2020**, *137*, 109705. [CrossRef]
37. Agarwal, N.; Kolba, N.; Jung, Y.; Cheng, J.; Tako, E. Saffron (*Crocus sativus* L.) Flower Water Extract Disrupts the Cecal Microbiome, Brush Border Membrane Functionality, and Morphology In Vivo (*Gallus gallus*). *Nutrients* **2022**, *14*, 220. [CrossRef]
38. Dias, D.M.; Kolba, N.; Hart, J.J.; Ma, M.; Sha, S.T.; Lakshmanan, N.; Nutti, M.R.; Martino, H.S.D.; Glahn, R.P.; Tako, E. Soluble Extracts from Carioca Beans (*Phaseolus vulgaris* L.) Affect the Gut Microbiota and Iron Related Brush Border Membrane Protein Expression in Vivo (*Gallus gallus*). *Food Res. Int.* **2019**, *123*, 172–180. [CrossRef]
39. Gomes, M.J.C.; Kolba, N.; Agarwal, N.; Kim, D.; Eshel, A.; Koren, O.; Tako, E. Modifications in the Intestinal Functionality, Morphology and Microbiome Following Intra-Amniotic Administration (*Gallus gallus*) of Grape (*Vitis vinifera*) Stilbenes (Resveratrol and Pterostilbene). *Nutrients* **2021**, *13*, 3247. [CrossRef]

40. Tako, E.; Glahn, R.P.; Welch, R.M.; Lei, X.; Yasuda, K.; Miller, D.D. Dietary Inulin Affects the Expression of Intestinal Enterocyte Iron Transporters, Receptors and Storage Protein and Alters the Microbiota in the Pig Intestine. *Br. J. Nutr.* **2008**, *99*, 472–480. [CrossRef]
41. Tako, E.; Glahn, R.P.; Knez, M.; Stangoulis, J.C. The Effect of Wheat Prebiotics on the Gut Bacterial Population and Iron Status of Iron Deficient Broiler Chickens. *Nutr. J.* **2014**, *13*, 58. [CrossRef]
42. Baker, R.G.; Hayden, M.S.; Ghosh, S. NF-KB, Inflammation, and Metabolic Disease. *Cell Metab.* **2011**, *13*, 11–22. [CrossRef]
43. Yu, H.; Lin, L.; Zhang, Z.; Zhang, H.; Hu, H. Targeting NF-KB Pathway for the Therapy of Diseases: Mechanism and Clinical Study. *Signal Transduct. Target. Ther.* **2020**, *5*, 209. [CrossRef]
44. Grancieri, M.; Martino, H.S.D.; de Mejia, E.G. Chia (*Salvia hispanica* L.) Seed Total Protein and Protein Fractions Digests Reduce Biomarkers of Inflammation and Atherosclerosis in Macrophages In Vitro. *Mol. Nutr. Food Res.* **2019**, *63*, 1900021. [CrossRef] [PubMed]
45. Yeşilyurt, N.; Yılmaz, B.; Ağagündüz, D.; Capasso, R. Involvement of Probiotics and Postbiotics in the Immune System Modulation. *Biologics* **2021**, *1*, 89–110. [CrossRef]
46. Brandelli, A.; Daroit, D.J.; Corrêa, A.P.F. Whey as a Source of Peptides with Remarkable Biological Activities. *Food Res. Int.* **2015**, *73*, 149–161. [CrossRef]
47. Kou, X.; Gao, J.; Zhang, Z.; Wang, H.; Wang, X. Purification and Identification of Antioxidant Peptides from Chickpea (*Cicer arietinum* L.) Albumin Hydrolysates. *LWT-Food Sci. Technol.* **2013**, *50*, 591–598. [CrossRef]
48. Sacks, F.M.; Lichtenstein, A.H.; Wu, J.H.Y.; Appel, L.J.; Creager, M.A.; Kris-Etherton, P.M.; Miller, M.; Rimm, E.B.; Rudel, L.L.; Robinson, J.G.; et al. Dietary Fats and Cardiovascular Disease: A Presidential Advisory from the American Heart Association. *Circulation* **2017**, *136*, e1–e23. [CrossRef]
49. Pedersen, J.; Coskun, M.; Soendergaard, C.; Salem, M.; Nielsen, O.H. Inflammatory Pathways of Importance for Management of Inflammatory Bowel Disease. *World J. Gastroenterol.* **2014**, *20*, 64–77. [CrossRef]
50. Kaminsky, L.W.; Al-sadi, R.; Ma, T.Y. IL-1 b and the Intestinal Epithelial Tight Junction Barrier. *Front. Immunol.* **2021**, *12*, 6–9. [CrossRef]
51. Suzuki, T. Regulation of the Intestinal Barrier by Nutrients: The Role of Tight Junctions. *Anim. Sci. J.* **2020**, *91*, e13357. [CrossRef]
52. Kang, Y.; Park, H.; Choe, B.H.; Kang, B. The Role and Function of Mucins and Its Relationship to Inflammatory Bowel Disease. *Front. Med.* **2022**, *9*, 848344. [CrossRef]
53. Pawłowska, B.; Sobieszczańska, B.M. Intestinal Epithelial Barrier: The Target for Pathogenic *Escherichia coli*. *Adv. Clin. Exp. Med.* **2017**, *26*, 1437–1445. [CrossRef]
54. Tarabova, L.; Makova, Z.; Piesova, E.; Szaboova, R.; Faixova, Z. Intestinal Mucus Layer and Mucins (A Review). *Folia Vet.* **2016**, *60*, 21–25. [CrossRef]
55. Yao, D.; Dai, W.; Dong, M.; Dai, C.; Wu, S. MUC2 and Related Bacterial Factors: Therapeutic Targets for Ulcerative Colitis. *EBioMedicine* **2021**, *74*, 103751. [CrossRef] [PubMed]
56. Da Silva, B.P.; Martino, H.S.D.; Tako, E. Plant Origin Prebiotics Affect Duodenal Brush Border Membrane Functionality and Morphology, in Vivo (*Gallus gallus*). *Food Funct.* **2021**, *12*, 6157–6166. [CrossRef]
57. Potten, C.S.; Loeffler, M. A Comprehensive Model of the Crypts of the Small Intestine of the Mouse Provides Insight into the Mechanisms of Cell Migration and the Proliferation Hierarchy. *J. Theor. Biol.* **1987**, *127*, 381–391. [CrossRef] [PubMed]
58. Wallaeys, C.; Garcia-Gonzalez, N.; Libert, C. Paneth Cells as the Cornerstones of Intestinal and Organismal Health: A Primer. *EMBO Mol. Med.* **2022**, *15*, e16427. [CrossRef]
59. Witten, J.; Samad, T.; Ribbeck, K. Selective Permeability of Mucus Barriers. *Curr. Opin. Biotechnol.* **2018**, *52*, 124–133. [CrossRef]
60. Verediano, T.A.; Agarwal, N.; Martino, H.S.D.; Grancieri, M.; Paes, M.C.D.; Tako, E. Effect of Black Corn Anthocyanin-Rich Extract (*Zea mays* L.) on Cecal Microbial Populations In Vivo (*Gallus gallus*). *Nutrients* **2022**, *14*, 4679. [CrossRef]

nutrients

MDPI

Article

Macauba (*Acrocomia aculeata*) Pulp Oil Prevents Adipogenesis, Inflammation and Oxidative Stress in Mice Fed a High-Fat Diet

Cíntia Tomaz Sant' Ana [1], Thaísa Agrizzi Verediano [2], Mariana Grancieri [2], Renata Celi Lopes Toledo [2], Elad Tako [3,*], Neuza Maria Brunoro Costa [4], Hércia Stampini Duarte Martino [2] and Frederico Augusto Ribeiro de Barros [1]

1 Department of Food Technology, Federal University of Viçosa, Viçosa 36570-900, MG, Brazil
2 Department of Nutrition and Health, Federal University of Viçosa, Viçosa 36570-900, MG, Brazil
3 Department of Food Science, Cornell University, Stocking Hall, Ithaca, NY 14850, USA
4 Department of Pharmacy and Nutrition, Federal University of Espírito Santo, Alegre 29500-000, ES, Brazil
* Correspondence: et79@cornell.edu

Abstract: Macauba is a palm tree native to Brazil, which fruits are rich in oil. Macauba pulp oil has high contents of oleic acid, carotenoids, and tocopherol, but its effect on health is unknown. We hypothesized that macauba pulp oil would prevent adipogenesis and inflammation in mice. Thus, the purpose of this study was to evaluate the effects of macauba pulp oil on the metabolic changes in C57Bl/6 mice fed a high-fat diet. Three experimental groups were used (n = 10): control diet (CD), high-fat diet (HFD), and high-fat diet with macauba pulp oil (HFM). The HFM reduced malondialdehyde and increased SOD activity and antioxidant capacity (TAC), showing high positive correlations between total tocopherol, oleic acid, and carotenoid intakes and SOD activity (r = 0.9642, r = 0.8770, and r = 0.8585, respectively). The animals fed the HFM had lower levels of PPAR-γ and NF-κB, which were negatively correlated with oleic acid intake (r = −0.7809 and r = −0.7831, respectively). Moreover, the consumption of macauba pulp oil reduced inflammatory infiltrate, adipocyte number and length, (mRNA) *TNF-α*, and (mRNA) *SREBP-1c* in the adipose tissue, and it increased (mRNA) *Adiponectin*. Therefore, macauba pulp oil prevents oxidative stress, inflammation, and adipogenesis and increases antioxidant capacity; these results highlight its potential against metabolic changes induced by an HFD.

Keywords: bioactive compounds; metabolic changes, oleic acid; carotenoid; tocopherol

Citation: Sant' Ana, C.T.; Agrizzi Verediano, T.; Grancieri, M.; Toledo, R.C.L.; Tako, E.; Costa, N.M.B.; Martino, H.S.D.; de Barros, F.A.R. Macauba (*Acrocomia aculeata*) Pulp Oil Prevents Adipogenesis, Inflammation and Oxidative Stress in Mice Fed a High-Fat Diet. *Nutrients* **2023**, *15*, 1252. https://doi.org/10.3390/nu15051252

Academic Editor: Antoni Sureda

Received: 3 February 2023
Revised: 23 February 2023
Accepted: 28 February 2023
Published: 2 March 2023

1. Introduction

Current eating habits characterized by elevated consumption of saturated fats and simple carbohydrates and low vitamins are one of the most important causes for the emergence of chronic non-communicable diseases, such as obesity and nonalcoholic fatty liver disease [1,2]. Obesity is characterized by adipogenesis, which favors the induction of metabolic changes, including changes in cytokine concentrations, activation of inflammatory pathways, and lipotoxic effects in tissues such as the liver. As a consequence of these changes, reactive oxygen species (ROS) production may increase and may, thus, result in an exacerbation of inflammation, oxidative stress, and cell alterations [3].

Regulation and control of adipogenesis and metabolic changes are performed by specific transcriptional regulators, such as peroxisome proliferator-activated receptor gamma (PPAR-γ), sterol regulatory element-binding protein 1 (SREBP-1), and nuclear factor kappa B (NF-κB) [4]. SREBP-1 controls fatty acid biosynthesis by favoring the transcription of specific enzymes and activating PPAR-γ, which controls the expression of genes that regulate adipocyte differentiation. NF-κB controls the expression of inflammatory genes. In obesity, there is an increase in these transcription factors, resulting in increased lipogenesis, which leads to an increase in triacylglycerol and a reduction in lipolysis, thereby favoring the development of inflammation and oxidative stress [3,4]. As such, research studies that demonstrate new dietary strategies with the purpose of preventing or controlling obesity and metabolic alterations become very important, and dietary fatty acid composition demonstrates a significant impact on disease development [5]. Thus, nutritional strategies that aim to treat or prevent these metabolic alterations are of great importance.

Macauba (*Acrocomia aculeata*) is a palm tree that is naturally present in almost all Brazilian territories, and it is considered a promising alternative to vegetable oil for fuel and for the cosmetic and food industries due to its high oil production and specific characteristics [6]. Two types of oils are obtained from macauba: pulp and kernel oils. Both have important chemical and economical characteristics, highlighting their nutritional action and applications in the food industry [7]. Similar to olive oil, macauba pulp oil is rich in oleic acid [6]. Oleic acid has been shown to reduce the expression of transcription factors related to the adipogenesis signaling pathway, such as PPAR-γ, and reduce oxidative stress markers [8]. Moreover, macauba pulp oil has a high content of carotenoids, which can act to reduce inflammation through NF-κB modulation [6,9]. Additionally, this oil contains tocopherol, which is an important antioxidant that has been shown to improve inflammation and oxidative stress [10]. Thus, macauba pulp oil consumption may result in an improvement in metabolic changes, which is associated with the bioactive compounds in its composition [6].

We hypothesized that macauba pulp oil would prevent adipogenesis and inflammation in mice. However, to the best of our knowledge, no research has been performed to provide evidence of the health benefits of macauba pulp oil. Thus, the objective of this work was to evaluate the effect of macauba pulp oil on the adipogenesis pathways and metabolic changes in mice fed a high-fat diet. This study is the first to explore the health benefits of this promising vegetable oil.

2. Material and Methods

2.1. Materials

Macauba fruits were harvested in Araponga, Minas Gerais (Brazil), in the mature stage, and then they were peeled and pulped to obtain the macauba pulp. The pulp was dried at 65 °C (CIENLAB CE220, Brazil) for 15 h. Oil was extracted using a manual hydraulic press (Laboratory Press, Fred S. Carver Inc., Summit, NJ, USA), centrifuged (5000 rpm/20 min), and then placed in a freezer (−80 °C).

2.2. Chemical Characterization of Macauba Pulp Oil

The fatty acid profile of the macauba pulp oil was determined using a gas chromatography equipped with a flame ionization detector (GC-FID) (Shimadzu, GC-2010, Kyoto, Japan) and a capillary column of 100 m × 0.25 mm (SP-2560, Sigma-Aldrich, San Luis, MO, USA) [11]. Helium gas was used as the dragging gas and maintained at a constant flow rate of 363 kPa. Fatty acid methyl esters (FAMEs) were separated using a linear heating ramp from 100 °C to 270 °C, at a heating rate of 20 °C mim^{-1} and with a high linear velocity for better peak resolution. Peak identification was confirmed by comparison with the standard FAME mix (Supelco 37 FAME mix, Sigma-Aldrich, San Luis, MO, USA). Moreover, the oleic acid content (mg/g) of the oil was also determined using a standard (Sigma-Aldrich).

Carotenoid analysis was carried out by a high-performance liquid chromatography (HPLC) with detection at 450 nm, using the following chromatographic conditions: a HPLC system (Shimadzu, SCL 10AT VP, Kyoto, Japan) and a chromatographic column Phenomenex Gemini RP-18 (250 mm × 4.6 mm, 5 mm) fitted with a guard column RP-18 Phenomenex ODS column (4 mm × 3 mm). The mobile phase consisted of methanol:ethylacetate:acetonitrile (70:20:10, $v/v/v$) with a flow rate of 2.0 mL·min^{-1} and a run time of 15 min. Total carotenoid content (μg/g) was expressed as the sum of the major carotenoids present in the macauba pulp oil [12].

Total tocopherol content was determined following the AOCS method, using a HPLC with fluorescence detection at 450 nm and the following chromatographic conditions: a silica column of 4.6 × 250 mm with a pore of 5 μm, a flow rate of 1.0 mL min^{-1}, and as the mobile phase, a mixture of 99.5% of *n*-hexane and 0.5% of isopropanol. The concentration of total tocopherols (μg/g) was expressed as the sum of the major tocopherols present in the macauba pulp oil [13].

2.3. Animals and Experimental Design

Black male mice C57Bl/6 (30 animals), which were 8 weeks old and had an average weight of 24.34 ± 0.18 g, were allocated into 3 groups, with 10 animals in each group, based on the homogeneity of body weight. The sample calculation equation determined how many animals should be in each group, using the following variables: α-error type I = 1.96, α-level = 5%, and data of fat mass mean reported by Schoemaker et al. in 2017 [14,15]. Individual stainless steel cages were used to keep the animals in a temperature-controlled room (light–dark cycles of 12 h and temperature of 22 ± 2 °C). Water and the respective experimental diets were supplied ad libitum.

The experimental diets were formulated according to AIN-93M and high-fat diet, using lard in the high-fat diet [16]. Each experimental group consumed the following diet: control diet—AIN93M (CD); high-fat diet (HFD); high-fat diet with macauba pulp oil (HFM). In the HFM, macauba pulp oil was added in a proportion of 40 g/kg (4%), replacing the soybean oil used in the AIN-93M diet (Table 1). The objective was to verify the effect of macauba pulp oil as a replacement of soybean oil, which is commonly used in control diets, and not as a supplementation. The formulated diets were stored at a low temperature (−20 °C) and offered to the animals every day.

Table 1. Composition of experimental diets (g/kg of diet).

Ingredients (g/kg)	CD	HFD	HFM
Albumin *	179.71	179.71	179.71
Dextrinized starch	155	155	155
Sucrose	100	100	100
Soybean oil	40	40	-
Lard	0	312	312
Cellulose	50	50	50
Mineral mix	35	35	35
Vitamin mix	10	10	10
L-cystine	1.8	1.8	1.8
Choline bitartrate	2.5	2.5	2.5
Corn starch	425.99	113.99	113.99
Macauba pulp oil	-	-	40
Carbohydrate (%)	76.9	44.1	44.1
Protein (%)	18.9	18.9	18.9
Lipids (%)	4.20	37	37
Caloric density (kcal g^{-1})	3.85	5.41	5.41

* Purity of 78%. CD: control diet (AIN93M); HFD: high-fat diet; HFM: high-fat diet with macauba pulp oil.

At the end of the 8 weeks, the animals were anesthetized after 12 h of fasting using isoflurane (Isoforine, Cristália), in accordance with the bodyweight of the mice. Using the methodology of cardiac puncture, blood was collected and centrifuged (4 °C at 800× *g* for 10 min using Fanem-204, São Paulo, Brazil), and the serum was collected and stored at −80 °C. The liver and adipose (epididymal and subcutaneous) tissues were extracted and stored at (−80 °C) until analysis, and another part was fixed in formaldehyde (10%) for the analysis of histological markers. Bodyweight gain and feed consumption were measured on a weekly basis throughout the experiment to calculate the feed efficiency ratio (weight gain/consumption × 100), and the percentage of adiposity was measured based on the weight of the adipose tissue (g) in relation to the total body weight. Body mass index (BMI) was measured using the ratio between weight and naso-anal length (cm) squared [17]. The hepatosomatic index was also determined (liver weight/body weight × 100) [18]. Carotenoid, oleic acid, and tocopherol intakes were determined by the total amount of diet consumed by the mice. Ethical principles for animal experimentation were implemented for all processes performed on the animals [19]. The Ethics Committee of the Federal University of Viçosa approved this research (Protocol 09/2019; date of approval: 28 May 2019).

2.4. Biochemical Analysis

The biochemical parameters were determined using the serum. Glucose concentration, total cholesterol (TC), high-density lipoprotein cholesterol (HDL-c), low-density lipoprotein cholesterol (LDL-c), triacylglycerides (TGL), aspartate aminotransferase (AST), and alanine aminotransferase (ALT) were determined based on the colorimetric method using commercial kits (Bioclin®, Belo Horizonte, Brazil).

2.5. Homogenate Preparation and Oxidative Stress Levels

Liver homogenate was prepared with 200 mg of the liver. The liver was mixed with 1 mM of EDTA (pH 7.4) and 1000 μL of phosphate buffer (50 mM). The content was macerated and centrifuged (1200× g/8 min/4 °C), and the supernatant was collected for the analysis of antioxidant enzymes.

For the quantification of the enzyme superoxide dismutase (SOD), 249 μL of 50 mM of Tris-HCl buffer (pH 8.2) (1 mM of EDTA, 6 μL of MTT (1.25 mM), 15 μL of pyrogallol (10 mM), and 279 μL of buffer) was mixed into the aliquoted homogenate. To determine the blank, 6 μL of MTT and 294 μL of buffer were added to the wells, which were incubated for 5 min at 37 °C, and the reading was performed on a spectrophotometer at 570 nm (Thermo Scientific Multiskan GO, Waltham, MA, USA). The SOD quantification was expressed as units of SOD/mg protein [20].

Malondialdehyde (MDA) was determined using the samples of the homogenate. A total of 400 μL of trichloroacetic acid solution (15%) and thiobarbituric acid (0.375%) was added into 400 μL of the sample. It was placed in a water bath (90 °C/40 min) and 600 μL of n-Butanol was added; then, the mixture was centrifuged (3500 rpm/ 5 min). The supernatant was removed, and the absorbance was read at 535 nm (Multiskan GO—Thermo Scientific). The MDA level was expressed as MDA/mg protein [21].

Catalase was performed on the samples of the homogenate as described above. At 0, 30, and 60 s after the reaction was initiated, the absorbance was determined at 240 nm (T70 + UV/VIS Spectrometer). Enzyme activity was reported as μmol per mL of sample, and the data were expressed in U of catalase/mg protein. Catalase activity was calculated according to the Beer–Lambert law [22].

For the quantification of nitric oxide, 50 μL of the homogenate was used. Then, 1% sulfanilamide solution and 0.1% nafityl ethylene amide dihydrochloride were added. A 0.025 M sodium nitrite standard curve was used, and the absorbance was determined at 570 nm (Multiskan GO—Thermo Scientific) [23].

2.6. Total Antioxidant Capacity of Serum and Liver

The total antioxidant capacity (TAC) of the serum and the liver was determined with an antioxidant assay kit (Cayman Chem Corp, Ann Arbor, MI, USA) Sigma Aldrich®. The absorbance reading was performed at 405 nm (Multiskan GO—Thermo Scientific).

2.7. PPAR-γ, PPAR-α, NF-κB, and TLR-4 Quantification

The adipose tissue and liver samples were homogenized using the NE-PER™ Nuclear and Cytoplasmic Extraction Kit reagents (Thermo Scientific Fisher, Waltham, MA, USA). The nuclear fractions were analyzed using an immunoassay with the Mouse PPAR-γ (Peroxisome Proliferator-Activated Receptor Gamma—E-EL-M0893, Elabscience, Houston, TX, USA), Mouse NF-κB p65 (Factor Nuclear Kappa B—E-EL-M0838, Elabscience, Houston, TX, USA), Rat PPAR-α (Peroxisome Proliferator-Activated Receptor Alfa—E-EL-R0725, Elabscience, Houston, TX, USA), Rat NF-κB p65 (Factor Nuclear Kappa B—E-EL-R0674, Elabscience, USA) and Rat TLR-4 (Toll-like Receptor 4—E-EL-R0990, Elabscience, USA)

ELISA kits, respectively. The microplates were, respectively, precoated with anti-PPAR-γ, anti-NF-κB p65, anti-PPAR-α, and anti-TLR-4 antibodies. The concentrations were calculated by comparison to the corresponding standard curves.

2.8. Determination of Gene Expression in Adipose Tissue and Liver by Reverse Transcriptase Quantitative Polymerase Chain Reaction (RT-qPCR)

TRIzol reagent (Invitrogen, CA, USA) was used to extract total RNA from the liver, and a specific kit (mirVana™ miRNA Isolation Kit, Life Technologies, Carlsbad, CA, USA) was used to extract RNA from the adipose tissue, according to the manufacturer's protocols. RNA concentration and purity were evaluated using a Microdrop plate spectrophotometer Multiskan™ GO (Thermo Scientific, Waltham, MA, USA). To create cDNA synthesis, the M-MLV Reverse Transcriptase Kit (Invitrogen, CA, USA) was used. RT-qPCR was used for the gene expression relative quantification using the AB StepOne Real-Time PCR System equipment and Fast SYBR Green Master Mix (Applied Biosystems, Carlsbad, CA, USA) reagent. The initial parameters used were 20 s at 95 °C and then 40 cycles at 95 °C (3 s), 60 °C (30 s), followed by the melting curve analysis. A melting point analysis was performed to improve the specificity and sensitivity of the amplification reactions detected. All primers were designed by using the Primer 3 Plus program and obtained from Sigma-Aldrich Brazil Ltda. (Table 2). The 2-Delta-Delta C (T) method was used to calculate the gene expression, by using GAPDH and β-actin as the references and the high-fat diet group as the control, which was normalized to 1 [24].

Table 2. Sequence of primers used in the RT-qPCR analyses.

Genes	Forward	Reverse
SREBP-1c	CGC TAC CGT TCC TCT ATC AAT GAC	AGT TTC TGG TTG CTG TGC TGT AAG
ADIPOR2	CAT GTT TGC CAC CCC TCA GTA	ATG CAA GGT AGG GAT TCC A
ACC-1	TCA AGA CGG CTC AGG TCA TCA	AGG CGC CAA ACT TCA GCA TC
CPT-1α	GTA AGG CCA CTG ATG AAG GAA GA	ATT TGG GTC CGA GGT TGA CA
LPL	TCA ACC ACA GCA GCA AGA	CCG ATA CAA CCA GTC TAC TAC AA
Adiponectin	ATG AGT ACC AGA CTA ATG AGA C	GGC AGG ATT AAG AGG AAC A
TNF-α	TAT GGC TCA GGG TCC AAC TC	GCT CCA GTG AAT TCG GAA AG
SREBP-1c	GCC GAG ATG TGC GAA CTG	GGA AGT CAC TGT CTT GGT TGT T
β-actin	TTC GTT GCC GGT CCA CC	GCT TTG CAC ATG CCG GAG CC
GAPDH	AGG TTG TCT CCT GTC ACT TC	CTG TTG CTG TAG CCA TAT TC

SREBP-1c: Sterol regulatory element-binding proteins 1c; *ADIPOR2*: adiponectin receptor 2; *ACC-1*: acetyl CoA carboxylase 1; *CPT-1α*: carnitine palmitoyl transferase 1 alpha; *LPL*: Lipoprotein lipase; *TNF-α*: Tumor necrosis factor alpha; *GAPDH*: Glyceraldehyde-3-phosphate dehydrogenase.

2.9. Histomorphometric Analysis of Adipose and Liver Tissues

Paraffin was used to fix the samples of adipose tissue and liver. Ten cuts per animal were performed (3 μm thick), and the samples were mounted on glass slides and stained with hematoxylin and eosin. Analyses were performed under a light microscope (Leica DM750®). The histological sections of the images were captured in a 20× objective. Inflammatory infiltrate number and length of adipocytes were evaluated using the adipose tissue (Image-Pro Plus® 4.5). Liver cellular components (fat vesicles, inflammatory infiltrate, cytoplasm, and nucleus), for 10 histological fields per animal, were analyzed using a test system with 266 points, obtaining 2660 total points for each animal analyzed (Image J®, Wayne Rasband). The following formula was used to calculate the parameters: $V_v = P_p/P_T$ (P_p = number of points located on the structure of interest, and P_T = total test points in the histological area) [25]. The steatosis degree was determined semi-quantitatively according to a 5° scale and the fat percentage: degree 0 (<5%), grade 1 (≥5% and <25%), grade 2 (≥25% and <50%), grade 3 (≥50% and <75%), grade 4 (≥75%) [26].

2.10. Statistical Analyses

Kolmogorov–Smirnov normality test was initially applied, and then an analysis of variance (ANOVA) test was performed, followed by the Newman–Keuls test for parametric variables. For the correlation analysis, Pearson's correlation was used. The results with a p-value $\leq$ 0.05 were considered statistically significant. The statistical analyses were performed using the GraphPad Prism® version 8.0 (GraphPad Software, San Diego, CA, USA).

3. Results

3.1. Chemical Characterization of Macauba Pulp Oil

The macauba pulp oil shows a high content of monounsaturated fatty acids (55%), with significant oleic acid content (49.32%), as shown in Table 3. In addition, it has high contents of carotenoids and tocopherol (Table 3).

Table 3. Fatty acid profile, carotenoids, and tocopherol contents in macauba pulp oil.

Components	
Palmitic (C16:0)	22.84%
Palmitoleic (C16:1)	5.93%
Stearic (C18:0)	1.23%
Oleic (C18:1n9c)	49.32%
Linoleic (C18:2n6c)	19.63%
Linolenic (C18:3n6c)	1.05%
Oleic acid (mg/g)	199.00
Tocopherol (µg/g)	40.80
Total carotenoids (µg/g)	207.52
β-carotene (µg/g)	163.63
α-carotene (µg/g)	21.03
Lutein (µg/g)	8.75
Lycopene (µg/g)	14.11

Caprylic, capric, lauric, and myristic acids are not detected.

3.2. Effects of Macauba Pulp Oil on Biometric Measures, Food Intake, and Lipid Profile

Weight gain, body mass index (BMI), and food efficiency ratio (FER) did not differ among the experimental groups ($p > 0.05$; Table 4). The CD group had higher food consumption compared to the HFD and HFM groups, which was associated with the reduced caloric density of the AIN93M diet ($p < 0.0001$; Table 4). The CD group had a lower percentage of adiposity compared to the HFD and HFM groups ($p = 0.0018$; Table 4).

Table 4. Biometric measures, food intake, and serum biochemical values of the mice after consuming the experimental diets for 8 weeks.

	CD	HFD	HFM
Weight gain (g)	4.01 ± 1.74 [a]	4.14 ± 2.23 [a]	3.66 ± 2.23 [a]
BMI (g/cm^2)	0.34 ± 0.02 [a]	0.33 ± 0.01 [a]	0.33 ± 0.02 [a]
Adiposity (%)	0.71 ± 0.24 [b]	2.43 ± 1.28 [a]	2.26 ± 1.25 [a]
Food consumption (g/day)	4.07 ± 0.16 [a]	2.53 ± 0.41 [b]	2.63 ± 0.42 [b]
Food efficiency (%)	1.69 ± 0.60 [b]	2.67 ± 1.46 [a]	2.42 ± 1.55 [a]
Hepatosomatic index (%)	3.61 ± 0.29 [a]	3.72 ± 0.28 [a]	3.48 ± 0.29 [a]
Oleic acid intake (mg/day)	-	-	0.52 ± 0.08
Carotenoid intake (µg/day)	-	-	21.96 ± 4.11
Tocopherol intake (mg/day)	0.46 ± 0.01 [a]	0.29 ± 0.03 [b]	0.31 ± 0.03 [b]
Total cholesterol (mg dL^{-1})	151.48 ± 13.79 [b]	166.49 ± 15.51 [a]	179.91 ± 6.87 [a]
Total triglycerides (mg dL^{-1})	79.91 ± 4.71 [a]	84.83 ± 5.63 [a]	83.06 ± 5.09 [a]
HDL-c (mg dL^{-1})	38.13 ± 4.29 [a]	37.35 ± 5.79 [a]	43.07 ± 5.64 [a]
LDL-c (mg dL^{-1})	12.80 ± 2.11 [b]	20.64 ± 5.25 [a]	20.80 ± 5.20 [a]
Glucose (mg dL^{-1})	160.67 ± 44.23 [a]	182.58 ± 30.09 [a]	197.31 ± 36.68 [a]
AST (mg dL^{-1})	88.14 ± 21.88 [a]	71.66 ± 21.30 [a]	73.39 ± 19.04 [a]
ALT (mg dL^{-1})	18.74 ± 9.88 [a]	15.71 ± 5.63 [a]	18.84 ± 9.26 [a]

Data are expressed as mean ± standard deviation (n = 10). Different lowercase letters in the same row indicate a statistical difference based on the Newman–Keuls test ($p \leq 0.05$). CD: control diet—AIN93M; HFD: high-fat diet; HFM: high-fat diet with macauba pulp oil; BMI: body mass index; HDL-c: high-density lipoprotein; LDL-c: low-density lipoprotein; ALT: alanine aminotransferase; AST: aspartate aminotransferase.

The group that consumed macauba pulp oil (HFM) did not differ from the HFD group in terms of glucose, triglyceride, TC, LDL, and HDL values, as well as hepatic enzymes AST and ALT, and hepatosomatic index ($p > 0.05$; Table 4).

3.3. Total Antioxidant Capacity and Oxidative Stress Marker Levels in Mice

The HFM group had a high SOD activity ($p = 0.0078$; Figure 1A) and showed a positive correlation with carotenoid (r = 0.8585, $p = 0.004$), oleic acid (r = 0.8770, $p = 0.009$) and tocopherol (r = 0.9642, $p < 0.0001$) intakes (Figure 1B–D). Macauba pulp oil decreased malondialdehyde ($p = 0.0057$; Figure 1E), showing a negative correlation between this parameter and oleic acid (r = −0.9401, $p < 0.001$) and tocopherol (r = −0.9021, $p = 0.0004$) (Figure 1G,H). Catalase and nitric oxide did not differ among the groups ($p > 0.05$; Figure 1M,N).

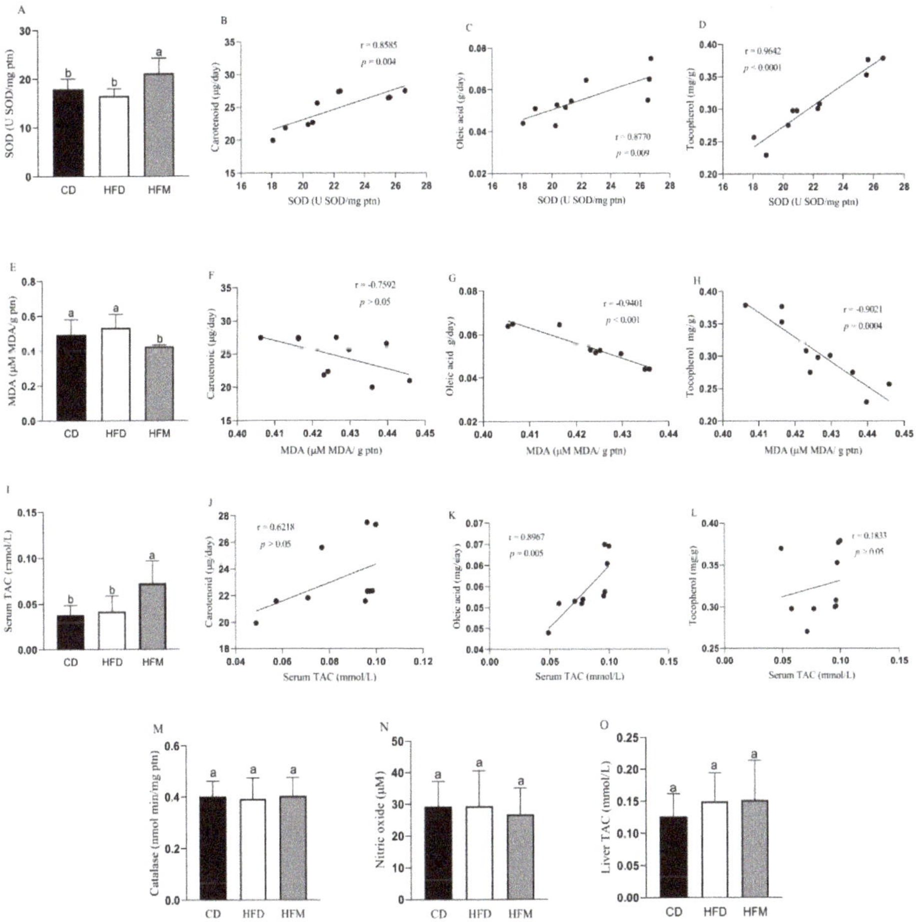

Figure 1. Oxidative stress level and antioxidant capacity of mice after consuming the experimental diets for 8 weeks, and correlation with carotenoid, oleic acid, and tocopherol intakes. (**A**) Superoxide dismutase level. (**B**) Correlation between SOD level and carotenoid intake. (**C**) Correlation between SOD level and oleic acid intake. (**D**) Correlation between SOD level and tocopherol intake. (**E**) Malondialdehyde level. (**F**) Correlation between MDA level and carotenoid intake. (**G**) Correlation between MDA level and oleic acid intake. (**H**) Correlation between MDA level and tocopherol intake.

(**I**) Serum total antioxidant capacity level. (**J**) Correlation between serum TAC level and carotenoid intake. (**K**) Correlation between serum TAC level and oleic acid intake. (**L**) Correlation between serum TAC level and tocopherol intake. (**M**) Catalase level. (**N**) Nitric oxide level. (**O**) Liver total antioxidant capacity level. Data are expressed as mean ± standard deviation (n = 10). Different letters indicate a statistical difference based on the Newman–Keuls test ($p \leq 0.05$). CD: control diet—AIN93M; HFD: high-fat diet; HFM: high-fat diet with macauba pulp oil; SOD: superoxide dismutase; MDA: malondialdehyde; TAC: total antioxidant capacity.

The HFM group had a higher serum TAC compared to the HFD and CD groups (p = 0.0058; Figure 1I), showing a positive correlation between serum TAC and oleic acid (r = 0.8967, p = 0.005) intake from macauba pulp oil (Figure 1K). Liver TAC did not differ among the experimental groups (p > 0.05; Figure 1O).

3.4. Effects of Macauba Pulp Oil on NF-κB, TLR-4, and PPAR-(α, γ) Quantification

The HFM group had a lower nuclear quantification of NF-κB in the adipose tissue compared to the HFD and CD groups (p = 0.0179; Figure 2A), showing a negative correlation with oleic acid (r = −0.7831, p = 0.037) and tocopherol (r = −0.8134, p = 0.0261) intakes (Figure 2C,D). Macauba pulp oil reduced the PPAR-γ quantification (p = 0.056; Figure 2E), showing a negative correlation with carotenoid (r = −0.7301, p = 0.021) and oleic acid (r = −0.7809, p = 0.022) (Figure 2F,G).

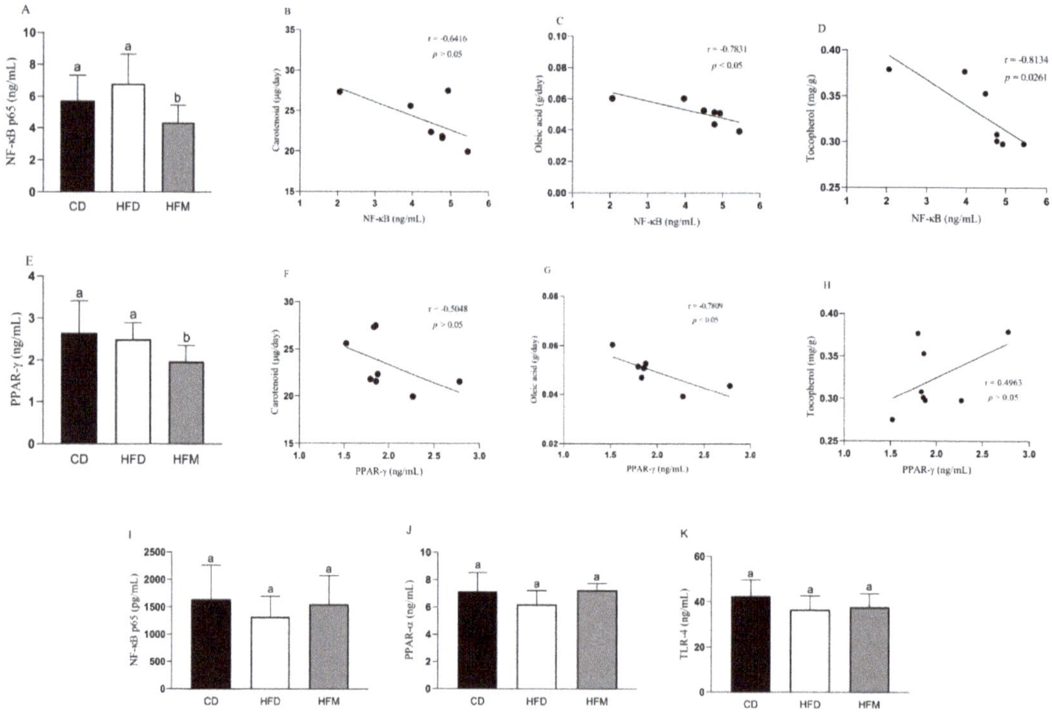

Figure 2. Levels of proteins in the adipose tissue (**A–H**) and liver (**I–K**) of mice after consuming the experimental diets for 8 weeks, and correlation with carotenoid, oleic acid, and tocopherol intakes. Data are expressed as mean ± standard deviation (n = 8). Different letters indicate a statistical difference based on the Newman–Keuls test ($p \leq 0.05$). CD: control diet—AIN93M; HFD: high-fat diet; HFM: high-fat diet with macauba pulp oil; NF-κB p65: nuclear factor kappa B subunit p65; PPAR-α: peroxisome proliferator-activated receptor alpha; PPAR-γ: peroxisome proliferator-activated receptor gamma; TLR-4: toll-like receptor 4.

NF-κB, PPAR-α, and TLR-4, as present in the nuclear fraction in the liver, did not differ among the experimental groups ($p > 0.05$; Figure 2I–K).

3.5. Effects of Macauba Pulp Oil on Gene Expression in Adipose and Hepatic Tissues

In the liver, in the HFM group, the mRNA expression of *SREBP-1c* was significantly increased compared to the control and HFD groups ($p < 0.0001$; Figure 3A), whereas (mRNA) *CPT-1α* was decreased ($p = 0.0031$; Figure 3B). The mRNA expression of *ACC-1α* and *AdipoR2* did not differ from the HFD group ($p > 0.05$; Figure 3C,D).

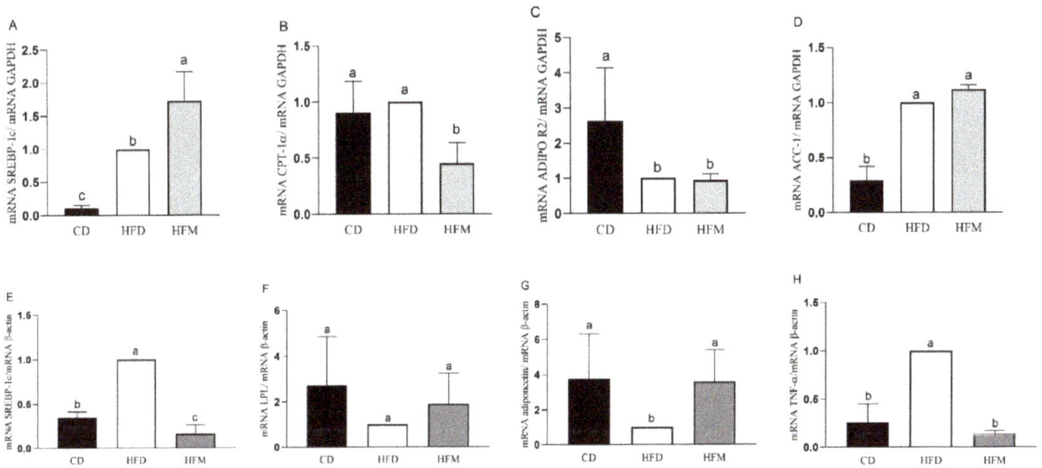

Figure 3. Gene expression in the liver (**A–D**) and adipose tissue (**E–H**) of mice after consuming the experimental diets for 8 weeks. Data are expressed as mean ± standard deviation ($n = 8$). Different letters indicate a statistical difference based on the Newman–Keuls test ($p \leq 0.05$). CD: control diet—AIN93M; HFD: high-fat diet; HFM: high-fat diet with macauba pulp oil; *SREBP-1c*: sterol regulatory element-binding proteins 1c; *ADIPOR2*: adiponectin receptor 2; *ACC-1*: acetyl CoA carboxilase 1; *CPT-1α*: carnitine palmitoyl transferase 1 alpha; *GAPDH*: glyceraldehyde 3-phosphate dehydrogenase; *LPL*: lipoprotein lipase; *TNF-α*: tumor necrosis factor alpha.

In the adipose tissue, in the HFM group, the mRNA expression of *SREBP-1c* ($p < 0.0001$; Figure 3E) and (mRNA) *TNF-α* ($p < 0.0001$; Figure 3H) were significantly decreased compared to the HFD group, and the mRNA expression of *Adiponectin* was similar between the HFM and CD groups ($p > 0.05$; Figure 3G). The mRNA expression of *LPL* was similar among the groups ($p > 0.05$; Figure 3F). The correlation analysis showed a negative correlation between mRNA *SREBP-1c* and carotenoid intake ($r = -0.8991$, $p = 0.012$), a positive correlation between mRNA *Adiponectin* and carotenoid intake ($r = 0.9130$, $p < 0.001$), and negative correlation between mRNA *TNF-α* and oleic acid intake ($r = -0.9057$, $p = 0.0009$).

3.6. Effects of Macauba Pulp Oil on Histological Morphometrics of Liver and Adipose Tissues

The percentage of the nucleus, cytoplasm, inflammatory infiltrate, and fat deposition in the hepatocytes did not differ among the groups ($p > 0.05$, Figure 4A). The control group was identified as steatosis grade 0, whereas the HFD and HFM groups increased the steatosis to grade 1 and had similar values between them (Figure 4B). The HFM group had lower inflammatory infiltrate ($p < 0.0001$) and adipocyte number ($p = 0.0027$) and length ($p = 0.0088$) in the adipose tissue compared to the HFD group, but its values were similar to the CD group (Figure 4C,D).

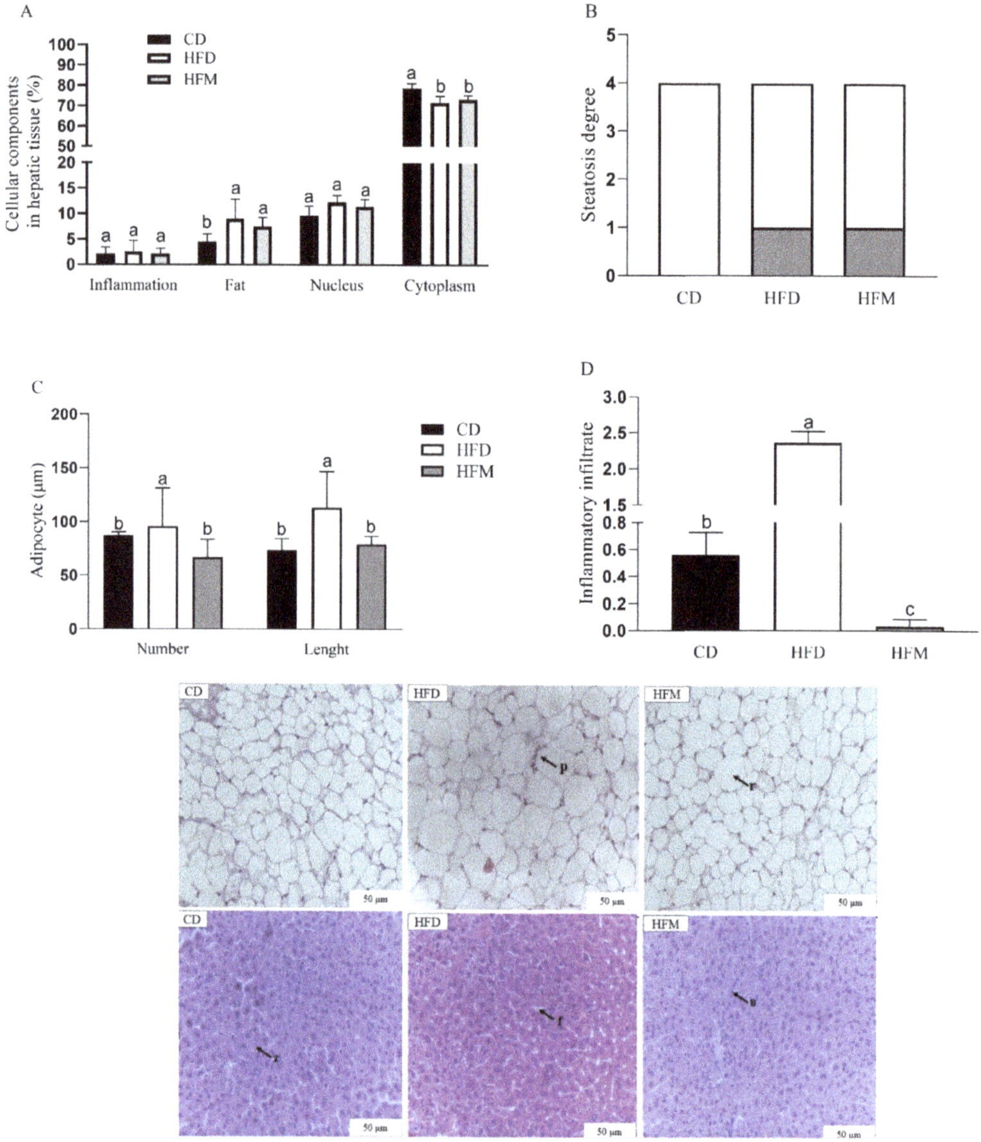

Figure 4. Cellular components: percentage in hepatic tissue (**A**), steatosis degree (**B**), number and length of adipocyte (**C**), and inflammatory infiltrate (**D**). The black arrows represent the following: z: cytoplasm, f: fat vesicles, n: nucleus, p: inflammatory infiltrate, and r: adiposity. Data are expressed as mean ± standard deviation (n = 10). Different letters indicate a statistical difference based on the Newman–Keuls test ($p \leq 0.05$). CD: control diet; HFD: high-fat diet; HFM: high-fat diet with macauba pulp oil.

4. Discussion

This is the first work that evaluated the influence of macauba pulp oil on undesirable metabolic changes in mice fed a high-fat diet. The present research focused on the effects of macauba pulp oil since there is evidence that oleic acid, carotenoid, and tocopherol present in this oil would trigger anti-inflammatory, anti-obesity, and antioxidant effects [27,28]. In this study, macauba pulp oil intake prevented the adipogenesis pathway, inflammation, and oxidative stress in mice fed a high-fat diet. In order to stimulate metabolic changes in animals, high-saturated fat diet consumption is extensively applied. The time to verify the effect of a specific food or compound on metabolic changes usually begins after seven or eight weeks of receiving the diet. In a different way, in our study, to determine the effects of macauba pulp oil as a preventive treatment, macauba pulp oil was added in the diet since the beginning of the experiment, along with the high-fat diet, to examine its mechanism of action and metabolic alterations.

In the present study, the consumption of macauba pulp oil reflected a higher total antioxidant capacity (TAC), which might be associated with the oleic acid content, and this was confirmed by the correlation analysis, which demonstrated a significant positive correlation between this compound consumption and TAC. Oleic acid is well documented for its anti-inflammatory properties, possibly associated with its chemical configuration with a double bond, thereby causing less chance of oxidation and resulting in the antioxidant property against a high oxidative load [10,29]. In addition, higher SOD activity and lower malondialdehyde levels were observed with the macauba pulp oil consumption. SODs are oxidoreductase enzymes that have a role in protecting cells against superoxide anions, performing the dismutation of $O_2^{\bullet -}$ into oxygen and H_2O_2, and providing antioxidant defense for the organism, while malondialdehyde is an important marker of lipid peroxidation [30,31]. it is shown that macauba pulp oil consumption can improve antioxidant defenses, with these results being attributed to the oleic acid, carotenoids, and tocopherol present in macauba pulp oil, as demonstrated in other studies that examined the relationship between these components and the improvement of the body's antioxidant defenses [32–34]. Additionally, there was a positive correlation between SOD and these compounds and a negative correlation between MDA and oleic acid and tocopherol.

The consumption of macauba pulp oil prevented the adipogenesis pathway by decreasing the expression of PPAR-γ and (mRNA) *SREBP-1c* and increasing the expression of (mRNA) *Adiponectin* in the adipose tissue. This effect was confirmed by the result of the histomorphometric analysis, which demonstrated that the animals that consumed the macauba pulp oil had smaller adipocyte number and length even with a high-fat diet consumption, that is, the macauba pulp oil caused less hypertrophy and hyperplasia of the adipocytes. Thus, the lower translocation of PPAR-γ in the present research could be associated with the high content of oleic acid and carotenoids in the macauba pulp oil and was confirmed by the significant negative correlation between the consumption of these compounds and PPAR-γ quantification. Oleic acid has been shown to act in PPAR-γ repression, resulting in less differentiation of pre-adipocytes into mature adipocytes and reducing adipogenesis [3,35]. Similar to our results, a previous study found a relationship between oleic acid consumption and reduction in PPAR-γ and (mRNA) *SREBP-1c* in an obese animal model [35]. Research shows that carotenoids can affect adipocyte function through the interaction with PPAR-γ, thereby interfering with adipocyte differentiation, as demonstrated in a study using experimental animals, which found an association between carotenoids and lower adipose tissue gain related to lower PPAR-γ expression [36]. Still, this result is related to the increased expression of adiponectin since PPAR-γ is tightly regulated by adiponectin [37]. Moreover, the observed results of a reduction in the genes related to the adipogenesis pathway, with a concomitant reduction in the histological markers of adipose tissue, could be related to the presence of β-carotene, which was the main carotenoid found in the macauba pulp oil that could suppress PPAR-γ, resulting in lower total lipid in adipocytes [38,39].

Related to this, macauba pulp oil was efficient in reducing inflammation in the adipose tissue since it reduced NF-κB in the nuclear fraction, and this indicates a reduction in the inflammation cascade, leading to a significant reduction in (mRNA) *TNF-α* gene expression. Corroborating this result, the histomorphometric analysis of the adipose tissue showed less inflammatory infiltrate with the consumption of macauba pulp oil. A hypertrophy of adipose tissue initiates the emission of chemotactic signals that recruit immune cells and lead to the infiltration of macrophages into the adipose tissue, contributing to systemic subclinical inflammation [3]. This result may be associated with a lower amount of PPAR-γ and higher adiponectin since PPAR-γ interferes with the differentiation of adipocytes and is consequently related to the inflammatory process. Obesity is an inflammatory condition: one of the complications related to obesity is the development of reactive oxygen species (ROS), and adiponectin is an anti-inflammatory adipokine with a negative correlation between the degree of obesity and the level of this adipokine [40,41]. These results were supported by the present study since there was a positive correlation between carotenoid consumption and an increase in the expression of adiponectin, indicating that the macauba pulp oil, which is high in carotenoids, may contribute to the reduction of inflammation. Additionally, there was a significant negative correlation between oleic acid and tocopherol and NF-κB, that is, an increase in oleic acid and tocopherol consumption was correlated with a decrease in the quantification of NF-κB. The study by Rosillo et al., with a mouse model, also demonstrated that the administration of oleic acid is able to suppress NF-κB activation [42]. Oleic acid is able to activate PGC-1α by forming a dimer with the protein called c-MAF, migrate to the nucleus, and then transcribe the gene responsible for IL-10, which dismantles the activation signaling of NF-κB due to its potent anti-inflammatory action [43]. Tocopherol can block NF-kB activation through its action on enzymes that regulate the NF-kB signaling pathway [44]. Despite the lack of a correlation between the reduction in NF-κB and the consumption of carotenoids in the present study, this compound presents interference with the NF-κB pathway, resulting in the modulation of their interacting proteins and interacting with the cysteine residues of IκB kinase, thereby suppressing NF-κB activation or inhibiting IκBα degradation [45,46].

Although macauba pulp oil prevented the adipogenesis pathway and inflammation in the adipose tissue, significant effects in the hepatic markers were not observed after eight weeks of the high-fat diet. The current study was carried out as a prevention model, and for this reason, it might not be able to verify alterations in the liver. Thus, in the current research, the consumption of the diets for eight weeks, even with a high concentration of saturated fats, was not able to cause metabolic changes in the liver. These results were confirmed by histomorphometric analyses, which showed that there was no alteration of the cellular components evaluated, such as fat and inflammation in the liver.

Despite the decreased expression of (mRNA) *CPT-1α* gene, the quantification of PPAR-α did not change with the consumption of macauba pulp oil, which might be because ADIPOR2 did not change either. The increase in the sensitization of ADIPOR2 receptor triggers the activation of PPAR-α, which regulates fatty acid oxidation [47]. Moreover, the high traffic of free fatty acids due to a high-fat diet has the ability to trigger SREBP-1c, which controls the expression of enzymes essential in triacylglycerol synthesis and storage, and restricts lipogenic genes, such as ACC-1, that are responsible for the transformation of ACC-1 to malonyl CoA [48]. However, despite the overexpression of this gene in the fatty acid synthesis pathway, there was no change in the proportion of fat and steatosis degree in the liver. This might be due to the increased antioxidant capacity, which decreased the expression of this gene in relation to fatty acid synthesis.

The strain of mice used in this study was chosen since they are prone to metabolic disturbances generated by a high-fat diet. However, it is known that experiments with mice do not fully reflect the effects in humans due to differences in the organs and metabolism of these two species. However, taking into account the macauba pulp oil intake per animal weight, a human with 70 kg needs a consumption of a small amount per day (approximately 8 g/day of macauba pulp oil—similar to one teaspoon) to have the same improvements observed in this research in the prevention of metabolic changes. Thus, further studies are needed to verify the real effects of macauba pulp oil in human.

The influence of a high-fat diet on the body and the mechanism of macauba pulp oil, which was demonstrated in our study, are summarized in Figure 5. The consumption of macauba pulp oil prevents inflammation and adipogenesis, as demonstrated by a reduction in the expression of PPAR-γ, (mRNA) *SREBP-1c*, NF-κB, and (mRNA) *TNF-*

α, and an increase in adiponectin in adipose tissue. In the liver, despite triggering the *SREBP-1c* expression and a lower (mRNA) *CPT-1α* level, it does not lead to liver changes, according to the histomorphometric analysis, due to an increased antioxidant capacity. These modes of action may be related to macauba pulp oil, which has a good composition of carotenoids, oleic acid, and tocopherol and improves the total antioxidant capacity, resulting in adipogenesis even with a high level of saturated fat consumption.

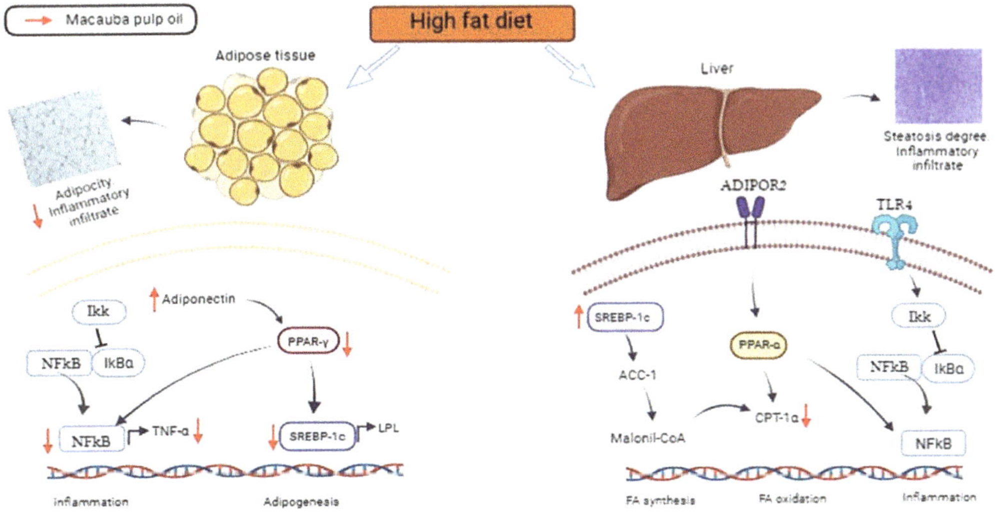

Figure 5. Effects of a high-fat diet on the adipose tissue and liver and potential action mechanism of macauba pulp oil. TLR-4: toll-like receptor 4; NF-κB: nuclear factor kappa B; *SREBP-1c*: sterol regulatory element-binding protein; *ACC-1*: acetyl-CoA carboxylase 1; *ADIPOR2*: adiponectin receptor 2; *CPT-1α*: carnitine palmitoyl transferase 1 alpha; *PPAR-γ*: peroxisome proliferator-activated receptor gamma; PPAR-α: peroxisome proliferator-activated receptor alpha; *LPL*: lipoprotein lipase; *TNF-α*: tumor necrosis factor alpha; Ikk: IkB kinase complex; IkBα nuclear factor of kappa light polypeptide gene enhancer in B-cells inhibitor alpha; Malonil-CoA: malonyl coenzyme A; FA: fatty acid.

5. Conclusions

Consumption of macauba pulp oil increases antioxidant capacity and prevents oxidative stress, inflammation, and the adipogenesis pathway. Therefore, macauba pulp oil has a great potential for inclusion in human foods to improve health, assisting in the prevention of risk factors for chronic non-communicable diseases.

Author Contributions: C.T.S.A.: data curation, formal analysis, methodology, investigation, validation, writing—original draft, and writing—review and editing. T.A.V.: formal analysis and methodology. M.G.: formal analysis and methodology. R.C.L.T.: formal analysis and methodology. E.T.: writing—review and editing. N.M.B.C.: supervision and writing—review and editing. H.S.D.M.: conceptualization, project administration, supervision, and writing—review and editing. F.A.R.d.B.: conceptualization, funding acquisition, project administration, supervision, and writing—review and editing. All authors have read and agreed to the published version of the manuscript.

Funding: This work was supported by the Conselho Nacional de Desenvolvimento Científico e Tecnológico (CNPq, Brazil). Cíntia Tomaz Sant' Ana was a recipient of a scholarship from CNPq (call # 23/2028-DAI/CNPq).

Institutional Review Board Statement: This study was conducted according to the guidelines of the Declaration of Helsinki and approved by the Ethics Committee of the Federal University of Viçosa (protocol code 09/2019 and date of approval 28 May 2019).

Informed Consent Statement: Not applicable.

Data Availability Statement: Data are contained within the article.

Acknowledgments: We gratefully acknowledge the *Soleá Brasil Óleos Vegetais Ltd.a* (Brazil) for providing the macauba used in this study and financial support, and Ceres Mattos Della Lucia (Federal University of Viçosa, Brazil) for her assistance with the quantification of tocopherols and carotenoids.

Conflicts of Interest: The authors declare no conflict of interest.

References

1. Koliaki, C.; Liatis, S.; Kokkinos, A. Obesity and cardiovascular disease: Revisiting an old relationship. *Metabolism* **2019**, *92*, 98–107. [CrossRef] [PubMed]
2. World Health Organization, WHO. Noncommunicable Diseases. Available online: https://www.who.int/news-room/fact-sheets/detail/noncommunicable-diseases (accessed on 27 July 2020).
3. Ambele, M.A.; Dhanraj, P.; Giles, R.; Pepper, M.S. Adipogenesis: A complex interplay of multiple molecular determinants and pathways. *Int. J. Mol. Sci.* **2020**, *21*, 4283. [CrossRef] [PubMed]
4. Lee, J.E.; Schmidt, H.; Lai, B.; Ge, K. Transcriptional and epigenomic regulation of adipogenesis. *Mol. Cell. Biol.* **2019**, *39*, e11. [CrossRef] [PubMed]
5. Hariri, N.; Gougeon, R.; Thibault, L. A highly saturated fat-rich diet is more obesogenic than diets with lower saturated fat content. *Nutr. Res.* **2010**, *30*, 632–643. [CrossRef]
6. Coimbra, M.C.; Jorge, N. Characterization of the pulp and kernel oils from *Syagrus oleracea*, *Syagrus romanzoffiana*, and *Acrocomia aculeata*. *J. Food Sci.* **2011**, *76*, C1156–C1161. [CrossRef]
7. Lieb, V.M.; Schex, R.; Esquivel, P.; Jiménez, V.M.; Schmarr, H.G.; Carle, R.; Steingass, C.B. Fatty acids and triacylglycerols in the mesocarp and kernel oils of maturing Costa Rican *Acrocomia aculeata* fruits. *NFS J.* **2019**, *14–15*, 6–13. [CrossRef]
8. Díaz, A.C.; Fiñana, I.T.; Granados, J.M.M.; Méndez, M.V.R.; Dorado, G.; Sánchez, M.C.R.; Valverde, C.N.; Gómez, J.M.Q. Serum from postmenopausal women treated with a by-product of olive-oil extraction process stimulates osteoblastogenesis and inhibits adipogenesis in human mesenchymal stem-cells (MSC). *Exp. Gerontol.* **2017**, *90*, 71–78. [CrossRef]
9. Bonet, M.L.; Canas, J.A.; Ribot, J.; Palou, A. Carotenoids and their conversion products in the control of adipocyte function, adiposity and obesity. *Arch. Biochem. Biophys.* **2015**, *572*, 112–125. [CrossRef]
10. Müller, L.; Theile, K.; Böhm, V. In vitro antioxidant activity of tocopherols and tocotrienols and comparison of vitamin E concentration and lipophilic antioxidant capacity in human plasma. *Mol. Nutr. Food Res.* **2010**, *54*, 731–742. [CrossRef]
11. Ichihara, K.; Fukubayashi, Y. Preparation of fatty acid methyl esters for gas-liquid chromatography. *J. Lipid Res.* **2010**, *51*, 635–640. [CrossRef]
12. Rodriguez-Amaya, D.B. *A Guide to Carotenoid Analysis in Foods*; ILSI Press: Washington, DC, USA, 2001; ISBN 1578810728.
13. Pinheiro-Sant'Ana, H.M.; Guinazi, M.; Da Silva, D.O.; Della Lucia, C.M.; De Lazzari, B.R.; Brandão, S.C.C. Method for simultaneous analysis of eight vitamin E isomers in various foods by high performance liquid chromatography and fluorescence detection. *J. Chromatogr. A* **2011**, *1218*, 8496–8502. [CrossRef]
14. Fontelles, M.J. Metodologias da pesquisa: Diretrizes para o cálculo do tamanho da amostra. *Rev. Para. Med.* **2010**, *24*, 57–64.
15. Schoemaker, M.H.; Kleemann, R.; Morrison, M.C.; Verheij, J.; Salic, K.; Van Tol, E.A.F.; Kooistra, T.; Weilinga, P.Y. A casein hydrolysate based formulation attenuates obesity and associated non-alcoholic fatty liver disease and atherosclerosis in LDLr-/-Leiden mice. *PLoS ONE* **2017**, *12*, e0180648. [CrossRef]
16. Reeves, P.G.; Nielsen, F.H.; Fahey, G.C., Jr. AIN-93 Purified diets for laboratory rodents: Final report of the American Institute of Nutrition ad hoc writing Committee on the Reformulation of the AIN-76A Rodent Diet. *J. Nutr.* **1993**, *123*, 1939–1951. [CrossRef]
17. Novelli, E.L.B.; Diniz, Y.S.; Galhardi, C.M.; Ebaid, G.M.X.; Rodrigues, H.G.; Mani, F.; Novelli Filho, J.L.V.B. Anthropometrical parameters and markers of obesity in rats. *Lab. Anim.* **2007**, *41*, 111–119. [CrossRef]
18. Kim, S.; Hong, J.; Jeon, R.; Kim, H.S. Adzuki bean ameliorates hepatic lipogenesis and proinflammatory mediator expression in mice fed a high-cholesterol and high-fat diet to induce nonalcoholic fatty liver disease. *Nutr. Res.* **2016**, *36*, 90–100. [CrossRef]
19. Percie du Sert, N.; Hurst, V.; Ahluwalia, A.; Alam, S.; Avey, M.T.; Baker, M.; Browne, W.J.; Clark, A.; Cuthill, I.C.; Dirnagl, U.; et al. The ARRIVE guidelines 2.0: Updated guidelines for reporting animal research. *PLoS Biol.* **2020**, *18*, e3000410. [CrossRef]
20. Marklund, S. Pyrogallol autooxidation. In *Handbook of Methods for Oxygen Radical Research*; Greenwald, R.A., Ed.; CRC Press Inc.: Boca Raton, FL, USA, 1985; pp. 243–247.
21. Santos-López, J.A.; Garcimartín, A.; López-Oliva, M.E.; Bautista-Ávila, M.; González-Muñoz, M.J.; Bastida, S.; Benedí, J.; Sánchez-Muniz, F.J. Chia Oil–Enriched Restructured Pork Effects on Oxidative and Inflammatory Status of Aged Rats Fed High Cholesterol/High Fat Diets. *J. Med. Food.* **2017**, *20*, 526–534. [CrossRef]
22. Aebi, H. Catalase in vitro. *Methods Enzymol.* **1984**, *105*, 121–126. [CrossRef]
23. Green, L.C.; Wagner, D.A.; Glogowski, J.; Skipper, P.L.; Wishnok, J.S.; Tannenbaum, S.R. Analyses of nitrate, nitrite and [^{15}N] nitrate in biological fluids. *Anal. Biochem.* **1982**, *126*, 131–138. [CrossRef]
24. Livak, K.J.; Schmittgen, T.D. Analysis of relative gene expression data using real-time quantitative PCR and the 2-DeltaCT method. *Methods* **2001**, *25*, 402–408. [CrossRef] [PubMed]

25. Cupertino, M.C.; Costa, K.L.C.; Santos, D.C.M.; Novaes, R.D.; Condessa, S.S.; Neves, A.C.; Oliveira, J.A.; Matta, S.L.P. Long-lasting morphofunctional remodelling of liver parenchyma and stroma after a single exposure to low and moderate doses of cadmium in rats. *Int. J. Exp. Pathol.* **2013**, *94*, 343–351. [CrossRef] [PubMed]
26. Turlin, B.; Mendler, M.H.; Moirand, R.; Guyader, G.; Guillygomarc'h, A.; Deugnier, Y. Histologic features of the liver in insulin resistance–associated Iron overload: A study of 139 patients. *Am. J. Clin. Pathol.* **2001**, *116*, 263–270. [CrossRef] [PubMed]
27. Elvira-Torales, L.I.; García-Alonso, J.; Piriago-Castón, M.J. Nutritional importance of carotenoids and their effect on liver health: A review. *Antioxidants* **2019**, *8*, 229. [CrossRef] [PubMed]
28. Chen, X.; Li, L.; Liu, X.; Luo, R.; Chen, Y. Oleic acid protects saturated fatty acid mediated lipotoxicity in hepatocytes and rat of non-alcoholic steatohepatitis. *Life Sci.* **2018**, *203*, 291–304. [CrossRef]
29. Bhattacharjee, B.; Pal, P.K.; Chattopadhyay, A.; Bandyopadhyay, D. Oleic acid protects against cadmium induced cardiac and hepatic tissue injury in male Wistar rats: A mechanistic study. *Life Sci.* **2020**, *244*, 117324. [CrossRef]
30. Wang, Y.; Branicky, R.; Noe, A.; Hekimi, S. Superoxide dismutases: Dual roles in controlling ROS damage and regulating ROS signaling. *J. Cell Biol.* **2018**, *217*, 1915–1928. [CrossRef]
31. Pizzino, G.; Irrera, N.; Cucinotta, M.; Pallio, G.; Mannino, F.; Arcoraci, V.; Squadrito, F.; Altavilla, D.; Bitto, A. Oxidative Stress: Harms and Benefits for Human Health. *Oxid. Med. Cell. Longev.* **2017**, *2017*, 8416763. [CrossRef]
32. Naguibi, Y.M. Antioxidant activities of astaxanthin and related carotenoids. *J. Agric. Food Chem.* **2000**, *48*, 1150–1154. [CrossRef]
33. Schnorr, C.E.; Morrone, M.S.; Simões-Pires, A.; Bittencourt, L.D.; Zeidán-Chuliá, F.; Moreira, J.C.F. Supplementation of adult rats with moderate amounts of β–carotene modulates the redox status in plasma without exerting pro-oxidant effects in the brain: A safer alternative to food fortification with vitamin A? *Nutrients* **2014**, *6*, 5572–5582. [CrossRef]
34. Sarada, S.; Dipti, P.; Anju, B.; Pauline, T.; Kain, A.; Sairam, M.; Sharma, S.; Ilavazhagan, G.; Kumar, D.; Selvamurthy, W. Antioxidant effect of β-carotene on hypoxia induced oxidative stress in male albino rats. *J. Ethnopharmacol.* **2002**, *79*, 149–153. [CrossRef]
35. Pan, J.H.; Kim, M.J.; Kim, J.H.; Cho, Y.J.; Shin, H.S.; Sung, J.S.; Park, T.S.; Yoon, H.G.; Park, S.; Kim, Y.J. Inhibition of the lipogenesis in liver and adipose tissue of diet-induced obese C57BL/6 mice by feeding oleic acid-rich sesame oil. *Food Sci. Biotechnol.* **2015**, *24*, 1115–1121. [CrossRef]
36. Ribot, J.; Felipe, F.; Bonet, M.L.; Palou, A. Changes of adiposity in response to Vitamin A status correlate with changes of PPARγ2 expression. *Obes. Res.* **2001**, *9*, 500–509. [CrossRef]
37. Arankumar, E.A.; Sushil, K.J. Adiponectin, a therapeutic target for obesity, diabetes and endothelial dysfunction. *Int. J. Mol. Sci.* **2017**, *18*, 1321. [CrossRef]
38. Lobo, G.P.; Amengual, J.; Li, H.N.; Golczak, M.; Bonet, M.L.; Palczewski, K.; von Lintig, J. β-carotene decreases peroxisome proliferator gamma activity and reduces lipid storage capacity of adipocytes in a β-carotene oxygenase 1-dependent manner. *J. Biol. Chen.* **2010**, *285*, 27891–27899. [CrossRef]
39. Amengual, J.; Gouranton, E.; Van Helden, Y.G.; Hessel, S.; Ribot, J.; Kramer, E.; Kiec-Wilk, B.; Razny, U.; Lietz, G.; Wyss, A.; et al. Beta-carotene reduces body adiposity of mice via BCMO1. *PLoS ONE* **2011**, *6*, e20644. [CrossRef]
40. Granados, N.; Amengual, J.; Ribot, J.; Palou, A.; Bonet, M.L. Distinct effects of oleic acid and its trans-isomer elaidic acid on the expression of myokines and adipokines in cell models. *Br. J. Nutr.* **2011**, *105*, 1226–1234. [CrossRef]
41. Włodarczyk, M.; Nowicka, G. Obesity, DNA damage, and development of obesity-related diseases. *Int. J. Mol. Sci.* **2019**, *20*, 1146. [CrossRef]
42. Rosillo, M.A.; Sanchez-Hidalgo, M.; Gonzalez-Benjumea, A.; Fernandez-Bolanos, J.G.; Lubberts, E.; Alarcon-De-La-Lastra, C. Preventive effects of dietary hydroxytyrosol acetate, an extra virgin olive oil polyphenol in murine collagen-induced arthritis. *Mol. Nutr. Food Res.* **2015**, *59*, 2537–2546. [CrossRef]
43. Morari, J.; Torsoni, A.S.; Anhê, G.F.; Roman, E.A.; Cintra, D.E.; Ward, L.S.; Bordin, S.; Velloso, L.A. The role of proliferator-activated receptor gamma coactivator-1alpha in the fatty-acid-dependent transcriptional control of interleukin-10 in hepatic cells of rodents. *Metabolism* **2010**, *59*, 215–223. [CrossRef]
44. Wang, Y.; Park, N.Y.; Jang, Y.; Ma, A.; Jiang, Q. Vitamin E γ-tocotrienol inhibits cytokine-stimulated NF-κB activation by induction of anti-inflammatory A20 via stress adaptive response due to modulation of sphingolipids. *J. Immunol.* **2015**, *1195*, 126–133. [CrossRef] [PubMed]
45. Kaulmann, A.; Bohn, T. Carotenoids, inflammation, and oxidative stress—Implications of cellular signaling pathways and relation to chronic disease prevention. *Nutr. Res.* **2014**, *34*, 907–929. [CrossRef] [PubMed]
46. Bai, S.-K.; Lee, S.-J.; Na, H.-J.; Ha, K.-S.; Han, J.-A.; Lee, H.; Kwon, Y.-G.; Chung, C.-K.; Kim, Y.-M. β-Carotene inhibits inflammatory gene expression in lipopolysaccharide-stimulated macrophages by suppressing redox-based NF-κB activation. *Exp. Mol. Med.* **2005**, *37*, 323–334. [CrossRef] [PubMed]
47. Abdelmegeed, M.A.; Moon, K.H.; Hardwick, J.P.; Gonzalez, F.J.; Song, B.J. Role of peroxisome proliferator-activated receptor-α in fasting-mediated oxidative stress. *Free Radic. Biol. Med.* **2010**, *47*, 767–778. [CrossRef]
48. Jeon, T.L.; Osborne, T.F. SREBPs: Metabolic integrators in physiology and metabolism. *Trends Endocrinol. Metab.* **2012**, *23*, 65–72. [CrossRef]

nutrients

MDPI

Article

Kombuchas from Green and Black Tea Modulate the Gut Microbiota and Improve the Intestinal Health of Wistar Rats Fed a High-Fat High-Fructose Diet

Mirian Aparecida de Campos Costa [1], Luiza de Paula Dias Moreira [2], Vinícius da Silva Duarte [3], Rodrigo Rezende Cardoso [1], Vinícius Parzanini Brilhante de São José [4], Bárbara Pereira da Silva [4], Mariana Grancieri [4], Viviana Corich [2], Alessio Giacomini [2], Josefina Bressan [4], Hércia Stampini Duarte Martino [4] and Frederico Augusto Ribeiro de Barros [1,*]

[1] Department of Food Technology, Universidade Federal de Viçosa, Avenida Peter Henry Rolfs, s/n, Viçosa 36570-900, MG, Brazil
[2] Department of Agronomy, Food Natural Resources, Animals, and Environment (DAFNAE), Università degli Studi di Padova, Via dell'Università 16, 35020 Legnaro, PD, Italy
[3] Faculty of Chemistry, Biotechnology, and Food Science, The Norwegian University of Life Sciences, P.O. Box 5003, 1432 Ås, Norway
[4] Department of Nutrition and Health, Universidade Federal de Viçosa, Avenida Peter Henry Rolfs, s/n, Viçosa 36570-000, MG, Brazil
* Correspondence: fredbarros@ufv.br

Citation: Costa, M.A.d.C.; Dias Moreira, L.d.P.; Duarte, V.d.S.; Cardoso, R.R.; São José, V.P.B.d.; Silva, B.P.d.; Grancieri, M.; Corich, V.; Giacomini, A.; Bressan, J.; et al. Kombuchas from Green and Black Tea Modulate the Gut Microbiota and Improve the Intestinal Health of Wistar Rats Fed a High-Fat High-Fructose Diet. *Nutrients* **2022**, *14*, 5234. https://doi.org/10.3390/nu14245234

Academic Editor: Elad Tako

Received: 10 November 2022
Accepted: 30 November 2022
Published: 8 December 2022

Abstract: The Western diet can negatively affect the gut microbiota and is associated with metabolic disorders. Kombucha, a tea fermented by a symbiotic culture of bacteria and yeast (SCOBY), is known for its bioactive properties and has become popular in the last years. In this study, we evaluated the effects of regular kombucha consumption on the gut microbiota and on outcomes related to the intestinal health of Wistar rats fed a high-fat high-fructose diet. After eight weeks receiving a standard diet (AIN-93M) (n = 10) or a high-fat and high-fructose diet (HFHF) (n = 30) to induce metabolic disorders, the animals were subdivided into four groups: AIN-93M (n = 10); HFHF (n = 10); GTK (HFHF + green tea kombucha (n = 10); and BTK (HFHF + black tea kombucha; n = 10) for 10 weeks. Although body composition did not differ among the groups, the HFHF diet was associated with metabolic alterations, and stimulated the growth of gram-negative bacteria such as *Proteobacteria* and *Bacteroides*. Kombucha ingestion could somewhat modulate the gut microbiota, attenuating the effects of a Western diet by increasing propionate production and favoring the growth of beneficial bacteria, such as *Adlercreutzia* in the GTK group. Our results suggest that regular kombucha consumption may be beneficial to intestinal health, which can be mostly attributed to its high content and diversity of phenolic compounds.

Keywords: experimental study; gut microbiome; intestinal permeability; obesity; polyphenols; probiotic; short-chain fatty acids

1. Introduction

The Western diet style, despite not having a specific definition, is an unhealthy diet generally composed of a high amount of saturated fat and fructose [1], which is involved in metabolic disorders such as obesity and metabolic syndrome [2,3], as well as in alterations in the gut microbiota profile [4]. Although the mechanisms by which dietary fat can modulate the gut microbiota are not completely understood yet, it is known that the small amount of this nutrient that is not absorbed in the small intestine can be fermented by the gut microbiota. Free fatty acids (FFA) resulting from the lipid metabolism can be utilized as substrates by the microorganisms present in the gut microbiota, influencing its composition [5] by increasing the *Bacteroidetes*:*Firmicutes* ratio [6] and the proportion of *Proteobacteria*, which are a major source of lipopolysaccharides [7]. Similarly, a high-fructose diet has also been linked with alterations in the gut microbiota. Studies have suggested that it can affect the morphology and function of the intestine by altering the structure of the tight junction proteins, leading to increased intestinal permeability and inflammation [8,9].

Kombucha, a fermented beverage usually produced from green or black tea, has been highlighted as a promising alternative to minimize the impact of the Western diet or even for those who wish for a healthier lifestyle [10]. Kombucha presents in its composition several microorganisms as a result of the fermentation process performed by microorganisms known as SCOBY (symbiotic culture of bacteria and yeast) [11,12]. These microorganisms include lactic and acetic bacteria, particularly from the genera *Acetobacter* and *Gluconobacter*, and yeasts [12–15].

Beyond the microorganisms, kombucha also presents in its composition organic acids such as acetic, gluconic, and glucuronic; vitamins C and B complex; minerals; and amino acids [10,11,16]. However, it seems that the main benefits associated with kombucha intake are due to the presence of bioactive compounds. Previous studies conducted by our research group have shown that kombuchas from green and black tea present a high antioxidant capacity due to a high amount and diversity of phenolic compounds. Of the 127 phenolic compounds that we have identified, 103 were reported for the first time in the literature [17]. Nonetheless, its nutritional composition is influenced by many factors, including the tea type and quality, the amount of substrate, and the time and temperature used in the fermentation process [17–19]. SCOBYs also present differences in their composition and can influence the microorganism profile [20,21].

Regardless of the differences obtained in the manufacturing process, kombucha intake has been associated with health benefits through the modulation of the gut microbiota in mice. Black tea kombucha was associated with a decreased abundance of *Allobaculum*, *Turicibacter*, and *Clostridium* genera and an increase in *Mucispirillum*, a genus positively correlated to circulating leptin, which is a hormone involved in the regulation of appetite and food intake [22]. In another study, green tea kombucha supplementation was associated with an increase in alpha-diversity as well as favored the growth of bacteria involved in butyrate production [23]. Additionally, a recent systematic review has pointed out that kombucha consumption was able to reduce intestinal dysbiosis in vivo, being suggested as a potential alternative for the control and treatment of obesity and its associated comorbidities [24].

Although there is evidence that kombucha intake can bring benefits to health, there is still no consensus in the literature, especially when associated with the Western diet. Even though many commercial kombuchas have an appeal as a probiotic product, there is no evidence to support it so far. Thus, we aimed to investigate the effects of regular kombucha consumption on the gut microbiota and on outcomes related to the intestinal health of Wistar rats fed a high-fat high-fructose diet. Based on the chemical and microbiological composition of the beverages, as well as on previous studies [22,23], we hypothesized that green and black tea kombuchas would be able to modulate the gut microbiota and improve the intestinal health of those animals.

2. Material and Methods

2.1. Kombuchas Preparation

Kombuchas from green and black tea were prepared as previously described [17,25]. In summary, green tea (Lung Ching) and black tea (Darjeeling Gielle FTGFOP1) were obtained in a certified store (Tea Shop®) located in Belo Horizonte, Minas Gerais, Brazil.

Both beverages were prepared using 12 g of tea leaves and 50 g of sugar per liter of mineral water. Green tea infusion was performed at 75 °C for 3 min and the black tea at 95 °C for 5 min, according to the manufacturer. The beverages were cooled in an ice bath and when they reached room temperature, they were added to a SCOBY (3% w/v) (Enziquímica®, Gravataí, Brazil). A previously prepared kombucha (10% v/v) was also added to the beverages to decrease the pH and inhibit the proliferation of pathogenic microorganisms [11].

Fermentation occurred for ten days at 25 °C and then the SCOBY was removed and the kombuchas were filtered (Whatman #1 qualitative filter paper). The beverages were stored at 4°C for up to two weeks before being offered to the animals.

2.2. *Animal Study*

2.2.1. Study Design

Forty Wistar rats (*Rattus norvegicus*) aged between 45 and 50 days old were obtained at the Central Animal Facility from the Center of Biological Sciences and Health at Universidade Federal de Vicosa, Brazil. The animals were allocated to individual stainless-steel cages and kept in a light-dark cycle (12 h/12 h) at room temperature at 22 ± 2 °C.

The experiment was divided into two phases. Phase I lasted eight weeks and the animals were separated into two groups: group 1 (n = 10); received a standard control diet (AIN-93M) [26] and group 2 (n = 30); received a high-fat and high-fructose (HFHF) diet to induce metabolic alterations [27]. In phase II, which lasted ten weeks, group 1 continued receiving a standard diet while the HFHF group was subdivided into three other groups: HFHF group (positive control) (n = 10); green tea kombucha (GTK group), which received HFHF diet + green tea kombucha diluted in water (30% v/v) (n = 10); and black tea kombucha (BTK group), which received HFHF diet + black tea kombucha diluted in water (30% v/v) (n = 10) (Figure 1). Both diets and water were consumed *ad libitum* during the whole experimental period.

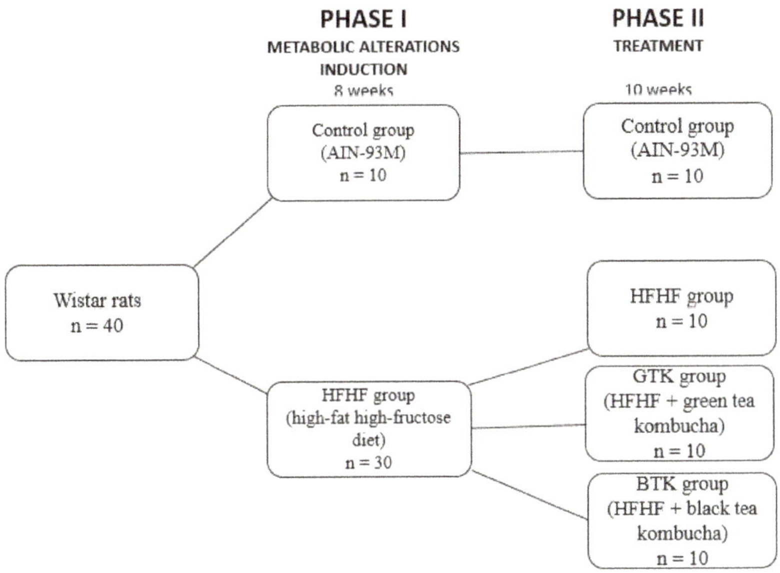

Figure 1. Experimental study design.

All procedures were performed following the ethical principles for animal use in experimental studies. The study protocol was approved by the Ethics Committee on Animal Use (CEUA—Universidade Federal de Vicosa, Protocol 06/2019; date of approval: 28 May 2019).

2.2.2. Kombuchas Characterization and Dosage

This study is a follow-up to our previous work and details about the experimental diet and the physical-chemical analyses of the kombuchas used in this experiment have been extensively described [17,25]. Briefly, sugars (sucrose, glucose, and fructose), organic acids (acetic, glucuronic, and lactic), and ethanol were identified and quantified by high-performance liquid chromatography (HPLC) (Shimadzu, model LC-10A VP) coupled to a refractive index detector (RID 6A). The total acidity was determined according to the methodology proposed by IAL (2008) using phenolphthalein as an indicator and the results expressed as % acetic acid (*w*/*v*) [28]. The pH was measured by a calibrated pH meter (Kasvi, K-39,1014B, China). The concentrations of theaflavins and thearubigins were determined by a spectrophotometer according to the methodology proposed by Jayabalan et al. (2007), and the results were expressed as % (*w*/*v*) [29]. Regarding the microbiological characterization, serial kombucha samples were used to determine acetic bacteria, lactic bacteria, and yeasts using plates with GYC agar and ethanol, MRS, and PDA agar, respectively. The results were expressed as CFU/mL.

Kombucha dosage was determined according to preliminary tests that indicated the total phenolic content of beverages. Calculations were performed based on a previous study that recommends a daily total phenolic intake of 17 mg/kg/body weight [30].

2.2.3. Euthanasia and Samples Collection

As previously described [25], at the end of the experimental period, the animals were anesthetized by inhalation (Isoforine, Cristália®, São Paulo, Brazil), and euthanized by exsanguination by cardiac puncture. The tissues were immediately collected, weighted, and frozen in liquid nitrogen, and stored at −80 °C for further analysis. Feces were collected from the cecum, weighted, and stored at −80 °C for future analyses. Colon fragments were collected and fixed in formaldehyde (10% *v*/*v*) for the first 24 h and then stored in ethanol (70% *v*/*v*) and embedded in paraffin for histological analysis.

2.2.4. Histological Analysis

Serial sections of the colon, with a thickness of 5 µm, were collected and subsequently deparaffinized in xylene, rehydrated with different alcohol solutions, and stained with hematoxylin and eosin.

Histological sections were visualized in an Olympus AX70 photomicroscope, and the images were captured in a 20X objective with an AxioCam HRc—Zeiss digital camera. The following features were analyzed: crypt depth, crypt width, and the number of goblet cells. For that, we randomly selected six animals per group and twenty random fields per animal, analyzing one crypt per field. Only crypts with a well-defined and visible structure were used.

The measurements of the crypts were performed using the ImagePro-Plus® version 4.5 software (Media Cybernetics Inc., 1700 Rockville Pike, Suite 240, Rockville, MD 20852, USA) and the goblet cell count was performed using the Image J® 1.48v software (Research Services Branch, National Institute of Mental Health, Bethesda, MD, USA).

2.2.5. Intestinal Permeability

Intestinal permeability analysis was performed at the end of the 10th week of treatment. After 12 h of fasting, the animals received 1 mL of a solution containing 100 mg lactulose and 50 mg mannitol by gavage. Then, they were kept in metabolic cages, fasting for 5 h. The urine was collected for 24 h and stored at −80 °C.

The urine was centrifuged (Hermle centrifuge, model Z326K, Wehingen, Germany), filtered on 0.45 µm membrane filters (Millipore, São Paulo, Brazil), and transferred to vials for high-performance liquid chromatography (HPLC). The mobile phase consisted of water in sulfuric acid (0.005 mM) with an injection volume of 20 µL and a mobile phase flow of 0.6 mL/min [31]. Lactulose® and Mannitol® were used as internal standards (Sigma-Aldrich, São Paulo, Brazil) and the concentrations were transformed to g/L to calculate the percentage of urinary excretion. The lactulose/mannitol ratio was calculated by dividing lactulose concentration by mannitol concentration [32].

2.2.6. Fecal pH and Short-Chain Fatty Acids Analysis

For pH analysis, approximately 1 g of cecum stool was homogenized in 10 mL of distilled water and vortexed with glass beads. Subsequently, the glass electrode of the pH meter was inserted, and the pH was measured in duplicate [33].

The short-chain fatty acids (SCFA) analysis was performed according to Siegfried et al. (1984) with modifications [34]. Briefly, approximately 500 mg of stool samples was homogenized in 1 mL of Milli-Q® water in a vortex and centrifuged at 12,000× *g* for 10 min. The supernatant was removed, and the samples were injected on a high-performance liquid chromatography (HPLC) (injection volume: 20 μL; Dionex Corporation, Sunnyvale, CA, USA). The SCFA were separated on a Phenomenex Rezex ROA ion exclusion column (300 × 7.8 mm) (Phenomenex Inc. Torrance, CA, USA) coupled to a Shodex RI-101 refractive index (IR) maintained at 45 °C. Sulfuric acid 5 mM with a flow of 0.7 mL/min was used as a mobile phase. Stock solutions were prepared using acetic, propionic, and butyric acids as standards with a final concentration of 10 mmol/L (Sigma-Aldrich, Sao Paulo, Brazil). Stock solutions were diluted 2-, 4-, 8-, and 16-fold in 5 mmol/L^{-1} sulfuric acid (0.08–10 mM) to be used as standards in the HPLC analysis.

2.2.7. DNA Extraction and Microbiota Profile

DNA extraction of stool samples collected from the cecum of the animals was performed according to the methodology proposed by Steveron and Weimer (2007) [35]. Briefly, mechanical cellular lysis was performed using glass beads and phenol chloroform to promote the partitioning of lipids and cellular debris into the organic phase. After DNA extraction, a total of 39 samples encompassing all the animals of each experimental group were sequenced at the Argonne National Laboratory, Illinois, USA (AIN-93M n = 10; HFHF: n = 10; GTK: n = 9; and BTK: n = 10). The V4 region of the 16S rRNA genes was amplified by PCR using 515f/806r primers and amplicons sequenced using Illumina MiSeq desktop sequencer producing 150 bp paired-end (PE) reads.

The demultiplexed raw paired-end reads obtained after sequencing were uploaded and processed into QIIME2 (version 2020.2) via the Casava 1.8 paired-end pipeline [36]. DADA2, which allows improved taxonomic resolution based on the exact identification and error correction of sample sequences that differ as little as a single nucleotide, was chosen to assess the quality of the reads in sequential steps such as filtering, trimming, denoising, dereplicating, merging paired reads, as well as chimeric sequences removal [37]. Afterward, amplicon sequence variants (ASV) were forwarded to generate a phylogenetic tree using the align-to-tree-mafft-fasttree pipeline from the q2-phylogeny plugin [38]. When convenient, samples were rarefied to an appropriate sampling depth of 15,349. Taxonomy was assigned to the 16S data using a Naïve Bayes pre-trained Greengenes 13_8 99% OTUs classifier [39].

With regards to the DNA obtained from the kombuchas and their respective SCOBYs, samples used during the experiment were mixed and lyophilized at −62 °C for 24 h under a pressure of 35 uHg (Liotop, model L101, serial no. 01610, Liobras, São Carlos, Brazil). Microbial DNA was extracted from frozen pellets using the Qiagen Powersoil Pro kit with bead beating, according to the manufacturer's protocol. Then, the samples were forwarded to the company Molecular Research LP (MR DNA, Shallowater, TX, USA) where amplicon preparation and sequencing were performed considering the bacterial V4 region of the 16S rRNA gene (515f/806r primers) and the variable internal transcribed spacer (ITS)-1 of the fungal rRNA region (ITS1F and ITS2 reverse primers). Amplicon libraries were prepared and sequenced with the Illumina MiSeq desktop sequencer producing 250 bp paired-end (PE) reads. For microbiota profiling, sequence data were processed using an MR DNA analysis pipeline (MR DNA, Shallowater, TX, USA). Briefly, sequences were joined, and short reads with ambiguous base calls were removed. Afterward, sequences were quality filtered using a maximum expected error threshold of 1.0, dereplicated, and denoised. Lastly, unique sequences identified with sequencing and/or PCR point errors were removed followed by chimera removal, thereby providing a denoised sequence or zero-radius OTU (zOTU). Final zOTUs were taxonomically classified using BLASTn against a curated database derived from NCBI (http://www.ncbi.nlm.nih.gov; accessed on 10 July 2020). Raw reads were deposited in the Sequence Read Archive (SRA) database (http://www.ncbi.nlm.nih.gov/sra; deposited on 23 November 2022) under the BioProject PRJNA904803.

2.3. Statistical Analysis

Statistical analysis was performed using the software GraphPad Prism®, version 6.01. The normality of the data was tested by the Shapiro–Wilk test. Groups with parametric distribution were analyzed by one-way ANOVA followed by Tukey post-hoc. Non-parametric data were evaluated by the Kruskal–Wallis test followed by Dunn's post-hoc. Data were expressed as mean ± standard deviation (SD) and the values were considered significant when $p < 0.05$.

For statistical analysis of the gut microbiota, qiime artifacts were imported into R (R Core Team 3.6.2, 2019) with the qiime2R package v.099.20 (https://github.com/jbisanz/qiime2R; accessed on 10 July 2020). Significant differences in alpha-diversity among the four groups (AIN-93M, GTK, BTK, and HFHF) were determined using the alpha function in microbiome R package v.2.1.24 adopting Kruskal–Wallis as a statistical test followed up by Wilcoxon's test to calculate pairwise comparisons between groups. For beta-diversity, weighted and unweighted UniFrac distances were subjected to permutational multivariate analysis of variance (PERMANOVA) to assess significant differences (pseudo-F test) in bacterial community composition and structure among the groups with a permutation number of 999. Principal coordinates analysis (PCoA) was chosen to explore and visualize the clustering of groups. All graphs were constructed and visualized with RStudio (v. 1.2.5033) using one or the combination of the following R packages: MicrobiomeR, dplyr, ggplot2, phyloseq [40], tidyr, and vegan.

To determine which bacterial taxa were differentially abundant among groups, taxonomy was firstly collapsed to the genus level and then analyzed via linear discriminant analysis (LDA) effect size (LEFSe) [41] (p-value cut-off of 0.05 and log LDA score of 2.0).

3. Results

3.1. Kombucha Chemical Characterization and Consumption

The main compounds found in kombuchas used in this study are presented in Table 1. As previously demonstrated, both green and black tea kombuchas presented high content of phenolic compounds; however, black tea kombucha presented a higher antioxidant capacity, probably due to its higher phenolic compound concentration [17,25].

Among the 127 phenolic compounds identified in the kombuchas, 14 were more abundant: gallocatechin 3-O-gallate/epigallocatechin 3-O-gallate; gallocatechin isomer 2/epigallocatechin; catechin; quercetin 3-O-rhamnosyl-rhamnosyl-glucoside isomer 2; quercetin 3-O-glucosyl-rhamnosylgalactoside-isomer 2; gallocatechin isomer 1/epigallocatechin; quercetin 3-O-rhamnosyl-rhamnosylglucoside-isomer 1; quercetin 3-O-glucosyl-rhamnosyl-galactoside isomer 1; catechin 3-O-gallate; and catechin 5-O-gallate, which belong to the flavonoids class; and 5-O-galloylquinic acid; 3-[2-(carboxymethyl)-3,4-dihydroxyphenyl] prop-2-enoic acid; 4-coumaroylquinic acid isomer 2; and 1-O-caffeoylquinic acid isomer 2/3-caffeoylquinic acid, which belong to the phenolic acids class [17,25].

We did not observe a difference between the GTK and BTK groups regarding daily, weekly, and total kombucha consumption. However, the BTK group ingested a higher amount of phenolic compounds, which can be explained by its higher concentration in the black tea kombucha compared to green tea kombucha, as mentioned (Figure 2).

Table 1. Green and black tea kombuchas chemical characterization.

	Green Tea Kombucha	Black Tea Kombucha	*p*-Value
Chemical composition			
Sucrose (g/L)	19.30 ± 2.73 [b]	34.98 ± 1.42 [a]	0.0382
Glucose (g/L)	3.19 ± 0.15 [a]	2.45 ± 0.96 [a]	0.4690
Fructose (g/L)	0.15 ± 0.01 [a]	0.05 ± 0.02 [a]	0.0583
Ethanol (g/L)	7.23 ± 0.03 [a]	4.91 ± 0.35 [a]	0.0653
Theaflavin (g/L)	0.28 ± 0.03 [b]	1.51 ± 0.06 [a]	0.0066
Thearubigin (g/L)	13.30 ± 0.67 [b]	19.99 ± 0.10 [a]	0.0416
pH	3.2 ± 0.1 [b]	3.5 ± 0.1 [a]	0.0078
Total acidity (% *w*/*v*)	0.36 ± 0.01 [a]	0.32 ± 0.01 [b]	0.0100
Organic acids			
Acetic acid (g/L)	3.22 ± 0.39 [a]	2.78 ± 0.16 [a]	0.3336
Glucuronic acid (g/L)	1.17 ± 0.06 [a]	0.47 ± 0.02 [b]	0.0323
Lactic acid (g/L)	0.01 ± 0.00 [a]	0.02 ± 0.00 [a]	0.2604
Microbiological characterization			
Acetic bacteria (log CFU/mL)	6.0 ± 0.30 [a]	5.30 ± 0.10 [a]	0.1071
Lactic bacteria (log CFU/mL)	6.50 ± 0.20 [a]	5.90 ± 0.60 [a]	0.3959
Yeast (log CFU/mL)	6.30 ± 0.40 [a]	5.50 ± 0.10 [a]	0.1690

Values are expressed as mean ± SD. Different letters in the same row indicate a significant difference ($p < 0.05$) according to unpaired *t* test followed by Welch's correction. Details about the methodology used for the analyses are described in Cardoso et al. (2020).

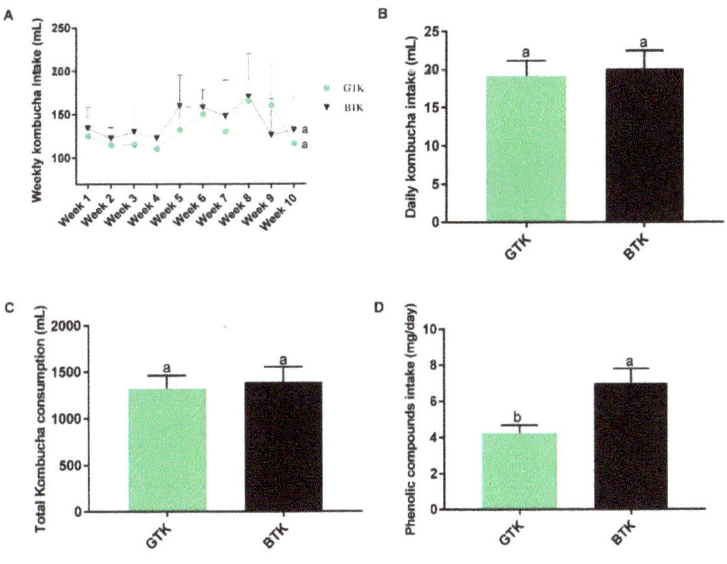

Figure 2. Weekly kombucha intake (**A**), daily kombucha intake (**B**), total kombucha consumption (**C**), and phenolic compounds intake (**D**) by the animals during the treatment. Data were expressed as mean ± SD. Different letters indicate a significant difference ($p < 0.05$) between groups according to t-test. GTK: HFHF diet + green tea kombucha diluted in water (30% *v*/*v*); BTK: HFHF diet + black tea kombucha diluted in water (30% *v*/*v*).

3.2. Biometric Parameters

Initial weight, final weight, weight gain, and BMI did not differ among groups, although some metabolic disorders were observed [25]. The group that received a standard diet (AIN-93M) presented a higher cecum weight than the HFHF and GTK groups, but no significant difference was observed compared to the BTK group. Regarding cecum weight:body weight ratio, the AIN93-M group presented a higher value when compared to the GTK group; however, this difference was not significant when compared to the other groups (Table 2).

Table 2. Body composition and intestinal parameters of the animals after 10 weeks of treatment.

Features	AIN93-M (n = 10)	HFHF (n = 10)	GTK (n = 9)	BTK (n = 10)
Body composition				
Initial weight (g)	349.90 ± 30.71 [a]	366.90 ± 36.90 [a]	370.40 ± 36.20 [a]	364.60 ± 36.05 [a]
Final weight (g)	415.00 ± 34.50 [a]	438.10 ± 66.65 [a]	415.30 ± 37.07 [a]	409.60 ± 50.08 [a]
Weight gain (g)	65.00 ± 22.70 [a]	71.25 ± 38.10 [a]	44.90 ± 30.69 [a]	44.90 ± 23.14 [a]
BMI (g/cm^2)	0.68 ± 0.08 [a]	0.61 ± 0.04 [a]	0.61 ± 0.08 [a]	0.61 ±0.08 [a]
Cecum weight (empty) (g)	1.01 ± 0.23 [a]	0.97 ± 0.13 [a]	0.96 ± 0.18 [a]	0.96 ± 0.04 [a]
Cecum weight (full) (g)	5.09 ± 1.15 [a]	3.92 ± 0.89 [b]	3.62 ± 0.75 [b]	4.03 ± 0.80 [ab]
Cecum weight:body weight ratio	1.23 ± 0.28 [a]	0.91 ± 0.24 [ab]	0.87 ± 0.17 [b]	0.98 ± 0.21 [ab]
Intestinal Permeability				
Lactulose:mannitol ratio	1.51 ± 0.57 [a]	1.56 ± 0.78 [a]	1.62 ± 0.94 [a]	2.17 ± 1.08 [a]
Histological Features				
Crypt depth (μM)	179.10 ± 43.20 [a]	223.10 ± 40.69 [a]	221.20 ± 24.92 [a]	209.30 ± 40.83 [a]
Crypt width (μM)	19.51 ± 2.66 [a]	18.85 ± 4.16 [a]	21.03 ± 1.46 [a]	19.46 ± 2.68 [a]
Number of goblet cells (units)	18.60 ± 3.27 [a]	16.59 ± 4.71 [a]	17.94 ± 2.93 [a]	22.38 ± 5.51 [a]
Fecal pH	9.01 ± 0.40 [a]	9.17 ± 0.25 [a]	9.27 ± 0.07 [a]	9.13 ± 0.13 [a]

Data are expressed as mean ± SD. Different letters in the same row indicate a significant difference ($p < 0.05$) according to one-way ANOVA followed by Tukey post-hoc (parametric data) or Kruskal–Wallis test followed by Dunn's post-hoc (non-parametric data). AIN-93M: standard diet (negative control group); HFHF: high-fat and high-fructose diet (positive control group); GTK: HFHF diet + green tea kombucha diluted in water (30% v/v); BTK: HFHF diet + black tea kombucha diluted in water (30% v/v).

3.3. Bioinformatics Analysis

To better comprehend the effects of the long-term intake of black tea kombucha (BTK) and green tea kombucha (GTK) on the gut microbiota of Wistar rats, we conducted a deep amplicon sequencing of the V4 region of the 16S rRNA genes. After the removal of low-quality and chimeric sequences from 39 datasets (AIN-93M, n = 10; BTK, n = 10; GTK, n = 9; HFHF, n = 10), a total of 1,004,357 high-quality reads, with an average of 25,752 (minimum: 15,349; maximum: 34,422) sequences for each sample, were obtained and assigned to 1218 predicted ASVs (≥99% similarity).

Regarding the sequences obtained from the kombuchas and their respective SCOBYs, 153,322 high-quality reads were obtained based on the amplification of the V4 region of the 16S rRNA genes (mean: 38,330; minimum: 36,957; maximum: 39,631) for bacterial community analysis, whereas 138,569 reads were forwarded for metataxonomic analysis based on the ITS1–2 regions of fungal ribosomal DNA (mean: 34,642; minimum: 34,566; maximum: 34,716).

3.4. Microbiota Profiling of GTK and BTK and Their Respective SCOBYs

In both kombuchas and their respective SCOBYs, the fungal community was greatly dominated by the species *Dekkera bruxellensis* (GTK SCOBY: 99.3%; BTK SCOBY: 99.6%; GTK: 99.6%; BTK: 99.9%) and, to a lesser extent, by the species *Saccharomyces bayanus* (GTK SCOBY: 0.7%; BTK SCOBY: 0.4%; GTK: 0.4%; BTK: 0.1%).

Regarding the bacterial community, microorganisms belonging to the phylum *Proteobacteria* dominated both tea and SCOBY samples (average of 77.07% considering the four groups); however, in the BTK group, this phylum accounted for only 37.24% of the sequences (Figure 3A). Notably, the phyla *Firmicutes*, *Bacteroidetes*, and *Actinobacteria* also represented an important part of the bacterial community found in kombucha samples, but not in their related SCOBYs. At the family level (Figure 3B), *Acetobacteraceae* stands as the dominant taxon across the groups, corresponding to almost 100% of the samples in the GTK and BTK SCOBYs groups. For kombucha samples, *Acetobacteraceae*, *Erysipelotrichaceae*, *Porphyromonadaceae*, *Rikenellaceae*, and *Streptococcaceae* encompass the top five families. Lastly, at the genus level (Figure 3C), *Gluconacetobacter* appears as the dominant taxon in samples obtained from both SCOBY samples, while *Acetobacter* comprehends the most abundant taxon in kombucha samples. In conjunction with the genus *Acetobacter*, *Allobaculum*, *Komagataeibacter*, and *Barnesiella*, they correspond to the top five genera. Interestingly, kombucha produced from black tea showed a higher number of taxa classified as low-abundant (13.05%), which is quite different from that obtained from the fermentation of green tea (6.85%), which may indicate greater bacterial diversity and richness in the GTK.

A

Phylum	SGTK	SBTK	GTK	BTK
Proteobacteria	99.67%	99.90%	71.46%	37.24%
Firmicutes	0.15%	0.04%	14.32%	32.85%
Bacteroidetes	0.12%	0.04%	8.16%	20.49%
Actinobacteria	0.04%	0.02%	2.37%	7.32%
Total	99.98%	99.99%	96.31%	97.90%

B

Family	SGTK	SBTK	GTK	BTK
Acetobacteraceae	99.38%	99.80%	66.96%	28.60%
Erysipelotrichaceae	0.06%	0.02%	7.57%	18.84%
Porphyromonadaceae	0.06%	0.03%	3.64%	10.93%
Rikenellaceae	0.03%	0.00%	2.32%	5.26%
Streptococcaceae	0.03%	0.00%	2.14%	3.81%
Bacteroidaceae	0.02%	0.01%	1.91%	3.76%
Clostridiaceae	0.02%	0.00%	1.61%	3.71%
Bifidobacteriaceae	0.01%	0.01%	1.02%	4.10%
Enterobacteriaceae	0.02%	0.00%	0.92%	2.63%
Lachnospiraceae	0.01%	0.00%	0.91%	1.86%
Oscillatoriales	0.02%	0.01%	2.66%	0.03%
Eubacteriaceae	0.01%	0.00%	0.77%	1.67%
Nocardiaceae	0.01%	0.00%	0.60%	1.50%
Helicobacteraceae	0.01%	0.00%	0.66%	1.40%
Pasteurellaceae	0.01%	0.00%	0.71%	1.06%
Xanthomonadaceae	0.01%	0.01%	0.41%	1.20%
Total	99.71%	99.87%	94.81%	90.35%

C

Genus	SGTK	SBTK	GTK	BTK
Gluconacetobacter	96.93%	94.21%	19.09%	9.97%
Acetobacter	1.67%	3.68%	38.18%	18.49%
Allobaculum	0.05%	0.01%	6.36%	15.92%
Komagataeibacter	0.79%	1.91%	9.89%	0.34%
Barnesiella	0.03%	0.01%	1.86%	5.53%
Bacteroides	0.02%	0.01%	1.92%	3.79%
Clostridium	0.02%	0.00%	1.62%	3.73%
Lactococcus	0.03%	0.00%	1.82%	3.44%
Bifidobacterium	0.01%	0.01%	1.03%	4.12%
Rikenella	0.01%	0.00%	1.61%	3.30%
Turicibacter	0.01%	0.01%	1.21%	2.87%
Odoribacter	0.01%	0.01%	0.59%	2.30%
Alistipes	0.02%	0.00%	0.71%	2.00%
Halospirulina	0.02%	0.01%	2.67%	0.03%
Eubacterium	0.01%	0.00%	0.78%	1.69%
Serratia	0.01%	0.00%	0.67%	1.74%
Rhodococcus	0.01%	0.00%	0.60%	1.51%
Helicobacter	0.01%	0.00%	0.66%	1.41%
Tannerella	0.02%	0.00%	0.54%	1.37%
Lachnoclostridium	0.01%	0.00%	0.62%	1.13%
Stenotrophomonas	0.01%	0.01%	0.41%	1.21%
Parabacteroides	0.00%	0.01%	0.31%	1.09%
Total	100%	100%	93.15%	86.95%

Figure 3. Heat map based on the relative abundance (>1.0% in at least one sample) of the most abundant bacterial taxa of phylum (**A**), family (**B**), and genus (**C**) identified in green tea kombucha (GTK), black tea kombucha (BTK), and their respective SCOBY (SGTK and SBTK). Green to red gradient indicates low to high relative levels of OTUs within the given taxonomic unit.

3.5. Alpha and Beta-Diversity Metrics of Gut Microbiota

The effects of the daily ingestion of green tea kombucha (GTK) and black tea kombucha (BTK) for 10 weeks on the gut microbiota of Wistar rats were firstly investigated using alpha- and beta-diversity indices and compared against the AIN-93M and HFHF groups. Considering the alpha-diversity analysis, the observed ASVs, Shannon, and Chao1 indices reached a plateau and are indicative that sequencing depth covered most of the microbial diversity and the majority of bacterial phylotypes were sampled (Figure 4).

A significant reduction in bacterial diversity, evaluated through Shannon's diversity index, was observed in the groups that received green and black tea kombucha when compared to both controls (Figure 4A). However, we did not observe significant differences among the control groups (AIN-93M vs. HFHF), as well as among the treatment groups (GTK vs. BTK) ($p > 0.05$). In terms of bacterial richness, a significantly lower Chao1 index was identified in the groups supplemented with GTK and BTK when compared to the AIN-93M group, but not when compared to the HFHF group (Figure 4B).

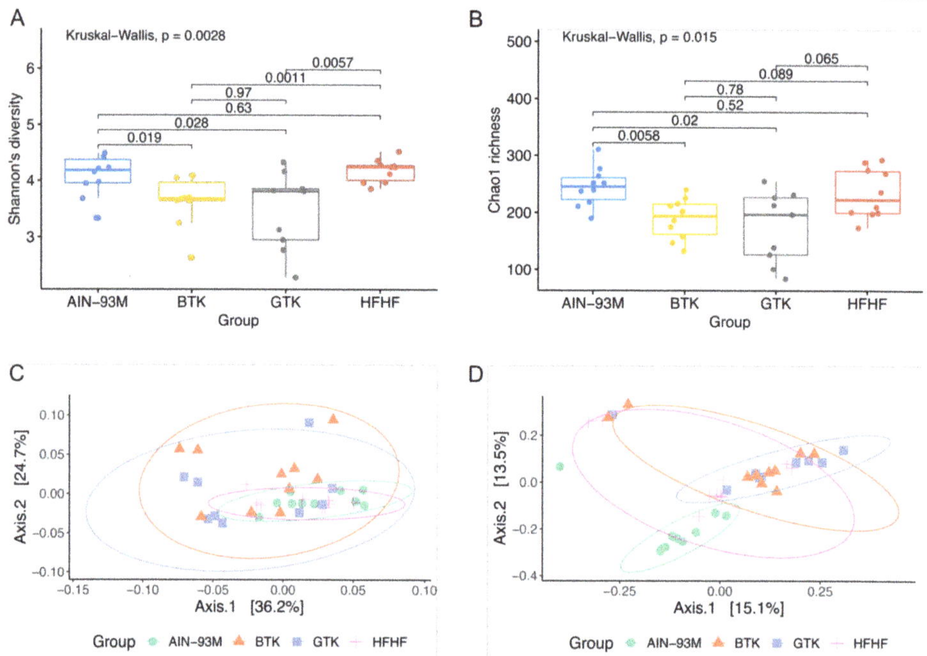

Figure 4. Box and whisker plots comparing species diversity (**A**) and richness (**B**) among the groups AIN-93M, BTK, GTK, and HFHF at the end of the experimental period. Horizontal bold lines show the median values. The bottom and top of the boxes show the 25th and the 75th percentiles, respectively. The whiskers extend up to the most extreme points within 1.5 times the interquartile ranges (IQR). Principal coordinate analysis (PCoA) based on weighted (**C**) and unweighted (**D**) UniFrac distances. PERMANOVA with 999 permutations was used to detect significant differences between microbial communities (dissimilarity) of different experimental groups. Standard error ellipses show 95% confidence areas.

Regarding the beta-diversity analysis, gut microbiota clustering on PCoA plots based on weighted and unweighted UniFrac distance metrics showed significant differences among the groups (Figure 4C,D). Pairwise comparisons using Qiime beta-group-significance command revealed that the gut composition of the groups that received the two different types of kombucha significantly differed from those animals receiving a standard (AIN-93M) or a high-fat and high-fructose (HFHF) diet. However, high similarity in terms of bacterial composition and abundance was observed among the GTK and BTK groups. Moreover, according to pairwise PERMANOVA results, the AIN-93M and HFHF groups differed only qualitatively regarding community dissimilarity (unweighted UniFrac: pseudo-F = 4.86, q = 0.0020). Taken together, alpha- and beta-diversity indices evidenced that after receiving an HFHF diet for eight weeks, long-term intake of green and black tea kombuchas was not able to establish a high diversity bacterial community as that observed in both control groups as well as a community as rich as that observed in the AIN-93M group.

3.6. Taxonomic Assignment and Gut Bacterial Composition

The Linear discriminant analysis Effect Size (LEfSe) was adopted to better characterize the gut bacterial composition of each group, as well as identify the differently abundant taxa. Moreover, inter-group comparisons were conducted based on the relative abundance of interest taxa. At the phylum level, *Firmicutes* (69.09%), *Bacteroidetes* (15.82%), *Proteobacteria* (8.13%), *Actinobacteria* (3.26%), and *Euryarchaeota* (2.15%) were the dominant taxa

(Figure 5A). The phylum *Firmicutes* showed higher relative abundance in the groups treated with kombucha prepared from black tea (BTK group, 75.21%) and green tea (GTK, 76.30%) when compared to both control groups (AIN-93M, 61.41%; HFHF, 63.45%), although a significant difference has been observed only between AIN-93M and BTK ($p = 0.019$). The same trend was observed for the phylum *Actinobacteria* (AIN-93M, 2.88%; BTK, 3.56%, GTK, 4.62%; HFHF, 1.97%), although no significant difference among the groups ($p > 0.05$) was observed. On the other hand, *Bacteroidetes* was significantly more abundant in the control groups AIN-93M (24.65%) and HFHF (22.20%) when compared to both kombucha groups (BTK, 8.72%; GTK, 7.69%), but not between AIN-93M and HFHF. Our results also indicate that the administration of both kombuchas increased the *Firmicutes*:*Bacteroidetes* ratio in these groups (BTK: F/B = 8.63; GTK: F/B = 9.93) when compared to the control groups AIN-93M (F/B = 2.49) and HFHF (F/B = 2.86). Regarding *Proteobacteria*, this phylum displayed the lowest proportion in the group GTK (5.38%), whereas its highest relative abundance was observed in the HFHF group (10.86%; Figure 6D, LDA > 3; $p = 0.035$). Interestingly, the phylum *Euryarchaeota* stands out in the BTK group (4.10%); it appears less abundant in the AIN-93M (1.08%) and GTK (2.53%) groups, while a very little abundance was observed in the group HFHF (0.88%) (BTK vs. HFHF, $p = 0.035$).

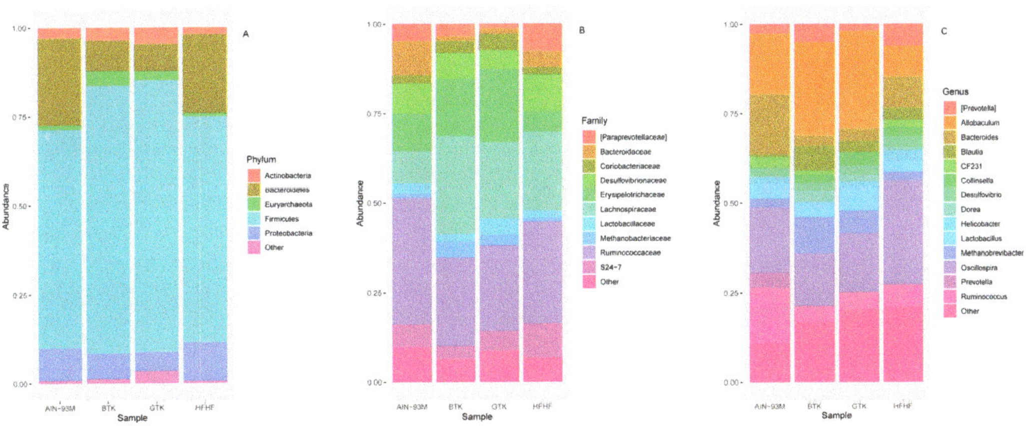

Figure 5. Stacked bar chart based on the relative abundance of major phyla (top-five) (**A**), families (top-ten) (**B**), and genera (top-fifteen) (**C**) across the groups AIN-93M, BTK, GTK, and HFHF.

At the family level, 12 taxa (relative abundance greater than 1%) accounted for approximately 95% of the total sequences in each group. Among them, *Ruminococcaceae* was the most abundant family in the AIN-93M, GTK, and HFHF groups, whereas *Lachnospiraceae* was the most prevalent in the BTK group (Figure 5B). Although *Lachnospiraceae* was the most abundant in the BTK group and differed from the AIN-93M group ($p = 0.00049$), this taxon appears as a biomarker of the HFHF group when compared to AIN93M (Figure 6A, LDA > 3, $p = 00013$). The third most abundant family among the groups, *Erysipelotrichaceae*, appears enriched in the groups that received both kombuchas (AIN-93M, 10.37%; BTK, 15.87%; GTK, 20.21%; HFHF, 5.74%) and was identified as a biomarker of the BTK group when compared to the group HFHF (Figure 6E LDA > 4; $p = 0.0068$). Interestingly, groups treated with green and black kombuchas showed a very low abundance (around 0.1%) of members belonging to the family *Bacteroidaceae*—a biomarker of the group HFHF—when compared to the group BTK (Figure 6E, LDA > 3), and when compared to both control groups ($p < 0.05$), which may justify the higher F/B ratio observed in kombucha treated groups. Positively correlated with diabetes and obesity [42,43], the families *S24–7* and *Desulfovibrionaceae* were significantly less abundant in the BTK and GTK groups, respectively, while in the HFHF group, these families reached their highest values. Indeed, LEfSe analysis revealed that the family *Desulfovibrionaceae*, in conjunction with the family

Paraprevotellaceae, were detected as biomarkers of the HFHF group when compared to the group that received green tea kombucha (Figure 6D, LDA >3). Lastly, as shown in Figure 6A–C, *Prevotellaceae* was identified as a biomarker of the AIN-93M group (LDA >3) when compared to the other experimental groups enrolled in this study (AIN-93M, 2.42%; BTK, 0.02%, GTK, 0.01%; HFHF, 0.004%).

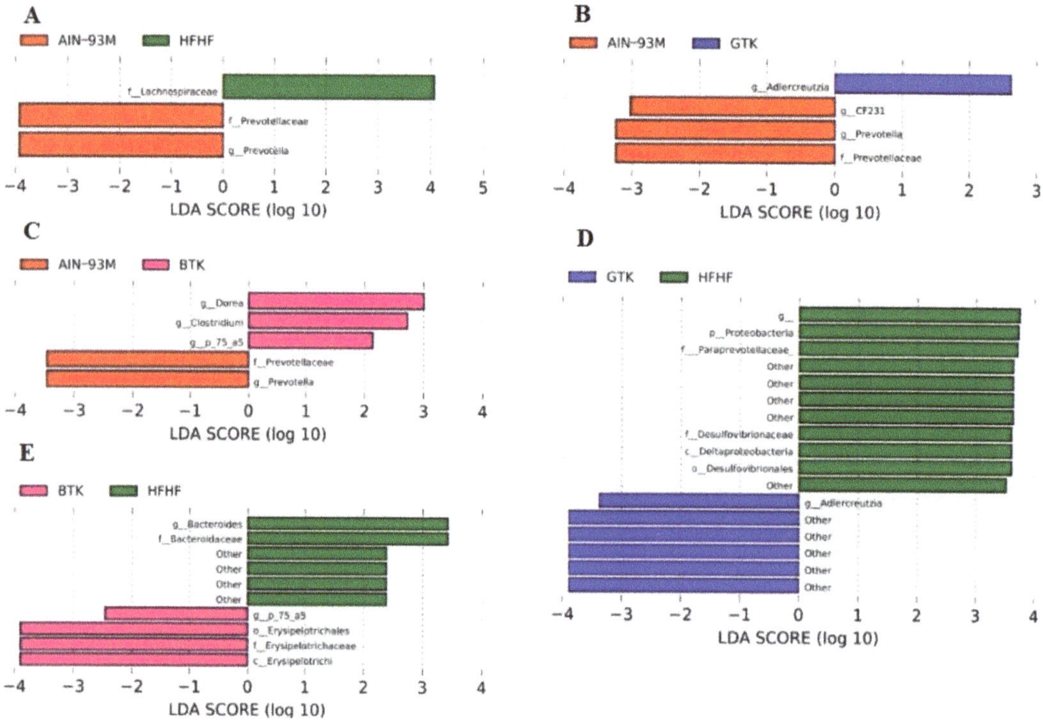

Figure 6. Differential abundance analysis was conducted with Linear discriminant analysis Effect Size (LEfSe) following the experimental period. Comparisons were made between AIN-93M with HFHF (**A**); GTK (**B**); and BTK (**C**) groups and between HFHF with GTK (**D**) and BTK (**E**) groups. Only biomarkers showing linear discriminant analysis (LDA) scores greater than 2.0 with a false discovery rate (FDR) $p < 0.05$ are depicted. Letters: p, phylum; c, class; o, order; f, family; g, genus.

After defining the most abundant genera, the top-ten taxa were selected and accounted for at least 70% of the total sequences in each group (Figure 5C). *Oscillospira* was the most abundant microorganism in the control group HFHF ($p < 0.05$). In the groups treated with kombuchas, *Allobaculum* was the dominant genus; however, this difference was significant just among the BTK and HFHF groups ($p = 0.0068$). Differential abundance analysis considering taxa at the genus level showed enrichment of *Prevotella* in the AIN-93M group (Figure 6A–C; LDA > 3). Although not considered biomarkers by LefSe analysis, but still included in the top-ten genera, it was possible to identify that *Bacteroides* and *Ruminococcus* are highly abundant in the group AIN-93M when compared to all other groups ($p < 0.05$). Regarding the BTK group, the genera *Dorea*, *Clostridium*, and *p-75-a5*, all of them belonging to the family *Erysipelotrichaceae*, appeared as biomarkers when compared to the AIN-93M group (Figure 6C, LDA > 3). However, when compared to the HFHF group, only the genus *p-75-a5* was identified as a biomarker in the BTK group (Figure 6E, LDA >3). Concerning the GTK, *Adlercreutzia* appeared as a biomarker in this group when compared to both control groups (Figure 6B,D; AIN-93M, LDA >2; HFHF, LDA >3). We did not identify biomarkers between the GTK and BTK groups regardless of the taxonomic level.

Lastly, we predicted and explored the structural basis (core taxa) of the bacterial communities of the groups enrolled in this study after the experimental period (Figure 7). Considering only ASVs with a prevalence of 75% across samples, the group AIN-93M contained 40 core taxa, while in the groups GTK, BTK, and HFHF, we noticed 14, 29, and 38 taxa, respectively. At a first inspection, five ASVs were identified as common to both groups and assigned to the following taxa: *Clostridiales*, *Allobaculum*, and *Oscillospira*. Secondly, considering only those ASVs specific to each group, 16 ASVs were identified in samples from the group AIN-93M and assigned to the following taxa: orders *Clostridiales* and *Bacteroidales*, class *Clostridia*, genera *Oscillospira*, *Bacteroides*, and *Helicobacter*, and species *Mucispirillum schaedleri*. Regarding the groups that underwent kombucha ingestion, only two ASVs stood out in the GTK group and were assigned to the genus *Lactobacillus* and the species *Ruminococcus flavefaciens*, whereas seven ASVs were typical for the BTK group and were assigned to the family *Lachnospiraceae*, genera *Dorea*, *Blautia*, *Allobaculum*, and *Mogibacteriaceae*, as well as the species *Collinsella stercoris*. Lastly, 12 ASVs were identified as specifically present in the HFHF group and were taxonomically assigned to the following taxa: families *Desulfovibrionaceae*, *S24-7*, *Ruminococcaceae*, and *Lachnospiraceae*, in addition to the genera *Roseburia* and *Oscillospira*.

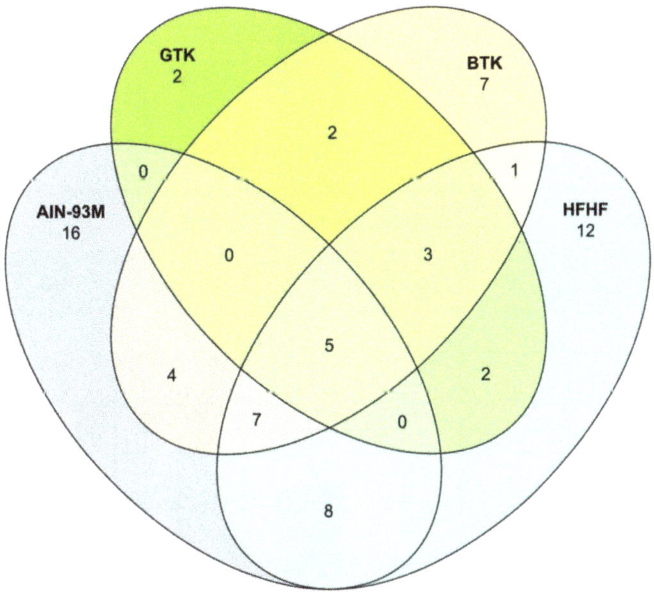

Figure 7. Venn diagram representing shared amplicon sequence variants (ASVs) of the core microbiome identified in the groups AIN-93M, GTK, BTK, and HFHF.

3.7. Fecal pH and Short-Chain Fatty Acids Content

We did not find a significant difference in fecal pH among the groups (Table 2). Regarding the SCFA, both treatment groups—GTK and BTK—presented a higher propionic acid concentration when compared to the AIN93-M and HFHF groups. Acetic acid concentration was significantly higher in the AIN93-M group when compared to the GTK and BTK groups, but no significant difference was noted when comparing the HFHF group to the other groups. Butyric acid concentrations did not differ significantly among groups (Figure 8).

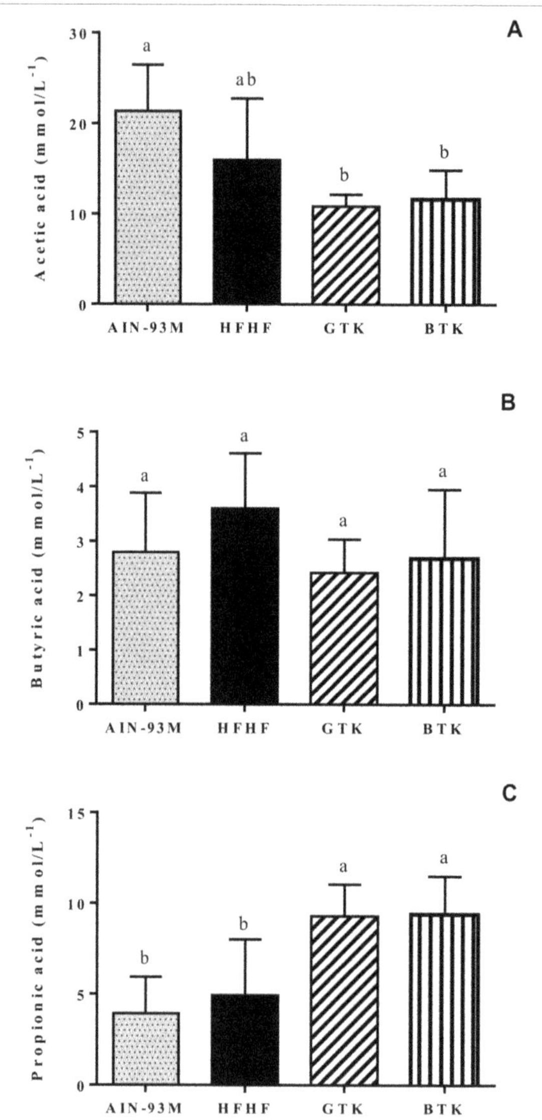

Figure 8. Acetic (**A**), butyric (**B**), and propionic (**C**) acids concentrations identified on stool samples from the animals. AIN-93M: standard diet (negative control group); HFHF: high-fat and high-fructose diet (positive control group); GTK: HFHF diet + green tea kombucha diluted in water (30% v/v); BTK: HFHF diet + black tea kombucha diluted in water (30% v/v). Values are expressed as means ± SD. Different letters indicate a significant difference between groups ($p < 0.05$) according to ANOVA one-way followed by Tukey post-hoc (acetic and butyric acids) and Kruskal–Wallis test followed by Dunn's post-hoc (propionic acid).

We also investigated whether there is an association between the SCFA content and the microorganisms found in the gut microbiota of the animals. For that, MaAsLin 2 [44] was performed to find significant multivariable associations between specific microbial genera, cecal SCFA (acetate, propionate, and butyrate), and phenolic intake. The compound Poisson linear model (CPLM) function was utilized on cumulative sum scaling (CSS) normalized data with minimum prevalence (1%). For analysis of changes across the different interventions, samples obtained from the HFHF group were assigned as references. All p-values were false discovery rate-adjusted (Benjamini–Hochberg, q-values), and features with $q < 0.25$ were considered significant (Table S1).

3.8. Intestinal Permeability and Histological Analysis

There was no difference between groups regarding the excretion of lactulose and mannitol in the urine, which was expressed as lactulose/mannitol ratio (Table 2).

Histological features are demonstrated in Figure 9. We did not observe differences among the groups in terms of crypt depth, crypt width, and the number of goblet cells (Table 2).

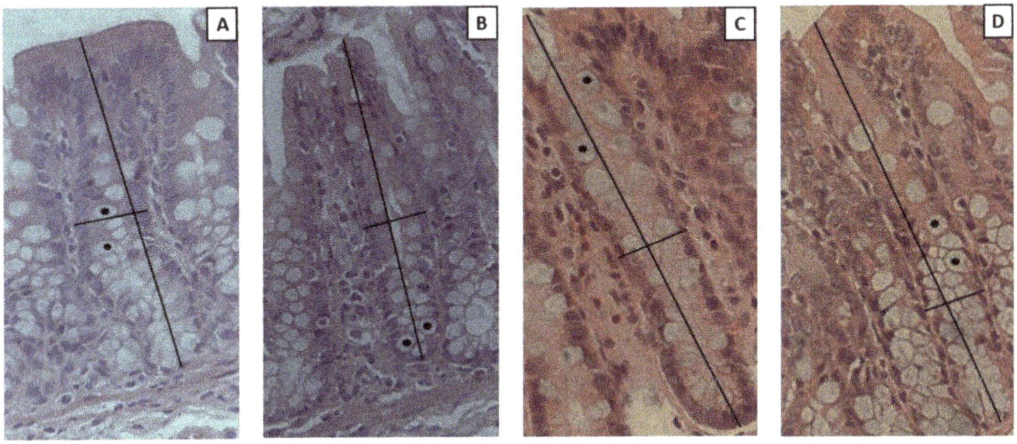

Figure 9. Representative photomicrographs of cecum sections after 10 weeks of treatment. (**A**) AIN-93M; (**B**) HFHF; (**C**) GTK; (**D**) BTK. Vertical lines indicate crypt depth; horizontal lines indicate crypt width; asterisks indicate goblet cells. All images were captured in a 20X objective. AIN-93M: standard diet (negative control group); HFHF: high-fat and high-fructose diet (positive control group); GTK: HFHF diet + green tea kombucha diluted in water (30% v/v); BTK: HFHF diet + black tea kombucha diluted in water (30% v/v).

4. Discussion

In this study, we evaluated the effects of regular kombucha consumption on the gut microbiota and on the intestinal health of Wistar rats fed a high-fat high-fructose diet. Our results show that both green tea (GTK) and black tea (BTK) kombuchas were able to modulate the gut microbiota, which corroborates our hypothesis.

We believe that our results can be attributed, in large part, to the high content and diversity of phenolic compounds present in kombuchas. In general, only a small portion—approximately 5–10%—of the dietary phenolic compound will be absorbed in the small intestine, mainly those with a monomeric or dimeric structure. The more complex ones—oligomeric and polymeric structures—reach the colon practically unchanged where they are metabolized by the gut microbiota, making them more bioactive [45–47].

Once biotransformed in less complex compounds such as phenolic acids, the generated metabolites and bioactive molecules will modulate the gut microbiota [45,47,48], exerting an effect similar to prebiotics by favoring the growth of beneficial bacteria and inhibiting the growth of pathogenic ones [47,49]. Studies have shown that the antioxidant and anti-inflammatory activities exerted by the polyphenols act against metabolic disorders such as cancer, obesity, and diabetes via modulation of the gut microbiota [24,48,50–52]. There is evidence that metabolic alterations induced by a high-fat diet can also be attenuated by polyphenols intake via activation of PPARα and GLUT4 [53].

Regarding the microbial composition of the kombuchas and their respective SCOBYs, we noticed that the most abundant microorganisms found in the SCOBYs were also found in the beverages, although the diversity was greater in the kombuchas. Interestingly, *Gluconacetobacter* was the predominant genus in both SCOBYs and was much less prevalent in the kombuchas. On the other hand, the genus *Acetobacter* was favored during the fermentation, reaching its maximum abundance in the GTK. Those differences can be explained by the metabolic adaptations of the microorganisms, which are capable of utilizing different substrates depending on the type of tea and consequently will generate different metabolites [17,18,54]. The genus *Acetobacter*, for example, belongs to the group of acetic acid bacteria and has the ability to oxidize ethanol and sugar to acetic acid [55]. Its higher prevalence in the GTK group is probably responsible for the lower pH observed in this beverage.

Our results were partially similar to other studies. Recently, it was evaluated, through metagenomics analysis, the microbial diversity of kombuchas whose fermentation time varied between 3 and 15 days. In all analyzed samples, the bacteria belong to eight phyla, and, likewise to our study, *Proteobacteria* was the predominant one, encompassing more than 99% of the species. Among the yeasts, the genus *Zygosaccharomyces* was predominant (>99%) [20]. In another study, green and black tea kombuchas produced on an industrial scale presented differences in microbiological composition. Lactic acid bacteria, especially *Oenococcus oeni*, was associated with the fermentation of green tea kombucha, while black tea kombucha showed a greater predominance of acetic bacteria. The presence of these bacteria was associated with a higher concentration of lactic acid in green tea kombucha and acetic acid in black tea kombucha. Yeast diversity was not influenced by the type of tea; in both kombuchas, the authors observed a predominance of the species *Dekkera bruxellensis*, D. anomala, Hanseniaspora valbyensis, and *Zygosaccharomyces bailli* [56].

The microbial composition of SCOBYs has also been investigated. In a recent study in which 103 samples obtained from commercial kombucha brewers were analyzed, the authors observed that the microorganisms' predominance changed according to their position at the SCOBY's surface. The fungi *Brettanomyces* and the bacteria *Gluconacetobacter*, which have a strong affinity for oxygen, were the main microorganisms found at the upper layer. On the other hand, a greater abundance of *Lactobacillus* was found at the bottom SCOBY side, which corroborates the fact that this genus prefers a low-oxygen environment [21].

Regarding the in vivo results, we should mention that although body composition was not significantly different among the groups, a high-fat and high-fructose diet was able to induce metabolic alterations, as previously reported [15]. As expected, the high-fat content stimulated the growth of gram-negative bacteria such as the phylum *Proteobacteria* and the genus *Bacteroides*, being a biomarker in the HFHF group when compared to the GTK and BTK groups, respectively. Gram-negative bacteria present lipopolysaccharides (LPS) in their cell wall [57], an endotoxin recognized by toll-like receptor 4 (TLR-4) that activates NF-κB (nuclear factor kappa-light-chain-enhancer of activated B cells) and induces the production of pro-inflammatory cytokines such as TNF-α, IL-6, and IL-1β [21]. A higher intestinal permeability allows LPS to migrate into the bloodstream, triggering an inflammatory response in the organism [58]. Our previous work demonstrated that the HFHF diet promoted an increase in the levels of inflammatory markers in the liver (TNF-α) and blood (NLR—neutrophil/lymphocyte ratio) as well as reduced the total antioxidant capacity in plasma and liver and increased oxide nitric concentrations, which were reverted in the groups that consumed both kombuchas [25]. A decrease in the *Proteobacteria* and *Bacteroides* abundance noted in the treatment groups suggest that green and black tea kombuchas may present activity against gram-negative bacteria, which may explain the attenuation of the systemic inflammation and oxidative stress markers observed.

The HFHF diet also favored the *Lachnospiraceae* family, which was a biomarker in this group when compared to AIN93-M, but not when compared to both treatments. Although bacteria that belong to the *Lachnospiraceae* family usually promote a good impact on the gut microbiota by being involved in SCFA production [59,60], it can also be a result of the bile acids' metabolism. The liver is the main organ responsible for lipid metabolism, where the primary bile acids are synthesized from cholesterol [61,62]. The liver–gut axis is not completely understood yet, but it is known that these bile acids can be used as substrates by the microorganisms in the colon and are converted into secondary bile acids, which are pro-inflammatory metabolites involved in steatosis and NAFLD (non-alcoholic fatty liver disease) [62,63]. Studies conducted with humans [64] and mice [65] have pointed out that the main microorganisms involved in this mechanism belong to the *Lachnospiraceae* family, particularly *Blautia* and *L. incertae sedis* [60,66], and elevated taxa of these bacteria in the gut microbiota have been related to liver diseases. Indeed, as reported in our previous work [25], the HFHF diet has induced liver steatosis in the animals, which was reverted from degree 2 to 1 in those treated with both kombuchas.

The genus *Prevotella* and its family *Prevotellaceae* were identified as biomarkers in the AIN-93M group when compared to the other groups, especially HFHF. They belong to the *Bacteroidetes* phylum and are involved in the metabolism of complex carbohydrates and cellulose. De Filippo et al. (2010) compared the gut microbiota of children living in a rural area in Africa who followed a low-fat and high-fiber diet versus children living in an industrialized city in Europe, whose diet has a high content of fat and protein. The African children presented a higher abundance of the genus *Prevotella* and a higher content of SCFA, which were attributed to a healthier diet [67]. *Erysipelotrichaceae* species are also related to diets rich in carbohydrates and negatively correlated with *Prevotella*. The genera *Dorea*, *Clostridium*, and *p-75-a5*, all belonging to the family *Erysipelotrichaceae*, were reported as biomarkers in the BTK group when compared to the AIN-93M group. The genus *Dorea* has been positively correlated to a Western diet [68] and *p-75-a5* is involved in protein and lipid digestion [69]. Interestingly, when comparing the BTK and HFHF groups, the genus *p-75-a5* and its family *Erysipelotrichaceae* were negatively associated with the BTK group, suggesting that BTK consumption was able to attenuate the negative impact of a HFHF diet. The same biomarkers were not observed in the GTK group, although it also presented a high abundance of the *Erysipelotrichaceae* family.

Considering the treatment groups, we observed that *Actinobacteria* was present in a higher amount in both of them. This phylum is associated with short-chain fatty acids (SCFA) production [70], which are by-products derived from microbial fermentation that are used as an energy source by the enterocytes, favoring intestinal homeostasis and metabolism [70,71]. SCFA also act in the activation of the hormones GLP-1 (glucagon-like peptide 1) and PYY (peptide YY). GLP-1 regulates appetite by inhibiting gastric emptying and stimulating insulin secretion. PYY, in turn, is involved in appetite reduction and gastric motility inhibition. Thus, SCFA act by regulating food consumption and satiety [72], acting on obesity control. Recent studies have suggested that SCFA can modify the epigenome, acting on tissues and organs besides the intestine [71]. Beyond *Actinobacteria*, other microorganisms are pointed out as SCFA producers as those belonging to the *Bifidobacterium* and *Lactobacillus* genera. Both have shown an increased abundance in treatment groups, while *Lactobacillus* was especially higher in the GTK group.

Among SCFA, butyrate is considered the main energy source of colonocytes and enterocytes, thus favoring their growth [73,74]. In our study, butyrate concentrations did not differ significantly among the groups. On the other hand, we observed a higher acetate production in the AIN-93M and HFHF groups, although this last one was not significantly different from the treatments. Since acetate is a metabolite produced especially by bacteria from the *Bacteroidetes* phylum [71], it may explain its elevated abundance in those groups. Finally, we observed a higher concentration of propionate in both treatment groups. This SCFA is produced by a few bacteria, especially those from the genus *Akkermansia*. Phenolic compounds can induce changes in microbial composition, favoring the growth of *Akkermansia muciniphila* [75]. Since both kombuchas present high amounts of phenolic compounds [17,76], it may explain the increase in the *Akkermansia* abundance observed in the treatment groups. The presence of *Akkermansia muciniphila* is associated with a decrease in intestinal permeability and beneficial effects on diabetes mellitus and obesity [77,78].

In both treatment groups, we also observed the presence of *Ocillospira*, a genus involved in glucuronic acid degradation that is positively associated with leanness and health [79]. However, its presence was predominant in the HFHF group, probably because *Ocillospira* can use metabolic products secreted by other microorganisms, including *Bacteroides*. Since a high-fat diet stimulates the growth of *Bacteroides*, it can indirectly favor *Ocillospira* [79].

Finally, when considering the LEfSe analysis, the genus *Adlercreutzia* was observed as a biomarker in the GTK group when compared to both controls. This genus, and more particularly, *Adlercreutzia equolifaciens*, exerts a fundamental role in the metabolism of polyphenols in conjunction with other bacteria such as *Flavonifractor plautii, Slackia equolifaciens, Slackia isoflavoniconvertens, Eubacterium ramulus, Eggerthella lenta,* and *Bifidobacterium* spp. [45].

Our study has several strengths. To our knowledge, this is the first that investigated the effects of regular kombucha consumption on the intestinal health of rats fed a high-fat high-fructose diet. Our methodology allowed us to compare both kombuchas and sequencing the beverages and their respective SCOBYs was crucial to analyze if those results reflect the ones found in vivo. The results will help on the understanding of the mechanisms involved in kombucha consumption, and certainly will contribute to filling out the lack of evidence about its impact on intestinal health. As the main limitation, we should mention that the gut microbiota analyses were performed using stool samples from the cecum, which probably has not allowed us to fully explore the results in the same way as if they were collected after undergoing the whole large intestine. The literature is still limited, and the results are controversial, so more studies are necessary to confirm those hypotheses.

5. Conclusions

Our results demonstrated that diets were able to modulate the gut microbiota in different ways. A high-fat high-fructose diet, as expected, was associated with the prevalence of pathobionts, such as *Proteobacteria* and *Bacteroides*. Even though a healthier diet will be always encouraged to prevent and attenuate metabolic disorders, we have noticed that kombucha intake could somewhat modulate the gut microbiota, mitigating the impairments provoked by a Western diet by increasing propionate production and favoring the growth of beneficial bacteria, such as *Adlercreutzia* in the GTK group. Thus, we conclude that regular kombucha intake may be beneficial to intestinal health, although more studies, especially clinical trials, are necessary.

Supplementary Materials: The following supporting information can be downloaded at: https://www.mdpi.com/article/10.3390/nu14245234/s1. Table S1. Multivariable associations between specific microbial genera, fecal short-chain fatty acids, and phenolic intake.

Author Contributions: M.A.d.C.C.: Data curation, formal analysis, investigation, and writing; L.d.P.D.M.: formal analysis, investigation, and writing; V.d.S.D.: formal analysis, investigation, and writing; R.R.C.: formal analysis, investigation, and writing; V.P.B.d.S.J.: formal analysis, investigation, and writing; B.P.d.S.: formal analysis, investigation, and writing; M.G.: investigation and writing; V.C.: investigation and writing; A.G.: investigation and writing; J.B.: investigation and writing; H.S.D.M.: conceptualization, data curation, investigation, methodology, and writing; F.A.R.d.B.: conceptualization, data curation, methodology, validation, writing—review and editing, supervision, and project administration. All authors have read and agreed to the published version of the manuscript.

Funding: The project has been funded in part by a grant from the Italian Ministry of Foreign Affairs and International Cooperation (MAECI) n.BR22GR06. Luiza de Paula Dias Moreira is recipient of a fellowship from the University of Padua (PhD course in Animal and Food Science).

Institutional Review Board Statement: The study was conducted according to the guidelines of the Declaration of Helsinki and approved by the Ethics Committee on Animal Use (CEUA) of Universidade Federal de Vicosa (Protocol code 06/2019; date of approval: 28 May 2019).

Informed Consent Statement: Not applicable.

Data Availability Statement: Not applicable.

Acknowledgments: The authors would like to thank Fundação de Amparo à Pesquisa do Estado de Minas Gerais (FAPEMIG, Brazil), Conselho Nacional de Desenvolvimento Científico e Tecnológico (CNPq, Brazil), and Coordenação de Aperfeiçoamento de Pessoal de Nível Superior (CAPES, Brazil—Finance Code 001) for financial support.

Conflicts of Interest: The authors declare no conflict of interest.

References

1. Malesza, I.J.; Malesza, M.; Walkowiak, J.; Mussin, N.; Walkowiak, D.; Aringazina, R.; Bartkowiak-Wieczorek, J.; Mądry, E. High-fat, western-style diet, systemic inflammation, and gut microbiota: A narrative review. *Cells* **2021**, *10*, 3164. [CrossRef] [PubMed]
2. Hasegawa, Y.; Chen, S.Y.; Sheng, L.; Jena, P.K.; Kalanetra, K.M.; Mills, D.A.; Wan, Y.J.Y.; Slupsky, C.M. Long-Term Effects of Western Diet Consumption in Male and Female Mice. *Sci. Rep.* **2020**, *10*, 14686. [CrossRef] [PubMed]
3. Kopp, W. How western diet and lifestyle drive the pandemic of obesity and civilization diseases. *Diabetes Metab. Syndr. Obes. Targets Ther.* **2019**, *12*, 2221–2236. [CrossRef] [PubMed]
4. Lam, Y.Y.; Ha, C.W.Y.; Campbell, C.R.; Mitchell, A.J.; Dinudom, A.; Oscarsson, J.; Cook, D.I.; Hunt, N.H.; Caterson, I.D.; Holmes, A.J.; et al. Increased gut permeability and microbiota change associate with mesenteric fat inflammation and metabolic dysfunction in diet-induced obese mice. *PLoS ONE* **2012**, *7*, e34233. [CrossRef]
5. Schoeler, M.; Caesar, R. Dietary Lipids, Gut microbiota and lipid metabolism. *Rev. Endocr. Metab. Disord.* **2019**, *20*, 461–472. [CrossRef]
6. Li, X.; Guo, J.; Ji, K.; Zhang, P. Bamboo shoot fiber prevents obesity in mice by modulating the gut microbiota. *Sci. Rep.* **2016**, *6*, 32953. [CrossRef]
7. Miura, K.; Ishioka, M.; Iijima, K. The roles of the gut microbiota and toll-like receptors in obesity and nonalcoholic fatty liver disease. *J. Obes. Metab. Syndr.* **2017**, *26*, 86–96. [CrossRef]
8. Lambertz, J.; Weiskirchen, S.; Landert, S.; Weiskirchen, R. Fructose: A dietary sugar in crosstalk with microbiota contributing to the development and progression of non-alcoholic liver disease. *Front. Immunol.* **2017**, *8*, 1159. [CrossRef]
9. Do, M.H.; Lee, E.; Oh, M.J.; Kim, Y.; Park, H.Y. High-glucose or-fructose diet cause changes of the gut microbiota and metabolic disorders in mice without body weight change. *Nutrients* **2018**, *10*, 761. [CrossRef]
10. Kapp, J.M.; Sumner, W. Kombucha: A systematic review of the empirical evidence of human health benefit. *Ann. Epidemiol.* **2019**, *30*, 66–70. [CrossRef]
11. Jayabalan, R.; Malbaša, R.V.; Lončar, E.S.; Vitas, J.S.; Sathishkumar, M. A review on kombucha tea-microbiology, composition, fermentation, beneficial effects, toxicity, and tea fungus. *Compr. Rev. Food Sci. Food Saf.* **2014**, *13*, 538–550. [CrossRef]
12. Villarreal-Soto, S.A.; Beaufort, S.; Bouajila, J.; Souchard, J.-P.; Taillandier, P. Understanding kombucha tea fermentation: A review. *J. Food Sci.* **2018**, *83*, 580–588. [CrossRef] [PubMed]
13. Greenwalt, C.J.; Steinkraus, K.H.; Ledford, R.A. Kombucha, the fermented tea: Microbiology, composition, and claimed health effects. *J. Food Prot.* **2000**, *63*, 976–981. [CrossRef] [PubMed]
14. Vina, I.; Semjonovs, P.; Linde, R.; Denina, I. Current evidence on physiological activity and expected health effects of kombucha fermented beverage. *J. Med. Food* **2014**, *17*, 179–188. [CrossRef] [PubMed]
15. Watawana, M.I.; Jayawardena, N.; Gunawardhana, C.B.; Waisundara, V.Y. Health, Wellness, and Safety Aspects of the Consumption of Kombucha. *J. Chem.* **2015**, *2015*, 1–11. [CrossRef]
16. Gaggìa, F.; Baffoni, L.; Galiano, M.; Nielsen, D.S.; Jakobsen, R.R.; Castro-Mejía, J.L.; Bosi, S.; Truzzi, F.; Musumeci, F.; Dinelli, G.; et al. Kombucha beverage from green, black and rooibos teas: A comparative study looking at microbiology, chemistry and antioxidant activity. *Nutrients* **2018**, *11*, 1. [CrossRef] [PubMed]
17. Cardoso, R.R.; Neto, R.O.; dos Santos D'Almeida, C.T.; do Nascimento, T.P.; Pressete, C.G.; Azevedo, L.; Martino, H.S.D.; Cameron, L.C.; Ferreira, M.S.L.; de Barros, F.A.R. Kombuchas from green and black teas have different phenolic profile, which impacts their antioxidant capacities, antibacterial and antiproliferative activities. *Food Res. Int.* **2020**, *128*, 108782. [CrossRef] [PubMed]
18. De Filippis, F.; Troise, A.D.; Vitaglione, P.; Ercolini, D. Different temperatures select distinctive acetic acid bacteria species and promotes organic acids production during kombucha tea fermentation. *Food Microbiol.* **2018**, *73*, 11–16. [CrossRef]
19. de Noronha, M.C.; Cardoso, R.R.; dos Santos D'Almeida, C.T.; do Carmo, M.A.V.; Azevedo, L.; Maltarollo, V.G.; Júnior, J.I.R.; Eller, M.R.; Cameron, L.C.; Ferreira, M.S.L.; et al. Black tea kombucha: Physicochemical, microbiological and comprehensive phenolic profile changes during fermentation, and antimalarial activity. *Food Chem.* **2022**, *384*, 132515. [CrossRef]
20. Arıkan, M.; Mitchell, A.L.; Finn, R.D.; Gürel, F. Microbial composition of kombucha determined using amplicon sequencing and shotgun metagenomics. *J. Food Sci.* **2020**, *85*, 455–464. [CrossRef]

21. Harrison, K.; Curtin, C.; Arıkan, M.; Mitchell, A.L.; Finn, R.D.; Gürel, F. Microbial composition of scoby starter cultures used by commercial kombucha brewers in north america. *J. Food Sci.* **2021**, *9*, 1060. [CrossRef] [PubMed]
22. Jung, Y.; Kim, I.; Mannaa, M.; Kim, J.; Wang, S.; Park, I.; Kim, J.; Seo, Y.S. Effect of kombucha on gut-microbiota in mouse having non-alcoholic fatty liver disease. *Food Sci. Biotechnol.* **2019**, *28*, 261–267. [CrossRef] [PubMed]
23. Wang, P.; Feng, Z.; Sang, X.; Chen, W.; Zhang, X.; Xiao, J.; Chen, Y.; Chen, Q.; Yang, M.; Su, J. Kombucha ameliorates LPS-induced sepsis in a mouse model. *Food Funct.* **2021**, *12*, 10263–10280. [CrossRef] [PubMed]
24. Costa, M.A.d.C.; Vilela, D.L.d.S.; Fraiz, G.M.; Lopes, I.L.; Coelho, A.I.M.; Castro, L.C.V.; Martin, J.G.P. Effect of kombucha intake on the gut microbiota and obesity-related comorbidities: A systematic review. *Crit. Rev. Food Sci. Nutr.* **2021**, 1–16. [CrossRef] [PubMed]
25. Cardoso, R.R.; Moreira, L.d.P.D.; Costa, M.A.d.C.; Toledo, R.C.L.; Grancieri, M.; Nascimento, T.P.d.; Ferreira, M.S.L.; da Matta, S.L.P.; Eller, M.R.; Martino, H.S.D.; et al. Kombuchas from green and black teas reduce oxidative stress, liver steatosis and inflammation, and improve glucose metabolism in wistar rats fed a high-fat high-fructose diet. *Food Funct.* **2021**, *12*, 10813–10827. [CrossRef]
26. Reeves, P.G.; Nielsen, F.H.; Fahey, G.C. AIN-93 Purified diets for laboratory rodents: Final report of the american institute of nutrition ad hoc writing committee on the reformulation of the AIN-76A rodent diet. *J. Nutr.* **1993**, *123*, 1939–1951. [CrossRef]
27. Martinez, O.D.M.; Theodoro, J.M.V.; Grancieri, M.; Toledo, R.C.L.; Queiroz, V.A.V.; de Barros, F.A.R.; Martino, H.S.D. Dry heated whole sorghum flour (BRS 305) with high tannin and resistant starch improves glucose metabolism, modulates adiposity, and reduces liver steatosis and lipogenesis in wistar rats fed with a high-fat high-fructose diet. *J. Cereal Sci.* **2021**, *99*. [CrossRef]
28. Zenebon, O.; Pascuet, N.S.; Tiglea, P. *Métodos fíSicos-Quimicos Para Análise de Alimentos*; Instituto Adolfo Lutz: São Paulo, Brazil, 2008; pp. 1–1020.
29. Jayabalan, R.; Chen, P.N.; Hsieh, Y.S.; Prabhakaran, K.; Pitchai, P.; Marimuthu, S.; Thangaraj, P.; Swaminathan, K.; Yun, S.E. Effect of solvent fractions of kombucha tea on viability and invasiveness of cancer cells-characterization of Dimethyl 2-(2-Hydroxy-2-Methoxypropylidine) malonate and vitexin. *Indian J. Biotechnol.* **2011**, *10*, 75–82.
30. Azevedo, L.; de Araujo Ribeiro, P.F.; de Carvalho Oliveira, J.A.; Correia, M.G.; Ramos, F.M.; de Oliveira, E.B.; Barros, F.; Stringheta, P.C. Camu-Camu (Myrciaria Dubia) from commercial cultivation has higher levels of bioactive compounds than native cultivation (amazon forest) and presents antimutagenic effects in vivo. *J. Sci. Food Agric.* **2019**, *99*, 624–631. [CrossRef]
31. De Sá, L.R.V.; De Oliveira, M.A.L.; Cammarota, M.C.; Matos, A.; Ferreira-Leitão, V.S. Simultaneous analysis of carbohydrates and volatile fatty acids by hplc for monitoring fermentative biohydrogen production. *Int. J. Hydrog. Energy* **2011**, *36*, 15177–15186. [CrossRef]
32. Song, P.; Zhang, R.; Wang, X.; He, P.; Tan, L.; Ma, X. Dietary grape-seed procyanidins decreased postweaning diarrhea by modulating intestinal permeability and suppressing oxidative stress in rats. *J. Agric. Food Chem.* **2011**, *59*, 6227–6232. [CrossRef] [PubMed]
33. Grancieri, M.; Costa, N.M.B.; Tostes, M.d.G.V.; de Oliveira, D.S.; de Carvalho Nunes, L.; de Nadai Marcon, L.; Veridiano, T.A.; Viana, M.L. Yacon flour (smallanthus sonchifolius) attenuates intestinal morbidity in rats with colon cancer. *J. Funct. Foods* **2017**, *37*, 666–675. [CrossRef]
34. Siegfried, R.; Ruckemann, H.; Stumpf, G. Method for the determination of organic acids in silage by high performance liquid chromatography. *Landwirt* **1984**, *37*, 298–304.
35. Stevenson, D.M.; Weimer, P.J. Dominance of prevotella and low abundance of classical ruminal bacterial species in the bovine rumen revealed by relative quantification real-time PCR. *Appl. Microbiol. Biotechnol.* **2007**, *75*, 165–174. [CrossRef] [PubMed]
36. Bolyen, E.; Rideout, J.R.; Dillon, M.R.; Bokulich, N.A.; Abnet, C.C.; Al-Ghalith, G.A.; Alexander, H.; Alm, E.J.; Arumugam, M.; Asnicar, F.; et al. Reproducible, interactive, scalable and extensible microbiome data science using QIIME 2. *Nat. Biotechnol.* **2019**, *37*, 852–857. [CrossRef]
37. Callahan, B.J.; McMurdie, P.J.; Rosen, M.J.; Han, A.W.; Johnson, A.J.A.; Holmes, S.P. DADA2: High-resolution sample inference from illumina amplicon data. *Nat. Methods* **2016**, *13*, 581–583. [CrossRef]
38. Katoh, K.; Misawa, K.; Kuma, K.I.; Miyata, T. MAFFT: A novel method for rapid multiple sequence alignment based on fast fourier transform. *Nucleic Acids Res.* **2002**, *30*, 3059–3066. [CrossRef]
39. DeSantis, T.Z.; Hugenholtz, P.; Larsen, N.; Rojas, M.; Brodie, E.L.; Keller, K.; Huber, T.; Dalevi, D.; Hu, P.; Andersen, G.L. Greengenes, a chimera-checked 16S RRNA gene database and workbench compatible with ARB. *Appl. Environ. Microbiol.* **2006**, *72*, 5069–5072. [CrossRef]
40. McMurdie, P.J.; Holmes, S. Phyloseq: An R Package for Reproducible Interactive Analysis and Graphics of Microbiome Census Data. *PLoS ONE* **2013**, *8*, e61217. [CrossRef]
41. Segata, N.; Izard, J.; Waldron, L.; Gevers, D.; Miropolsky, L.; Garrett, W.S.; Huttenhower, C. Metagenomic biomarker discovery and explanation. *Genome Biol.* **2011**, *12*, 1–18. [CrossRef]
42. Campbell, L.; Yu, R.; Li, F.; Zhou, Q.; Chen, D.; Qi, C.; Yin, Y.; Sun, J. Diabetes, Metabolic syndrome and obesity: Targets and therapy dovepress modulation of fat metabolism and gut microbiota by resveratrol on high-fat diet-induced obese mice. *Diabetes Metab. Syndr. Obes. Targets Ther.* **2019**, *12*, 97–107. [CrossRef] [PubMed]
43. Ormerod, K.L.; Wood, D.L.A.; Lachner, N.; Gellatly, S.L.; Daly, J.N.; Parsons, J.D.; Dal'Molin, C.G.O.; Palfreyman, R.W.; Nielsen, L.K.; Cooper, M.A.; et al. Genomic characterization of the uncultured bacteroidales family S24-7 inhabiting the guts of homeothermic animals. *Microbiome* **2016**, *4*, 1–17. [CrossRef] [PubMed]

44. Mallick, H.; Rahnavard, A.; McIver, L.J.; Ma, S.; Zhang, Y.; Nguyen, L.H.; Tickle, T.L.; Weingart, G.; Ren, B.; Schwager, E.H.; et al. Multivariable association discovery in population-scale meta-omics studies. *PLoS Comput. Biol.* **2021**, *17*, e1009442. [CrossRef] [PubMed]
45. Corrêa, T.A.F.; Rogero, M.M.; Hassimotto, N.M.A.; Lajolo, F.M. The two-way polyphenols-microbiota interactions and their effects on obesity and related metabolic diseases. *Front. Nutr.* **2019**, *6*, 188. [CrossRef]
46. Fava, F.; Rizzetto, L.; Tuohy, K.M. Gut microbiota and health: Connecting actors across the metabolic system. *Proc. Nutr. Soc.* **2019**, *78*, 177–188. [CrossRef]
47. Gowd, V.; Karim, N.; Shishir, M.R.I.; Xie, L.; Chen, W. Dietary polyphenols to combat the metabolic diseases via altering gut microbiota. *Trends Food Sci. Technol.* **2019**, *93*, 81–93. [CrossRef]
48. Sheng, S.; Chen, J.; Zhang, Y.; Qin, Q.; Li, W.; Yan, S.; Wang, Y.; Li, T.; Gao, X.; Tang, L.; et al. Structural and functional alterations of gut microbiota in males with hyperuricemia and high levels of liver enzymes. *Front. Med.* **2021**, *8*, 779994. [CrossRef]
49. Caponio, G.R.; Lippolis, T.; Tutino, V.; Gigante, I.; De Nunzio, V.; Milella, R.A.; Gasparro, M.; Notarnicola, M. Nutraceuticals: Focus on anti-inflammatory, anti-cancer, antioxidant properties in gastrointestinal tract. *Antioxidants* **2022**, *11*, 1274. [CrossRef]
50. Hurst, H.; Harborne, J. Plant polyphenols—XVI: Identification of flavonoids by reductive cleavage. *Phytochemistry* **1967**, *6*, 1111–1118. [CrossRef]
51. Rastmanesh, R. High Polyphenol, Low Probiotic diet for weight loss because of intestinal microbiota interaction. *Chem. Biol. Interact.* **2011**, *189*, 1–8. [CrossRef]
52. Caponio, G.R.; Cofano, M.; Lippolis, T.; Gigante, I.; De Nunzio, V.; Difonzo, G.; Noviello, M.; Tarricone, L.; Gambacorta, G.; Giannelli, G.; et al. Anti-proliferative and pro-apoptotic effects of digested aglianico grape pomace extract in human colorectal cancer cells. *Molecules* **2022**, *27*, 6791. [CrossRef] [PubMed]
53. Ding, X.; Guo, L.; Zhang, Y.; Fan, S.; Gu, M.; Lu, Y.; Jiang, D.; Li, Y.; Huang, C.; Zhou, Z. Extracts of pomelo peels prevent high-fat diet-induced metabolic disorders in C57BL/6 mice through activating the PPARα and GLUT4 pathway. *PLoS ONE* **2013**, *8*, e77915. [CrossRef] [PubMed]
54. Jakubczyk, K.; Kałduńska, J.; Kochman, J.; Janda, K. Chemical profile and antioxidant activity of the kombucha beverage derived from white, green, black and red tea. *Antioxidants* **2020**, *9*, 477. [CrossRef] [PubMed]
55. Gomes, R.J.; Borges, M.d.F.; Rosa, M.d.F.; Castro-Gómez, R.J.H.; Spinosa, W.A. Acetic acid bacteria in the food industry: Systematics, characteristics and applications. *Food Technol. Biotechnol.* **2018**, *56*, 139–151. [CrossRef]
56. Coton, M.; Pawtowski, A.; Taminiau, B.; Burgaud, G.; Deniel, F.; Coulloumme-Labarthe, L.; Fall, A.; Daube, G.; Coton, E. Unraveling microbial ecology of industrial-scale kombucha fermentations by metabarcoding and culture-based methods. *FEMS Microbiol. Ecol.* **2017**, *93*, 1–16. [CrossRef]
57. Maldonado, R.F.; Sá-Correia, I.; Valvano, M.A.; Gasmi, A.; Mujawdiya, P.K.; Pivina, L.; Doşa, A.; Semenova, Y.; Benahmed, A.G.; Bjørklund, G. Lipopolysaccharide modification in gram-negative bacteria during chronic infection. *FEMS Microbiol. Rev.* **2016**, *28*, 827–839. [CrossRef]
58. Gasmi, A.; Mujawdiya, P.K.; Pivina, L.; Doşa, A.; Semenova, Y.; Benahmed, A.G.; Bjørklund, G. Relationship between gut microbiota, gut hyperpermeability and obesity. *Curr. Med. Chem.* **2020**, *28*, 827–839. [CrossRef]
59. Kang, C.; Wang, B.; Kaliannan, K.; Wang, X.; Lang, H.; Hui, S.; Huang, L.; Zhang, Y.; Zhou, M.; Chen, M.; et al. Gut microbiota mediates the protective effects of dietary capsaicin against chronic low-grade inflammation and associated obesity induced by high-fat diet. *MBio* **2017**, *8*. [CrossRef]
60. Vacca, M.; Celano, G.; Calabrese, F.M.; Portincasa, P.; Gobbetti, M.; De Angelis, M. The controversial role of human gut lachnospiraceae. *Microorganisms* **2020**, *8*, 573. [CrossRef] [PubMed]
61. Staley, C.; Weingarden, A.R.; Khoruts, A.; Sadowsky, M.J. Interaction of gut microbiota with bile acid metabolism and its influence on disease states. *Appl. Microbiol. Biotechnol.* **2017**, *101*, 47–64. [CrossRef]
62. Wahlström, A.; Sayin, S.I.; Marschall, H.U.; Bäckhed, F. Intestinal Crosstalk between Bile Acids and Microbiota and Its Impact on Host Metabolism. *Cell Metab.* **2016**, *24*, 41–50. [CrossRef] [PubMed]
63. Calabrese, F.M.; Disciglio, V.; Franco, I.; Sorino, P.; Bonfiglio, C.; Bianco, A.; Campanella, A.; Lippolis, T.; Pesole, P.L.; Polignano, M.; et al. A Low glycemic index mediterranean diet combined with aerobic physical activity rearranges the gut microbiota signature in NAFLD patients. *Nutrients* **2022**, *14*, 1773. [CrossRef] [PubMed]
64. Shen, F.; Zheng, R.D.; Sun, X.Q.; Ding, W.J.; Wang, X.Y.; Fan, J.G. Gut microbiota dysbiosis in patients with non-alcoholic fatty liver disease. *Hepatobiliary Pancreat. Dis. Int.* **2017**, *16*, 375–381. [CrossRef]
65. Zeng, H.; Larson, K.J.; Cheng, W.H.; Bukowski, M.R.; Safratowich, B.D.; Liu, Z.; Hakkak, R. Advanced liver steatosis accompanies an increase in hepatic inflammation, colonic, secondary bile acids and Lactobacillaceae/Lachnospiraceae bacteria in C57BL/6 mice fed a high-fat diet. *J. Nutr. Biochem.* **2020**, *78*, 108336. [CrossRef] [PubMed]
66. Vojinovic, D.; Radjabzadeh, D.; Kurilshikov, A.; Amin, N.; Wijmenga, C.; Franke, L.; Ikram, M.A.; Uitterlinden, A.G.; Zhernakova, A.; Fu, J.; et al. Relationship between gut microbiota and circulating metabolites in population-based cohorts. *Nat. Commun.* **2019**, *10*, 5813. [CrossRef] [PubMed]
67. De Filippo, C.; Cavalieri, D.; Di Paola, M.; Ramazzotti, M.; Poullet, J.B.; Massart, S.; Collini, S.; Pieraccini, G.; Lionetti, P. Impact of diet in shaping gut microbiota revealed by a comparative study in children from Europe and rural Africa. *Proc. Natl. Acad. Sci. USA* **2010**, *107*, 14691–14696. [CrossRef] [PubMed]

68. Precup, G.; Vodnar, D.C. Gut Prevotella as a possible biomarker of diet and its Eubiotic versus dysbiotic roles: A Comprehensive literature review. *Br. J. Nutr.* **2019**, *122*, 131–140. [CrossRef]
69. Ke, S.; Fang, S.; He, M.; Huang, X.; Yang, H.; Yang, B.; Chen, C.; Huang, L. Age-Based dynamic changes of phylogenetic composition and interaction networks of health pig gut microbiome feeding in a uniformed condition. *BMC Vet. Res.* **2019**, *15*, 1–13. [CrossRef]
70. Illescas, O.; Rodríguez-Sosa, M.; Gariboldi, M. Mediterranean Diet to Prevent the Development of Colon Diseases: A Meta-Analysis of Gut Microbiota Studies. *Nutrients* **2021**, *13*, 2234. [CrossRef]
71. van der Hee, B.; Wells, J.M. Microbial regulation of host physiology by short-chain fatty acids. *Trends Microbiol.* **2021**, *29*, 700–712. [CrossRef]
72. Dahiya, D.K.; Renuka; Puniya, M.; Shandilya, U.K.; Dhewa, T.; Kumar, N.; Kumar, S.; Puniya, A.K.; Shukla, P. Gut microbiota modulation and its relationship with obesity using prebiotic fibers and probiotics: A review. *Front. Microbiol.* **2017**, *8*, 563. [CrossRef] [PubMed]
73. Maslowski, K.M.; MacKay, C.R. Diet, gut microbiota and immune responses. *Nat. Immunol.* **2011**, *12*, 5–9. [CrossRef] [PubMed]
74. Holscher, H.D. Dietary fiber and prebiotics and the gastrointestinal microbiota. *Gut Microbes* **2017**, *8*, 172–184. [CrossRef]
75. Van Hul, M.; Cani, P.D. Targeting carbohydrates and polyphenols for a healthy microbiome and healthy weight. *Curr. Nutr. Rep.* **2019**, *8*, 307–316. [CrossRef]
76. Bhattacharya, S.; Gachhui, R.; Sil, P.C. Effect of kombucha, a fermented black tea in attenuating oxidative stress mediated tissue damage in alloxan induced diabetic rats. *Food Chem. Toxicol.* **2013**, *60*, 328–340. [CrossRef] [PubMed]
77. Xu, Y.; Wang, N.; Tan, H.Y.; Li, S.; Zhang, C.; Feng, Y. Function of Akkermansia Muciniphila in obesity: Interactions with lipid metabolism, immune response and gut systems. *Front. Microbiol.* **2020**, *11*, 219. [CrossRef] [PubMed]
78. Chelakkot, C.; Choi, Y.; Kim, D.K.; Park, H.T.; Ghim, J.; Kwon, Y.; Jeon, J.; Kim, M.S.; Jee, Y.K.; Gho, Y.S.; et al. Akkermansia Muciniphila-derived extracellular vesicles influence gut permeability through the regulation of tight junctions. *Exp. Mol. Med.* **2018**, *50*, e450-11. [CrossRef]
79. Konikoff, T.; Gophna, U. Oscillospira: A central, enigmatic component of the human gut microbiota. *Trends Microbiol.* **2016**, *24*, 523–524. [CrossRef]

nutrients

MDPI

Article

Preparation, Characterization, Wound Healing, and Cytotoxicity Assay of PEGylated Nanophytosomes Loaded with 6-Gingerol

Ali Al-Samydai [1,*], Moath Al Qaraleh [1], Walhan Alshaer [2], Lidia K. Al-Halaseh [3], Reem Issa [1], Fatima Alshaikh [1], Aseel Abu-Rumman [1], Hayat Al-Ali [4] and Emad A. S. Al-Dujaili [5,*]

1 Pharmacological and Diagnostic Research Centre, Department Pharmacological, Faculty of Pharmacy, Al-Ahliyya Amman University, Amman 19328, Jordan
2 Cell Therapy Center, The University of Jordan, Amman 11942, Jordan
3 Department of Pharmaceutical Chemistry, Faculty of Pharmacy, Mutah University, Al-Karak 61710, Jordan
4 Pharmacological and Diagnostic Research Center, Faculty of Allied Medical Sciences, Al-Ahliyya Amman University, Al-Salt 19328, Jordan
5 Centre for Cardiovascular Science, Queen's Medical Research Institute, University of Edinburgh, 47 Little France Crescent, Edinburgh EH16 4TJ, UK
* Correspondence: a.alsamydai@ammanu.edu.jo (A.A.-S.); ealduja1@exseed.ed.ac.uk (E.A.S.A.-D.)

Abstract: Background: Nutrients are widely used for treating illnesses in traditional medicine. Ginger has long been used in folk medicine to treat motion sickness and other minor health disorders. Chronic non-healing wounds might elicit an inflammation response and cancerous mutation. Few clinical studies have investigated 6-gingerol's wound-healing activity due to its poor pharmacokinetic properties. However, nanotechnology can deliver 6-gingerol while possibly enhancing these properties. Our study aimed to develop a nanophytosome system loaded with 6-gingerol molecules to investigate the delivery system's influence on wound healing and anti-cancer activities. Methods: We adopted the thin-film hydration method to synthesize nanophytosomes. We used lipids in a ratio of 70:25:5 for DOPC(dioleoyl-sn-glycero-3-phosphocholine): cholesterol: DSPE/PEG2000, respectively. We loaded the 6-gingerol molecules in a concentration of 1.67 mg/mL and achieved size reduction via the extrusion technique. We determined cytotoxicity using lung, breast, and pancreatic cancer cell lines. We performed gene expression of inflammation markers and cytokines according to international protocols. Results: The synthesized nanophytosome particle sizes were 150.16 ± 1.65, the total charge was −13.36 ± 1.266, and the polydispersity index was 0.060 ± 0.050. Transmission electron microscopy determined the synthesized particles' spherical shape and uniform size. The encapsulation efficiency was 34.54% ± 0.035. Our biological tests showed that 6-gingerol nanophytosomes displayed selective antiproliferative activity, considerable downregulation of inflammatory markers and cytokines, and an enhanced wound-healing process. Conclusions: Our results confirm the anti-cancer activity of PEGylated nanophytosome 6-gingerol, with superior activity exhibited in accelerating wound healing.

Keywords: cytotoxicity; 6-gingerol; liposomes; nanoparticles; phytosomes; wound healing

Citation: Al-Samydai, A.; Al Qaraleh, M.; Alshaer, W.; Al-Halaseh, L.K.; Issa, R.; Alshaikh, F.; Abu-Rumman, A.; Al-Ali, H.; Al-Dujaili, E.A.S. Preparation, Characterization, Wound Healing, and Cytotoxicity Assay of PEGylated Nanophytosomes Loaded with 6-Gingerol. *Nutrients* **2022**, *14*, 5170. https://doi.org/10.3390/nu14235170

Academic Editor: Elad Tako

Received: 24 October 2022
Accepted: 2 December 2022
Published: 5 December 2022

1. Introduction

Human skin, the body's largest organ, is considered the first protective barrier against environmental changes besides its other functions, such as thermoregulation and maintaining aesthetic appearance [1]. Wounded skin undergoes repairing and regenerating processes to avoid traumatized tissues and painful sensations [2]. The healing process involves the contraction of injured tissues, biosynthesizing collagen, and epithelization [3]. Untreated and unhealed wounds can develop into chronic wounds, which might lead to further complications, such as septic infection, and even the need for organ transplantation [1,4]. The internal self-healing process lasts from 5 to 10 days depending on the severity of the wound and other factors. Sometimes, the process might take up to 30 days until complete healing, with tissue hypoxia, exudation, and necrosis among the delaying factors. Furthermore, infectious wounds are very likely to develop into chronic wounds [5]. Bioactive molecules originating from natural sources have been extensively studied for their wound-healing capabilities. These include plant polyphenols, such as lupeol, curcumin, and catechin, and animal sources, such as bee venom [6].

Ginger, *Zingiber officinale*, is rich in bioactive compounds that have confirmed activities against inflammation, oxidative stress, and cancerous cells. Alongside its hypoglycemic and hypolipidemic activities, it also contributes to relieving motion sickness [1–3]. Its bioactive compounds include terpenes and numerous phenolic compounds, such as gingerols, shogaols, and paradols [3]. Indeed, 8- and 10-gingerol, 6-gingerol is an active principle with known therapeutic activity that is found in abundance in fresh ginger rhizomes [7].

Reports published on 6-gingerol highlight its capability in promoting the epithelization process and suppressing the production of macrophages and cytokines, thus contributing to a decrease in the inflammatory response [1,8]. This anti-inflammatory effect is advantageous because it stimulates vascularization and the wound-healing process. Moreover, in combination with vitamin D, it enhances wound healing in diabetic patients by increasing expression levels of hemopexin-like domains (HPX), fibrillogenesis (FN), and collagen, accelerating the healing process and shortening epithelialization time [9].

Various vehicles and delivery systems have been utilized to deliver therapeutic agents to their targets in optimal conditions. Nanoparticle carriers have shown promising results in treating acute and chronic wounds due to their physiochemical, optical, and biological properties. Nanocomposites are smart materials that have been synthesized by incorporating the nanoparticles into scaffolds, and they have accelerated wound-healing activity in addition to their antimicrobial, anti-inflammatory, and angiogenic properties [10–12].

Nanoparticles affect the healing mechanism by influencing collagen deposition and realignment, thus initiating skin regeneration. Nanotechnology is being increasingly assimilated in wound therapy because it provides flexibility, reasonable mechanical strength, and large porosity, and ensures non-adherence to the wound surface. Additionally, its cooling sensations, moistening effects, and protection from microbial contamination are desirable properties [12–15].

Buflomedil hydrochloride, a topical vasoactive agent, is an example of a therapeutic agent incorporated into nanocarriers, and it has shown improved nutritive perfusion within the lesion's vicinity, increasing wound-healing properties [16].

Uploading high concentrations of the therapeutic agent, in addition to its direct delivery to the wounded tissue, should decrease systemic side effects; therefore, the system is a promising delivery model. Similar findings were obtained after treating calvarial bone wounds with gallic acid in the liposomal system of an animal model [17]. Quercetin-loaded liposomes exerted a sustained release and displayed effective wound-healing properties [18]. Nanoscale liposomes provide benefits when treating various illnesses, such as rheumatoid arthritis, namely reduced toxicity and enhanced delivery [13,19].

The active constituent of ginger, 6-gingerol, is known to exhibit wound healing and anti-inflammatory activities; however, it has poor pharmacokinetic properties regarding its slight solubility in aqueous media and its low bioavailability after oral administration [14]. To the best of our knowledge, there are few clinical studies on 6-gingerol activity regarding these biological effects. Therefore, in our research, we loaded 6-gingerol particles into PEGylated nanophytosomes to investigate their wound healing, anti-inflammatory, and antiproliferative properties using lung, pancreatic, and breast cancer cell lines.

2. Materials and Methods

2.1. Chemical Reagents

We purchased dried ginger rhizomes from a local market in Amman, Jordan, and we kept them at room temperature. We purchased 6-gingerol, quercetin, and gallic acid from Santa Cruz Biotechnology, CA, San Juan, USA. We purchased analytical- and HPLC-grade solvents from Sigma Aldrich, St. Louis, MO, USA, unless stated otherwise. We purchased chemicals from, Jordan Genome store, Amman, Jordan. We obtained deionized water from Millipore Waters, California, CA, USA. We obtained PANC1 pancreatic (ATCC number: CRL-1469), A549 lung (ATCC number; CCL-185), and MDA-MB-231 breast (ATCC number: HTB-26) cancer cell lines from PDRC (Pharmacological and Diagnostic Research Center at Al-Ahliyya Amman University). We purchased cell culture plates from TPP, Zollstraße, Switzerland.

2.2. Plant Preparation, Extraction, and Phytochemical Analysis

We crushed the dried rhizomes to reduce their size and then macerated them in methanol at a ratio of 2:3 w/v. We filtered, concentrated, and kept the mixture at 4 °C. We performed phytochemical analysis and quantification of the polyphenol and flavonoid content [15]. We tested the dried extract for its antioxidant properties by following optimized methods [20].

2.3. HPLC Analysis of the Ginger Crude Extracts

Sample Preparation

We prepared a stock solution of the authentic 6-gingerol by initially dissolving in DMSO and then diluting it with acetonitrile. We centrifuged the mixture (4000 rpm, 2 min) using a benchtop centrifuge (Eppendorf®, Germany). We took the supernatant and injected the 4 µL sample size into the autosampler.

2.4. PEGylated Nanophytosome Formation

We prepared PEGylated nanophytosomes using the thin-film hydration method (TFHM) as previously described [11,16], concentrating cholesterol (25%), 1,2-dioleoyl-sn-glycero-3-phosphocholine (DOPC) (70%), and 1, 2-distearoyl-sn-glycero-3-phosphoethanolamine-Poly (ethylene glycol) (5%) into 6-gingerol fixed at 1.67 mg/mL.

2.5. Characterization of PEGylated Nanophytosomes

2.5.1. Size and Charge

We measured the average size, charge, and polydispersity index (PDI) for both the free and 6-gingerol-encapsulated PEGylated nanophytosomes using dynamic light scattering (DLS) on a Zetasizer Nano-ZS® (Malvern Instruments Ltd., Malvern, UK). We diluted samples (1:20) with deionized water. We placed all samples in the Zetasizer specimen holder for approximately 1–2 min before measuring to equilibrate with room temperature [19].

2.5.2. Encapsulation Efficiency (%EE) and Drug Loading (%DL)

We determined the encapsulated 6-gingerol's percentage as follows. First, we degraded the loaded PEGylated nanophytosomes by adding acetonitrile at a ratio of 4:1 acetonitrile: phytosomes. Second, we centrifuged the mixture (7000, 10 min). Third, we performed the sonication step (10 min, 35 °C), then centrifuged (12,000 rpm, 10 min) and filtered with the aid of 0.45 µm syringe filters [12,19].

We measured the encapsulated 6-gingerol's concentration using an HPLC system (UV Detector at 280 nm, HPLC Nucleodur-C18 Column; 250 mm × 4.6 mm, 5 µm), at a slow flow rate of 1 mL/min and column temperature of 40 °C using a 10 µL injection volume connected to Labsolution® computer software, Weiswampach, Luxembourg. The mobile phase comprised 60:40 HPLC-grade methanol: water [14]. Equations (1) and (2) show the calculations of % encapsulation efficiency (EE) and % drug loading (DL), respectively:

$$(\%EE) = \frac{[\text{Encapsulated6} - \text{Gingerol}]}{[\text{Total6} - \text{Gingerol}]} \times 100 \quad (1)$$

Equation (1): The encapsulation efficiency of 6-gingerol.

$$(\%DL) = \frac{[\text{WeightofLoadeddrug}]}{[\text{Weightoflipids}]} \times 100 \quad (2)$$

Equation (2): The drug loading of 6-gingerol.

2.5.3. Transmission Electron Microscopy (TEM)

We analyzed the structure and morphology of the empty and loaded PEGylated nanophytosomes using the TEM analysis negative staining method. First, we coated 200-mesh formvar copper grids (SPI Supplies, West Chester, PA, USA) with carbon under a low-vacuum Leica EM ACE200 glow discharge coating system (Leica, Westbahnstraße, Austria). Then, we further coated the carbon-coated grids with 1.5% Vinylec K in a chloroform solution. We placed a drop of deionized water-diluted liposome suspension on the 200-mesh copper grid, followed by an air-drying step. Then, we stained the loaded grids with a 3% v/v aqueous solution of uranyl acetate for 20 min at room temperature. After incubation, we washed the grids with distilled water and dried them at room temperature before imaging with a Versa 3D (FEI, Hoofstraat, The Netherlands) TEM operating system at an acceleration voltage of 30 kV [19].

2.6. Stability Study (Lyophilization of Loaded PEGylated Nanophytosomes)

We dissolved the lyoprotectant (dextrose) in phosphate-buffered saline (10% w/v). We freeze-dried PEGylated nanophytosome suspensions in a buffer with the lyoprotectant (the liquid froze at −195 °C). The dehydration step lasted for two days at −40 °C until we obtained dried powders. We stored the resulting lyophilized powders for one month and rehydrated them (when needed) to their original volume at room temperature with phosphate-buffered saline (PBS). After adding PBS, we equilibrated the samples at room temperature for 30 min. Then, we subjected the samples to the size PDI and encapsulation efficiency test [21].

2.7. Viability Assay

We determined cytotoxicity measurements using a 3-(4,5-dimethylthiazol-2-yl)-2,5-diphenyltetrazolium bromide (MTT) assay for cytotoxicity [22,23]. We seeded lung cancer A549, breast cancer MDA-MB-231, and pancreatic cancer Panc1 cell lines in 96-well plates at 1×10^4 cells/well and cultured them in a medium containing *Zingiber officinale* extract, 6-gingerol, and PEGylated nanophytosomes loaded at concentrations of 1.5–50 µg/mL. After 48 h, we performed the MTT assay [20]. Human periodontal ligament fibroblasts (PDL) are a primary cell culture used for verifying selective cytotoxicity.

2.8. Gene Expression Level Assay

We plated MDA cells at a density of 1×10^6 cells/well in 6-well plates [24]. After 24 h, we incubated the selected concentration of ginger extract and liposomes for another 24 h to measure the fold change in gene expressions of IL-10, IL-1beta, IL-6, IL-8, TNF-alpha, and IRAK1.

2.9. Ribonucleic Acid (RNA) Extraction and Analysis

We extracted total RNA using a RNeasy Mini kit®, QIAGEN, USA. We transferred the cell lysate to the provided spin column before centrifugation (10,000 rpm, 15 s). We performed the assay reactions following the manufacturer's instructions. We synthesized cDNA using a reverse transcription system (Applied Biosystem, Waltham, MA, USA). We measured the optical densities at 260 and 280 nm, respectively. The ratio (A260/A280) was 1.6–1.8 for most of the cDNA-extracted samples. We performed relative-quantitative analyses of our tested genes' mRNA expression levels using a fast SYBR green kappa master mix (Biosystem, USA). We investigated GAPDH, IL-1, beta IL-10, IL-6, IL-8, TNF-alpha, and IRAK1. We used glyceraldehyde-3-phosphate dehydrogenase (GAPDH) as an internal reference gene to normalize the expression of the tested genes [24,25].

2.10. Statistical Analysis

We determined statistical differences between control and treated groups using GraphPad Prism ANOVA followed by Dunnett's post hoc test. For all statistical analyses, we considered a *p*-value of less than 0.05 as statistically significant.

3. Results

3.1. Extraction and Total Phenol, Flavonoid, and Antioxidant Contents of Crude Extract

The methanolic extraction yield was 46.32 mg/g. We measured the total flavonoids based on the quercetin standard curve as 18.52 ± 1.48 mg/g ± SD, and the total phenol content based on the gallic acid standard was 39.17 ± 1.24 SD. Furthermore, we measured the antioxidant activity of the methanolic extract compared with vitamin C in IC50 as 3.325 ± 0.21 mg/mL ± SD.

3.2. Effect of 6-Gingerol on Particle Size, PDI, and Zeta Potential

We formulated the free and the 6-gingerol-loaded PEGylated nanophytosomes in lipid ratios of DOPC: Cholesterol: DSPE/PEG2000 (70:25:5). Table 1 and Figure 1A show the particle sizes, PDIs, and zeta potentials of the different formulations. All the nanoliposome data were within the range of optimum formulation [25]. There were no significant differences in the average particle sizes and PDIs between the free and loaded nanoparticles; furthermore, we calculated the *p* values as 0.96 and 0.71, respectively. However, we measured significant differences in the zeta potential (charge) ($p < 0.05$), which could arise from 6-gingerol molecules' phenolic oxygen. Accordingly, we included more dispersed negative charges [26].

Table 1. Effect of 6-gingerol on nanoparticle characterization.

Description						Two-Sample *t*-Test
Test	Group	Mean	Std. Deviation	Minimum	Maximum	Sig.
Size	1	126.01	0.435	125.70	126.50	
	2	150.16	1.650	148.80	152.00	0.96 ns
Charge	1	−11.13	0.152	−11.00	−11.30	
	2	−13.36	1.266	−12.40	−14.80	0.04 *
PDI	1	0.127	0.017	0.11	0.14	
	2	0.060	0.050	0.004	0.09	0.71 ns

1: Free PEGylated nanophytosomes. 2: PEGylated nanophytosomes loaded with 6-gingerol. All data are normally distributed according to the Shapiro–Wilk normality test. $n = 3$; * Statistically different at significance level $p < 0.05$. ns: Not statistically different at significance level $p < 0.05$.

Our TEM study revealed the successful loading of 6-gingerol molecules into PEGylated nanophytosomes. Our results represent the loaded nanophytosomes' spherical shape and uniform size. We measured the average particle size as 23.46 ± 40.21 nm ($n = 9$). Figure 1A–C shows the particle size of the product and that 6-gingerol has been successfully encapsulated in the system.

3.2.1. Stability after Lyophilization

We measured the loaded system's encapsulation efficiency as 34.54 ± 0.035%. A month later, after storing the lyophilized system, we found that the retained formula's mean of encapsulated efficiency was 33.94%.

Lyophilization should increase the nanophytosomes' shelf-life by preserving them dry in a lyophilized cake until needed; then, they can be reconstituted in water and used. However, we observed no significant differences in our other studied factors after lyophilization, such as PDI and size, compared with data obtained before the freezing process ($p > 0.05$). We added a lyoprotectant to the final formula to maintain the particles' distribution during the lyophilization–rehydration cycle (Table 2).

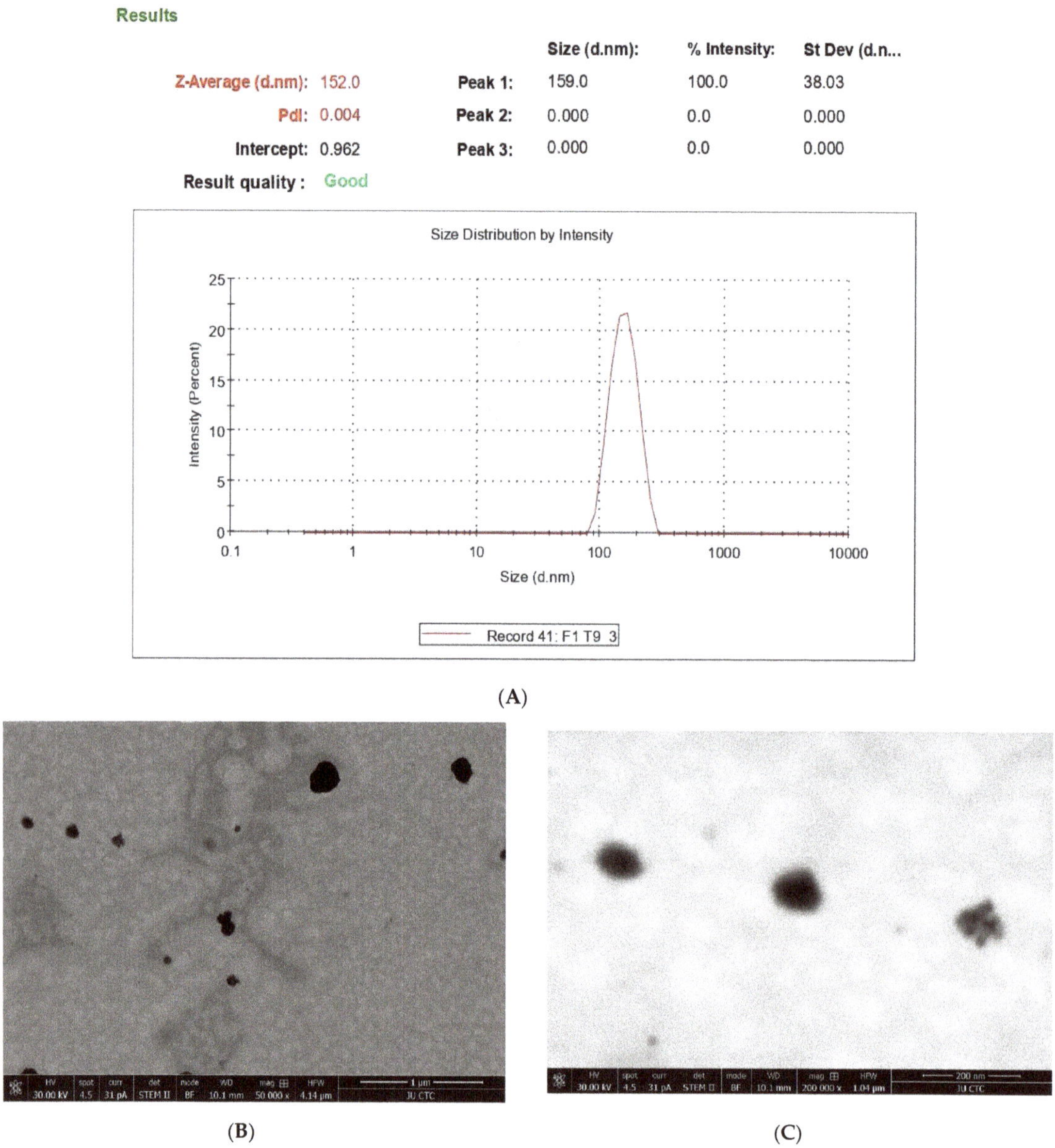

Figure 1. (**A**) Size distribution of 6-gingerol-loaded PEGylated nanophytosomes by DLS; (**B**) TEM images of shape and size of loaded PEGylated nanophytosomes; (**C**) morphology (6-gingerol-loaded PEGylated nanophytosomes).

Table 2. Effect of lyophilization process on characterization of PEGylated nanophytosomes loaded with 6-gingerol molecules.

Description				Two-Sample *t*-Test
Test	Group	Mean	Std. Deviation	Sig.
Size	1	150.1667	1.65025	0.102 ns
	2	160.0667	9.31254	
PDI	1	0.0623	0.05054	0.628 ns
	2	0.2400	0.06207	

1: The 6-gingerol-loaded PEGylated nanophytosomes before lyophilization. 2: The 6-gingerol-loaded PEGylated nanophytosomes after lyophilization. All data are normally distributed according to the Shapiro–Wilk normality test, $n = 3$; ns: Not statistically different at significance level $p < 0.05$.

3.2.2. Modulation of Proliferation of Lung and Pancreatic Cancer Cell Lines, as Well as Fibroblasts, by Ginger Extract, 6-Gingerol, and 6-Gingerol Particles Carried by PEGylated Nanophytosomes

Excluding fibroblasts, 6-gingerol-loaded PEGylated nanophytosomes exerted considerable antiproliferative efficacy against A549 and Panc1 cell lines over 48 h incubations. Furthermore, the crude extract of rhizomes and the authentic 6-gingerol sustained their antiproliferative efficacies against fibroblasts, A549, and Panc1 cell lines (Figures 2–4).

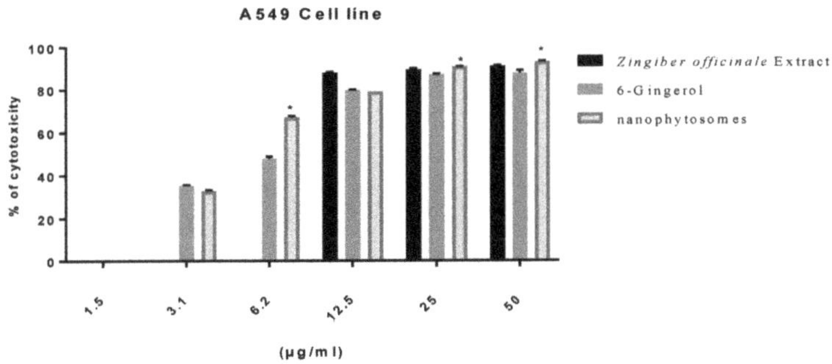

Figure 2. Effect of *Zingiber officinale* extract, 6-gingerol, and 6-gingerol-loaded PEGylated nanophytosomes on cytotoxicity of A549 cells compared with control (6-gingerol). * Statistically different at significance level $p < 0.05$.

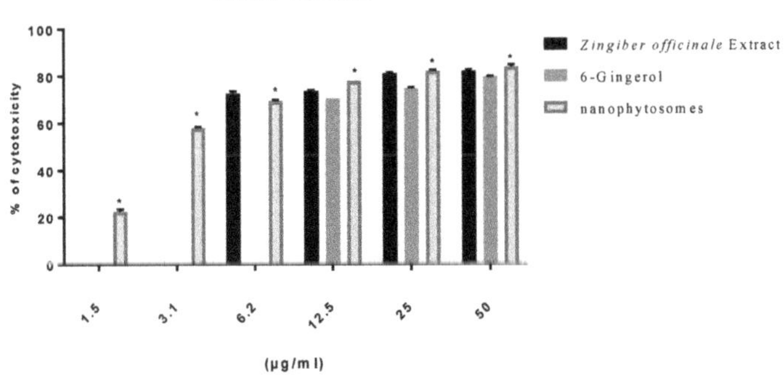

Figure 3. Effect of *Zingiber officinale* extract, 6-gingerol, and 6-gingerol-loaded PEGylated nanophytosomes on cytotoxicity of Panc1 cells compared with control (6-gingerol). * Statistically different at significance level $p < 0.05$.

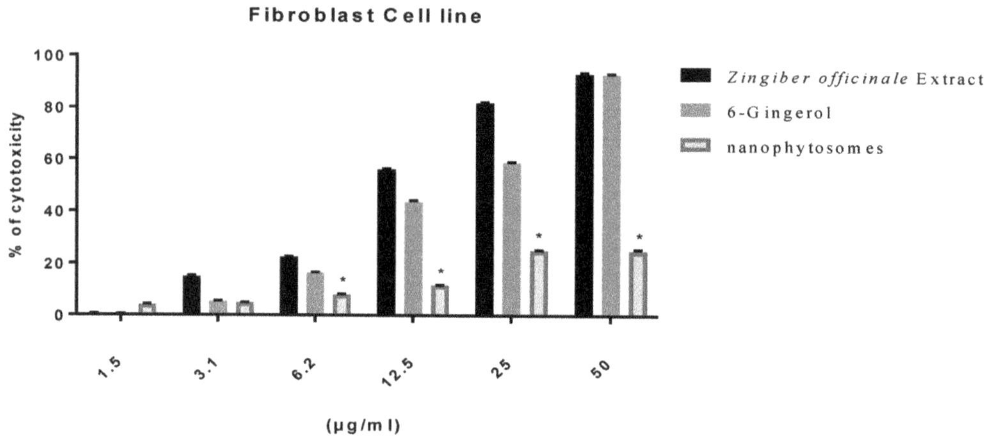

Figure 4. Effect of *Zingiber officinale* extract, 6-gingerol, and 6-gingerol-loaded PEGylated nanophytosomes on cytotoxicity of human periodontal ligament fibroblast cells compared with control (6-gingerol). * Statistically different at significance level $p < 0.05$.

3.2.3. Gene Expression of Pro-Inflammatory Cytokines

We determined the gene expression levels of the cytokines in ginger crude extract and the loaded PEGylated nanophytosomes (1.5 µg/mL) with the authentic 6-gingerol using RT-PCR technology. The cytokines we examined were TNF-α, IL-1beta, IL-6, IL-8, IL-10, and IRAK1.

The breast cancer MDA-MB-231 cell line elicits the production of proinflammatory cytokines TNF-α, IL-1beta, IL-6, IL-8, and IL-10. Our analysis of IRAK1.RT-PCR expression levels revealed that the crude extract as well as the pure compound 6-gingerol have significantly ($p < 0.05$) downregulated factors (TNF-α, IL-1beta, IL-6, IL-8, and IRAK1) in MDA after being separately loaded in PEGylated nanophytosomes (Figure 5).

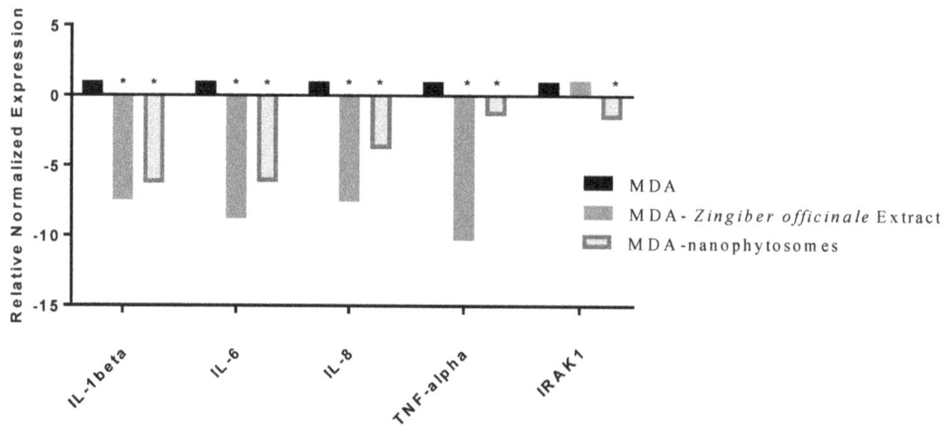

Figure 5. Effect of *Zingiber officinale* extract and 6-gingerol-loaded PEGylated nanophytosomes on the expression of IL-1beta, IL-6, IL-8, TNF-alpha, and IRAK1 on MDA cells compared with control (cells only). The crude extract as well as the pure compound 6-gingerol have significantly ($p < 0.05$) downregulated factors (TNF-α, IL-1beta, IL-6, IL-8, and IRAK1) in MDA after being separately loaded in PEGylated nanophytosomes. * Statistically different at significance level $p < 0.05$.

3.2.4. Wound Healing

The effects of ginger crude extract, 6-gingerol, and PEGylated nanophytosomes loaded with 6-gingerol on the wound-healing process are shown in Figure 6. We observed a marked and considerable increase in the wound-healing process of the human periodontal ligament fibroblasts in the medicated cells compared with those left untreated.

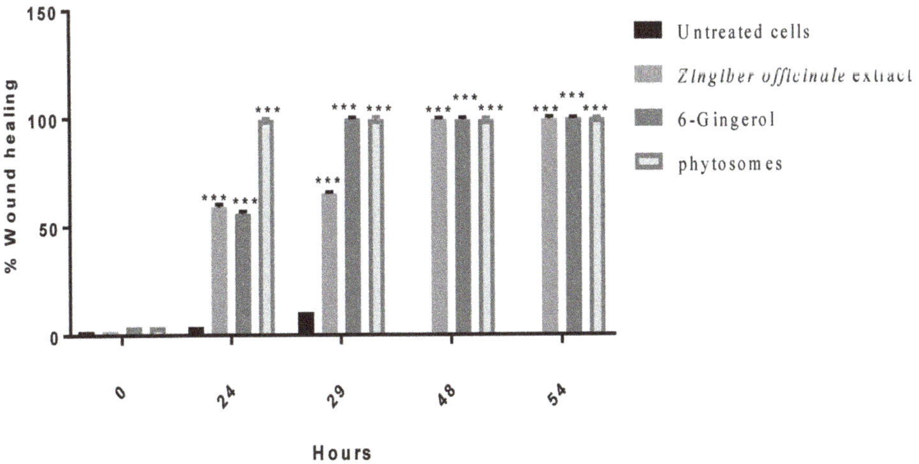

Figure 6. Effect of *Zingiber officinale* extract, 6-gingerol, and 6-gingerol-loaded PEGylated nanophytosomes on wound healing of human periodontal ligament fibroblasts cells compared with control (untreated cells). ***: $p < 0.001$.

4. Discussion

Inflammatory responses to harmful stimuli are among the leading causes of variable health issues in the cardiovascular and pulmonary systems, contributing to autoimmune illnesses. Furthermore, the chemicals released by inflamed and injured tissues could initiate and exaggerate the proliferation and metastasis of cancerous cells. Collectively, cytokines, which are pro-inflammatory signaling agents, enhance the survival of tumor cells through activating the NFκB pathway [27]. Considering the harmful consequences of inflammatory reactions, large quantities of research have been conducted to discover and formulate agents with anti-inflammatory potential. Ameliorating inflammation should accelerate wound-healing cascades, possibly reducing the growth of numerous cancerous cells. Several natural and synthetic agents have been investigated for their anti-inflammatory activity, and researchers have discovered that many compounds have such effects [28–30]. The results of previous reports have elucidated the relationship between medicinal plants and the NFκB pathway. Ginger, *Zingiber officinale*, is among the studied plants, and was approved as an alternative anti-inflammatory agent against the production of cyclooxygenase (COX)-2 and nitric oxide (NO) [31–33].

According to our results, we confirmed the anti-inflammatory properties of ginger crude extract and its active constituents in concomitance with previously published data. We determined this biological activity based on the expression levels of pro-inflammatory cytokines in human breast cancerous cells. This potential activity is promising for limiting breast cancer metastasis. According to our obtained data, the crude ginger extract, 6-gingerol, and 6-gingerol formulated into nanocarriers have shown that they can accelerate the wound-healing process. This activity may be due to the strong relationship between reductions in the biosynthesis of pro-inflammatory markers and enhancements in the wound-healing process [33]. Acute and chronic wounds occur after any disruption in skin integrity, either in the epidermis or dermis layers. While the skin serves as the body's mechanical barrier to the outer environment, maintaining its integrity should protect it from infections, inflammation, and electrolyte imbalance [34]. Usually, acute wounds heal faster than chronic wounds, and we observed a delay of up to three months when comparing chronic and acute wounds. Treating chronic wounds is one of the most challenging issues facing the regenerative medicine field, and there have been extensive scientific efforts to develop therapeutic strategies in a fast, qualified, and cost-effective therapy modality [34].

A technology to utilize nanocarrier particles to deliver therapeutic agents would increase the effectiveness of treatment and revolutionize medicinal therapy. Wound-healing processes benefit from incorporating nanotechnology through minimization of the time required to achieve complete healing and improving the response rate to medicinal therapy [35]. Nanomaterials, nanoscaffolds, and nanofibers, in addition to biomaterials, are used for topical drug delivery in wound-healing processes [35]. The use of nanomaterials for biomedical and pharmaceutical applications has gained considerable attention, especially in recent years, and several nanomaterials are used in biomedical applications for wound healing, drug delivery, etc. [4,6]. We adopted a thin-film hydration method to load 6-gingerol particles because it is a simple, globally accepted, and well-optimized technique [36]. Encapsulating medicinal agents from their natural origin into nanoliposomes is a promising delivery system by which to improve overall therapeutic efficacy. For example, capsaicin loaded into nanoliposome carriers has shown enhanced stability, selectivity, and improvement in activity against human cancerous cells. The same system considerably decreased the bioproduction of IL-6 and IL-8 compared with the control [13]. Our compound of interest, 6-gingerol, has been studied in animal models for its healing activity. In these studies, 6-gingerol was incorporated in an ointment dosage form at a concentration of 10% of the total formula. Then, it was applied to wounded rats (*Rattus norvegicus*). The incision wounds showed an accelerated healing process compared with those untreated or treated with Oxyfresh Soothing Pet Gel, which was used as a positive control. The average recovery took 16 days [37]. According to Rahayu et al., applying ginger extract to incision-wounded rats reduces neutrophil concentrations in the proliferation and maturation phases [38].

Our research has produced a novel formula with remarkable stability and selectivity. Our system of PEGylated nanophytosomes loaded with 6-gingerol showed promising biological activities and good physiochemical properties. However, this study has some limitations. Flow cytometry technique should have been used and correlated with immunohistochemistry data, and dose-response at different concentrations of 6-gingerol should have been performed to determine maximum drug load ability. In addition to the MTT assay, we should have tested the effect of the compound on a macrophage cell line.

5. Conclusions

Our novel formulation of 6-gingerol-loaded PEGylated nanophytosomes ensured considerable stability, selectivity, and safety and offered higher wound-healing activity levels than 6-gingerol alone and the crude extract. Our results are comparable with other natural compounds using encapsulated nanoliposome technology. In the future, 6-gingerol-loaded PEGylated nanophytosomes may provide a promising approach for managing wounds and inflammatory illnesses, including cancers. Future studies would look at carrying out experiments on in-vivo wound healing and optimizing the dose of 6-gingerol loaded on the PEGylated nanophytosomes. In addition, flow cytometry will be used to study in detail the induction of apoptosis and the effect of synthesized 6-gingerol phytosomes on the cell cycle.

Author Contributions: Conceptualization, A.A.-S.; research design, methodology and data curation, writing—original draft A.A.-S. and M.A.Q.; formal analysis, M.A.Q., L.K.A.-H. and W.A.; funding acquisition, and resources, F.A., A.A.-R. and H.A.-A.; writing original draft, A.A.-S. and E.A.S.A.-D.; writing, revision and editing, R.I., L.K.A.-H. and E.A.S.A.-D.; design extraction of plant, A.A.-S. All authors have read and agreed to the published version of the manuscript.

Funding: This research received no external funding.

Institutional Review Board Statement: Not applicable.

Informed Consent Statement: Not applicable.

Data Availability Statement: Study data are available from the authors upon request.

Acknowledgments: We would like to thank the Pharmacological and Diagnostic Research Centre, Al-Ahliyya Amman University, and the Cell Therapy Centre (Jordan University), for allowing us to use their facilities. All individuals have consented to the acknowledgement.

Conflicts of Interest: The authors declare no conflict of interest.

Abbreviations

A549	Lung cancer cells
CAP	Capsaicin
DDs	Drug delivery systems
DLS	Dynamic light scattering
DMSO	Dimethyl sulfoxide
DSPE-PEG2000	1,2-distearoyl-*sn*-glycero-3-phosphoethanolamine-NPEG2000
DOPC	1,2-dioleoyl-sn-glycero-3-phosphocholine
EE	Encapsulation efficiency
EGFR	Epidermal growth factor receptor
EMEM	Eagle's minimum essential medium
FBS	Fetal bovine serum
FDA	Food and Drug Administration
HPLC	High-performance liquid chromatography
IC_{50}	Half-maximal inhibitory concentration
(COX)-2	Cyclooxygenase
IL-1	Active human IL-1 beta protein
IL-6	Interleukin 6
IL-8	Interleukin 8
IL-10	Interleukin 10
IRAK1	Interleukin-1 receptor-associated kinase 1
MEM	Minimum essential medium
MTT	3-(4,5-dimethyl-2-thiazolyl)-2,5-diphenyltetrazolium bromide solution
NP	Nanoparticle
NO	Nitric oxide
Panc1	Pancreatic cancer cells
PBS	Phosphate-buffered saline
PANC1	Pancreas cancer
PDI	Polydispersity index
RPMI	RPMI 1640 growth medium
RNA	Ribonucleic acid
TEM	Transmission electron microscopy
TNBC	Triple-negative breast cancer
TNFα	Tumor necrosis factor-alpha
Z	Zeta

References

1. Percival, N.J. Classification of Wounds and their Management. *Surgery* **2002**, *20*, 114–117. [CrossRef]
2. Friedstat, J.; Brown, D.A.; Levi, B. Chemical, Electrical, and Radiation Injuries. *Clin. Plast. Surg.* **2017**, *44*, 657–669. [CrossRef] [PubMed]
3. Fentahun, N.; Anteneh, Y.; Menber, Y. Malnutrition in the Outcome of Wound Healing at Public Hospitals in Bahir Dar City, Northwest Ethiopia: A Prospective Cohort Study. *J. Nutr. Metab.* **2021**, *2021*, 8824951. [CrossRef] [PubMed]
4. Herman, T.F.; Bordoni, B. Wound Classification. In *StatPearl*; StatPearl Publishing: Orlando, FL, USA, 2022.
5. Guo, S.; DiPietro, L.A. Factors Affecting Wound Healing. *J. Dent. Res.* **2010**, *89*, 219–229. [CrossRef] [PubMed]
6. Kurek-Górecka, A.; Komosinska-Vassev, K.; Rzepecka-Stojko, A.; Olczyk, P. Bee venom in wound healing. *Molecules* **2020**, *26*, 148. [CrossRef]

7. Baliga, M.S.; Latheef, L.; Haniadka, R.; Fazal, F.; Chacko, J.; Arora, R. Ginger (*Zingiber officinale* Roscoe) in the Treatment and Prevention of Arthritis. In *Bioactive Food as Dietary Interventions for Arthritis and Related Inflammatory Diseases*; Elsevier Inc.: Amsterdam, The Netherlands, 2013; pp. 529–544. [CrossRef]
8. Tripathi, S.; Maier, K.G.; Bruch, D.; Kittur, D.S. Effect of 6-Gingerol on Pro-Inflammatory Cytokine Production and Costim-ulatory Molecule Expression in Murine Peritoneal Macrophages. *J. Surg. Res.* **2007**, *138*, 209–213. [CrossRef]
9. Al-Rawaf, H.A.; Gabr, S.A.; Alghadir, A.H. Molecular Changes in Diabetic Wound Healing following Administration of Vitamin D and Ginger Supplements: Biochemical and Molecular Experimental Study. *Evid.-Based Complement. Altern. Med.* **2019**, *2019*, 4352470. [CrossRef]
10. Naderi, N.; Karponis, D.; Mosahebi, A.; Seifalian, A.M. Nanoparticles in wound healing; from hope to promise, from prom-ise to routine. *Front. Biosci.-Landmark* **2018**, *23*, 1038–1059.
11. Abu Hajleh, M.N.; Abu-Huwaij, R.; AL-Samydai, A.; Al-Halaseh, L.K.; Al-Dujaili, E.A. The Revolution of Cosmeceuti-cals Delivery by Using Nanotechnology: A Narrative Review of Advantages and Side Effects. *J. Cosmet. Dermatol.* **2021**, *20*, 3818–3828. [CrossRef]
12. Abdelnabi, H.; Alshaer, W.; Azzam, H.; Alqudah, D.; Al-Samydai, A.; Aburjai, T. Loading of capsaicin-in-cyclodextrin inclu-sion complexes into PEGylated liposomes and the inhibitory effect on IL-8 production by MDA-MB-231 and A549 cancer cell lines. *Z. Naturforsch. C J. Biosci.* **2021**, *76*, 503–514. [CrossRef]
13. Al-Samydai, A.; Alshaer, W.; Al-Dujaili, E.A.S.; Azzam, H.; Aburjai, T. Preparation, Characterization, and Anticancer Effects of Capsaicin-Loaded Nanoliposomes. *Nutrients* **2021**, *13*, 3995. [CrossRef] [PubMed]
14. Wei, Q.; Yang, Q.; Wang, Q.; Sun, C.; Zhu, Y.; Niu, Y.; Yu, J.; Xu, X. Formulation, Characterization, and Pharmacokinetic Studies of 6-Gingerol-Loaded Nanostructured Lipid Carriers. *AAPS PharmSciTech* **2018**, *19*, 3661–3669. [CrossRef] [PubMed]
15. Pawar, N.; Pai, S.; Nimbalkar, M.; Dixit, G. RP-HPLC analysis of phenolic antioxidant compound 6-gingerol from different ginger cultivars. *Food Chem.* **2011**, *26*, 1330–1336. [CrossRef]
16. Roesken, F.; Uhl, E.; Curri, S.B.; Menger, M.D.; Messmer, K. Acceleration of wound healing by topical drug delivery via lip-osomes. *Langenbeck's Arch. Surg.* **2000**, *385*, 42–49. [CrossRef]
17. Altan, A.; Yuce, H.B.; Karataş, Ő.; Taşkan, M.M.; Gevrek, F.; Çolak, S.; Akbulut, N. Free and liposome form of gallic acid improves calvarial bone wound healing in Wistar rats. *Asian Pac. J. Trop. Biomed.* **2020**, *10*, 156. [CrossRef]
18. Jangde, R.; Singh, D. Preparation and optimization of quercetin-loaded liposomes for wound healing, using response surface methodology. *Artif. Cells Nanomed. Biotechnol.* **2014**, *44*, 635–641. [CrossRef]
19. Kapoor, B.; Singh, S.; Gulati, M.; Gupta, R.; Vaidya, Y. Application of Liposomes in Treatment of Rheumatoid Arthritis: Quo Vadis. *Sci. World J.* **2014**, *2014*, 978351. [CrossRef]
20. Cafino, E.; Lirazan, M.; Marfori, E. A simple HPLC method for the analysis of [6]-gingerol produced by multiple shoot culture of ginger (*Zingiber officinale*). *Int. J. Pharmacogn. Phytochem. Res.* **2016**, *8*, 38–42.
21. Ghanbarzadeh, S.; Valizadeh, H.; Zakerimilani, P. The Effects of Lyophilization on the Physico-Chemical Stability of Sirolimus Liposomes. *Adv. Pharm. Bull.* **2013**, *3*, 25–29. [CrossRef]
22. Yousef, I.; Oran, S.; Alqaraleh, M.; Bustanji, M. Evaluation of cytotoxic, antioxidant and antibacterial activities of Origanum dayi, Salvia palaestina and Bongardia chrysogonum plants growing wild in Jordan. *Trop. J. Nat. Prod. Res.* **2021**, *5*, 66–70.
23. Al-Tawarah, N.M.; Qaralleh, H.; Khlaifat, A.M.; Nofal, M.N.; Khleifat, K.M.; Al-Limoun, M.O.; Alqaraleh, M.; Al Shhab, M.A. Anticancer and Antibacterial Properties of Verthemia Iphionides Essential Oil/Silver Nanoparticles. *Biomed. Pharmacol. J.* **2020**, *13*, 1175–1184. [CrossRef]
24. Alqaraleh, M.; Kasabri, V.; AL-Majali, I.; Aljaafreh, A.; AL-Othman, N.; Khleifat, K.; AL-Tawarah, N.; Qaralleh, H.; Khwaldeh, A.; Alalawi, S. Branched chain amino Acids as in vitro and in vivo Anti-Oxidation Compounds. *Res. J. Pharm. Technol.* **2021**, *14*, 3899–3904. [CrossRef]
25. Trinh, X.-T.; Long, N.-V.; Van Anh, L.T.; Nga, P.T.; Giang, N.N.; Chien, P.N.; Nam, S.-Y.; Heo, C.-Y. A Comprehensive Review of Natural Compounds for Wound Healing: Targeting Bioactivity Perspective. *Int. J. Mol. Sci.* **2022**, *23*, 9573. [CrossRef]
26. Furlan, V.; Bren, U. Protective Effects of [6]-Gingerol against Chemical Carcinogens: Mechanistic Insights. *Int. J. Mol. Sci.* **2020**, *21*, 695. [CrossRef]
27. Vasarri, M.; Degl'innocenti, D. Antioxidant and Anti-Inflammatory Agents from the Sea: A Molecular Treasure for New Po-tential Drugs. *Mar. Drugs* **2022**, *20*, 132. [CrossRef]
28. Kaur, B.; Singh, P. Inflammation: Biochemistry, cellular targets, anti-inflammatory agents and challenges with special empha-sis on cyclooxygenase 2. *Bioorg. Chem.* **2022**, *121*, 105663. [CrossRef] [PubMed]
29. Da Cruz Nascimento, S.; Carvalho DE Queiroz, J.; Fernandes De Medeiros, A.; De França Nunes, A.; Piuvezam, G.; Lima Maciel, B.; Souza Passos, T.; Morais, A. Anti-inflammatory agents as modulators of the inflammation in adipose tissue: A systematic review. *PLoS ONE* **2022**, *17*, e0273942. [CrossRef] [PubMed]
30. Morvaridzadeh, M.; Fazelian, S.; Agah, S.; Khazdouz, M.; Rahimlou, M.; Agh, F.; Potter, E.; Heshmati, S.; Heshmati, J. Effect of ginger (*Zingiber officinale*) on inflammatory markers: A systematic review and meta-analysis of randomized controlled trials. *Cytokine* **2020**, *135*, 155224. [CrossRef]
31. Al-Radadi, N.S.; Abdullah; Faisal, S.; Alotaibi, A.; Ullah, R.; Hussain, T.; Rizwan, M.; Saira; Zaman, N.; Iqbal, M.; et al. Zingiber officinale driven bioproduction of ZnO nanoparticles and their anti-inflammatory, anti-diabetic, anti-Alzheimer, anti-oxidant, and anti-microbial applications. *Inorg. Chem. Commun.* **2022**, *140*, 109274. [CrossRef]

32. Daniels, C.; Isaacs, Z.; Finelli, R.; Leisegang, K. The Efficacy of Zingiber officinale on Dyslipidaemia, Blood Pressure, and In-flammation as Cardiovascular Risk Factors: A Systematic Review. *Clin. Nutr. ESPEN* **2022**, *51*, 72–82. [CrossRef]
33. Yeshi, K.; Turpin, G.; Jamtsho, T.; Wangchuk, P. Indigenous Uses, Phytochemical Analysis, and Anti-Inflammatory Properties of Australian Tropical Medicinal Plants. *Molecules* **2022**, *27*, 3849. [CrossRef] [PubMed]
34. Rajendran, N.K.; Kumar, S.S.D.; Houreld, N.N.; Abrahamse, H. A review on nanoparticle based treatment for wound healing. *J. Drug Deliv. Sci. Technol.* **2018**, *44*, 421–430. [CrossRef]
35. Basnet, P.; Hussain, H.; Tho, I.; Skalko-Basnet, N. Liposomal delivery system enhances anti-inflammatory properties of cur-cumin. *J. Pharm. Sci.* **2012**, *101*, 598–609. [CrossRef]
36. Lombardo, D.; Kiselev, M.A. Methods of Liposomes Preparation: Formation and Control Factors of Versatile Nanocarriers for Biomedical and Nanomedicine Application. *Pharmaceutics* **2022**, *14*, 543. [CrossRef]
37. Jamaluddin, A.; Lestari, A. Effectiveness of Ginger Ointment (*Zingiber officinale* roscoe) on Incision Wound Healing in White Rats (*Rattus norvegicus*). *J. Indones. Vet. Res.* **2021**, *5*, 20–26.
38. Rahayu, K.I.N.; Suharto, I.P.S.; Etika, A.N.; Nurseskasatmata, S.E. The Effect of Ginger Extract (*Zingiber officinale* roscoe) on the Number of Neutrophil Cells, Fibroblast and Epithelialization on Incision Wound. *J. Phys. Conf. Ser.* **2020**, *1569*, 032063. [CrossRef]

nutrients MDPI

Article

Empire Apple (*Malus domestica*) Juice, Pomace, and Pulp Modulate Intestinal Functionality, Morphology, and Bacterial Populations In Vivo (*Gallus gallus*)

Cydney Jackson, Viral Shukla, Nikolai Kolba, Nikita Agarwal, Olga I. Padilla-Zakour and Elad Tako *

Department of Food Science, Cornell University, Ithaca, NY 14850, USA
* Correspondence: et79@cornell.edu; Tel.: +1-607-255-0884

Abstract: Approximately $20 billion of apple sales are generated annually in the United States. With an estimated 5 million tons produced yearly in the U.S. within the last decade, apple consumption is considered ubiquitous. Apples are comprised of bioactive constituents such as phytochemicals and prebiotics that may potentiate intestinal health and the gut microbiome. This study aimed to evaluate the effects of Empire apple juice, pomace, and pulp soluble extracts on intestinal functionality, morphology, and the microbiome in vivo (*Gallus gallus*). There were five treatment groups: non-injected (NI); 18 MΩ H_2O (H_2O); 6% apple juice (AJ); 6% apple pomace (APo); 6% apple pulp (APu). The eggs were treated by intra-amniotic administration of the samples on day 17 of incubation. After hatching, the blood, tissue, and cecum samples were collected for further analyses—including duodenal histomorphology, hepatic and duodenal mRNA expression, and cecal bacterial populations. Crypt depth was significantly ($p < 0.5$) shortest in AJ when compared to APo and APu. APo and APu soluble extracts significantly improved villi surface area compared to NI and H_2O control groups. The highest count of Paneth cells per crypt was observed in APo as compared to all groups. In addition, the expression of brush border membrane micronutrient metabolism and functional proteins varied between treatments. Lastly, *Lactobacillus* cecal microbial populations increased significantly in the AJ group, while AJ, APu, and APu increased the abundance of *Clostridium* ($p < 0.5$). Ultimately, these results indicate the potential of Empire apple pomace to improve host intestinal health and the gut microbiome.

Keywords: apples; intra amniotic administration; gene expression; microbiome; intestinal morphology; *Gallus gallus*

Citation: Jackson, C.; Shukla, V.; Kolba, N.; Agarwal, N.; Padilla-Zakour, O.I.; Tako, E. Empire Apple (*Malus domestica*) Juice, Pomace, and Pulp Modulate Intestinal Functionality, Morphology, and Bacterial Populations In Vivo (*Gallus gallus*). *Nutrients* **2022**, *14*, 4955. https://doi.org/10.3390/nu14234955

Academic Editor: Maria Traka

Received: 24 October 2022
Accepted: 18 November 2022
Published: 22 November 2022

1. Introduction

Apples (*Malus domestica*) are a well-established, domesticated fruit worldwide, with approximately 86 million metric tons produced annually [1]. On a fresh basis, apples are among the top fruit varieties grown in the United States—second only to grapes [2]. Fresh apple consumption is most common among consumers, however, 35% of the apples consumed are processed [3]. While processed apple products generally include jams, jellies, cider, vinegar, and dried products, most apples are processed for apple juice. There is a 75% juice extraction efficiency in the apple juice industry; thus, 25–30% of the fruit waste remains. Known as the pomace, this leftover fraction is a heterogeneous mixture of skin, flesh, seeds, stems, core, and calyx [4,5]. According to the U.S. Apple Association, approximately 33.4 million bushels (701.4 million kg) of apples were produced in 2021–2022 for juice and cider production [6]. Based on the juice extraction efficiency (75%), an estimated 175,350 metric tons of apple pomace were produced annually [4]. Apple waste is typically disposed of in landfills, leading to several damaging environmental effects, as disruption to the carbon:nitrogen ratio of soil can occur due to sugar content, organic acids, and microbial fermentation of apple pomace [7]. The high-water content of apple pomace is also an issue as it can cause water pollution. Given the complications of pomace disposal, various industries have taken advantage of the by-product as it is a rich source of nutrients such as carbohydrates, micronutrients, and phytochemicals [8]. Further utilization includes pectin extraction, production of enzymes and aroma compounds, cultivation of microbial strains and edible mushrooms, and incorporation into animal feed [5]. However, the burden remains to prevent the disposal of apple waste into landfills and, ultimately, avoid environmental pollution.

Apples contain health-promoting bioactive constituents. Quercetin derivatives (galactoside, glucoside, rhamnoside), catechin, gallic acid, phloretin, and chlorogenic acid are polyphenolic compounds that are reportedly found in apples [8]. Studies have reported the antioxidative [9–13], antiproliferative [11,14–17], and anti-inflammatory [18–20] potential of apple polyphenols. It is important to note that apple phenolic compounds differ in concentration throughout the fruit matrix. For example, while apple flesh and apple peels contain chlorogenic acid, the flesh contains higher concentrations [21,22]. Research has also shown apple peels to contain greater amounts of antioxidants and, therefore, greater antioxidant potential compared to apple flesh. In addition, interest in apple seeds as a source of polyphenolic compounds has been investigated and found to be a rich source of quercetin derivatives, phenolic acids, catechin, and phloridzin [23]. Another constituent with health-promoting properties within the apple is dietary fiber, specifically, the non-digestible soluble polysaccharide pectin. Historically utilized as a commercial thickening and gelling agent, pectin holds potential functional properties which may improve intestinal health. The resistance to gastric digestion enables pectin to reach the host gut and undergo fermentation by microbiota, and ultimately produces short-chain fatty acid (SCFA) metabolites [24–27]. Previous studies have shown SCFAs to be beneficial to the gut by promoting enteric epithelial cell proliferation, enhancing gut barrier function, enhancing micronutrient absorption, and favoring the growth of beneficial bacteria over potentially pathogenic bacteria [28–32].

This study aimed to evaluate the in vivo effects of apple juice, pomace, and pulp soluble extracts on intestinal morphology, functionality, and the microbiome using *Gallus gallus* as our model. The broiler chicken is an established model employed to evaluate the effects of plant-origin bioactives on intestinal health and the microbiome [33]. The broiler chicken model exhibits genetic homology, a complex gut microbiota, and notable microbial similarity at the phylum level to human gut microbiota [34].

2. Materials and Methods

2.1. Apple Preparation

Empire apples (*Malus domestica*) were harvested (>2 tons) during the fall of 2021 from multiple trees from the Cornell AgriTech Orchards and processed at the Cornell Food Venture Center Pilot Plant (Geneva, NY, USA). Before processing, Empire apples were destemmed and washed. Apple pulp was made by removing the core and seeds and dicing. The apple pieces were then freeze-dried (Max53, Millrock Technology, Kingston, NY, USA) for 24 h. The dried apple was then ground into a fine powder using a bench-scale processor (Robo Coue; Jackson, MI, USA) at 1500 RPM. Apple juice was made to typical industry standards. For apple juice, whole apples were ground in a hammer mill in a blade configuration. The pulp was then pressed using a pilot-scale hydraulic press (Orchard Equipment Co., Conway, MA, USA) at 1200–1400 PSI. The juice was commercially sterilized by hot-packing into PET bottles at 85 °C and keeping it hot for 2 min. Apple pomace was the resulting pomace from the juice pressing. After juice pressing, the pomace was freeze-dried for 24 h. The dried pomace was ground into a fine powder using a bench-scale grinder. The powders were vacuumed-sealed, and all samples were kept frozen until use.

2.2. Apple Analysis Sample Preparation

Apple samples were extracted under dark conditions utilizing absolute methanol and constant agitation for 2 h. The resulting slurry was centrifuged and decanted to acquire the supernatant. The subsequent isolate and washings were diluted to attain an extract (15% w/v) which was ultimately utilized for the further analysis below.

2.2.1. Polyphenol Analysis

The Folin-Ciocalteu method previously detailed by Waterhouse was utilized to quantify total polyphenol content (TPC) [35]. Essentially, the Folin-Ciocalteu reagent and the extract were allowed to incubate at room temperature. The sodium carbonate solution was used to quench the reaction and sample absorbance was measured immediately using a UV-visible spectrophotometer (Thermo Fisher; Waltham, MA, USA) at 765 nm. Therefore, TPC was calculated as gallic equivalents (GE) using a standard curve prepared under the same conditions.

2.2.2. Fibrous and Non-Fibrous Carbohydrate Analysis

According to AOAC 962.09, the non-fibrous carbohydrate analysis (NFC) was completed. Acid detergent fiber (ADF) and neutral detergent fiber (NDF) analyses were conducted according to AOAC 973.18. The analysis was performed by Dairy One Co-Op Inc. (Ithaca, NY, USA).

2.3. Extraction of Soluble Apple Contents

Apple powders and juice samples were dissolved and diluted in distilled water to create 6% concentrations. All apple samples were heated via water bath for 1 h at 60 °C, centrifuged (3500 RPM) for 10 min at room temperature, and the supernatant was collected only in the case of the pomace and pulp.

2.4. Animals and Design

Fertile Cornish-cross broiler chicken eggs (n = 55) were provided by a hatchery (Moyer's chicks, Quakertown, PA, USA). All animal protocols were approved by Cornell University Institutional Animal Care and Use Committee (ethic approval code: 2020-0077). Apple extract powders and juice were diluted with 18 MΩ H_2O to acquire the necessary concentration to maintain an osmolarity value of less than 320 Osm. On day 17 of embryonic incubation, viable eggs were weighed and randomly allocated into five groups (n = 12) with a similar weight frequency distribution. After identifying the amniotic fluid by candling, the treatment solution (1 mL) was injected with a 21-gauge needle. Subsequent to injection, the injection site was sealed with cellophane tape. Eggs were placed in hatching baskets for each treatment and equal representation at each incubator location. The five treatment groups consisted as follows: non-injected (NI); 18 MΩ H_2O (H_2O); 6% apple juice (AJ); 6% apple pomace (APo); and 6% apple pulp (APu). On day 21, exposure to CO_2 was used to euthanize hatchlings, and the blood, pectoral muscle, liver, duodenum, and cecum were collected for further analysis.

2.5. Blood Analysis

Blood was collected from the heart using micro-hematocrit heparinized capillary tubes (Fisher Scientific Waltham, MA, USA). Blood glucose concentrations were determined using the Accu-Chek® blood glucose monitor following the manufacturer's protocol.

2.6. Pectoral Glycogen

The pectoral muscle was collected on the day of the hatch. Glycogen analysis was completed as previously described [36–38]. Briefly, 20 mg of the sample was homogenized in 8% perchloric acid and centrifuged at 12,000 rpm (4 °C) for 15 min. The supernatant was removed, and 1.0 mL of petroleum ether was added. Following mixing, the petroleum ether fraction was discarded, and the remaining sample layer was transferred to a new container with the color reagent (300 μL). Samples were read in an ELISA reader at 450 nm, and glycogen content was analyzed based on the standard curve. Total glycogen content in the pectoral sample was identified as the product of multiplying tissue weight by the amount of glycogen per 1 g of wet tissue.

2.7. Total RNA Extraction from Duodenum and Liver Tissue Samples

Total RNA was extracted from 30 mg of duodenal and liver tissue samples (n = 5) as previously described [39–43]. Briefly, the Qiagen Rneasy Mini Kit (Rneasy Mini Kit, Qiagen Inc., Valencia, CA, USA) was used. All protocols were carried out according to the manufacturer and under Rnase-free conditions. RNA was quantified by absorbance at A 260/280. The integrity of 18S ribosomal rRNA was verified by 1.5% agarose gel electrophoresis, followed by ethidium bromide staining. RNA was stored at −80 °C until further use.

2.8. Real-Time Polymerase Chain Reaction (RT-PCR)

cDNA was created from the extracted RNA by a 20 μL reverse transcriptase (RT) reaction. To complete the reaction, the BioRad C1000 touch thermocycler using the Improm-II Reverse Transcriptase Kit (Catalog #A1250; Promega, Madison, WI, USA) was utilized. The cDNA concentration was measured by Nanodrop (Thermo Fisher Scientific, Waltham, MA, USA) at an absorbance of 260 nm and 280 nm using an extinction coefficient of 33 (for single-stranded DNA). Genomic DNA contamination was assessed by a real-time RT-PCR assay for the reference gene samples.

The RT-PCR primers were designed based on relevant gene sequences from the GenBank database using the Real-Time Primer Design Tool software (IDT DNA, Coralvilla, IA, USA), as detailed previously [44]. Table 1 indicates the primer sequences used in accordance with iron, zinc, and vitamin A metabolism, immune response, and brush border membrane functionality. The reference gene used was the *Gallus gallus* primer 18S rRNA. BLAST searches against the genomic National Center for Biotechnology Information (NCBI) database were applied to verify primer specificity.

Table 1. DNA primer sequences used in this study.

Analyte	Forward Primer (5′-3′)	Reverse Primer (5′-3′)	Base Pair	GI Identifier
		Iron Metabolism		
DcytB	CATGTGCATTCTCTTCCAAAGTC	CTCCTTGGTGACCGCATTAT	103	20380692
DMT1	TTGATTCAGAGCCTCCCATTAG	GCGAGGAGTAGGCTTGTATTT	101	206597489
Ferroportin	CTCAGCAATCACTGGCATCA	ACTGGGCAACTCCAGAAATAAG	98	423984
Hepcidin	AGACGACAATGCAGACTAACC	CTGCAGCAATCCCACATTTC	132	SAMN08056490
		Zinc Metabolism		
ZnT1	GGTAACAGAGCTGCCTTAACT	GGTAACAGAGCTGCCTTAACT	105	54109718
ZnT7	GGAAGATGTCAGGATGGTTCA	CGAAGGACAAATTGAGGCAAAG	87	56555152
ZIP4	TCTCCTTAGCAGACAATTGAG	GTGACAAACAAGTAGGCGAAAC	95	107050877
ZIP1	TGCCTCAGTTTCCCTCAC	GGCTCTTAAGGGCACTTCT	144	121112053
		Vitamin A Metabolism		
CRBP2	GGCTACATGGTTGCACTAGACA	AACCACCCGGTTATCGAGTC	195	NM_001277417.1
LRAT	GATTTTGCCTATGGCGGCAG	TTGTCGGTCTGGAAGCTGAC	197	XM_420371.7
RBP4	TGCCACCAACACAGAACTCTC	CTTTGAAGCTGCTCACACGG	149	NM_205238.2
STRA6	GTGCGCTGAACTTTGTCTGC	TTCTTCCTGCTCCCGACCT	116	NM_001293202.2

Table 1. *Cont.*

Analyte	Forward Primer (5′-3′)	Reverse Primer (5′-3′)	Base Pair	GI Identifier
		Inflammatory Response		
NF-κB	CACAGCTGGAGGGAAGTAAAT	TTGAGTAAGGAAGTGAGGTTGAG	100	2130627
TNF-α	GACAGCCTATGCCAACAAGTA	TTACAGGAAGGGCAACTCATC	109	53854909
IL6	ACCTCATCCTCCGAGACTTTA	GCACTGAAACTCCTGGTCTT	105	302315692
		Brush Border Membrane Functionality		
VDAC2	CAGCACTCGCTTTGGAATTG	GTGTAACCCACTCCAACTAGAC	99	395498
OCLN	GTCTGTGGGTTCCTCATCGT	GTTCTTCACCCACTCCTCCA	124	396026
SI	CCAGCAATGCCAGCATATTG	CGGTTTCTCCTTACCACTTCTT	95	2246388
MUC6	CCAACTTGCAGTGTTCCAAAG	CTGACAGTGTAGAGCAAGTACAG	106	XM_015286750.1
18S rRNA	GCAAGACGAACTAAAGCGAAAG	TCGGAACTACGACGGTATCT	100	7262899

DcytB, Duodenal cytochrome B; DMT1, Divalent metal transport 1; ZnT1, Zinc transporter 1; ZnT7, Zinc transporter 7; CRBP2, Cellular retinol-binding protein 2; LRAT, Lecithin; Retinol Acyltransferase; RBP4, Retinol binding protein 4; STRA6, Stimulated by Retinoic aid 6; NF-kB, Nuclear factor kappa beta; TNF-α, Tumor necrosis factor-alpha; IL6, Interleukin 6; VDAC2, Voltage-dependent anion channel 2; OCLN, Occludin; SI, Sucrase isomaltase; MUC6, Mucin 6.

2.9. Microbial Samples and Intestinal Contents DNA Isolation

Cecum samples were weighed and placed in sterile tubes containing PBS. Subsequently, the samples were vortexed with sterile glass beads for 3 min. All protocols were completed as previously described [42,45–49].

2.10. Primer Design and PCR Amplification of Bacterial 16S rRNA

Bifidobacterium, *Lactobacillus*, *Escherichia coli*, *Clostridium*, *Klebsiella*, and *L. plantarum* primers were used. 16S rRNA was the universal primer and internal standard. Therefore, the proportions of each bacterial group are presented. PCR products were applied to 1.5% agarose gel with ethidium bromide stain and quantified with Gel-Pro analyzer version 3.0 (Media Cybernetics LP, Rockville, MD, USA).

2.11. Histomorphological Examination

On the day of the hatch, proximal duodenal samples were collected. Subsequently, the samples were soaked in 4% (*v*/*v*) buffered formaldehyde, dehydrated, cleared, and embedded in paraffin. Several sections were cut with a 5 μm thickness and placed on glass slides. Intestinal sections were then deparaffinized in xylene and rehydrated in a series of graded alcohol. Ultimately, the slides were stained with Alcian Blue/Periodic acid-Schiff and investigated by light microscopy using EPIX XCAP software (Standard version, Olympus, Waltham, MA, USA). The following features were measured in the duodenum: villus surface area, crypt depth, villus and crypt goblet diameter, crypt goblet cell number and type, and Paneth cell number and diameter within the crypt, as previously described [42,43,49–51]. Per treatment group, five biological samples (n = 5) (four segments each) were analyzed. Ten randomly selected villi and crypts were measured and analyzed, and cell measurements and counts were completed in ten randomly selected villi or crypts per segment. The following equation was utilized to calculate the villus surface area:

$$Villus\ surface\ area = 2\pi \times \frac{VW}{2} \times VL \quad (1)$$

in which VW is the mean of three villus width measurements, and VL is the villus length.

2.12. Statistical Analysis

In this paper, values are portrayed as the mean values ± standard error means. Experimental treatments and controls for the intra-amniotic administration were assigned with approximately equal weight distribution. Tested parameters were analyzed for normal distribution and equal variance through a Shapiro-Wilk test. If the test was accepted, a one-way analysis for variance (ANOVA) was utilized. ANOVA was used to analyze the results, followed by a Duncan post-hoc test to determine significance based on *p*-values ($p < 0.05$). For all statistical evaluations, software SPSS version 20.0 was utilized.

3. Results

3.1. Apple Polyphenol and Fiber Content

Empire apple pulp had the highest ($p < 0.05$) TPC content, followed by apple pomace (APo) and apple juice (AJ) (Table 2). Apple pomace had the greatest acid detergent fiber (ADF) and neutral detergent fiber (NDF) compared to apple pulp and apple juice. Lastly, non-fiber carbohydrate content was greatest in the apple pulp.

Table 2. Total polyphenolic (mg/g GAE) and fiber content of apple products.

Sample	TPC (mg/g GAE)	ADF (%/DM)	NDF (%/DM)	NFC (%/DM)
Pomace	0.834 ± 0.059 [b]	22.6	25.6	62
Juice	0.300 ± 0.029 [c]	NA	NA	NA
Pulp	1.57 ± 0.074 [a]	6.7	7.9	85.4

Values are the means ± SEM, [a–c] Treatment groups not indicated by the same letter in the same column are significantly different ($p < 0.05$) by Duncan's post-hoc test. ADF: cellulose, lignin, and insoluble minerals; NDF: cellulose, lignin, insoluble minerals, and hemicellulose; NFC: sugars, starches, organic acids, and pectin.; GAE: gallic acid equivalents; DM: dry matter; NA: not applicable.

3.2. Body Weight, Blood Glucose, and Glycogen Content

No significant differences were observed between the treatment and control groups for body weight, blood glucose, and glycogen content ($p < 0.05$, Table 3).

Table 3. Effect of the intra-amniotic administration of apple fraction soluble extracts on body weight (g), blood glucose (mg/dL), and glycogen (mg/g).

Treatment Group	Body Weight (g)	Blood Glucose (mg/dL)	Glycogen (mg/g)
NI	40.83 ± 1.24 [a]	254.11 ± 23.83 [a]	0.396 ± 0.101 [a]
H_2O	38.29 ± 4.31 [a]	234.4 ± 11.16 [a]	0.294 ± 0.093 [a]
AJ	40 ± 0.99 [a]	225.5 ± 11.43 [a]	0.431 ± 0.092 [a]
APo	35.7 ± 0.67 [a]	230.56 ± 21.28 [a]	0.271 ± 0.054 [a]
APu	36.44 ± 0.85 [a]	205.5 ± 32.05 [a]	0.442 ± 0.077 [a]

Values are the means ± SEM, $n = 5$. [a] Treatment groups not indicated by the same letter in the same column are significantly different ($p < 0.05$) by Duncan's post-hoc test.

3.3. Duodenal and Hepatic Gene Expression of Related Proteins

Figures 1–3 depict the gene expression of proteins relevant to micronutrients (iron, zinc, and vitamin A) metabolism, immune response, and functionality.

3.3.1. Iron and Zinc-Related Proteins

DcytB expression in treatment groups AJ and Apo were downregulated ($p < 0.05$) compared to the NI and H_2O controls (Figure 1). AJ reduced ($p < 0.05$) gene expression of DMT1 relative to the NI control, but no difference was observed relative to the H_2O control. The apple groups did not alter ($p < 0.05$) the gene expression of ferroportin compared to the NI and H_2O groups. Hepcidin, an iron-related protein located in the liver, was upregulated by the AJ treatment relative to the NI control. APo and APu were not significantly altered compared to NI and H_2O controls.

A)

	DcytB	DMT1	Ferroportin	Hepcidin
NI	a 5.01 ± 0.091	a 1.021 ± 0.002	a 7.831 ± 0.074	b 0.999 ± 0.008
H_2O	a 5.067 ± 0.157	ab 1.016 ± 0.002	a 7.622 ± 0.04	a 1.024 ± 0.007
AJ	b 4.705 ± 0.05	b 1.011 ± 0.001	a 7.729 ± 0.111	a 1.024 ± 0.003
APo	b 4.633 ± 0.064	a 1.019 ± 0.002	a 7.795 ± 0.194	ab 1.00 8 ± 0.005
APu	ab 4.905 ± 0.055	ab 1.016 ± 0.001	a 8.032 ± 0.24	ab 1.017 ± 0.006

B)

	ZnT1	ZnT7	ZIP4	ZIP1
NI	a 7.799 ± 0.11	a 1.391 ± 0.005	a 0.898 ± 0.002	a 0.212 ± 0.004
H_2O	a 7.756 ± 0.176	a 1.381 ± 0.004	a 0.897 ± 0. 004	a 0.211 ± 0.006
AJ	a 7.851 ± 0.527	a 1.386 ± 0.005	a 0.89 ± 0.001	a 0.214 ± 0.003
APo	a 8.449 ± 0.526	a 1.398 ± 0.01	a 0.899 ± 0.004	a 0.183 ± 0.04
APu	a 7.88 ± 0.228	a 1.395 ± 0.006	a 0.896 ± 0.001	a 0.21 ± 0.006

Low AU — High AU

Figure 1. Effect of intra-amniotic administration of apple fraction soluble extracts on iron (**A**) and zinc (**B**) intestinal and liver (Hepcidin) gene expression. Values are the means ± SEM, $n = 5$. ^{a,b} Per gene, treatment groups not indicated by the same letter in the same column are significantly different ($p < 0.05$) by Duncan's post-hoc test. DcytB, Duodenal cytochrome B; DMT1, Divalent metal transporter 1; Zinc transporters: ZnT1, ZnT7, ZIP4, ZIP1; AU, Arbitrary units.

	CRBP2	LRAT	RBP4	STRA6
NI	a 8.237 ± 0.477	a 14.164 ± 0.352	b 1.081 ± 0.008	a 288.809 ± 52.715
H_2O	a 8.877 ± 0.647	a 13.83 ± 0.334	ab 1.099 ± 0.006	ab 253.056 ± 29.29
AJ	a 8.863 ± 0.446	a 14.112 ± 0.362	a 1.105 ± 0.002	b 190.933 ± 10.375
APo	a 9.576 ± 0.458	a 14.756 ± 1.39	ab 1.088 ± 0.005	ab 210.5 ± 14.971
APu	a 8.876 ± 0.557	a 13.949 ± 0.553	ab 1.095 ± 0.005	ab 231.638 ± 9.739

Low AU — High AU

Figure 2. Effect of intra-amniotic administration of apple fraction soluble extracts on vitamin A intestinal (CRBP2 and LRAT) gene expression and hepatic (RBP4 and STRA6) metabolic proteins. Values are the means ± SEM, $n = 5$. ^{a,b} Treatment groups not indicated by the same letter in the same column are significantly different ($p < 0.05$) by Duncan's post-hoc test. CRBP2, Cellular retinol-binding protein 2; LRAT, Lecithin: Retinol Acyltransferase; RBP4, Retinol binding protein 4; STRA6, Stimulated by Retinoic acid 6; AU, Arbitrary units.

	NF-κB	TNF-α	IL6	VDAC2	OCLN	SI	MUC6
NI	a 11.932 ± 0.344	a 2.698 ± 0.021	a 6.26 ± 0.141	a 1.705 ± 0.016	a 1.074 ± 0.003	a 2.896 ± 0.008	ab 7.507 ± 0.099
H_2O	a 11.406 ± 0.283	a 2.661 ± 0.041	a 5.9 ± 0.299	a 1.689 ± 0.012	a 1.069 ± 0.002	ab 2.847 ± 0.022	b 7.214 ± 0.057
AJ	a 12.455 ± 0.429	a 2.707 ± 0.021	a 6.15 ± 0.067	a 1.671 ± 0.018	b 1.063 ± 0.001	b 2.823 ± 0.02	a 7.739 ± 0.14
APo	a 12.624 ± 0.334	a 2.711 ± 0.02	a 6.702 ± 0.453	a 1.693 ± 0.035	a 1.071 ± 0.002	ab 2.848 ± 0.022	ab 7.707 ± 0.121
APu	a 12.283 ± 0.494	a 2.737 ± 0.065	a 6.121 ± 0.157	a 1.7 ± 0.009	a 1.068 ± 0.001	ab 2.864 ± 0.006	a 7.998 ± 0.284

Low AU — High AU

Figure 3. Effect of intra-amniotic administration of apple fraction soluble extracts on inflammatory and functional intestinal protein gene expression. Values are the means ± SEM, $n = 5$. [a, b] Treatment groups not indicated by the same letter in the same column are significantly different ($p < 0.05$) by Duncan's post-hoc test. NF-kB, Nuclear factor kappa beta; TNF-α, Tumor necrosis factor-alpha; IL6, Interleukin 6, VDAC2, Voltage-dependent anion channel 2; OCLN, Occludin; SI, Sucrase isomaltase; MUC6, Mucin 6; AU, Arbitrary units.

The administration of apple juice, pomace, and pulp extractions did not alter ($p < 0.05$) the gene expression of zinc-related proteins at the brush border membrane relative to the NI and H_2O controls.

3.3.2. Vitamin A-Related Proteins

No significant differences ($p < 0.05$) were observed in the gene expression of CRBP2 and LRAT between the treatment and control groups (Figure 2). AJ had the greatest ($p < 0.05$) expression of RBP4 compared to the NI control, which had the lowest ($p < 0.05$) expression of RBP4, however, there were no differences ($p < 0.05$) between the H_2O control and the apple treatment groups.

3.3.3. Inflammatory and Functionality-Related Proteins

The gene expression of proteins related to inflammation—NF-κB, TNF-α, and IL6—was not significantly different ($p < 0.05$) between the control and experimental groups (Figure 3). Further, gene expression of functional proteins VDAC2, SI, and MUC6 were not significantly ($p < 0.05$) different between treatment and control groups. However, the gene expression of OCLN was significantly ($p < 0.05$) lowered in the AJ group compared to all other treatment groups.

3.4. Morphometric Analysis

Groups APo and APu have significantly ($p < 0.05$) greater villi surface area compared to both control and experimental groups (Table 4). Crypt depth is significantly shortest ($p < 0.05$) in APu, APo, and AJ, respectively, related to both controls.

The intra-amniotic administration of AJ increased ($p < 0.05$) goblet cell diameter within the villi relative to the other experimental groups and H_2O control, but not the NI control (Table 5). Goblet cell diameter within the crypts was significantly ($p < 0.05$) lower in the APo group. Further, the count of goblet cells within the intestinal crypts was highest ($p < 0.05$) in APu relative to the other treatment and control groups. The abundance of acidic goblet cells per unit area was more significant ($p < 0.05$) in the APu treatment compared to the other apple fraction treatments and NI and H_2O controls. Neutral goblet cell abundance was significantly ($p < 0.05$) lowered in all apple fraction treatment groups relative to the

H_2O control but was similar to the NI control. Mixed goblet cells were most numerous ($p < 0.05$) in APu relative to the APo, NI, and H_2O groups.

Table 4. Effect of the intra-amniotic administration of apple fraction soluble extracts on villi surface area and crypt depth.

Treatment Group	Villi Surface Area (μm^2)	Crypt Depth (μm)
NI	16,458.04 ± 771.84 ^{b}	22.1 ± 0.81 ^{a}
H_2O	16,101.54 ± 383.07 ^{b}	21.93 ± 0.72 ^{a}
AJ	17,470.91 ± 444.08 ^{b}	14.45 ± 0.54 ^{c}
APo	23,116.65 ± 509.84 ^{a}	16.85 ± 0.79 ^{b}
APu	23,520.69 ± 739.04 ^{a}	17.38 ± 0.71 ^{b}

Values are the means ± SEM, $n = 5$. $^{a-c}$ Treatment groups not indicated by the same letter in the same column are significantly different ($p < 0.05$) by Duncan's post-hoc test.

Table 5. Effect of the intra-amniotic administration of apple fraction soluble extracts on villi and crypt goblet cells.

Treatment Group	Villi Goblet Diameter (μm)	Crypt Goblet Diameter (μm)	Crypt Goblet Cell Number	Crypt Goblet Cell Number		
				Acidic	Neutral	Mixed
NI	3.48 ± 0.07 ^{a}	3.01 ± 0.05 ^{a}	7.01 ± 0.24 ^{c}	5.79 ± 0.2 ^{c}	0.02 ± 0.02 ^{b}	1.21 ± 0.13 ^{c}
H_2O	3.17 ± 0.06 ^{b}	2.89 ± 0.05 ^{a}	8.55 ± 0.32 ^{b}	6.92 ± 0.27 ^{b}	0.13 ± 0.03 ^{a}	1.51 ± 0.12 ^{bc}
AJ	3.55 ± 0.07 ^{a}	2.88 ± 0.05 ^{a}	7.51 ± 0.26 ^{c}	5.81 ± 0.22 ^{c}	0.02 ± 0.01 ^{b}	1.69 ± 0.11 ^{ab}
APo	3.17 ± 0.06 ^{b}	2.71 ± 0.06 ^{b}	8.36 ± 0.27 ^{b}	6.75 ± 0.23 ^{b}	0.01 ± 0.01 ^{b}	1.60 ± 0.10 ^{b}
APu	3.08 ± 0.08 ^{b}	2.84 ± 0.05 ^{a}	9.14 ± 0.31 ^{a}	7.35 ± 0.26 ^{a}	0.00 ± 0.00 ^{b}	1.79 ± 0.11 ^{a}

Values are the means ± SEM, $n = 5$. $^{a-c}$ Treatment groups not indicated by the same letter in the same column are significantly different ($p < 0.05$) by Duncan's post-hoc test.

With regard to the morphometric measurements of Paneth cells within the crypt, APo and APu groups had the greatest ($p < 0.05$) crypt Paneth cell count per unit area, respectively (Table 6). The H_2O group had the lowest ($p < 0.05$) observed Paneth cell count per unit area relative to all other groups, yet had the greatest Paneth cell diameter.

Table 6. Effect of the intra-amniotic administration of apple fraction soluble extracts on Paneth cells.

Treatment Group	Crypt Paneth Cell Number	Paneth Cell Diameter (μm)
NI	1.22 ± 0.03 ^{c}	1.37 ± 0.02 ^{c}
H_2O	1.04 ± 0.01 ^{d}	1.5 ± 0.02 ^{a}
AJ	1.31 ± 0.04 ^{bc}	1.45 ± 0.02 ^{ab}
APo	1.44 ± 0.04 ^{a}	1.43 ± 0.02 ^{b}
APu	1.38 ± 0.04 ^{b}	1.45 ± 0.02 ^{ab}

Values are the means ± SEM, $n = 5$. $^{a-d}$ Treatment groups not indicated by the same letter in the same column are significantly different ($p < 0.05$) by Duncan's post-hoc test.

3.5. Microbial Analysis

The AJ group reduced the relative abundance of Bifidobacterium ($p < 0.05$) compared to the control and experimental groups (Figure 4). Furthermore, the greatest increase ($p < 0.05$) of *Lactobacillus* abundance relative to NI and water-only control groups occurred in the AJ group, while APo also increased ($p < 0.05$) *Lactobacillus* abundance relative only to the water injection control. All apple treatment groups have an increased ($p < 0.05$) abundance of *Clostridium* relative to the NI control, whereas only AJ and APu abundance is significantly higher than both controls. No differences were observed between the apple treatments relative to all controls for *Klebsiella* and *L. plantarum* microbial abundance.

	Bifidobacterium	*Lactobacillus*	*E. coli*	*Clostridium*	*Klebsiella*	*L. plantarum*
NI	a 0.958 ± 0.016	c 0.824 ± 0.045	b 0.882 ± 0.033	c 0.213 ± 0.015	b 0.709 ± 0.066	a 0.780 ± 0.015
H_2O	ab 0.926 ± 0.029	bc 0.873 ± 0.048	ab 0.872 ± 0.033	bc 0.324 ± 0.097	ab 0.800 ± 0.045	ab 0.695 ± 0.013
AJ	b 0.719 ± 0.021	a 1.021 ± 0.009	ab 0.940 ± 0.011	a 0.589 ± 0.062	a 0.849 ± 0.015	b 0.583 ± 0.075
APo	a 0.788 ± 0.021	ab 0.981 ± 0.035	ab 0.888 ± 0.035	ab 0.432 ± 0.020	ab 0.797 ± 0.026	a 0.808 ± 0.025
APu	a 0.738 ± 0.048	bc 0.846 ± 0.075	a 0.955 ± 0.075	a 0.556 ± 0.040	ab 0.739 ± 0.018	a 0.736 ± 0.064

Low Density (INT/mm^2) — High Density (INT/mm^2)

Figure 4. Effect of intra-amniotic administration of apple fraction soluble extracts on genera- and species-level bacterial populations from lower intestine contents on the day of hatch. Values are the means ± SEM, $n = 5$. $^{a–c}$ Treatment groups not indicated by the same letter in the same column are significantly different ($p < 0.05$) by Duncan's post-hoc test.

4. Discussion

The Empire apple variety is a cross between McIntosh (*Malus domestica* "McIntosh") and Red Delicious (*Malus domestica* "Red Delicious") cultivars and is native to New York state [52]. According to the United States Apple Association, Empire apples are among the top-produced apples in the nation [6,53]. Here, we have investigated the effects of Empire apple juice (AJ), pomace (APo), and pulp (APu) extracts via intra-amniotic administration on micronutrient absorption, intestinal immune response, gut morphology, and cecal bacterial populations. To our knowledge, this is the first study to examine such physiological effects of the Empire apple cultivar.

Body weight, blood glucose, and glycogen content (Table 3) did not differ throughout the treatment and control groups. As apples contain select macro- and micronutrients, consumption of this fruit and weight gain prevention have been studied in previous animal trials [54]. While Cho et al. (2013) reported apple pomace and juice supplementation reduced ($p < 0.05$) body weight gain in Sprague-Dawley rats [55], and Samout et al. (2016) found apple pectin to exert anti-obesity effects in Wistar rats [56], changes in body weight did not occur in our study. We hypothesize that this may be the case as treatment groups were a single dose in a naïve system, while the aforementioned studies utilized overweight models over a prolonged period.

The intra-amniotic administration of apple juice and pomace extracts reduced the gene expression of DcytB reductase relative to the controls (Figure 1A). Located in the proximal duodenum, DcytB functions to reduce ferric dietary iron to the bioavailable form (Fe^{2+}) for uptake into the enterocyte [57]. No significant changes occurred in the remaining iron metabolism proteins (DMT1, Ferroportin, and Hepcidin). Nonetheless, the reduced expression of DcytB potentially suggests an improvement in iron absorption into the enterocyte [58,59]. Shah et al. (2003) reported that apple juice enhances iron bioavailability in American children of 3 to 6 years of age [60]. The soluble apple fractions did not enhance zinc transporter proteins and vitamin A metabolism proteins, yet these results revealed no adverse effects of a single-dose administration. We expected to observe an anti-inflammatory effect of the apple treatments due to the naturally occurring phenolics in apples (represented in Table 2), which possess antioxidative properties (e.g., chlorogenic acid, quercetin glycosides, catechin) [21–23]. However, Figure 3 reveals no effect of the apple treatments on inflammatory cytokine expression (NF-κB, TNF-α, IL6). Given the reported anti-inflammatory potential of apples [10,18–20], the lack of agreement with this study's results can be attributed to the short exposure time and concentrations administered. Conversely, our results reveal that apple pomace extract did not stimulate a negative intestinal immune response. Apple seeds are known to generate toxic cyanogenic glycosides upon grinding and have been a factor of concern when upscaling apple pomace for consumption. In a recent study, using Fisher rats, Ravn-Haren et al. investigated the effects of a different cultivar (Shampion) apple pomace with and without seeds [61]. It was reported that apple pomace, regardless of seed content, did not elevate alanine aminotransferase, a liver toxicity biomarker [61].

Intestinal barrier integrity is vital to gastrointestinal functionality and health, and tight junction proteins play a crucial role in maintaining the luminal structure [62,63]. Located in luminal epithelial cells, tight junctions, such as claudin and occludin regulate the permeability of ions, water, and macronutrients [64–66]. Expression of the tight junction protein occludin (OCLN) was significantly reduced by apple juice administration (Figure 3), which suggests an increase in epithelial permeability. Apple juice is known to naturally contain high amounts of simple sugars such as glucose and fructose [67,68]. High-sugar diets have reportedly increased intestinal barrier permeability, although the direct mechanism is unclear [69]. One proposed mechanism is through the intestinal microbiome, as diets rich in simple sugars have been linked to disrupting the balance of gut microbes, causing dysbiosis [69]. Dysbiosis can be characterized by an increase in the *Firmicutes/Bacteroidetes* ratio [69–71]. AJ increased the abundance of *Clostridium* and *Klebsiella*, as depicted in Figure 4. The *Clostridium* genus comprises commensal bacteria within the *Firmicutes* phylum that can exert pathogenic effects under dysbiosis conditions. Essentially, the overgrowth of opportunistic species within the *Clostridium* and *Klebsiella* may lead to the degradation of the intestinal barrier [72–74] or render severe infection [75]. Conversely, while the beneficial bacteria *Bifidobacterium* decreased abundance in AJ, *Lactobacillus* increased abundance. *Lactobacillus* is a lactic acid-producing bacteria within the *Firmicutes* phylum; thus, our results suggest a possible selective stimulation of *Firmicute* proliferation by AJ.

Figure 4 reveals the increased abundance of *Clostridium* within APo and APu relative to the NI control. According to our results in Figure 3 and Tables 4–6, it is possible that the *Clostridium* genera exerted a beneficial effect as an induced effect on intestinal permeability and inflammatory cytokine expression was not observed. Therefore, we hypothesize that valuable species of *Clostridia* were increased. *Clostridium* is an SCFA-producing genus that has been reported to grow in abundance through pectin fermentation [76–78]. APo and APu had the greatest amount of non-fiber carbohydrates (including pectin) (Table 2), villi surface area (Table 4), and acidic goblet cells within the crypt (Table 5). A previous study reported that apple-derived pectin-fed rats increased *Clostridium* species abundance four-fold, whilst also increasing butyrate levels [79]. Butyrate is a short-chain fatty acid produced upon carbohydrate fermentation by *Clostridium* that can lower intestinal pH [78], which could explain the increased count of acidic goblet cells for APo and APu. Dufourny et al. (2021) previously assessed the effects of apple pomace on intestinal morphology and microbiota in weaned piglets. They found the pomace to increase *Clostridia* abundance and duodenal and ileal villi length [80]. Our results agree with the significant findings of this study. This further establishes the role of apple pomace to modulate Clostridia groups in both animal models which leads to improvements in gut health and intestinal homeostasis.

Shortened crypt depth was observed in all apple treatment groups (Table 4). Shortened crypt depths are morphological evidence for improved intestinal health as it suggests a slower intestinal epithelial cell turnover rate, allowing sufficient time for enterocytes to differentiate and function at capacity [81–83]. Within intestinal crypts, goblet cells play a role in maintaining the gut epithelial layer as they secrete mucus and mucin glycoproteins that function as a protective layer along the intestinal lumen [84]. A previous study found apple polysaccharide isolated from Fuji apple pomace to stimulate the enhancement of gut epithelial integrity by goblet cell autophagy [85]. Our study observed that APu and APo had the greatest crypt goblet cell count per unit area, respectively. In addition, APu had a significantly lower crypt goblet cell diameter. This finding likely suggests lower mucus content since intestinal crypt goblet cells only secrete mucus upon stimulation [84] Our findings suggest that the Empire apple has a level of effect on goblet cells located in the crypt. However, further studies should be completed to elucidate the potential impact. Moreover, Paneth cells are in the small intestinal crypts and function in the gut immunological response—secreting antimicrobial peptides and immunomodulating proteins to maintain intestinal homeostasis [86–88]. Crypt Paneth cell count per unit area was greatest ($p < 0.05$) in APo, APu, and AJ, respectively, compared to H_2O control (Table 6). Yet, the Paneth cell diameter of APo treatment was similar ($p > 0.05$) to APu and AJ, and lower ($p > 0.05$) than the H_2O control. Based on these observations and a recent review that summarized the impacts of dietary fiber on host gastrointestinal immune response, it seems that the Empire apple may stimulate Paneth cell function and improve intestinal immune response [89].

5. Conclusions

The intra-amniotic administration of Empire apple soluble extracts from various fractions was completed in this study. The data suggests that each apple fraction can alter duodenal brush border membrane functionality, morphology, and the cecal microbial populations. More specifically, the potential health benefits of apple pomace are revealed in this study, evident by reducing iron metabolism protein gene expression (DcytB), increasing villi surface area and decreasing crypt depth, increasing Paneth cell count per intestinal crypts, and increasing potentially beneficial gut bacteria (*Clostridium* spp.). Additional long-term studies should be completed to further establish potential health benefits.

Author Contributions: Conceptualization, O.I.P.-Z. and E.T.; Methodology, N.K., N.A., V.S., O.I.P.-Z. and E.T.; Investigation, C.J., N.A., V.S. and N.K.; Data curation, C.J., N.A., V.S. and N.K.; Writing—original draft preparation, C.J. and E.T.; Writing—review and editing, V.S., N.K., O.I.P.-Z. and E.T.; Supervision, O.I.P.-Z. and E.T. All authors have read and agreed to the published version of the manuscript.

Funding: This research was partially funded by the National Institute of Food and Agriculture, U.S. Department of Agriculture Federal Capacity Funds Multistate Project (NC1023).

Institutional Review Board Statement: Animal protocol used in this study was conducted according to the guidelines of the Declaration of Helsinki and were approved by the Cornell University Institutional Animal Care and Use committee by the ethic approval code: 2020-0077.

Conflicts of Interest: The authors declare that they have no conflict of interest.

References

1. FAO Apple Production Worldwide from 2010 to 2020 (in Million Metric Tons). Available online: https://www.statista.com/statistics/961248/production-of-apples-worldwide/ (accessed on 1 June 2022).
2. US Department of Agriculture. Economic Research Service Leading Fruits in the United States in 2018, Based on Production Volume (in 1000 Tons). Available online: https://www.statista.com/statistics/631886/leading-fruits-united-states-based-on-production-quantity/ (accessed on 9 June 2022).
3. Nicklas, T.A.; O'Neil, C.E.; Fulgoni, V.L. Consumption of Various Forms of Apples Is Associated with a Better Nutrient Intake and Improved Nutrient Adequacy in Diets of Children: National Health and Nutrition Examination Survey 2003–2010. *Food Nutr. Res.* **2015**, *59*, 25948. [CrossRef]
4. Lyu, F.; Luiz, S.F.; Azeredo, D.R.P.; Cruz, A.G.; Ajlouni, S.; Ranadheera, C.S. Apple Pomace as a Functional and Healthy Ingredient in Food Products: A Review. *Processes* **2020**, *8*, 319. [CrossRef]
5. Vendruscolo, F.; Albuquerque, P.M.; Streit, F.; Esposito, E.; Ninow, J.L. Apple Pomace: A Versatile Substrate for Biotechnological Applications. *Crit. Rev. Biotechnol.* **2008**, *28*, 1–12. [CrossRef]
6. U.S. Apple Association. US Apple Industry Outlook 2021. 2021. Available online: https://usapple.org/wp-content/uploads/2021/08/USAppleIndustryOutlook2021.pdf (accessed on 22 October 2022).
7. Gołębiewska, E.; Kalinowska, M.; Yildiz, G. Sustainable Use of Apple Pomace (AP) in Different Industrial Sectors. *Materials* **2022**, *15*, 1788. [CrossRef]
8. Patocka, J.; Bhardwaj, K.; Klimova, B.; Nepovimova, E.; Wu, Q.; Landi, M.; Kuca, K.; Valis, M.; Wu, W. Malus Domestica: A Review on Nutritional Features, Chemical Composition, Traditional and Medicinal Value. *Plants* **2020**, *9*, 1408. [CrossRef]

9. Fernandes, P.A.R.; Ferreira, S.S.; Bastos, R.; Ferreira, I.; Crus, M.T.; Pinto, A.; Coelho, E.; Passos, C.P.; Coimbra, M.A.; Cardoso, S.M.; et al. Apple Pomace Extract as a Sustainable Food Ingredient. *Antioxidants* **2019**, *8*, 189. [CrossRef]
10. Zielinska, D.; Laparra-Llopis, J.M.; Zielinski, H.; Szawara-Nowak, D.; Gimenez-Bastida, J.A. Role of Apple Phytochemicals, Phloretin and Phloridzin, in Modulating Processes Related to Intestinal Inflammation. *Nutrients* **2019**, *11*, 1173. [CrossRef]
11. Wolfe, K.; Wu, X.; Liu, R.H. Antioxidant Activity of Apple Peels. *J. Agric. Food Chem.* **2003**, *51*, 609–614. [CrossRef]
12. Vieira, F.G.K.; Borges, G.D.S.C.; Copetti, C.; Di Pietro, P.F.; da Costa Nunes, E.; Fett, R. Phenolic Compounds and Antioxidant Activity of the Apple Flesh and Peel of Eleven Cultivars Grown in Brazil. *Sci. Hortic.* **2011**, *128*, 261–266. [CrossRef]
13. Wojdylo, A.; Oszmianski, J.; Laskowski, P. Polyphenolic Compounds and Antioxidant Activity of New and Old Apple Varieties. *J. Agric. Food Chem.* **2008**, *56*, 6520–6530. [CrossRef]
14. Reagan-Shaw, S.; Eggert, D.; Mukhtar, H.; Ahmad, N. Antiproliferative Effects of Apple Peel Extract against Cancer Cells. *Nutr. Cancer* **2010**, *62*, 517–524. [CrossRef]
15. Luo, J.; Zhang, P.; Li, S.; Shah, N.P. Antioxidant, Antibacterial, and Antiproliferative Activities of Free and Bound Phenolics from Peel and Flesh of Fuji Apple. *J. Food Sci.* **2016**, *81*, M1735–M1742. [CrossRef]
16. Li, C.X.; Zhao, X.H.; Zuo, W.F.; Zhang, T.L.; Zhang, Z.Y.; Chen, X.S. Phytochemical Profiles, Antioxidant, and Antiproliferative Activities of Red-Fleshed Apple as Affected by in Vitro Digestion. *J. Food Sci.* **2020**, *85*, 2952–2959. [CrossRef]
17. He, X.; Liu, R.H. Phytochemicals of Apple Peels: Isolation, Structure Elucidation, and Their Antiproliferative and Antioxidant Activities. *J. Agric. Food Chem.* **2008**, *56*, 9905–9910. [CrossRef]
18. Andre, C.M.; Greenwood, J.M.; Walker, E.G.; Rassam, M.; Sullivan, M.; Evers, D.; Perry, N.B.; Laing, W.A. Anti-Inflammatory Procyanidins and Triterpenes in 109 Apple Varieties. *J. Agric. Food Chem.* **2012**, *60*, 10546–10554. [CrossRef]
19. Jung, M.; Triebel, S.; Anke, T.; Richling, E.; Erkel, G. Influence of Apple Polyphenols on Inflammatory Gene Expression. *Mol. Nutr. Food Res.* **2009**, *53*, 1263–1280. [CrossRef]
20. Carrasco-Pozo, C.; Speisky, H.; Brunser, O.; Pastene, E.; Gotteland, M. Apple Peel Polyphenols Protect against Gastrointestinal Mucosa Alterations Induced by Indomethacin in Rats. *J. Agric. Food Chem.* **2011**, *59*, 6459–6466. [CrossRef]
21. Kschonsek, J.; Wolfram, T.; Stockl, A.; Bohm, V. Polyphenolic Compounds Analysis of Old and New Apple Cultivars and Contribution of Polyphenolic Profile to the in Vitro Antioxidant Capacity. *Antioxidants* **2018**, *7*, 20. [CrossRef]
22. Boyer, J.; Liu, R.H. Apple Phytochemicals and Their Health Benefits. *Nutr. J.* **2004**, *3*, 5. [CrossRef]
23. Fidelis, M.; de Moura, C.; Kabbas, T.; Pap, N.; Mattila, P.; Mäkinen, S.; Putnik, P.; Kovačević, D.B.; Tian, Y.; Yang, B.; et al. Fruit Seeds as Sources of Bioactive Compounds: Sustainable Production of High Value-Added Ingredients from by-Products within Circular Economy. *Molecules* **2019**, *24*, 3854. [CrossRef]
24. Fu, X.; Liu, Z.; Zhu, C.; Mou, H.; Kong, Q. Nondigestible Carbohydrates, Butyrate, and Butyrate-Producing Bacteria. *Crit. Rev. Food Sci. Nutr.* **2019**, *59*, S130–S152. [CrossRef]
25. Elshahed, M.S.; Miron, A.; Aprotosoaie, A.C.; Farag, M.A. Pectin in Diet: Interactions with the Human Microbiome, Role in Gut Homeostasis, and Nutrient-Drug Interactions. *Carbohydr. Polym.* **2021**, *255*, 117388. [CrossRef]
26. Koh, A.; de Vadder, F.; Kovatcheva-Datchary, P.; Bäckhed, F. From Dietary Fiber to Host Physiology: Short-Chain Fatty Acids as Key Bacterial Metabolites. *Cell* **2016**, *165*, 1332–1345. [CrossRef]
27. Macfarlane, G.T.; Macfarlane, S. Bacteria, Colonic Fermentation, and Gastrointestinal Health. *J. AOAC Int.* **2012**, *95*, 50–60. [CrossRef]
28. Campbell, J.M.; Fahey, G.C., Jr.; Wolf, B.W. Selected Indigestible Oligosaccharides Affect Large Bowel Mass, Cecal and Fecal Short-Chain Fatty Acids, PH, and Microflora in Rats. *J. Nutr.* **1997**, *127*, 130–136. [CrossRef]
29. Venegas, D.P.; de La Fuente, M.K.; Landskron, G.; González, M.J.; Quera, R.; Dijkstra, G.; Harmsen, H.J.M.; Faber, K.N.; Hermoso, M.A. Short Chain Fatty Acids (SCFAs)Mediated Gut Epithelial and Immune Regulation and Its Relevance for Inflammatory Bowel Diseases. *Front. Immunol.* **2019**, *10*, 277. [CrossRef]
30. Markowiak-Kopeć, P.; Śliżewska, K. The Effect of Probiotics on the Production of Short-Chain Fatty Acids by Human Intestinal Microbiome. *Nutrients* **2020**, *12*, 1107. [CrossRef]
31. Alexander, C.; Swanson, K.S.; Fahey, G.C., Jr.; Garleb, K.A. Perspective: Physiologic Importance of Short-Chain Fatty Acids from Nondigestible Carbohydrtae Fermentation. *Adv. Nutr.* **2019**, *10*, 576–589. [CrossRef]
32. Pearce, S.C.; Weber, G.J.; van Sambeek, D.M.; Soares, J.W.; Racicot, K.; Breault, D.T. Intestinal Enteroids Recapitulate the Effects of Short-Chain Fatty Acids on the Intestinal Epithelium. *PLoS ONE* **2020**, *15*, e0230231. [CrossRef]
33. Hou, T.; Tako, E. The in Ovo Feeding Administration (Gallus Gallus)—An Emerging in Vivo Approach to Assess Bioactive Compounds with Potential Nutritional Benefits. *Nutrients* **2018**, *10*, 418. [CrossRef]
34. Yegani, M.; Korver, D.R. Factors Affecting Intestinal Health in Poultry. *Poult. Sci.* **2008**, *87*, 2052–2063. [CrossRef]
35. Waterhouse, A.L. Determination of Total Phenolics. *Curr. Protoc. Food Anal. Chem.* **2002**, *6*, 11.1.1–11.1.8. [CrossRef]
36. Kornasio, R.; Halevy, O.; Kedar, O.; Uni, Z. Effect of in Ovo Feeding and Its Interaction with Timing of First Feed on Glycogen Reserves, Muscle Growth, and Body Weight. *Poult. Sci.* **2011**, *90*, 1467–1477. [CrossRef]
37. Uni, Z.; Ferket, P.R.; Tako, E.; Kedar, O. In Ovo Feeding Improves Energy Status of Late-Term Chicken Embryos. *Poult. Sci.* **2005**, *84*, 764–770. [CrossRef]
38. Dreiling, C.E.; Brown, D.E.; Casale, L.; Kelly, L. Muscle Glycogen: Comparison of Iodine Binding and Enzyme Digestion Assays and Application to Meat Samples. *Meat Sci.* **1987**, *20*, 167–177. [CrossRef]

39. Tako, E.; Glahn, R.P. Intra-Amniotic Administration and Dietary Inulin Affect the Iron Status and Intestinal Functionality of Iron-Deficient Broiler Chickens. *Poult. Sci.* **2012**, *91*, 1361–1370. [CrossRef]
40. Tako, E.; Beebe, S.; Reed, S.; Hart, J.; Glahn, R.P. Polyphenolic Compounds Appear to Limit the Nutritional Benefit of Biofortified Higher Iron Black Bean (*Phaseolus vulgaris* L.). *Nutr. J.* **2014**, *13*, 28. [CrossRef]
41. Agrizzi Verediano, T.; Stampini Duarte Martino, H.; Kolba, N.; Fu, Y.; Cristina Dias Paes, M.; Tako, E. Black Corn (*Zea mays* L.) Soluble Extract Showed Anti-Inflammatory Effects and Improved the Intestinal Barrier Integrity in Vivo (Gallus Gallus). *Food Res. Int.* **2022**, *157*, 111227. [CrossRef]
42. Agarwal, N.; Kolba, N.; Jung, Y.; Cheng, J.; Tako, E. Saffron (*Crocus sativus* L.) Flower Water Extract Disrupts the Cecal Microbiome, Brush Border Membrane Functionality, and Morphology in Vivo (Gallus Gallus). *Nutrients* **2022**, *14*, 220. [CrossRef]
43. da Silva, B.P.; Kolba, N.; Martino, H.S.D.; Hart, J.; Tako, E. Soluble Extracts from Chia Seed (*Salvia hispanica* L.) Affect Brush Border Membrane Functionality, Morphology and Intestinal Bacterial Populations in Vivo (Gallus Gallus). *Nutrients* **2019**, *11*, 2457. [CrossRef]
44. Agarwal, N.; Kolba, N.; Khen, N.; Even, C.; Turjeman, S.; Koren, O.; Tako, E. Quinoa Soluble Fiber and Quercetin Alter the Composition of the Gut Microbiome and Improve Brush Border Membrane Morphology In Vivo (Gallus Gallus). *Nutrients* **2022**, *14*, 448. [CrossRef]
45. Wang, X.; Kolba, N.; Liang, J.; Tako, E. Aleterations in Gut Microflora Populations and Brush Border Functionality Following Intra-Amniotic Administration (Gallus Gallus) of Wheat Bran Prebiotic Extracts. *Food Funct.* **2019**, *10*, 4834–4843. [CrossRef]
46. Tako, E.; Glahn, R.P.; Welch, R.M.; Lei, X.; Yasuda, K.; Miller, D.D. Dietary Inulin Affects the Expression of Intestinal Enterocyte Iron Transporters, Receptors and Storage Protein and Alters the Microbiota in the Pig Intestine. *Br. J. Nutr.* **2008**, *99*, 472–480. [CrossRef]
47. Gomes, M.J.C.; Martino, H.S.D.; Kolba, N.; Cheng, J.; Agarwal, N.; de Moura Rocha, M.; Tako, E. Zinc Biofortified Cowpea (*Vigna unguiculata* L. Walp.) Soluble Extracts Modulate Assessed Cecal Bacterial Populations and Gut Morphology In Vivo (Gallus Gallus). *Front. Biosci.-Landmark* **2022**, *27*, 140. [CrossRef]
48. Dias, D.M.; Kolba, N.; Hart, J.J.; Ma, M.; Sha, S.T.; Lakshmanan, N.; Nutti, M.R.; Martino, H.S.D.; Glahn, R.P.; Tako, E. Soluble Extracts from Carioca Beans (*Phaseolus vulgaris* L.) Affect the Gut Microbiota and Iron Related Brush Border Membrane Protein Expression in Vivo (Gallus Gallus). *Food Res. Int.* **2019**, *123*, 172–180. [CrossRef]
49. Martino, H.S.D.; Kolba, N.; Tako, E. Yacon (*Smallanthus sonchifolius*) Flour Soluble Extract Improve Intestinal Bacterial Populations, Brush Border Membrane Functionality and Morphology in Vivo (Gallus Gallus). *Food Res. Int.* **2020**, *137*, 109705. [CrossRef]
50. Tako, E.; Ferket, P.R.; Uni, Z. Changes in Chicken Intestinal Zinc Exporter MRNA Expression and Small Intestinal Functionality Following Intra-Amniotic Zinc-Methionine Administration. *J. Nutr. Biochem.* **2005**, *16*, 339–346. [CrossRef]
51. Uni, Z.; Noy, Y.; Sklan, D. Posthatch Development of Small Intestinal Function in the Poult. *Poult. Sci.* **1999**, *78*, 215–222. [CrossRef]
52. New York Apple Association Apples from New York: Varieties-Empire. Available online: https://www.applesfromny.com/varieties/empire/ (accessed on 17 July 2022).
53. US Apple Association. U.S. Apple Association: Apple Varieties. Available online: https://usapple.org/apple-varieties (accessed on 12 July 2022).
54. Asgary, S.; Rastqar, A.; Keshvari, M. Weight Loss Associated With Consumption of Apples: A Review. *J. Am. Coll. Nutr.* **2018**, *37*, 627–639. [CrossRef]
55. Cho, K.-D.; Han, C.-K.; Lee, B.-H. Loss of Body Weight and Fat and Improved Lipid Profiles in Obese Rats Fed Apple Pomace or Apple Juice Concentrate. *J. Med. Food* **2013**, *16*, 823–830. [CrossRef]
56. Samout, N.; Bouzenna, H.; Dhibi, S.; Ncib, S.; ElFeki, A.; Hfaiedh, N. Therapeutic Effect of Apple Pectin in Obese Rats. *Biomed. Pharmacother.* **2016**, *83*, 1233–1238. [CrossRef]
57. McKie, A.T. The Role of Dcytb in Iron Metabolism: An Update. *Biochem. Soc. Trans.* **2008**, *36*, 1239–1241. [CrossRef]
58. Lane, D.; Bae, D.-H.; Merlot, A.; Sahni, S.; Richardson, D. Duodenal Cytochrome b (Dcytb) in Iron Metabolism: An Update on Function and Regulation. *Nutrients* **2015**, *7*, 2274–2296. [CrossRef]
59. Warkentin, T.; Kolba, N.; Tako, E. Low Phytate Peas (*Pisum sativum* L.) Improve Iron Status, Gut Microbiome, and Brush Border Membrane Functionality in Vivo (Gallus Gallus). *Nutrients* **2020**, *12*, 2563. [CrossRef]
60. Shah, M.; Griffin, I.J.; Lifschitz, C.H.; Abrams, S.A. Effect of Orange and Apple Juices on Iron Absorption in Children. *Arch. Pediatr. Adolesc. Med.* **2003**, *157*, 1232–1236. [CrossRef]
61. Ravn-Haren, G.; Krath, B.N.; Markowski, J.; Poulsen, M.; Hansen, M.; Kolodziejczyk, K.; Kosmala, M.; Dragsted, L.O. Apple Pomace Improves Gut Health in Fisher Rats Independent of Seed Content. *Food. Funct.* **2018**, *9*, 2931–2941. [CrossRef]
62. Lee, B.; Moon, K.M.; Kim, C.Y. Tight Junction in the Intestinal Epithelium: Its Association with Diseases and Regulation by Phytochemicals. *J. Immunol. Res.* **2018**, *2018*, 1–11. [CrossRef]
63. Suzuki, T. Regulation of the Intestinal Barrier by Nutrients: The Role of Tight Junctions. *Anim. Sci. J.* **2020**, *91*, e13357. [CrossRef]
64. Suzuki, T. Regulation of Intestinal Epithelial Permeability by Tight Junctions. *Cell Mol. Life Sci.* **2013**, *70*, 631–659. [CrossRef]
65. Furuse, M.; Fujita, K.; Hiiragi, T.; Fujimoto, K.; Tsukita, S. Claudin-1 and -2: Novel Integral Membrane Proteins Localizing at Tight Junctions with No Sequence Similarity to Occludin. *J. Cell Iology* **1998**, *141*, 1539–1550. [CrossRef]
66. Furuse, M.; Hirase, T.; Itoh, M.; Nagafuchi, A.; Yonemura, S.; Tsukita, S.; Tsukita, S. Occludin: A Novel Integral Membrane Protein Localizing at Tight Junctions. *J. Cell Biol.* **1993**, *123*, 1777–1788. [CrossRef] [PubMed]

67. Ma, C.; Sun, Z.; Chen, C.; Zhang, L.; Zhu, S. Simultaneous Separation and Determination of Fructose, Sorbitol, Glucose and Sucrose in Fruits by Hplc-Elsd. *Food. Chem.* **2014**, *145*, 784–788. [CrossRef] [PubMed]
68. Yang, S.; Meng, Z.; Li, Y.; Chen, R.; Yang, Y.; Zhao, Z. Evaluation of Physiological Characteristics, Soluble Sugars, Organic Acids and Volatile Compounds in "Orin" Apples (*Malus domestica*) at Different Ripening Stages. *Molecules* **2021**, *26*, 807. [CrossRef] [PubMed]
69. Bischoff, S.C.; Kaden-Volynets, V.; Filipe Rosa, L.; Guseva, D.; Seethaler, B. Regulation of the Gut Barrier by Carbohydrates from Diet–Underlying Mechanisms and Possible Clinical Implications. *Int. J. Med. Microbiol.* **2021**, *311*, 151499. [CrossRef] [PubMed]
70. Volynets, V.; Louis, S.; Pretz, D.; Lang, L.; Ostaff, M.J.; Wehkamp, J.; Bischoff, S.C. Intestinal Barrier Function and the Gut Microbiome Are Differentially Affected in Mice Fed a Western-Style Diet or Drinking Water Supplemented with Fructose. *J. Nutr.* **2017**, *147*, 770–780. [CrossRef]
71. Laffin, M.; Fedorak, R.; Zalasky, A.; Park, H.; Gill, A.; Agarwal, A.; Keshteli, A.; Hotte, N.; Madsen, K.L. A High-Sugar Diet Rapidly Enhances Susceptibility to Colitis via Depletion of Luminal Short-Chain Fatty Acids in Mice. *Sci. Rep.* **2019**, *9*, 1–11. [CrossRef]
72. Goldstein, J.; Morris, W.E.; Loidl, C.F.; Tironi-Farinatti, C.; McClane, B.A.; Uzal, F.A.; Fernandez Miyakawa, M.E. Clostridium Perfringens Epsilon Toxin Increases the Small Intestinal Permeability in Mice and Rats. *PLoS ONE* **2009**, *4*, e7065. [CrossRef]
73. Lopetuso, L.R.; Scaldaferri, F.; Petito, V.; Gasbarrini, A. Commensal Clostridia: Leading Players in the Maintenance of Gut Homeostasis. *Gut Pathog.* **2013**, *5*, 23. [CrossRef]
74. Moore, R.; Pothoulakis, C.; LaMont, J.T.; Carlson, S.; Madara, J.L.C. Difficile Toxin A Increases Intestinal Permeability and Induces Cl- Secretion. *Am. J. Physiol.* **1990**, *259*, G165–G172. [CrossRef]
75. Paczosa, M.K.; Mecsas, J. Klebsiella Pneumoniae: Going on the Offense with a Strong Defense. *Microbiol. Mol. Biol. Rev.* **2016**, *80*, 629–661. [CrossRef]
76. Bang, S.-J.; Kim, G.; Lim, M.Y.; Song, E.-J.; Jung, D.-H.; Kum, J.-S.; Nam, Y.-D.; Park, C.-S.; Seo, D.-H. The Influence of in Vitro Pectin Fermentation on the Human Fecal Microbiome. *AMB Express* **2018**, *8*, 10–1186. [CrossRef] [PubMed]
77. Larsen, N.; Bussolo de Souza, C.; Krych, L.; Barbosa Cahu, T.; Wiese, M.; Kot, W.; Meyer Hansen, K.; Blennow, A.; Venema, K.; Jespersen, L. Potential of Pectins to Beneficially Modulate the Gut Microbiota Depends on Their Structural Properties. *Front. Microbiol.* **2019**, *10*, 223. [CrossRef] [PubMed]
78. Guo, P.; Zhang, K.; Ma, X.; He, P. Clostridium Species as Probiotics: Potentials and Challenges. *J. Anim. Sci. Biotechnol.* **2020**, *11*, 24. [CrossRef] [PubMed]
79. Licht, T.R.; Hansen, M.; Bergström, A.; Poulsen, M.; Krath, B.N.; Markowski, J.; Dragsted, L.O.; Wilcks, A. Effects of Apples and Specific Apple Components on the Cecal Environment of Conventional Rats: Role of Apple Pectin. *BMC Microbiol.* **2010**, *10*, 13. [CrossRef] [PubMed]
80. Dufourny, S.; Antoine, N.; Pitchugina, E.; Delcenserie, V.; Godbout, S.; Douny, C.; Scippo, M.-L.; Froidmont, E.; Rondia, P.; Wavreille, J.; et al. Apple Pomace and Performance, Intestinal Morphology and Microbiota of Weaned Piglets—A Weaning Strategy for Gut Health? *Microorganisms* **2021**, *9*, 572. [CrossRef]
81. Laudadio, V.; Passantino, L.; Perillo, A.; Lopresti, G.; Passantino, A.; Khan, R.U.; Tufarelli, V. Productive Performance and Histological Features of Intestinal Mucosa of Broiler Chickens Fed Different Dietary Protein Levels. *Poult. Sci.* **2012**, *91*, 265–270. [CrossRef]
82. Pluske, J.R.; Thompson, M.J.; Atwood, C.S.; Bird, P.H.; Williams, I.H.; Hartmann, P.E. Maintenance of Villus Height and Crypt Depth, and Enhancement of Disaccharide Digestion and Monosaccharide Absorption, in Piglets Fed on Cows' Whole Milk after Weaning. *Br. J. Nutr.* **1996**, *76*, 409–422. [CrossRef]
83. Oliveira, M.C.; Rodrigues, E.A.; Marques, R.H.; Gravena, R.A.; Guandolini, G.C.; Moraes, V.M.B. Performance and Morphology of Intestinal Mucosa of Broilers Fed Mannan-Oligosaccharides and Enzymes [Desempenho e Morfologia Da Mucosa Intestinal de Frangos de Corte Alimentados Com Mananoligossacarídeos e Enzimas]. *Arq. Bras. Med. Vet. Zootec.* **2008**, *60*, 442–448. [CrossRef]
84. Birchenough, G.M.H.; Johansson, M.E.V.; Gustafsson, J.K.; Bergström, J.H.; Hansson, G.C. New Developments in Goblet Cell Mucus Secretion and Function. *Mucosal. Immunol.* **2015**, *8*, 712–719. [CrossRef]
85. Wang, S.; Li, Q.; Zang, Y.; Zhao, Y.; Liu, N.; Wang, Y.; Xu, X.; Liu, L.; Mei, Q. Apple Polysaccharide Inhibits Microbial Dysbiosis and Chronic Inflammation and Modulates Gut Permeability in HFD-Fed Rats. *Int. J. Biol. Macromol.* **2017**, *99*, 282–292. [CrossRef]
86. Lueschow, S.R.; McElroy, S.J. The Paneth Cell: The Curator and Defender of the Immature Small Intestine. *Front. Immunol.* **2020**, *11*, 587. [CrossRef] [PubMed]
87. Salzman, N.H.; Bevins, C.L. Dysbiosis-A Consequence of Paneth Cell Dysfunction. *Semin. Immunol.* **2013**, *25*, 334–341. [CrossRef] [PubMed]
88. Bevins, C.; Salzman, N. Paneth Cells, Antimicrobial Peptides and Maintenance of Intestinal Homeostasis. *Nat. Rev. Microbiol.* **2011**, *9*, 356–368. [CrossRef] [PubMed]
89. Beukema, M.; Faas, M.; de Vos, P. The Effects of Different Dietary Fiber Pectin Structures on the Gastrointestinal Immune Barrier: Impact via Gut Microbiota and Direct Effects on Immune Cells. *Exp. Mol. Med.* **2020**, *52*, 1364–1376. [CrossRef] [PubMed]

nutrients

MDPI

Article

Effect of Chia (*Salvia hispanica* L.) Associated with High-Fat Diet on the Intestinal Health of *Wistar* Rats

Marcella Duarte Villas Mishima [1], Bárbara Pereira Da Silva [1], Mariana Juste Contin Gomes [1], Renata Celi Lopes Toledo [1], Hilário Cuquetto Mantovani [2], Vinícius Parzanini Brilhante de São José [1], Neuza Maria Brunoro Costa [3], Elad Tako [4] and Hércia Stampini Duarte Martino [1,*]

1 Department of Nutrition and Health, Federal University of Viçosa, Av. Purdue, s/n, Campus Universitário, Viçosa 36570-900, MG, Brazil
2 Department of Microbiology, Federal University of Viçosa, Av. Peter Henry Rolfs, s/n, Campus Universitário, Viçosa 36570-000, MG, Brazil
3 Department of Pharmacy and Nutrition, Federal University of Espírito Santo, Alto Universitário, s/n, Alegre 29500-000, ES, Brazil
4 Department of Food Science, Cornell University, Stocking Hall, Ithaca, NY 14850, USA
* Correspondence: hercia72@gmail.com; Tel.: +55-31-3612-5207

Abstract: A direct correlation has been reported between excessive fat intake and the development and progression of various enteropathies. Plant foods may contain bioactive compounds and non-digestible dietary fiber, with potential to improve intestinal health. Chia is a good source of dietary fiber and bioactive compounds. Our study evaluated the role of chia flour associated with a high-fat diet (HFD) on colon histomorphometry, intestinal functionality and intestinal microbiome composition and function in *Wistar* rats. The study used 32 young male rats separated into four groups to receive a standard diet (SD) or HFD, with or without chia, for 35 days. At the end of the study, the cecum, cecal content and duodenum were collected. The consumption of chia increased the production of short-chain fatty acids and improved fecal moisture. Chia consumption improved the circular muscle layer in the SD group. The diversity and abundance of intestinal bacteria were not affected, but increased richness was observed in the microbiome of the SD+chia group. Moreover, chia consumption decreased the expression of proteins involved in intestinal functionality. Chia consumption improved intestinal morphology and functionality in young *Wistar* rats but was insufficient to promote significant changes in the intestinal microbiome in a short term of 35 days.

Keywords: chia seed; western diet; intestinal microbiota; intestinal morphology; SCFA; intestinal functionality

Citation: Mishima, M.D.V.; Da Silva, B.P.; Gomes, M.J.C.; Toledo, R.C.L.; Mantovani, H.C.; José, V.P.B.d.S.; Costa, N.M.B.; Tako, E.; Martino, H.S.D. Effect of Chia (*Salvia hispanica* L.) Associated with High-Fat Diet on the Intestinal Health of *Wistar* Rats. *Nutrients* **2022**, *14*, 4924. https://doi.org/10.3390/nu14224924

Academic Editor: Susanna Iossa

Received: 31 October 2022
Accepted: 18 November 2022
Published: 21 November 2022

1. Introduction

The large consumption of a western diet, which is increased in fat and reduced in vegetables and fiber, is increasingly becoming a big risk factor for several chronic metabolic and inflammatory disorders involving different organs. The chronic high-fat diet (HFD) consumption can induce or aggravate many diseases that affect a wide range of organs besides obesity and metabolic syndrome [1,2]. In addition, a direct correlation has been reported between excessive fat intake and the development and progression of various enteropathies, with conditions that lead to the secretion of pro-inflammatory cytokines, which initiate and maintain an inflammatory process, thus causing alteration in the intestinal microbiota, dysfunction of the epithelial barrier and reduced permeability and integrity of the intestinal mucosal barrier [2–4].

By manipulating external factors, the plasticity of the microbiota allows the reshaping of the architecture and biological outputs of intestinal microbes to improve human health. A relation between diet and the intestinal microbiota is observed, which leads to the conclusion that dietary factors are the most potent modulators of microbiota composition and function. Intestinal microorganisms, in turn, influence the absorption, metabolism and storage of nutrients, with effects on host physiology [5].

When it reaches the colon, the fiber diet is anaerobically fermented by the intestinal bacteria. Some nutritionally specialized bacteria in the phyla Firmicutes and Actinobacteria are considered to be important in initiating the degradation of the fiber diet, and the continued breakdown is attributed to certain abundant species within the phylum Bacteroidetes, and then they can produce short-chain fatty acids (SCFAs). SCFAs play a significant role in intestinal homeostasis and affect the body [5,6]. The most abundant SCFAs in the intestine are acetic acid, propionic acid and butyric acid [6,7]. Microbiota-accessible carbohydrates provide an energy source for intestinal bacteria, some of which are nutritionally specialized in degrading these carbohydrates and producing certain SCFA [6]. The consequent production of SCFAs benefits the host by serving as both recovered energy from otherwise inaccessible carbohydrates as well as potent regulatory molecules, with vast physiological effects, including energy homeostasis, lipid and carbohydrate metabolism, and suppression of inflammatory signals [5,6].

Plant foods may contain bioactive compounds and dietary fibers with potential effects on intestinal bacterial populations, gastrointestinal motility improvement, intestinal functionality and morphology, increased mucus production, number/diameter of goblet cells, surface area of villi and crypt depth. Such effects seem to result from the increased motility of the digestive tract, which leads to hyperplasia and/or hypertrophy of muscle cells [7–9].

Chia (*Salvia hispanica* L.) is a potentially bioactive food. Its consumption may reduce the risk and attenuate metabolic alterations, due to a series of health benefits, such as antidiabetic effects, antitumor potential, immunostimulant activity, antioxidant protection, cardiovascular and liver protection, remission of inflammation and reduced heart fat content [10–12]. Chia contains components with diverse actions, whose benefits are mainly due to their high nutritional value, high concentrations of lipids (32.16 g.100 g^{-1}), proteins (18.18 g.100 g^{-1}) and total dietary fiber (33.37 g.100 g^{-1}), besides insoluble fiber (30.47 g.100 g^{-1}), minerals and antioxidant compounds beneficial to health [13]. The intra-amniotic (in ovo) administration of a soluble chia extract improved intestinal morphology, affected the intestinal microbiota and positively regulated the gene expression of proteins associated to mineral metabolism [8]. Furthermore, the relationship between chia seed peptides and the regulation of adipogenesis and inflammation has already been demonstrated [14]. Therefore, the objective of the present study was to evaluate the role of chia associated with a high-fat diet on colon histomorphometry, intestinal functionality and microbiome composition and function of young *Wistar* rats.

2. Materials and Methods

2.1. Sample Material

The seeds (*Salvia hispanica* L.) used in this study were grown in the state of Rio Grande do Sul (Brazil). The seeds were planted in January 2015, and the harvest was carried out in June 2015. The samples were packed and transported in cardboard boxes. Then, the samples were stored in hermetically sealed plastic bags until use, protected from light and frozen (-18 °C $\pm$ 1 °C). The seeds were ground up to a particle size of 850 µm to obtain flour, using a knife mill (Marconi Equipment, Brazil), in three replicates. Then, chia flour was packed in polyethylene aluminum bags and stored in a freezer (-18 °C $\pm$ 1 °C) until the chia was added to the diet.

2.2. Animals and Diets

Thirty-two newly weaned, 21-day-old male rats (Rattus norvegicus, *Wistar*, albinus variation) from the Central Animal Facility of the Center for Biological Sciences and Health, at the Federal University of Viçosa, Minas Gerais, Brazil, were separated into 4 groups (n = 8) randomized by body weight, using the WinPepi Program version 11.65 [15]. The rats were distributed into individual stainless steel cages in a temperature-controlled environment (22 °C) and automatically controlled light and dark cycles of 12 h. The experimental diets were based either on the standard diet (AIN-93G) [16] or high-fat diet (Research Diets, New Brunswick, NJ, USA), with modifications. The animals received their respective experimental diets and deionized water ad libitum.

The standard diet was composed of 20% protein, 30% fat and 50% carbohydrate. The high-fat diet was prepared with the following proportions: 64% fat, 16% protein and 20% carbohydrate. The amount of chia was determined based on the human consumption of 40 g of chia/day (2 tablespoons). The chia seed was added to diets based on its composition, described by Da Silva (2017) [13], of lipids (32.16 g.100 g^{-1}), proteins (18.18 g.100 g^{-1}), total dietary fiber (33.37 g.100 g^{-1}) and carbohydrates (4.59 g.100 g^{-1}). The other ingredients were added in quantities sufficient to provide the planned amounts of proteins, lipids, carbohydrates, fiber and calories (Table 1).

Table 1. Compositions of the experimental diets.

	Experimental Diets			
Ingredients (g/kg of Diet)	**SD**	**SD+chia**	**HFD**	**HFD+chia**
Albumin *	217.90	117.60	217.92	117.60
Chia flour	0.00	416.80	0.00	416.80
Dextrinized starch	132.00	132.00	132.00	115.10
Sucrose	100.00	100.00	100.00	100.00
Lard	0.00	0.00	200.00	200.00
Soybean oil (mL)	134.20	0.00	134.20	0.00
Microcrystalline cellulose	139.20	0.00	139.20	0.00
Mineral mix	35.00	35.00	35.00	35.00
Vitamin mix	10.00	10.00	10.00	10.00
L-cystine	3.00	3.00	3.00	3.00
Choline bitartrate	2.50	2.50	2.50	2.50
Corn starch	221.20	183.10	21.20	0.00
Total calories (kcal)	3700.42 ^{b}	3624.75 ^{b}	4700.42 ^{a}	4624.11 ^{a}
Caloric density (kcal g^{-1})	3.70 ^{b}	3.62 ^{b}	4.70 ^{a}	4.62 ^{a}

* Considering that albumin is 78% pure. SD: standard diet, HFD: high-fat diet. Means followed by different small letters in the same row differ significantly, according to Newman–Keuls post hoc test, at the 5% threshold of probability.

Each group received one of the following experimental diets: standard diet (SD), standard diet + chia flour (SD+chia), high-fat diet (HFD) or high-fat diet + chia flour (HFD+chia). Body weight gain and food consumption were monitored weekly during the experimental period. On the 35th day, after 12 h of fasting, the animals were anesthetized with isoflurane (Isoforine, Cristália®), and then euthanized by cardiac puncture. Cecum, cecal content and duodenum were collected, weighed and stored at −80 °C prior to analysis. The colon segment was collected, flushed with phosphate buffered saline solution, fixed in formaline for 24 h and kept in 70% ethanol for histological analysis. The following fat depots was weighed: visceral, retroperitoneal, epididymal and abdominal adipose tissues. The adiposity was calculated as a percentage using the following formula: (visceral + retroperitoneal + epididymal + abdominal adipose tissues)/total body weight × 100 [17].

All the experimental procedures with animals were performed in accordance with Directive 86/609/EEC of November 24, 1986, in compliance with the ethical principles for animal experimentation. The study protocol was approved by the Ethics Committee of the Federal University of Viçosa (protocol 20/2017; date of approval: 13 July 2017).

2.3. Fecal Moisture

The moisture content in the feces, collected at the end of the experiment, was determined by the gravimetric method. The samples were oven-dried at 105 °C for 24 h [18].

2.4. Cecal Content pH

About 1 g of cecum feces was homogenized in 10 mL of distilled water, with the aid of vortex glass spheres. Subsequently, the glass electrode of the pHmeter (Bel, Italy) was inserted. The measurements were performed in duplicate [19].

2.5. IgA Quantification

In order to determine sIgA, the standard procedures were followed to prepare the cecal content homogenates, 1:5 (w/v) . The cecal content samples were defrosted on ice. To prepare the suspensions, 200 mg feces was added to 800 μL of phos-phate-buffered saline and homogenized with the aid of a vortex. It was used an Immunochron enzyme-linked immunosorbent assay (ELISA) to measured the mucosal immunity, based on the cecal content sIgA concentration. [20].

2.6. Short Chain Fatty Acids (SCFA) Content

The analysis to determine the content of SCFA followed the methodology proposed by Siegfried and Ruckemmann (1984) [21], with modifications. Briefly, 500 mg of cecum feces were homogenized in MiliQ water, following a vortex shaking protocol. Next, the samples were centrifuged at 12,000× g for 10 min. The supernatants were transferred to new tubes, received the addition of calcium hydroxide and cupric sulfate and were vigorously shaken. The samples were frozen and then thawed at room temperature for centrifugation. The supernatant was added with concentrated sulfuric acid and then frozen. Then, the samples were thawed, refrozen and finally thawed for centrifugation at 12,000× g, for 10 min. The supernatant was transferred to vials for analysis by high-performance liquid chromatography (HPLC).

The SCFAs were determined in a Dionex Ultimate 3000 Dual detector HPLC apparatus (Dionex Corporation, Sunnyvale, CA, USA) equipped with a refractive index detector Shodex RI-101, using a Bio-Rad HPX-87H column, 300 mm × 4.6 mm, main-tained at 45 °C. The conditions that the analyses were performed were as follow: mobile phase sulfuric acid 5 mmol L^{-1}, flow rate 0.7 mL min^{-1} and injection volume 20 μL. Acetic, propionic and butyric acids were used to prepare the stock solutions of the standards. The SCFAs were prepared with a final concentration of 10 mmol/L.

2.7. Colon Histomorphometry

Semi-serialized histological fragments from proximal colon (3 μm thickness) were obtained on a semi-automated rotating microtome (Leica, Brazil) and stained by the hematoxylin/eosin technique. To measure the length and thickness of the crypts and the thickness of the circular and longitudinal muscle layers twenty random fields per animal were selected. The slides were examined under a AX70 photomicroscope (Olympus, Japan). The ImagePro-Plus® software sys-tem, version 4.5 (Media Cybernetics, Rockville, MD, USA) was used to processed the images.

2.8. DNA Extraction and Sequencing

The total genomic DNA was extracted from the cecum content samples, following a mechanical disruption and phenol/chloroform extraction protocol [22]. PCR amplicon libraries targeting the hypervariable V4 region of the 16S rRNA gene were produced using the primers 515F (5′GTGYCAGCMGCCGCGGTAA3′) and 806R (GGACTACNVGGGTWTCTAAT3′) and a barcoded primer set adapted for the Illumina MiSeq platform (Illumina, San Diego, California, CA, USA) [23,24]. Illumina MiSeq was used to load the samples onto an Illumina flow cell for paired-end sequencing reactions in the Environmental Sample Preparation and Sequencing Facility (ESPSF), at the Argonne National Laboratory (Lemont, IL, USA).

The sequences obtained for all samples in the present study were submitted to the Sequence Read Archive (SRA) of the National Center for Biotechnology Information (NCBI), under the accession number PRJNA805271.

The Mothur software system v.1.44.3 was used to data processing and analysis [25]. The R1 and R2 paired-end reads were joined, and sequences smaller than 150 or greater than 300 bp were removed. Sequences with homopolymers with at least 8 nucleotides or containing ambiguous base pairs were also eliminated. Chimera sequences were detected and removed using UCHIME [26]. After cleaning the sequences, they were aligned with the 16S rRNA gene, using the SILVA database v.138 [27].

The Operational Taxonomic Units (OTUs) were grouped according to the sequence similarity cutoff. The coverage of all samples was assessed by Good's coverage estimator (bacteria > 97%). A normalized data table was used for calculating alpha and beta diversity and the relative abundance of OTUs. Alpha diversity was estimated by using Chao1, Shannon and Simpson indices. Beta diversity between dietary groups was assessed by Principal Coordinate Analysis (PoA), based on the Jaccard dissimilarity index [28]. Metagenome functional predictive analysis was carried out using PICRUSt2 software [29].

2.9. Determination of Gene Expression of Proteins Involved in Intestinal Health by Quantitative Reverse Transcriptase Polymerase Chain Reaction (RT-qPCR)

mRNA expression levels of genes in the intestinal tissue (duodenum) which are involved in intestinal health were analyzed by RT-qPCR. The SYBR Green PCR master mix from Applied Biosystems (Foster City, CA, USA) was employed, and the analyses were performed on the StepOne™ Real-Time PCR System (Thermo Fisher Scientific) by means of the measurement system involving SYBR-Green Fluorescence and Primer Express software (Applied Biosystems, Foster City, CA, USA). Sense and antisense primer sequences were ordered (Choma Biotechnologies) to amplify aminopeptidase (AP) (ID: 301368687), sucrose isomaltase (SI) (ID: 301368688), peptide transporter 1 (PepT1) (ID: 301368693) and sodium–glucose transport protein 1 (SGLT1) (ID: 301368686). The relative expression levels of mRNA were normalized to the endogenous control (beta-actin; Table 2). All the steps were performed under open conditions with RNase.

Table 2. Sequencing primers used in the RT-qPCR analysis.

Genes	Oligonucleotide (5′-3′)	
	Forward	Reverse
Beta-actin	TTCGTGCCGGTCCACACCC	GCTTTGCACATGCCGGAGCC
AP	CTCTCTCCTCAAACACATGAA	AGTTCAGGGCCTTCTCATATTC
SI	CCTCCAGAACACAATCCCTATAC	GGAGAGGTGAGATGGATTAGA
PEPT1	CCTGGTCGTCTCATCATATT	TTCTTCTCATCCTCATCGAACTG
SGLT1	CATCCAGTCCATCACATTAC	CAATCAGGAAGCCGAGAATCA

AP: aminopeptidase; SGLT1: sodium–glucose transport protein 1; SI: sucrose isomaltase; PepT1: peptide transporter 1.

2.10. Statistical Analysis

Food consumption, body weight, colonic histomorphometry characteristics and concentrations of SCFA data were first submitted to a Kolmogorov–Smirnov normality test. Next, a one-way analysis of variance (ANOVA) was applied, followed by the Newman–Keuls post hoc test to compare all test groups. The experimental treatments were arranged in a completely randomized design, with eight repetitions. The data are presented as means $\pm$ standard deviation, and statistical significance was established at $p < 0.05$. Correlations between biological markers and intestinal parameters were assessed by Pearson's correlation test. The biochemical parameters and stress oxidative markers were previously carried out and published [30].

Regarding the microbiome results, the Chao1, Shannon and Simpson indexes were used to estimates differences between alpha diversity, ANOVA was used to analyse the differences between the groups. The differences between beta diversity were analyzed by the pairwise PERMANOVA test. Statistically significant p-values associated with microbial clades and functions were corrected for multiple comparisons using the Benjamini–Hochberg false discovery rate (FDR) correction. The statistical analysis was performed using SPSS version 20.0. The level of significance was established at $p < 0.05$.

3. Results

3.1. Data of Animals

The food consumption was higher in the rats fed the standard diet (SD and SD+chia) than in the rats fed HFD (HFD and HFD+chia). The animals that were fed with the HFD containing chia (HFD+chia) presented higher food consumption than those of the HFD group. The animals fed the SD and HFD without chia (SD and HFD) presented lower final weight than those in the groups fed chia. Adiposity, cecum weight and cecal content pH did not differ between the experimental groups. The consumption of chia (SD+chia × SD and HFD+chia × HFD) increased the moisture concentration in the feces of the animals and did not alter the IgA concentration, while the SD group presented the lowest IgA concentration (Table 3). Furthermore, chia consumption increased the concentration of acetic, propionic and butyric acids in the cecal content (Figure 1).

Table 3. Biometric data, consumption data, fecal moisture, cecal content pH and IgA concentration.

	SD	SD+chia	HFD	HFD+chia
Food consumption (g/day)	14.74 ± 0.86 [a]	15.46 ± 1.20 [a]	10.86 ± 0.49 [c]	12.29 ± 0.93 [b*]
Final weight (g)	221.69 ± 15.25 [b]	241.10 ± 15.65 [a*]	208.76 ± 11.95 [b]	240.11 ± 16.89 [a*]
Adiposity (% of body weight)	2.32 ± 0.60 [a]	2.66 ± 0.39 [a]	2.69 ± 0.49 [a]	2.96 ± 0.46 [a]
Cecum weight (g)	4.20 ± 0.71 [a]	4.76 ± 0.60 [a]	4.25 ± 0.40 [a]	4.06 ± 0.83 [a]
Cecal content pH	7.23 ± 1.28 [a]	7.33 ± 0.27 [a]	6.70 ± 0.21 [a]	7.13 ± 0.45 [a]
Fecal moisture (%)	16.36 ± 3.86 [c]	37.65 ± 5.37 [a*]	14.36 ± 6.41 [c]	28.01 ± 5.97 [b*]
IgA (ng)	271.09 ± 95.70 [b]	555.50 ± 193.67 [a]	751.83 ± 278.88 [a]	681.17 ± 155.91 [a]

Values referring to means ± SD, n = 8/group. SD: standard diet; SD+chia: standard diet + chia; HFD: high-fat diet; HFD+chia: high-fat diet + chia. The data were analyzed by one-way ANOVA. Means followed by different small letters in the same row differ significantly, according to Newman–Keuls post hoc test, at the 5% threshold of probability. * Indicates differences between the groups by the *t*-test, at 5% probability, in the comparison of the groups that received the same diet, either with or without chia (SD × SD+chia and HFD × HFD+chia).

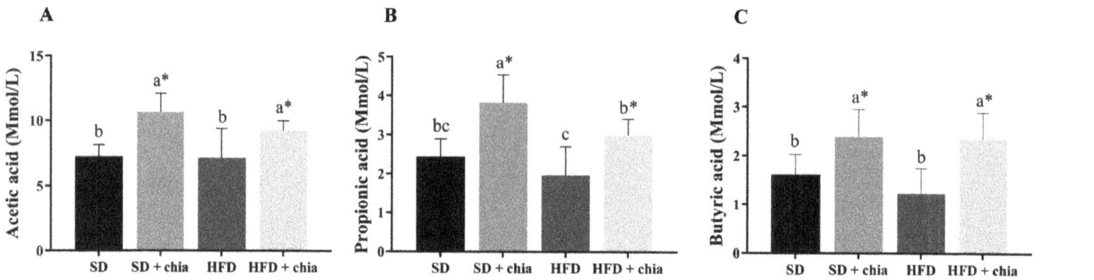

Figure 1. Short-chain fatty acid (SCFA) concentration in cecal contents of young *Wistar* rats, after 35 days of treatment (n = 8). (**A**) Concentration of acetic acid; (**B**) concentration of propionic acid; (**C**) concentration of butyric acid. SD: standard diet; SD+chia: standard diet + chia; HFD: high-fat diet; HFD+chia: high-fat diet + chia. The data were analyzed by one-way ANOVA. Means followed by different small letters in the same row differ significantly, according to Newman–Keuls post hoc test, at 5% threshold of probability. * Indicates differences between the groups by the *t*-test, at 5% probability, in the comparison of the groups that received the same diet, either with or without chia (SD × SD+chia and HFD × HFD+chia).

3.2. Colonic Histomorphometry Characteristics

The HFD consumption reduced the longitudinal muscle layer, circular muscle layer, crypt length and crypt thickness. On the other hand, when associated with high-fat diet, chia consumption reduced the longitudinal muscle layer. Chia consumption with a standard diet increased the circular muscle layer compared to all the other groups. However, in the high-fat groups, chia consumption reduced this measurement. The crypt length was increased by chia consumption in the group that received HFD, but in the group fed the standard diet, the crypt length and crypt thickness were reduced by chia consumption (Table 4).

Table 4. Colonic histomorphometry characteristics of young *Wistar* rats after 35 days of treatment.

	SD	SD+chia	HFD	HFD+chia
LML (μm)	49.16 ± 12.48 ^{a}	52.65 ± 26.90 ^{a}	26.02 ± 11.36 ^{b*}	22.68 ± 10.94 ^{b}
CML (μm)	111.88 ± 35.84 ^{b}	127.44 ± 70.86 ^{a}	62.32 ± 24.37 ^{c*}	50.01 ± 24.13 ^{d}
Crypt length (μm)	180.53 ± 41.33 ^{a*}	163.67 ± 44.52 ^{b}	103.39 ± 13.97 ^{d}	121.51 ± 30.96 ^{c*}
Crypt thickness (μm)	45.57 ± 7.64 ^{a*}	36.00 ± 13.53 ^{b}	26.72 ± 5.67 ^{c}	25.41 ± 6.08 ^{c}

Values referring to means ± SD, n = 8/group. SD: standard diet; SD+chia: standard diet + chia; HFD: high fat diet; HFD+chia: high-fat diet + chia; LML: longitudinal muscle layer; CML: circular muscle layer. The data were analyzed by one-way ANOVA. Means followed by different small letters in the same row differ significantly, according to the Newman–Keuls post hoc test, at the 5% threshold of probability. * Indicates differences between the groups by the *t*-test, at 5% probability, in the comparison of the groups that received the same diet, either with or without chia (SD × SD+chia and HFD × HFD+chia).

3.3. Intestinal Microbiota Analysis

Regarding the intestinal microbiota analysis, the sequencing of the 16S rRNA gene from the fecal samples generated 731.297 raw sequences. Following the filtering and cleaning of the sequences, 569.406 sequences of suitable quality were obtained. The Good's coverage obtained in the samples was >99%, which indicates adequate sequencing coverage. The summary of the sample sequencing data is shown in the Supplementary Table S1.

The richness and diversity indexes were used to evaluate the alpha diversity of the microbial samples. No difference was observed in the Chao1, Shannon and Simpson indexes among the experimental groups (Figure 2A–C). However, when assessed pairwise (SD × SD+chia and HFD × HFD+chia), the Chao1 index show an increase in richness in the intestinal microbiota of the animals fed SD+chia compared to the SD group (Figure 2A).

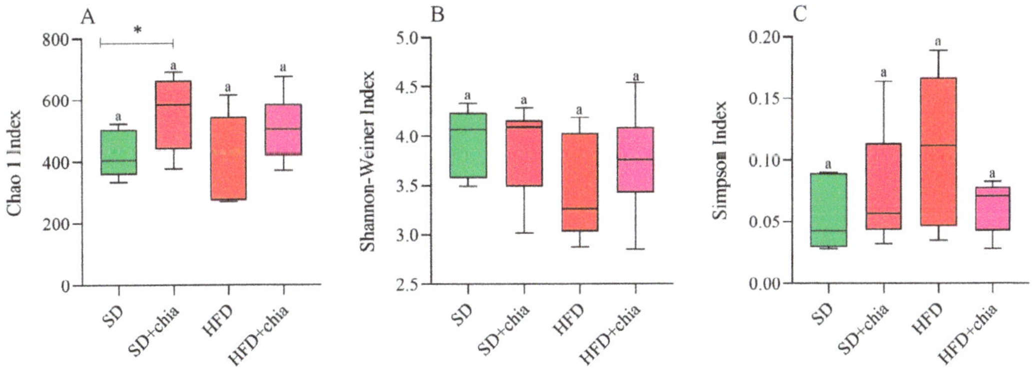

Figure 2. Alpha diversity metrics of bacterial communities in the cecum faces. n = 6 per group. (**A**) Chao1 index; (**B**) Shannon-Weiner Index; (**C**) Simpson Index; SD: standard diet; SD+chia: standard diet+chia; HFD: high-fat diet; HFD+chia: high-fat diet+chia. Treatment groups indicated by the same letter are not significantly different ($p < 0.05$). * Indicates differences between the groups by the *t*-test, at 5% probability, in the comparison of the groups that received the same diet, either with or without chia (SD × SD+chia and HFD × HFD+chia).

According to the beta diversity analyses, calculated by using Jaccard distances, there was some variation in bacterial communities in response to the consumption of the four types of diet at the level of OTU, phyla and genera, as indicated by the PCoA analysis and PERMANOVA (Figure 3A–C).

The spatial ordination of the OTUs (Figure 3A) indicated differences in the clustering of the samples of the different experimental groups. The data based on their collections of sequences presented differences in the distance metrics among the SD group compared to SD+chia, and between the HFD group and HFD+chia. However, after FDR correction, the clustering of OTU sequences presented no difference. Spatial ordination at the phylum level (Figure 3B) indicated no changes between all treatment groups or between the HFD and HFD+chia groups. Although PERMANOVA identified a significant difference between the clustering of phyla from SD and SD+chia groups, no difference was verified after FDR correction.

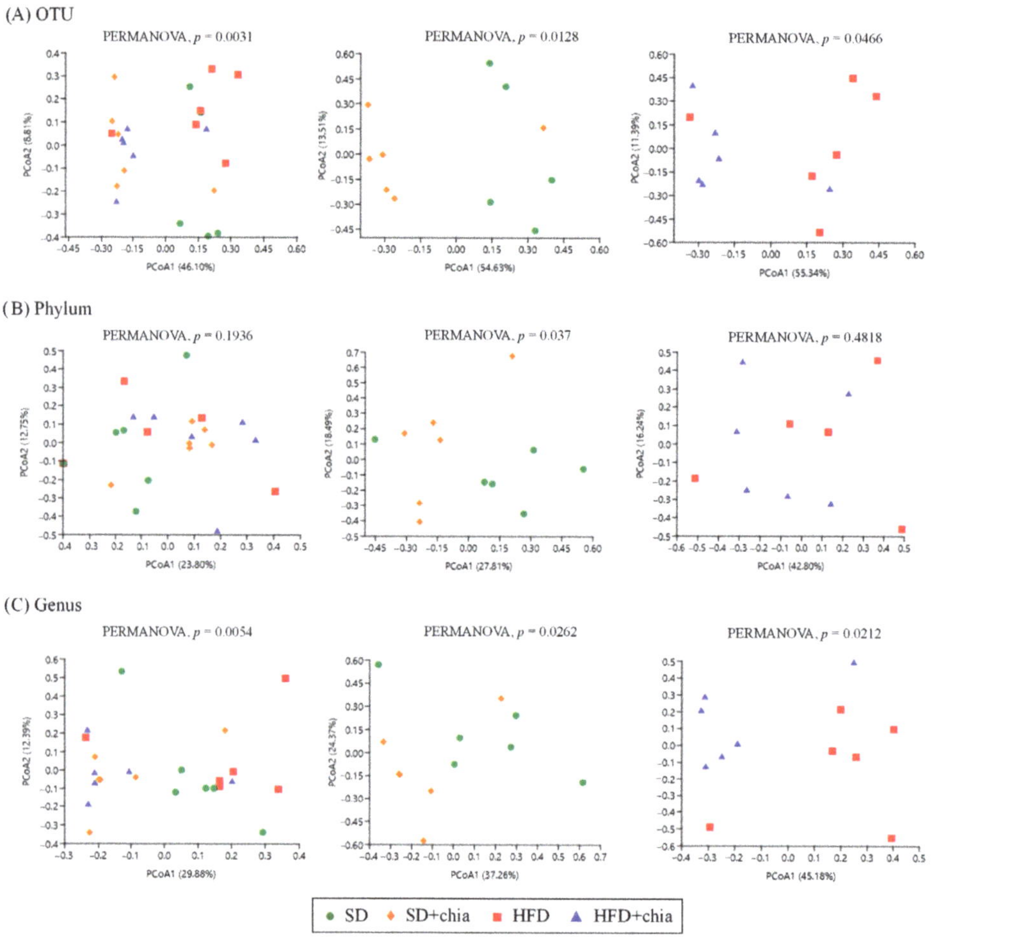

Figure 3. Changes in β-diversity of the cecal microbiome of young *Wistar* rats after 35 days of treatment. (**A**) Principal Coordinate Analysis (PCoA) based on the Jaccard distance at OTU level. (**B**) Principal Coordinate Analysis (PCoA) based on Jaccard distance at phylum level. (**C**) Principal Coordinate Analysis (PCoA) based on Jaccard distance at genus level.

The statistical data revealed that before FDR correction, the Lachnospiraceae family was enriched in the group fed HFD+chia compared to the HFD group. In addition, the Muribaculaceae family and the genus *Roseburia* were enriched in the group fed SD+chia compared to the SD group, which indicates a beneficial effect of chia consumption on the SCFA-producing bacteria (Supplementary Table S2).

The spatial ordination at the genus level (Figure 3C) presented differences in the distance metrics among the experimental groups. However, as previously observed at the OTU and phylum levels, after FDR correction, these differences were lost. It is important to highlight that the *Corynebacterium* genera concentration differs between the groups fed with HFD and HFD+chia.

The taxonomic analysis of the bacterial community in response to chia flour consumption revealed the existence of 18 phyla, 30 classes, 65 orders, 93 families and 193 genera. The stratification of phyla, genus and the Firmicutes to Bacteroidetes ratio is plotted in Figure 4. Although no difference was found in the relative abundance of the identified phyla and genus between the experimental groups, after FDR correction, we observed the dominancy of phyla Firmicutes (63–68% of relative abundance) and Bacteroidetes (19–22% of relative abundance) composing the intestinal microbiota of young rats (Figure 4A), and the dominancy of genera from the Muribaculaceae family, and *Bacteroides* (Figure 4C), identified after 35 days of treatment. Furthermore, no difference was observed in the Firmicutes to Bacteroidetes ratio between the experimental groups (Figure 4B).

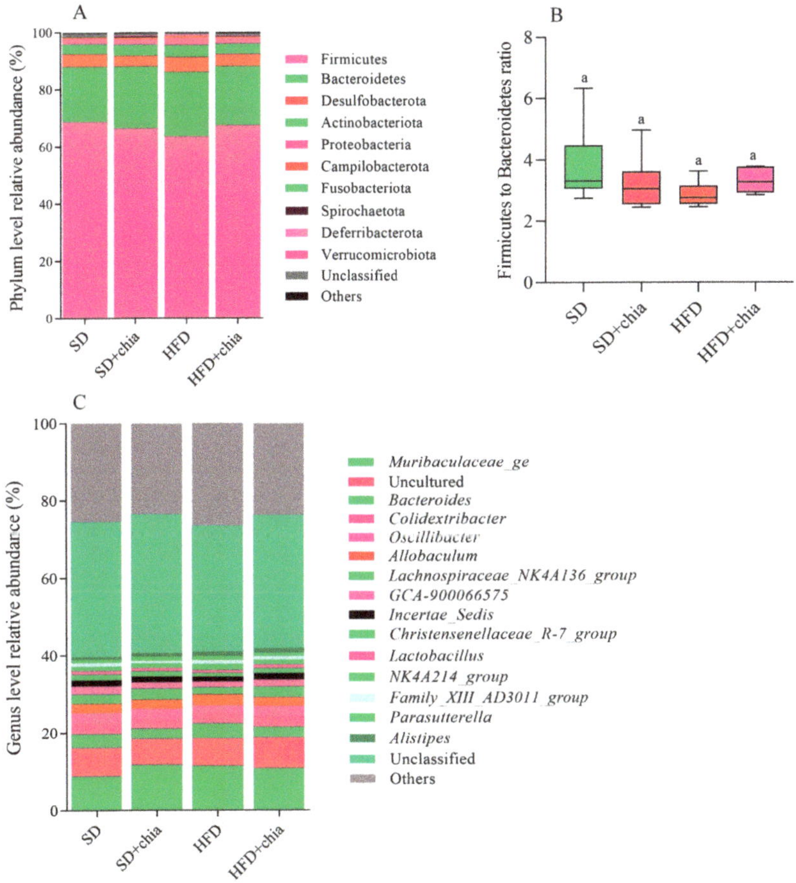

Figure 4. Relative abundances of bacterial microbiota composition at phylum and genera level of young *Wistar* rats, after 35 days of treatment. (**A**) Relative abundance of each identified phylum. (**B**) Firmicutes to Bacteroidetes ratio. (**C**) Genera relative abundance. n = 6/group. SD: standard diet; SD+chia: standard diet + chia; HFD: high-fat diet; HFD+chia: high-fat diet + chia. Only phyla with abundance >0.2% and genera with abundance >1% in at least one group are displayed. The data were analyzed with FDR correction. Treatment groups indicated by the same letter are not significantly different ($p < 0.05$). * Indicates differences between the groups by the *t*-test, at 5% probability, in the comparison of the groups that received the same diet, either with or without chia (SD × SD+chia and HFD × HFD+chia).

We investigated whether chia treatments affected the genetic capacity of the microbiota and explored possible functional changes. PICRSt2 was used for the functional predictive analysis of the metagenome and revealed the most abundant pathways (Figure 5), but no difference was detected between the treatments.

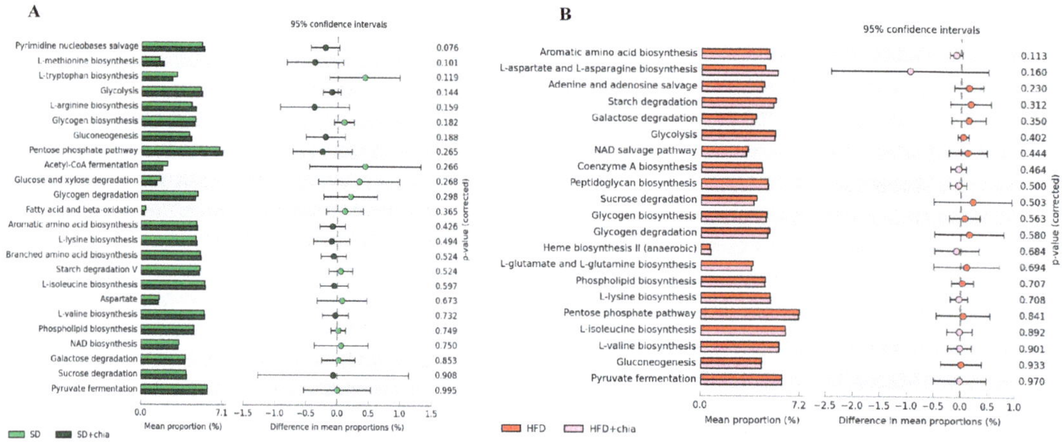

Figure 5. Difference in the relative abundance of the most abundant KEGG microbial metabolic pathways in the microbiota of young *Wistar* rats, after 35 days of treatment. (**A**) Enriched pathways between the SD and SD+chia treatment groups. (**B**) Enriched pathways between the HFD and HFD+chia treatment groups. Extended error bar plot was performed by bioinformatic software (STAMP). Welch's two-sided test was used and Welch's inverted was 0.95. n = 6/group. SD: standard diet; SD+chia: standard diet + chia; HFD: high-fat diet; HFD+chia: high-fat diet + chia.

3.4. BBM Functional Proteins

Regarding intestinal functionality, we observed that the group that consumed HFD presented higher gene expression for all intestinal genes evaluated. Furthermore, chia consumption (SD+chia and HFD+chia) reduced PepT1, AP, SGLT1 and SI mRNA gene expression in relation to the control diets (SD and HFD) (Figure 6).

3.5. Pearson Correlation Analysis

We performed the Pearson correlation analysis to assess the relationship between changes in intestinal microbial abundance, gut health parameters, oxidative stress markers and biochemical variables. When correlations were assessed, *Bacteroides* were negatively correlated with the Muribaculaceae family (r = −0.487), *Roseburia* (r = −0.480) and butyric acid (r = −0.496). The Lachnospiraceae family was negatively correlated with total cholesterol (TC) (r = −0.507). The Muribaculaceae family was positively correlated with IgA (r = 0.482). *Roseburia* was positively correlated with acetic acid (r = 0.602), propionic acid (r = 0.471), catalase (CAT) (r = 0.485) and malondialdehyde (MDA) (r = 0.495). Acetic acid was positively correlated with propionic acid (r = 0.836), butyric acid (r = 0.649), fecal moisture (r = 0.662) and total antioxidant capacity of plasma (TAC) (r = 0.475), and negatively correlated with TC (r = −0.426). Propionic acid was positively correlated with butyric acid (r = 0.664) and fecal moisture (r = 0.711), CAT (r = 0.590), and negatively correlated with TC (r = −0.515). Butyric acid was positively correlated with fecal moisture (r = 0.581) and TAC (r = 0.486), and negatively correlated with TC (r = −0.423). The fecal moisture was positively correlated with the cecal content pH (r = 0.785) and CAT (r = 0.576), and negatively correlated with TC (r = −0.598). Superoxide dismutase (SOD) was positively correlated with MDA (r = 0.597). Finally, CAT was positively correlated with MDA (r = 0.515) and glucose (r = 0.391), and negatively correlated with TC (r = −0.427) (Figure 7).

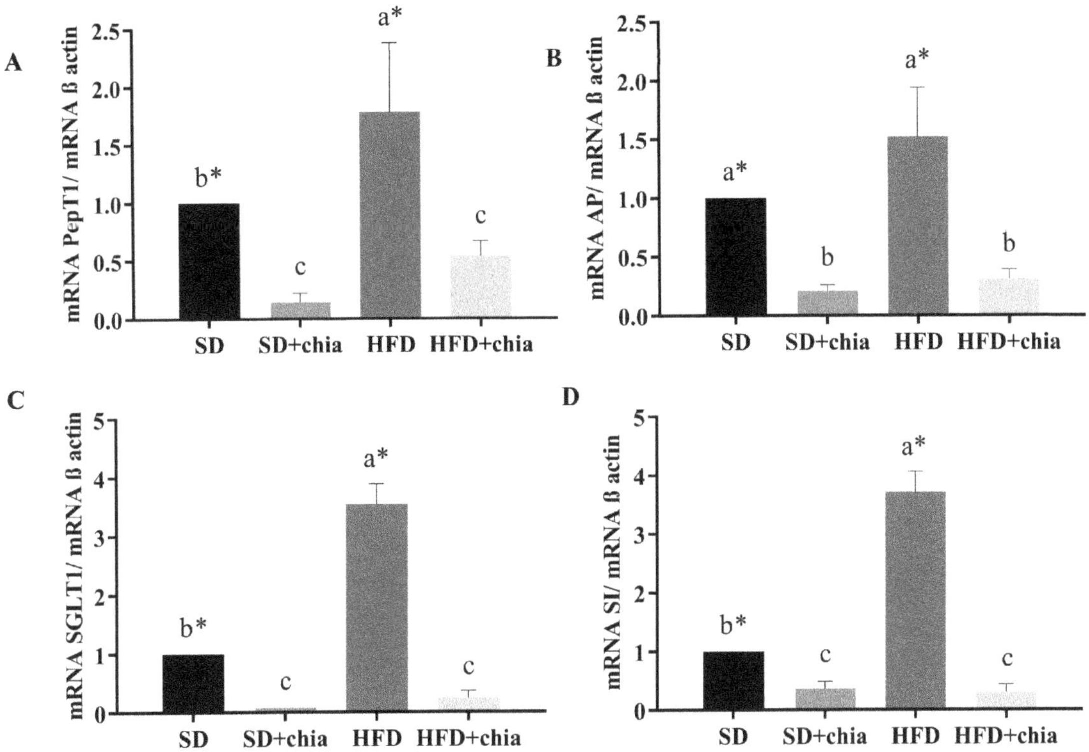

Figure 6. Gene expression of brush border membrane functional proteins in the intestinal tissue. RT-qPCR analysis. (**A**) PepT1 expression, (**B**) AP expression, (**C**) SGLT1 expression and (**D**) SI expression. ST: standard diet; ST+chia: standard diet + chia; HFD: high-fat diet; HF+chia: high-fat diet + chia; PepT1: peptide transporter 1; AP: aminopeptidase; SGLT1: sodium–glucose transport protein 1; SI: sucrose isomaltase. The data were analyzed by one-way ANOVA. Means followed by different small letters in the same row differ significantly, according to the Newman–Keuls post hoc test, at the 5% threshold of probability. * Indicates differences between the groups by the *t*-test, at 5% probability, in the comparison of the groups that received the same diet, either with or without chia (SD × SD+chia and HFD × HFD+chia).

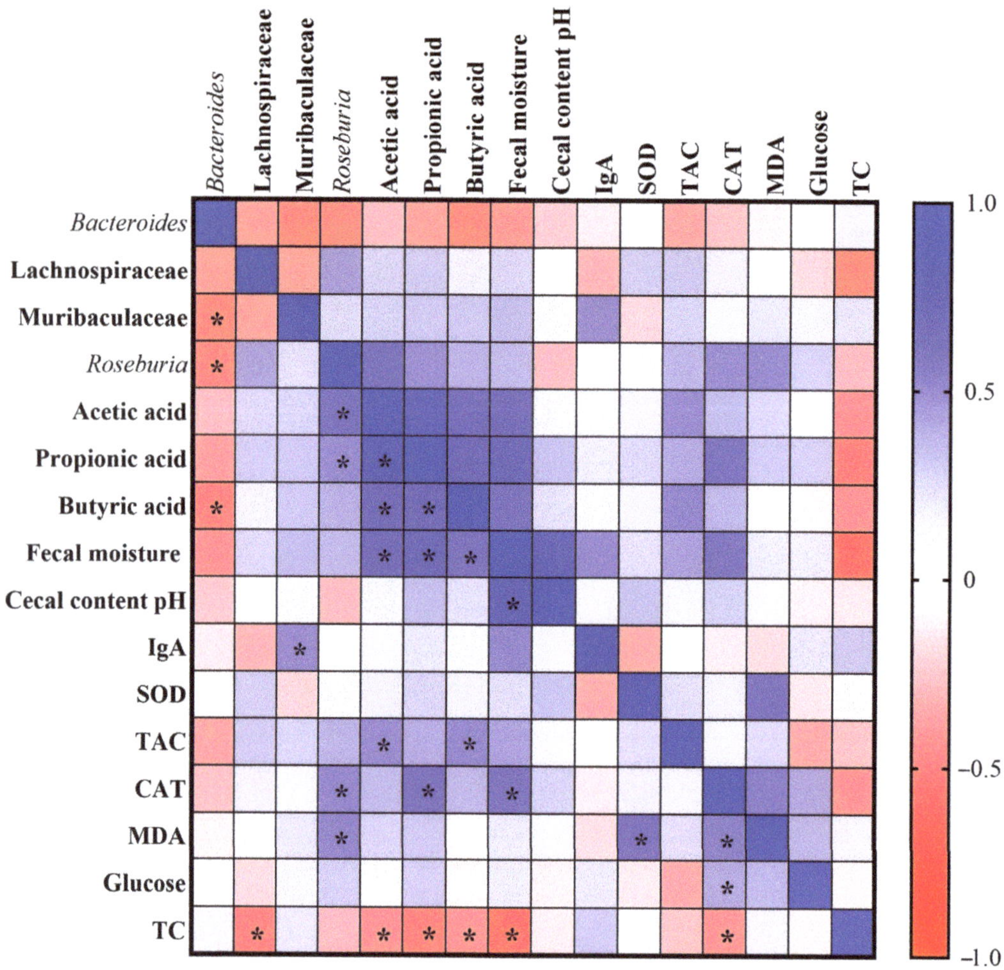

Figure 7. Heatmap of Pearson correlation analysis. IgA: immunoglobulin; SOD: superoxide dismutase; TAC: total antioxidant capacity of plasma; CAT: catalase; MDA: malondialdehyde; TC: total cholesterol. * Indicates statistically significant difference.

4. Discussion

Chia is a good source of bioactive compounds and dietary fiber [13,31]. However, the potential effects of chia associated with an inflammatory condition on the intestinal microbiota composition, intestinal morphology and intestinal functionality have not been investigated. Thus, the present study focused on evaluating the effect of chia consumption associated with HFD, for 35 days, on the gut health of young *Wistar* rats. Our study revealed that the intake of chia with the SD or HFD increased the production of acetic acid, propionic acid and butyric acid and increased fecal moisture. The consumption of HFD affects the intestinal morphology, reducing the longitudinal and circular muscle layer and the length and thickness of the crypts. Chia consumption increased the crypt length in the group that received HFD and improved circular muscle layer in the group that received SD. The diversity and abundance of intestinal bacteria were not affected, but increased richness was observed in the intestinal microbiome of animals fed SD+chia compared to the SD group. Additionally, chia consumption reduced the expression of proteins involved in intestinal functionality.

The lower food consumption observed in animals fed HFD can be attributed to higher energy density and greater satiety during the experimental period, as previously reported [12,30]. When chia was consumed with HFD, the food consumption was increased, as was the final weight. In the group that was fed SD, although the food consumption was not altered by chia consumption, the SD+chia group presented higher final weight, but no difference in adiposity was observed between groups. This leads us to assume that there was an increase in the muscle mass of these animals. The study of Grancieri et al. (2022) [32] observed that the incorporation of the digested protein of chia into a normal diet increased body weight and contributed to muscle mass gain in the animals, which impacted weight gain without increasing body fat. The skeletal muscle plays a major role in fatty acid uptake and oxidation, while HFD can increase the susceptibility to loss of muscle mass and the degradation of proteins [33]. Chia consumption, when associated with SD and HFD, increased fecal moisture, probably due to the higher content of dietary fibers present in food. The gel-forming dietary fibers present sufficient water-holding capacity and increase fecal bulking action [34]. The increased weight of the feces makes it easier for them to pass through the colon and be expelled from the body, thus alleviating problems such as constipation. Here, the increased moisture in groups that were fed chia can be attributed to chia's ability to absorb water, which adds bulk to the feces. Chia secrets a mucilage when it becomes wet, a gelatinous and viscous substance after water absorption, which can increase fecal moisture [34].

In general, a high amount of soluble and insoluble dietary fiber increases intestinal motility and fecal volume, which tends to increase the thickness of muscle layers. The circular muscle layer was higher in the group fed SD+chia than in the SD group. This increase was also observed in female ovariectomized *Wistar* rats [35]. This result is probably due to the increased motility of the digestive tract, enabled by the intact form of the dietary fiber found in chia, besides the formation of gel, promoted by the soluble dietary fiber fraction, which leads to the hypertrophy of muscle cells [36].

The consumption of dietary fiber provides a substrate for microbial activity and affects the intestinal microbiota by altering bacterial fermentation and fermentation products, such as SCFA [5,6]. In our study, increased SCFA production was observed in groups that were fed chia. These changes in total SCFAs indicate that the dietary fiber from chia seed could be utilized by the microbiota [37]. SCFAs, such as butyric acid, acetic acid and propionic acid, are the most abundant fecal metabolites. It is important to mention that acetic acid was the most abundant SCFA, which is in agreement with previous reports by Tamargo et al. (2018) [37]. Different SCFA production patterns are possibly dependent on the type of fiber. These fermentation products are extremely important for host health, since butyric acid, apart from serving as the primary energy source for colonocytes, also improves the integrity of intestinal epithelial cells by promoting tight junctions and cell proliferation. Both acetic acid and propionic acid also aid in anti-inflammatory processes and cytokine production [38]. It is suggested that acetic acid plays an important regulatory role in body weight control and insulin sensitivity, acting as a direct mediator in lipid metabolism and glucose homeostasis [39]. The production of SCFAs can reduce the intestinal pH, which may increase mineral solubility and absorption [8,40]. However, in the present study, the cecal content pH was not altered. In our study, the Pearson's correlation indicated that SCFAs were positively correlated with fecal moisture and antioxidant capacity, and negatively correlated with cholesterol.

In our study, the consumption of chia did not affect the immunoglobulin A (IgA) concentration. IgA is abundant in the intestine and plays an essential role in the defense of the intestinal mucosa against harmful pathogens. It is suggested that diet and intestinal microbiota are involved in the regulation of IgA production [41,42]. Our result is in agreement with another study that also evaluated the impact of chia consumption on intestinal health and found that there was no change in the concentration of IgA [35]. Nakajima et al. (2020) [43] revealed that the microbiota in the large intestine is involved in IgA induction by dietary fiber, and the amount of IgA was considerably higher in the intestinal contents of mice that were fed a diet with soluble dietary fiber than in those that were fed a diet with insoluble dietary fiber. The concentration of dietary fiber in chia is mainly formed by insoluble fiber [13], which can explain why the IgA concentration was not altered in our animals.

The diversity and abundance of intestinal bacteria were not affected when the Chao1, Shannon and Simpson indexes were calculated among the four experimental groups. However, the Chao1 index indicated increased richness in the intestinal microbiome of the animals fed a standard diet+chia (SD+chia) compared to the SD group, in agreement with our findings from another study [35]. Richness is related to the number of different species, and high richness of the intestinal microbiome is associated with healthy host metabolism, while their absence is aligned with unhealthy outcomes [38,44–46]. In the same group (SD+chia x SD), the Muribaculaceae family and the genus *Roseburia* were enriched before FDR correction. Studies suggest that the Muribaculaceae family have members with a functional potential in the intestine, which is the ability to degrade dietary carbohydrates and ferment polysaccharides into SCFAs [47,48]. *Roseburia* was positively correlated with acetic acid and propionic acid and has also been reported to produce SCFAs, mainly propionic acid [38], which is in accordance with our results. Therefore, our results indicated a beneficial effect of chia consumption on the SCFA-producing bacteria. It is important to highlight that *Corynebacterium* genus concentration varies between the groups fed with HFD and HFD+chia. Chia consumption was able to reduce the concentration of the genus. This is a positive event, since *Corynebacterium* can produce phospholipase D, an exotoxin that degrades lipids in cell membranes, which may increase cellular permeability and thereby facilitate the spread of the pathogen in the tissues [49]. In addition, according to the functional analysis of the microbiota, we observed the KEGG metabolic pathways that were more abundant, although no differences were detected between the treatments. This fact can be explained by the short time of chia consumption by the animals (35 days).

When we evaluated the intestinal functionality, we found that the HDF increased the expression of all genes evaluated, and chia consumption (HFD+chia) decreased the gene expression, which was similar to the group that consumed SD+chia. Chia consumption reduced the expression of SI and SGLT1, which are genes related to carbohydrate digestion and absorption. Studies have reported that fasting plasma glucose concentration is reduced after chia consumption [36,50,51]. Thus, chia intake may reduce carbohydrate absorption/digestion through the downregulation of SI and SGLT1 gene expression and the consequent reduction in SI, which suppresses the rapid production of glucose on the surface of the brush border membrane, in combination with decreased glucose absorption by reducing the relative expression of SGLT1 [52]. The same result was observed for AP and PepT1 protein gene expression. The function of PepT1 is to transport peptides to the enterocyte. AP is an exopeptidase responsible for cleaving amino acids from the N-terminus of peptides [53]. In our study, we observed that chia consumption reduced PepT1 and AP gene expression, probably due to the fact that the organism is in homeostasis, which does not require the increase in the expression of these genes to perform its function with regard to protein metabolism.

5. Conclusions

Chia consumption in standard and high-fat diet increased the production of short-chain fatty acids (acetic acid, propionic acid and butyric acid) and improved fecal moisture. Furthermore, chia consumption improved intestinal morphology, increasing the circular muscle layer in the SD group and the crypt length in the group that received HFD. Besides this, chia decreased the gene expression of SGLT1, SI, AP and PepT1 in all groups. The 35-day intervention in young male *Wistar* rats did not affect the diversity and abundance of intestinal bacteria but promoted an increase in richness in the SD+chia group. Therefore, further studies using a longer intervention period are needed to clarify the effects of chia on the intestinal microbiome.

Supplementary Materials: The following supporting information can be downloaded at: https://www.mdpi.com/article/10.3390/nu14224924/s1, Table S1: Sequencing data at the end of 35 days of treatment, according to each experimental group; Table S2: Significantly different families and genus identified prior to FDR correction, at the end of 35 days of treatments.

Author Contributions: Conceptualization, M.D.V.M., B.P.D.S., H.S.D.M.; methodology, M.D.V.M., B.P.D.S., M.J.C.G., R.C.L.T., V.P.B.d.S.J.; formal analysis, M.D.V.M., M.J.C.G., R.C.L.T., V.P.B.d.S.J.; investigation, M.D.V.M., B.P.D.S., R.C.L.T., H.C.M., V.P.B.d.S.J.; resources, H.C.M., N.M.B.C., H.S.D.M.; data curation, M.D.V.M., M.J.C.G.; writing—original draft preparation, M.D.V.M., B.P.D.S., M.J.C.G.; writing—review and editing, M.D.V.M., B.P.D.S., H.S.D.M., E.T.; supervision, H.S.D.M.; project administration, H.S.D.M.; funding acquisition, N.M.B.C., H.S.D.M. All authors have read and agreed to the published version of the manuscript.

Funding: This research received no external funding.

Institutional Review Board Statement: The animal study protocol was approved by the Ethics Committee of the Federal University of Viçosa (protocol 20/2017; date of approval: 13 July 2017).

Informed Consent Statement: Not applicable.

Data Availability Statement: Not applicable.

Acknowledgments: The authors are thankful to the Foundation for Research Support of Minas Gerais (FAPEMIG, Brazil, number APQ-02183-17), for the financial support; we are also grateful to the Coordination for the Improvement of Higher Education Personnel (CAPES, Brazil, grant number 88887.599144/2021-00), the National Counsel of Technological and Scientific Development (CNPq, Brazil, number 406517/2018-5, Research Productivity fellowships (PQ2—grant number 310910/2020-0)), and the Foundation for Research an Innovation Support of Espirito Santo (FAPES, Brazil—PRONEX—CNPq/FAPES, Public Notice 24/2018-TO 567/2018).

Conflicts of Interest: The authors declare no conflict of interest.

References

1. Wali, J.A.; Jarzebska, N.; Raubenheimer, D.; Simpson, S.J.; Rodionov, R.N.; Sullivan, J.F.O. Cardio-Metabolic Effects of High-Fat Diets and Their Underlying Mechanisms—A Narrative Review. *Nutrients* **2020**, *12*, 1505. [CrossRef] [PubMed]
2. Sferra, R.; Pompili, S.; Cappariello, A.; Gaudio, E.; Latella, G.; Vetuschi, A. Prolonged Chronic Consumption of a High Fat with Sucrose Diet Alters the Morphology of the Small Intestine. *Int. J. Mol. Sci.* **2021**, *22*, 7280. [CrossRef] [PubMed]
3. Araújo, J.R.; Tomas, J.; Brenner, C.; Sansonetti, P.J. Impact of High-Fat Diet on the Intestinal Microbiota and Small Intestinal Physiology before and after the Onset of Obesity. *Biochimie* **2017**, *141*, 97–106. [CrossRef] [PubMed]
4. Rohr, M.W.; Narasimhulu, C.A.; Rudeski-Rohr, T.A.; Parthasarathy, S. Negative Effects of a High-Fat Diet on Intestinal Permeability: A Review. *Adv. Nutr.* **2020**, *11*, 77–91. [CrossRef] [PubMed]
5. Gentile, C.L.; Weir, T.L. The Gut Microbiota at the Intersection of Diet and Human Health. *Science* **2018**, *362*, 776–780. [CrossRef]
6. Van Der Hee, B.; Wells, J.M. Microbial Regulation of Host Physiology by Short-Chain Fatty Acids. *Trends Microbiol.* **2021**, *29*, 700–712. [CrossRef]
7. Slavin, J. Fiber and Prebiotics: Mechanisms and Health Benefits. *Nutrients* **2013**, *5*, 1417–1435. [CrossRef]
8. Da Silva, B.P.; Kolba, N.; Martino, H.S.D.; Hart, J.; Tako, E. Soluble Extracts from Chia Seed (*Salvia Hispanica* L.) Affect Brush Border Membrane Functionality, Morphology and Intestinal Bacterial Populations in Vivo (*Gallus gallus*). *Nutrients* **2019**, *11*, 2457. [CrossRef]
9. Martino, H.S.D.; Kolba, N.; Tako, E. Yacon (Smallanthus Sonchifolius) Flour Soluble Extract Improve Intestinal Bacterial Populations, Brush Border Membrane Functionality and Morphology in Vivo (*Gallus gallus*). *Food Res. Int.* **2020**, *137*, 109705. [CrossRef]
10. Melo, D.; Machado, T.B.; Oliveira, M.B.P.P. Chia Seeds: An Ancient Grain Trending in Modern Human Diets. *Food Funct.* **2019**, *10*, 3068–3089. [CrossRef]
11. Enes, B.N.; Moreira, L.P.D.; Silva, B.P.; Grancieri, M.; Lúcio, H.G.; Venâncio, V.P.; Mertens-Talcott, S.U.; Rosa, C.O.B.; Martino, H.S.D. Chia Seed (*Salvia Hispanica* L.) Effects and Their Molecular Mechanisms on Unbalanced Diet Experimental Studies: A Systematic Review. *J. Food Sci.* **2020**, *85*, 226–239. [CrossRef] [PubMed]
12. Mishima, M.D.V.; Ladeira, L.C.M.; da Silva, B.P.; Toledo, R.C.L.; de Oliveira, T.V.; Costa, N.M.B.; Martino, H.S.D. Cardioprotective Action of Chia (*Salvia Hispanica* L.) in Ovariectomized Rats Fed a High Fat Diet. *Food Funct.* **2021**, *12*, 3069–3082. [CrossRef]
13. Da Silva, B.P.; Anunciação, P.C.; Matyelka, J.C.D.S.; Della Lucia, C.M.; Martino, H.S.D.; Pinheiro-Sant'Ana, H.M. Chemical Composition of Brazilian Chia Seeds Grown in Different Places. *Food Chem.* **2017**, *221*, 1709–1716. [CrossRef] [PubMed]
14. Grancieri, M.; Martino, H.S.D.; de Mejia, E.G. Protein Digests and Pure Peptides from Chia Seed Prevented Adipogenesis and Inflammation by Inhibiting Pparγ and Nf-Kb Pathways in 3t3l-1 Adipocytes. *Nutrients* **2021**, *13*, 176. [CrossRef]

15. Abramson, J.H. WINPEPI Updated: Computer Programs for Epidemiologists, and Their Teaching Potential. *Epidemiol. Perspect. Innov.* **2011**, *8*, 1. [CrossRef]
16. Reeves, P.G.; Nielsen, F.H.; Fahey, G.C. AIN-93 Purified Diets for Laboratory Rodents: Final Report of the American Institute of Nutrition Ad Hoc Writing Committee on the Reformulation of the AIN-76A Rodent Diet. *J. Nutr.* **1993**, *123*, 1939–1951. [CrossRef]
17. Pereira, S.S.; Teixeira, L.G.; Aguilar, E.C.; Matoso, R.O.; Soares, F.L.P.; Ferreira, A.V.M.; Alvarez-Leite, J.I. Differences in Adipose Tissue Inflammation and Oxidative Status in C57BL/6 and ApoE-/- Mice Fed High Fat Diet. *Anim. Sci. J.* **2012**, *83*, 549–555. [CrossRef]
18. *AOAC Official Methods of Analysis of AOAC International*, 20th ed.; Association of Official Analytical Chemistry: Gaithersburg, MD, USA, 2016.
19. Grancieri, M.; Costa, N.M.B.; Vaz Tostes, M.D.G.; de Oliveira, D.S.; Nunes, L.D.C.; Marcon, L.D.N.; Veridiano, T.A.; Viana, M.L. Yacon Flour (Smallanthus Sonchifolius) Attenuates Intestinal Morbidity in Rats with Colon Cancer. *J. Funct. Foods* **2017**, *37*, 666–675. [CrossRef]
20. das Graças Vaz-Tostes, M.; Viana, M.L.; Grancieri, M.; dos Santos Luz, T.C.; de Paula, H.; Pedrosa, R.G.; Costa, N.M.B. Yacon Effects in Immune Response and Nutritional Status of Iron and Zinc in Preschool Children. *Nutrition* **2014**, *30*, 666–672. [CrossRef]
21. Siegfried, V.R.; Ruckemmann, H.; Stumpf, G. Method for the Determination of Organic Acids in Silage by High Performance Liquid Chromatography. *Landwirtsch. Forsch.* **1984**, *37*, 298–304.
22. Stevenson, D.M.; Weimer, P.J. Dominance of Prevotella and Low Abundance of Classical Ruminal Bacterial Species in the Bovine Rumen Revealed by Relative Quantification Real-Time PCR. *Appl. Microbiol. Biotechnol.* **2007**, *75*, 165–174. [CrossRef] [PubMed]
23. Caporaso, J.G.; Lauber, C.L.; Walters, W.A.; Berg-Lyons, D.; Lozupone, C.A.; Turnbaugh, P.J.; Fierer, N.; Knight, R. Global Patterns of 16S RRNA Diversity at a Depth of Millions of Sequences per Sample. *Proc. Natl. Acad. Sci. USA* **2011**, *108*, 4516–4522. [CrossRef] [PubMed]
24. Caporaso, J.G.; Lauber, C.L.; Walters, W.A.; Berg-Lyons, D.; Huntley, J.; Fierer, N.; Owens, S.M.; Betley, J.; Fraser, L.; Bauer, M.; et al. Ultra-High-Throughput Microbial Community Analysis on the Illumina HiSeq and MiSeq Platforms. *ISME J.* **2012**, *6*, 1621–1624. [CrossRef]
25. Schloss, P.D.; Westcott, S.L.; Ryabin, T.; Hall, J.R.; Hartmann, M.; Hollister, E.B.; Lesniewski, R.A.; Oakley, B.B.; Parks, D.H.; Robinson, C.J.; et al. Introducing Mothur: Open-Source, Platform-Independent, Community-Supported Software for Describing and Comparing Microbial Communities. *Appl. Environ. Microbiol.* **2009**, *75*, 7537–7541. [CrossRef]
26. Edgar, R.C.; Haas, B.J.; Clemente, J.C.; Quince, C.; Knight, R. UCHIME Improves Sensitivity and Speed of Chimera Detection. *Bioinformatics* **2011**, *27*, 2194–2200. [CrossRef]
27. Quast, C.; Pruesse, E.; Yilmaz, P.; Gerken, J.; Schweer, T.; Yarza, P.; Peplies, J.; Glöckner, F.O. The SILVA Ribosomal RNA Gene Database Project: Improved Data Processing and Web-Based Tools. *Nucleic Acids Res.* **2013**, *41*, 590–596. [CrossRef]
28. Lozupone, C.; Knight, R. UniFrac: A New Phylogenetic Method for Comparing Microbial Communities. *Appl. Environ. Microbiol.* **2005**, *71*, 8228–8235. [CrossRef]
29. Caicedo, H.H.; Hashimoto, D.A.; Caicedo, J.C.; Pentland, A.; Pisano, G.P. Overcoming Barriers to Early Disease Intervention. *Nat. Biotechnol.* **2020**, *38*, 669–673. [CrossRef]
30. Da Silva, B.P.; Toledo, R.C.L.; Grancieri, M.; de Castro Moreira, M.E.; Medina, N.R.; Silva, R.R.; Costa, N.M.B.; Martino, H.S.D. Effects of Chia (*Salvia Hispanica* L.) on Calcium Bioavailability and Inflammation in Wistar Rats. *Food Res. Int.* **2019**, *116*, 592–599. [CrossRef]
31. Grancieri, M.; Stampini, H.; Martino, D.; De Mejia, E.G. Chia Seed (*Salvia Hispanica* L.) as a Source of Proteins and Bioactive Peptides with Health Benefits : A Review. *Compr. Rev. Food Sci. Food Saf.* **2019**, *18*, 480–499. [CrossRef]
32. Grancieri, M.; Verediano, T.A.; Sant'Ana, C.T.; de Assis, A.; Toledo, R.L.; de Mejia, E.G.; Martino, H.S.D. Digested Protein from Chia Seed (Salvia Hispanica L) Prevents Obesity and Associated Inflammation of Adipose Tissue in Mice Fed a High-Fat Diet. *Pharma Nutr.* **2022**, *21*, 100298. [CrossRef]
33. Marthandam Asokan, S.; Hung, T.H.; Chiang, W.D.; Lin, W.T. Lipolysis-Stimulating Peptide from Soybean Protects Against High Fat Diet-Induced Apoptosis in Skeletal Muscles. *J. Med. Food* **2018**, *21*, 225–232. [CrossRef] [PubMed]
34. Fu, X.; Li, R.; Zhang, T.; Li, M.; Mou, H. Study on the Ability of Partially Hydrolyzed Guar Gum to Modulate the Gut Microbiota and Relieve Constipation. *J. Food Biochem.* **2019**, *43*, e12715. [CrossRef] [PubMed]
35. Mishima, M.D.V.; da Silva, B.P.; Gomes, M.J.C.; Toledo, R.C.L.; Pereira, C.E.R.; Costa, N.M.B.; Martino, H.S.D. Effect of Chia Flour Associated with High Fat Diet on Intestinal Health in Female Ovariectomized Wistar Rats. *Eur. J. Nutr.* **2022**. [CrossRef] [PubMed]
36. Da Silva, B.P.; Matyelka, J.C.D.S.; Moreira, M.E.D.C.; Toledo, R.C.L.; Della Lucia, C.M.; Pinheiro-Sant'Ana, H.M.; Martino, H.S.D. A High Fat Diet Does Not Affect the Iron Bioavailability in Wistar Rats Fed with Chia and Increases Gene Expression of Iron Metabolism Proteins. *Food Funct.* **2016**, *7*, 4861–4868. [CrossRef] [PubMed]
37. Tamargo, A.; Cueva, C.; Laguna, L.; Moreno-Arribas, M.V.; Muñoz, L.A. Understanding the Impact of Chia Seed Mucilage on Human Gut Microbiota by Using the Dynamic Gastrointestinal Model Simgi®. *J. Funct. Foods* **2018**, *50*, 104–111. [CrossRef]
38. Ghosh, S.; Pramanik, S. Structural Diversity, Functional Aspects and Future Therapeutic Applications of Human Gut Microbiome. *Arch. Microbiol.* **2021**, *203*, 5281–5308. [CrossRef]
39. Hernández, M.A.G.; Canfora, E.E.; Jocken, J.W.E.; Blaak, E.E. The Short-Chain Fatty Acid Acetate in Body Weight Control and Insulin Sensitivity. *Nutrients* **2019**, *11*, 1943. [CrossRef]

40. Patterson, J.K.; Lei, X.G.; Miller, D.D. The Pig as an Experimental Model for Elucidating the Mechanisms Governing Dietary Influence on Mineral Absorption. *Exp. Biol.* **2008**, *233*, 651–664. [CrossRef]
41. Geva-zatorsky, N.; Sefik, E.; Kua, L.; Pasman, L.; Tan, T.G.; Ortiz-lopez, A.; Yanortsang, T.B.; Yang, L.; Jupp, R.; Mathis, D.; et al. Mining the Human Gut Microbiota for Immunomodulatory Organisms. *Cell* **2017**, *168*, 928–943. [CrossRef]
42. Takeuchi, T.; Ohno, H. Reciprocal Regulation of IgA and the Gut Microbiota: A Key Mutualism in the Intestine. *Int. Immunol.* **2021**, *33*, 781–786. [CrossRef]
43. Nakajima, A.; Sasaki, T.; Itoh, K.; Kitahara, T.; Takema, Y.; Hiramatsu, K.; Ishikawa, D. A Soluble Fiber Diet Increases Bacteroides Fragilis Group Abundance and Immunoglobulin A Production in the Gut. *Appl. Environ. Microbiol.* **2020**, *86*, e00405-20. [CrossRef]
44. Covasa, M.; Stephens, R.W.; Toderean, R.; Cobuz, C. Intestinal Sensing by Gut Microbiota: Targeting Gut Peptides. *Front. Endocrinol.* **2019**, *10*, 82. [CrossRef]
45. Pinart, M.; Dötsch, A.; Schlicht, K.; Laudes, M.; Bouwma, J.; Forslund, S.K.; Pischon, T.; Nimptsch, K. Gut Microbiome Composition in Obese and Non-Obese Persons: A Systematic Review and Meta-Analysis. *Nutrients* **2022**, *14*, 12. [CrossRef]
46. Siddiqui, M.T.; Cresci, G.A.M. The Immunomodulatory Functions of Butyrate. *J. Inflamm. Res.* **2021**, *14*, 6025–6041. [CrossRef]
47. Lagkouvardos, I.; Lesker, T.R.; Hitch, T.C.A.; Gálvez, E.J.C.; Smit, N.; Neuhaus, K.; Wang, J.; Baines, J.F.; Abt, B.; Stecher, B.; et al. Sequence and Cultivation Study of Muribaculaceae Reveals Novel Species, Host Preference, and Functional Potential of This yet Undescribed Family. *Microbiome* **2019**, *7*, 28. [CrossRef]
48. Smith, B.J.; Miller, R.A.; Ericsson, A.C.; Harrison, D.C.; Strong, R.; Schmidt, T.M. Changes in the Gut Microbiome and Fermentation Products Concurrent with Enhanced Longevity in Acarbose-Treated Mice. *BMC Microbiol.* **2019**, *19*, 130. [CrossRef]
49. Barros de Pinho, R.; de Oliveira Silva, M.T.; Brenner, G.; Dié Alves, M.S.; Azevedo, V.; Dias Portela, R.; Borsuk, S. A Novel Approach for an Immunogen against Corynebacterium Pseudotuberculosis Infection: An Escherichia Coli Bacterin Expressing Phospholipase D. *Microb. Pathog.* **2021**, *151*, 104746. [CrossRef]
50. da Silva, B.P.; Toledo, R.C.L.; Mishima, M.D.V.; Moreira, M.E.D.C.; Vasconcelos, C.M.; Pereira, C.E.R.; Favarato, L.S.C.; Costa, N.M.B.; Martino, H.S.D. Effects of Chia (*Salvia Hispanica* L.) on Oxidative Stress and Inflammation in Ovariectomized Adult Female Wistar Rats. *Food Funct.* **2019**, *10*, 4036–4045. [CrossRef]
51. Enes, B.N.; Moreira, L.D.P.D.; Toledo, R.C.L.; Moraes, É.A.; Moreira, M.E.D.C.; Hermsdorff, H.H.M.; Noratto, G.; Mertens-Talcott, S.U.; Talcott, S.; Martino, H.S.D.; et al. Effect of Different Fractions of Chia (*Salvia Hispanica* L.) on Glucose Metabolism, in Vivo and in Vitro. *J. Funct. Foods* **2020**, *71*, 157–164. [CrossRef]
52. Rosas-Pérez, A.M.; Honma, K.; Goda, T. Sustained Effects of Resistant Starch on the Expression of Genes Related to Carbohydrate Digestion/Absorption in the Small Intestine. *Int. J. Food Sci. Nutr.* **2020**, *71*, 572–580. [CrossRef] [PubMed]
53. da Silva, B.P.; Martino, H.S.D.; Tako, E. Plant Origin Prebiotics Affect Duodenal Brush Border Membrane Functionality and Morphology, in Vivo (*Gallus gallus*). *Food Funct.* **2021**, *12*, 6157–6166. [CrossRef] [PubMed]

nutrients

Article

Curcumin-Added Whey Protein Positively Modulates Skeletal Muscle Inflammation and Oxidative Damage after Exhaustive Exercise

Kelly Aparecida Dias [1], Aline Rosignoli da Conceição [1], Stephanie Michelin Santana Pereira [1], Lívya Alves Oliveira [1], João Vitor da Silva Rodrigues [2], Roberto Sousa Dias [2], Sérgio Oliveira de Paula [2], Antônio José Natali [3], Sérgio Luis Pinto da Matta [2], Reggiani Vilela Gonçalves [4], Elad Tako [5], Hercia Stampini Duarte Martino [1] and Ceres Mattos Della Lucia [1,*]

1 Department of Nutrition and Health, Universidade Federal de Viçosa, Viçosa 36570-900, MG, Brazil
2 Department of General Biology, Universidade Federal de Viçosa, Viçosa 36570-900, MG, Brazil
3 Department of Physical Education, Universidade Federal de Viçosa, Viçosa 36570-900, MG, Brazil
4 Department of Animal Biology, Universidade Federal de Viçosa, Viçosa 36570-900, MG, Brazil
5 Department of Food Science, Cornell University, Stocking Hall, Ithaca, NY 14850, USA
* Correspondence: cmdellalucia@ufv.br; Tel.: +55-31-986597011

Abstract: (1) Background: Exhaustive exercise can induce muscle damage. The consumption of nutritional compounds with the ability to positively influence the oxidative balance and an exacerbated inflammatory process has been previously studied. However, little is known about the nutritional value of curcumin (CCM) when mixed with whey protein concentrate (WPC). This study was developed to evaluate the effect of CCM-added WPC on inflammatory and oxidative process control and histopathological consequences in muscle tissue submitted to an exhaustive swimming test (ET). (2) Methods: 48 animals were randomly allocated to six groups (n = 8). An ET was performed 4 weeks after the start of the diet and animals were euthanized 24 h post ET. (3) Results: WPC + CCM and CCM groups reduced IL-6 and increased IL-10 expression in muscle tissue. CCM reduced carbonyl protein after ET compared to standard AIN-93M ET and WPC + CCM ET diets. Higher nitric oxide concentrations were observed in animals that consumed WPC + CCM and CCM. Consumption of WPC + CCM or isolated CCM reduced areas of inflammatory infiltrate and fibrotic tissue in the muscle. (4) Conclusions: WPC + CCM and isolated CCM contribute to the reduction in inflammation and oxidative damage caused by the exhaustive swimming test.

Keywords: exhaustive swimming test; curcuma; bioactive compounds; antioxidants; inflammatory process; oxidative stress; experimental study

Citation: Dias, K.A.; da Conceição, A.R.; Pereira, S.M.S.; Oliveira, L.A.; da Silva Rodrigues, J.V.; Dias, R.S.; de Paula, S.O.; Natali, A.J.; da Matta, S.L.P.; Gonçalves, R.V.; et al. Curcumin-Added Whey Protein Positively Modulates Skeletal Muscle Inflammation and Oxidative Damage after Exhaustive Exercise. *Nutrients* **2022**, *14*, 4905. https://doi.org/10.3390/nu14224905

Academic Editor: Stephen Ives

Received: 19 October 2022
Accepted: 17 November 2022
Published: 19 November 2022

1. Introduction

Exhaustive exercise can induce muscle damage [1,2], which, consequently, causes the development of inflammatory responses and the production of reactive oxygen species (ROS) and free radicals inducing oxidative stress in the muscle tissue [2]. Inflammation and oxidative stress induced by exhaustive exercise are indispensable for tissue recovery; however, when exacerbated and uncontrolled, they can cause secondary muscle damage [1,3]. Thus, strenuous exercise and secondary muscle damage can reduce muscle strength, impaired neuromuscular function, fatigue, a reduced range of motion, and muscle soreness [4–6]. In addition, this muscle damage can decrease muscle function, affecting physical activities and performance [3,7]. Therefore, strategies that control or minimize muscle damage, attenuate exacerbated inflammatory responses, and improve performance are increasingly studied [6,8].

In this context, consuming products with antioxidant and anti-inflammatory properties has become common and necessary to treat muscle damage [4,8]. Among the nutritional compounds that are capable of promoting a good response in the function muscle tissue is curcumin (CCM), the main curcuminoid extracted from *Curcuma longa* L, which has received considerable attention. Studies have shown that CCM can positively affect different illnesses due to its antioxidant, anti-inflammatory, and immunoregulatory effects, which highlight its cardioprotective, antineoplastic, and hepatoprotective potential [9–11]. In addition to its demonstrated health benefits, polyphenol curcumin has also been shown to improve performance in physical exercise by reducing muscle pain and damage [3,12–14], which is characterized by sarcolemma disruption, extracellular myofiber matrix abnormalities, swelling or disruption of the sarcotubular system, and distortion of the myofibril contractile components, to name a few [15].

Whey proteins (WP) have become widely consumed as a nutritional supplement because of their leucine content, digestibility, and ability to activate muscle protein synthesis [11,16]. In addition to their efficacy and safety, these proteins have high biological value and metabolic efficiency, optimizing recovery from strenuous exercise and sports performance, strength, and muscle hypertrophy [17,18]. Studies suggest that the use of WP can also attenuate oxidative stress and exercise-induced inflammation, as well as aid recovery after resistance training [16,19,20]. The benefits of WP are mainly related to its concentration of bioactive peptides [21]. In addition, it was previously demonstrated that WP rich in leucine acts as an anabolic signal, decreasing the levels of ubiquitinated proteins and inhibiting proteasome activity, decreasing muscle atrophy [22].

Previous studies have shown that CCM and whey protein ingestion alone can attenuate muscle damage induced by unusual exercise [3,18,20,23,24]. However, current scientific knowledge is lacking in regard to the potentiation of the beneficial effects of CCM-added WP, especially concerning the maintenance of muscle tissue integrity, the improvement of the oxidative balance, or the attenuation of inflammatory responses, and consequently, the histopathological changes in the muscle during the practice of strenuous exercise. Thus, and in order to bridge this gap, the aim of the present study was to evaluate the effects of curcumin-added whey protein concentrate (WPC) on the oxidative balance, inflammation process, and histopathology changes of the muscle of Wistar rats submitted to an acute protocol of exhaustive physical exercise.

2. Materials and Methods

2.1. Feedstock

2.1.1. Curcumin

The curcumin used in this study was Theracurmin® (Theravalues, Tokyo, Japan), which is highly absorbable. This product has more bioavailability than conventional CCM.

2.1.2. Whey Protein Concentrate (WPC)

The milk for the manufacture of whey protein concentrate was obtained from the stable of the Federal University of Viçosa-MG, being submitted to microfiltration. The whey-drying procedure was carried out in a LabMaq MSD 1.0 single-level spray dryer (Ribeirão Preto, SP, Brazil), according to the methodology of Perrone et al. (2013) [25].

2.1.3. Curcumin-Added Whey Protein Concentrate

CCM was mixture to WPC to obtain a product with a concentration of 0.8 g of curcumin in every 100 g of WPC, and an intake of 3.0 mg/kg/day of curcumin (maximum ADI value for humans [26]) was considered for a 70 kg individual [27]. Therefore, the protein supply to the animals through WPC corresponded to 30% of the requirements for rats in the maintenance phase [28].

2.2. Animals and Experimental Diets

Forty-eight male rats (young adults) of the Wistar strain (Rattus novergicus, variety albinus, Rodentia), 12 weeks old, were randomly divided into six experimental groups with 8 animals each ($n = 8$/group), as follows: (i) the control group that received a standard diet

(AIN-93M); (ii) the control group that received a standard diet submitted to the exhaustive test (AIN-93M ET); (iii) the curcumin-added whey protein concentrate group (WPC + CCM); (iv) the curcumin-added whey protein concentrate group submitted to the exhaustive test (WPC + CCM ET); (v) the curcumin group (CCM); and (vi) the curcumin group submitted to the exhaustive test (CCM ET) (Figure 1).

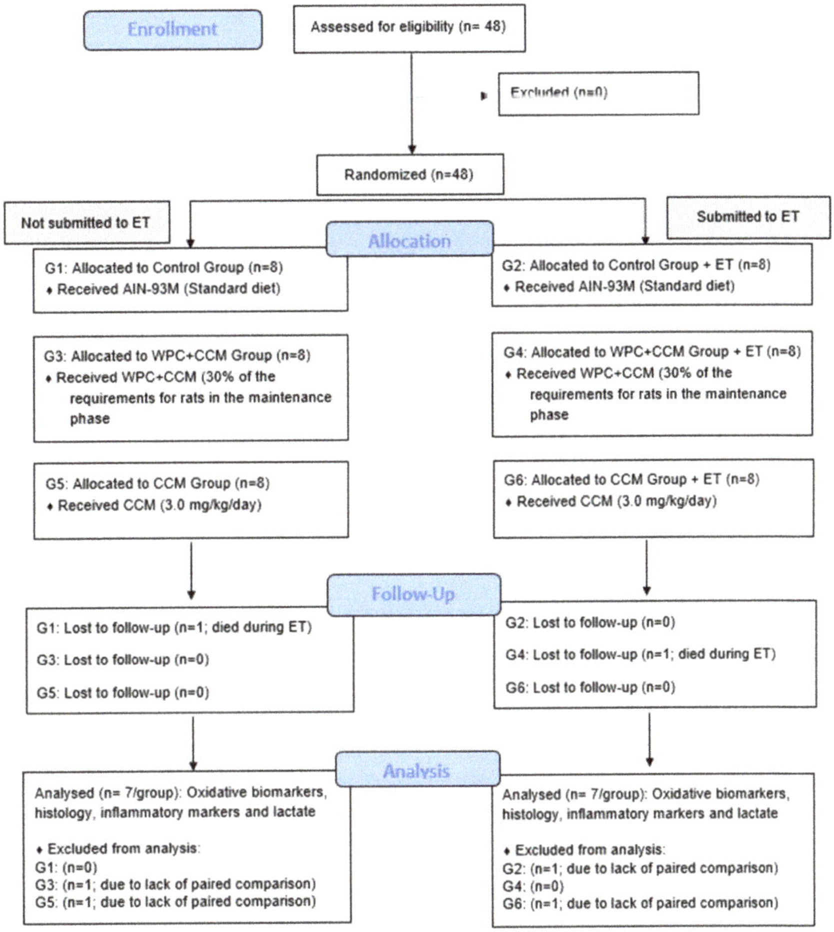

Figure 1. Flow diagram depicting the study design. AIN-93M: group that received a standard diet; WPC + CCM ET: group that received curcumin-added whey protein concentrate; CCM: group that received curcumin; ET: exhaustion test. Based on CONSORT [28].

The experimental diets specific to each group (formulated and produced in the laboratory of experimental nutrition at UFV, according to the AIN-93M recommendations proposed by the American Institute of Nutrition (1993) [29]) were introduced and maintained for 4 weeks (Table 1). The animals received water and their experimental diets ad libitum. Diets were normocaloric, normoproteic and normolipidic. The animals were distributed in individual cages in a temperature-controlled environment (22 °C ± 2 °C) and automatically controlled 12 h light/dark cycles.

Table 1. Composition of experimental diets.

Ingredients	g/kg of Diet		
	AIN-93M	WPC + CCM	CCM
Cornstarch	455.7	455.7	455.7
Albumin [1]	150.0	105.0	150.0
Maltodextrin	155.0	155.0	155.0
Sucrose	100.0	99.0	99.0
Soybean oil	40.0	40.0	40.0
Cellulose	50.0	50.0	50.0
Mineral blend	35.0	35.0	35.0
Vitamin blend	10.0	10.0	10.0
L-cystine	1.8	1.8	1.8
Choline bitartrate	2.5	2.5	2.5
WPC [2]	-	44.0	-
Theracurmin®	-	1.2	1.2

Reeves et al., 1993; AIN-93M: standard diet for maintenance phase rodents; WPC: whey protein concentrate; CCM: curcumin. [1] Albumin was considered to have 80% protein content. [2] WPC with 82.4% protein content was considered.

The animals from the groups AIN-93M ET, WPC + CCM ET, and CCM ET were submitted to the exhaustive test after 4 weeks of experiment. The rats swam to exhaustion, supporting a loading equivalent to 4% of the body weight fixed to the tail (the exhaustion protocol was modified) [30]. Exhaustion was defined as when the animal remained submerged for 10 s [31].

At the end of the experimental period and 24 h after the exhaustive test, the animals were anesthetized with isoflurane (Isoforine, Cristália®, Itapira, Brazil) and euthanized. Skeletal muscle tissues (gastrocnemius and biceps) were removed, weighed, frozen in liquid nitrogen, and stored at −80 °C until analysis. All experimental procedures were carried out following the ethical principles of animal experimentation.

2.3. Analysis of Oxidation Products and Activity of Antioxidant Enzymes

The muscle tissue samples (gastrocnemius) were weighed (300 mg) and homogenized in 1.5 mL of cold PBS (pH 7.4) using an Ultra-Turrax homogenizer (T10 basic Ultra-Turrax, IKA®, Campinas, SP, Brazil). The homogenate was centrifuged at 10,000× *g* for 10 min at 4 °C. The supernatant was used for enzymatic analysis and the evaluation of oxidation biomarkers.

Catalase (CAT), determined by its ability to cleave hydrogen peroxide (H_2O_2) into water and molecular oxygen, was evaluated as proposed by Aebi (1984) [32]. Superoxide dismutase (SOD) activity, defined as the amount of enzyme that causes a reduction in pyrogallol autooxidation, was performed according to the method proposed by Dieterich (2000) [33]. The activity of glutathione-S-transferase (GST) was obtained from its ability to metabolize 1-chloro-2,4-dinitrobenzene (CDNB), conjugated to reduced glutathione, following the methodology described by Habig (1981) [34].

The concentration of malondialdehyde (MDA) was determined as thiobarbituric acid reactive substances (TBARS) of lipid peroxidation, according to the methodology proposed by Buege and Aust (1978) [35]. The production of NO was estimated by the production of NO_2/NO_3 by the standard Griess reaction and followed the methodology of TSIKAS (2007) [36], and the oxidative damage of muscle proteins, indicated by the levels of carbonyl proteins (CP), was measured according with the method of Levine (1990) [37]. The results were normalized to the total protein concentration of the supernatant using the method described by Lowry (1951) [38].

2.4. Cytokine Profile in Muscle Homogenate

The muscle samples (200 mg) were homogenized using a tissue homogenizer (IKA WORKS GMBH and CO, Staufen, BW, Germany, model T10) in PBS buffer (pH 7.0), centrifuged (10,000× g, for 10 min at 4 °C), and the supernatant recovered. The cytokines Interleukin-2, Interleukin-10, and tumor necrosis factor were determined by a FACSVerse flow cytometer (BD, Franklin Lakes, NJ, USA) by the Cytometric Bead Array (CBA) following the manufacturer's recommendations (BD). Data were processed using FCAP Array v3.0 software, and the results were expressed in pg/g tissue.

2.5. Real-Time Polymerase Chain Reaction (RT-qPCR)

The relative expression level was assessed by a qPCR in an Illumina Eco® real-time polymerase chain reaction system (Illumina, San Diego, CA, USA), using the GoTaq® 1-Step RT-qPCR System (Promega, Madison, WI, USA). After extraction, the RNA from the muscle samples (gastrocnemius) was quantified using the Qubit 4 Fluorometer (Thermo Scientific). The sequences of the primers that were used are described in Table 2. The analysis was performed using the 2−ΔCt method in EcoStudy® software (Illumina, San Diego, CA, USA), using GAPDH as an endogenous control.

Table 2. Sequences of primers for qPCR analysis.

Genes	Oligonucleotides (5′–3′)	
	Forward	**Reverse**
GAPDH	CCCCCAATGTATCCGTTGTG	TAGCCCAGGATGCCCTTTAGT
IL-6	TCCTACCCCAACTTCCAATGCTC	TTGGATGGTCTTGGTCCTTAGCC
IL-10	TTGAACCACCCGGCATCTAC	CCAAGGAGTTGTTCCCGTTA
TNF-α	TGGGCTACGGGCTTGTCACTC	GGGGGCCACCACGCTCTT

GAPDH—Glyceraldehyde-3-Phosphate-Dehydrogenase; IL-6—Interleukin-6; IL-10—Interleukin-10; TNF-α—tumor necrosis factor-α.

2.6. Histopathological Analyzes

The muscle samples (biceps) were removed and fixed in formaldehyde. After fixation, the tissues were dehydrated in ethanol and embedded in resin containing Historesin® hardener (Leica, Wetzlar, HE, Germany). Histological sections were obtained using an automatic microtome (Leica). They were then submitted to staining using the hematoxylin and eosin (H&E) technique, mounted with Entellan (Merck), and analyzed under a light microscope (Leica DM750).

The images of the histological sections were obtained with a 20× objective using the LEICA MC170 HD digital camera. From 10 photos per histology slide, the areas containing muscle fiber, connective tissue (which was considered fibrosis when the quantity of collagen was exacerbated), and inflammatory infiltrate in the muscle were quantified using the ImagePro-Plus® application version 4.5 (Media Cybernetics, Rockville, MD, USA) by manually counting points on the tissue [39].

2.7. Blood Lactate Analysis during the Exhaustion Test

Blood samples (25 μL) were collected at two stages: before (rest) and after the end of the exhaustive swimming test (final 60 s of testing), following the protocol of Gobatto et al. (2001) [40], with modifications. The blood samples were collected from the tail end of the animal (n = 8). The animals were quickly dried with a towel before blood collection to avoid dilution with water. Lactate concentration was determined using the Accutrend Check automated blood lactate analyzer (Accusport—Roche, Basel, Switzerland).

2.8. Statistical Analysis

The results were expressed as mean and standard deviation (mean ± SD). Data normality was assessed using the Kolmogorov–Smirnov test. The *t*-test evaluated differences within the same group (not submitted to ET × submitted to ET). Intra-group differences (AIN-93M, WPC + CCM, and CCM) were submitted to an analysis of variance (ANOVA), followed by the Newman–Keuls mean test at 5% probability. All statistical analyses and graph construction were performed using the GraphPad Prism software, version 8 (GraphPad Prism Inc, La Jolla, CA, USA).

3. Results

3.1. Antioxidant Enzymes and Markers Oxidatives

The SOD activity did not differ among the AIN-93M, WPC + CCM, and CCM groups, regardless of whether or not ET was performed (Figure 2a). Among the animals not submitted to ET, the CAT activity was higher in the AIN-93M group when compared to the animals that consumed CCM ($p < 0.05$). The coefficient of variation (CV) among groups showed variation of 2.4 to 19% (Table 3). On the other hand, the CAT activity was not altered in the animals of the groups submitted to ET. The animals in the WPC + CCM ET group, as well as the animals in the CCM ET group, showed higher CAT activity compared to the animals in the WPC + CCM and CCM group, respectively ($p < 0.05$) (Figure 2b). The groups showed a CV range of 3.5 to 15.7% (Table 3). Glutathione activity, in turn, also did not differ among groups, regardless of whether the ET was performed. However, the animals in the CCM ET group showed higher GST activity when compared to the CCM group not submitted to ET ($p < 0.05$) (Figure 2c). The CV range was from 15.6 to 41.5% (Table 3).

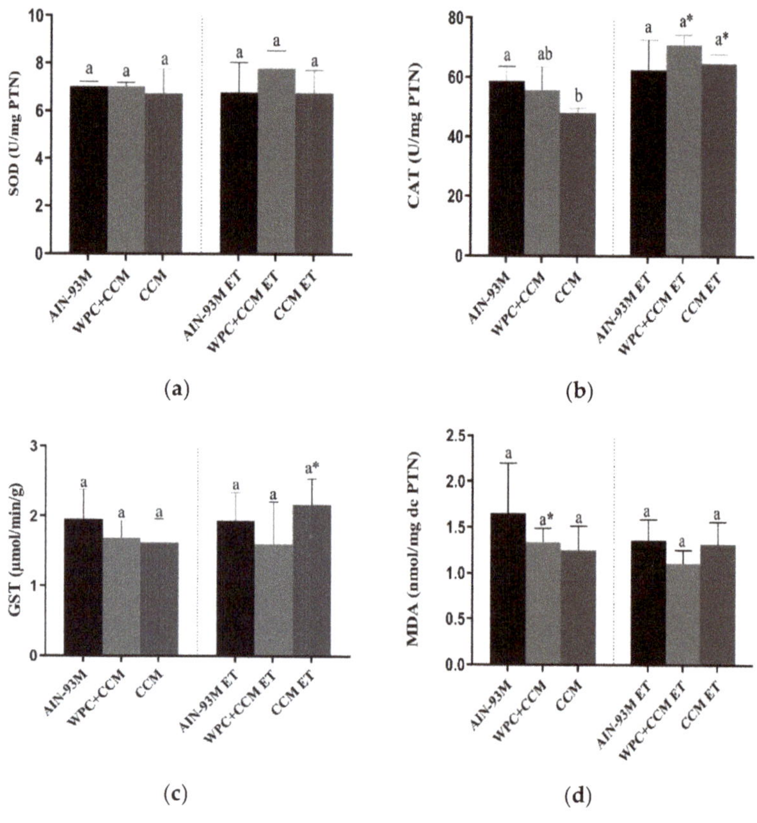

Figure 2. *Cont.*

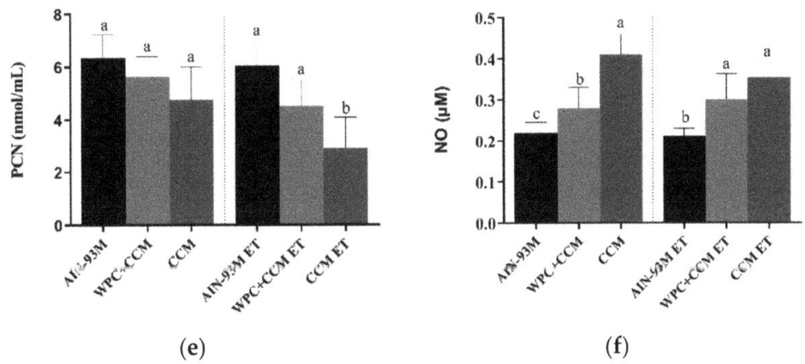

Figure 2. Analysis of oxidative balance in muscle. AIN-93M: group that received a standard diet; AIN-93M ET: group submitted to the exhaustion test that received a standard diet; WPC + CCM: group not submitted to ET that received curcumin-added whey protein concentrate; WPC + CCM ET: group submitted to the exhaustion test that received curcumin-added whey protein concentrate; CCM: group not submitted to ET that received curcumin; CCM ET: group submitted to the exhaustion test that received curcumin; ET: exhaustion test. The graphs show (**a**) superoxide dismutase, (**b**) catalase, (**c**) glutathione S-transferase, (**d**) malondialdehyde, (**e**) carbonylated protein, and (**f**) Nitric Oxide. * indicates significant differences between groups that were not submitted to ET and groups that were submitted to ET, according to the *t*-test ($p < 0.05$). According to ANOVA, different lowercase letters (**a–c**) indicate significant mean differences among the groups, followed by the Newman–Keuls test, at 5% probability. Same lowercase letters indicate that there was no significant mean difference among the groups. Data expressed as mean ± standard deviation.

Table 3. Variability expressed as coefficient of variation (CV%), tabulated within different groups.

Biomarkers	CV (%) AIN-93M	CV (%) AIN-93M ET	CV (%) WPC + CCM	CV (%) WPC + CCM ET	CV (%) CCM	CV (%) CCM ET
SOD	2.8	19	2.4	9.4	16.6	15.2
CAT	7.9	15.7	15	4.8	3.5	5
GST	21.2	21.7	15.6	41.5	21.6	15.9
MDA	33.7	18.3	11.2	12.9	22.3	19.5
PCN	14.2	15.8	13.7	23.4	25.5	42.8
NO	11.2	8.5	17.2	21.8	12.1	9.8
IL-6	-	2.4	13.9	69	11.1	26.5
TNF	-	16.2	12.3	58.8	37.9	28.9
IL-10	-	75.7	23.9	54.5	0.3	30.7
TNF (CBA)	48.6	9.2	44.5	18.7	8.8	25.2
Blood lactate	-	20.37	-	16.7	-	41.9
Swimming time	-	37.9	-	41.8	-	30
Histopathology: Inflammation	14.4	22.2	14.6	8.1	14.3	28.3
Point on muscle fiber	2.4	2.3	1.8	1.9	2	1.9
Point on connective tissue	13.6	16.4	7.8	8.2	18.3	9.6
Muscle fiber area	5.1	2.7	14.4	7.1	12.5	4.3
Muscle fiber diameter	3.5	4.8	0.2	4.4	9.2	0.1

AIN-93M: group not submitted to ET that received a standard diet for rodents in the maintenance phase; AIN-93M ET: group submitted to the exhaustion test that received a standard diet for rodents in the maintenance phase; WPC + CCM: group not submitted to ET that received curcumin-added whey protein concentrate; WPC + CCM ET: group submitted to the exhaustion test that received curcumin-added whey protein concentrate; CCM: group not submitted to ET that received curcumin; CCM ET: group submitted to the exhaustion test that received curcumin. CV: coefficient of variation.

MDA concentrations showed no difference among the AIN-93M, WPC + CCM, and CCM groups, regardless of whether ET was performed. However, the animals in the WPC + CCM ET group had lower concentrations of MDA when compared to the WPC + CCM group ($p < 0.05$) not submitted to ET (Figure 2d). The groups showed a CV range of 11.2 to 33.7% (Table 3). PCN concentrations were similar among the animals of the AIN-93M, WPC + CCM, and CCM groups not submitted to ET. However, the animals in the CCM ET group had lower concentrations of PCN ($p < 0.05$) when compared to the animals in the AIN-93M ET group and the WPC + CCM ET group (Figure 2e). The CV among groups showed variation of 13.7 to 42.8% (Table 3). Among animals not submitted to ET, NO levels were higher in the CCM group when compared to the AIN-93M and WPC + CCM groups ($p < 0.05$). Moreover, the WPC + CCM group had a higher amount of NO when compared to the AIN-93M group ($p < 0.05$). Concerning the animals submitted to ET, the animals of the CCM ET group, as well as the animals of the WPC + CCM ET group, showed higher levels of NO when compared to the animals of the AIN-93M ET group ($p < 0.05$) (Figure 2f). The groups showed a CV range of 8.5 to 21.8% (Table 3).

3.2. Gene Expression of TNF-α, IL-6, and IL-10

The TNF-α expression levels remained similar among groups (Figure 3a). The groups showed a CV range of 12.3 to 58.8% (Table 3). The gene expression of the cytokine IL-6 was higher in the AIN-93M ET group when compared to the other groups ($p < 0.05$) (Figure 3b). The CV range was from 2.4 to 69% (Table 3). The gene expression of the cytokine IL-10 was higher in the groups that received WPC + CCM and CCM, regardless of whether ET was performed, when compared to the AIN-93M ET group ($p < 0.05$) (Figure 3c). The CV range was from 0.3 to 75.7% (Table 3).

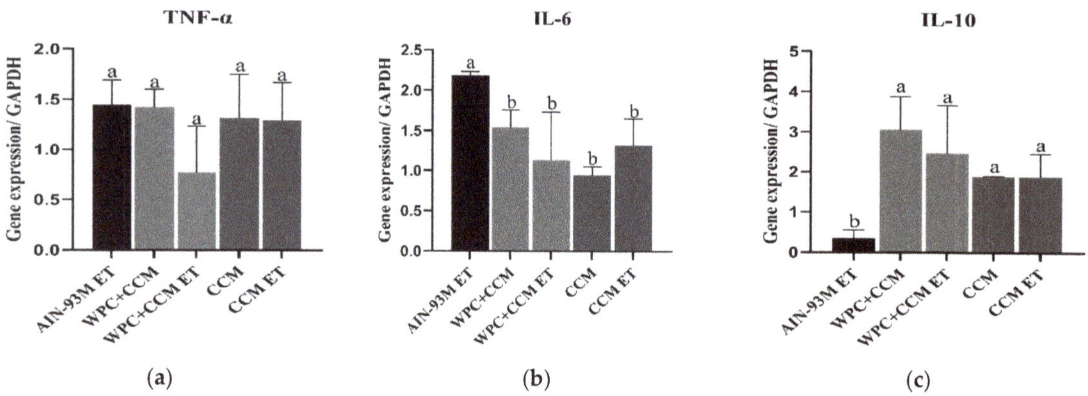

Figure 3. Gene expression of TNF-α, IL-6, and IL-10 in muscle. AIN-93M: group not submitted to ET that received a standard diet was used as the reference group, and the results expressed concerning it; AIN-93M ET: group submitted to the exhaustion test that received a standard diet; WPC + CCM: group not submitted to ET that received curcumin-added whey protein concentrate; WPC + CCM ET: group submitted to the exhaustion test that received curcumin-added whey protein concentrate; CCM: group not submitted to ET that received curcumin; CCM ET: group submitted to the exhaustion test that received curcumin; ET: exhaustion test. The graphs show (**a**) TNF-α, (**b**) IL-6, (**c**) IL-10. According to ANOVA, different lowercase letters (**a**,**b**) indicate significant mean differences among the groups, followed by the Newman–Keuls test, at 5% probability. Same lowercase letters indicate that there was no significant mean difference among the groups. Data expressed as mean ± standard deviation.

3.3. Cytokine Production Profile in Muscle Tissue

The cytometry-measured levels of cytokines IL-2, IL-10, and TNF-α were calculated in muscle. IL-2 and IL-10 were not detected, probably because they were not yet in their protein form or were below the detection limit.The TNF-α levels among animals not submitted to ET were similar among groups. The animals in the WPC + CCM and CCM group showed a tendency towards a reduction in TNF-α levels compared to the AIN-93M group, but without statistical significance. Among the animals submitted to ET, a trend in the reduction in TNF-α was also observed in the animals of the WPC + CCM ET group (130.8 ± 26.08) when compared to the AIN-93 ET group (196.0 ± 18, 16) and CCM ET (Figure 4). The groups showed a CV range of 8.8 to 48.6% (Table 3).

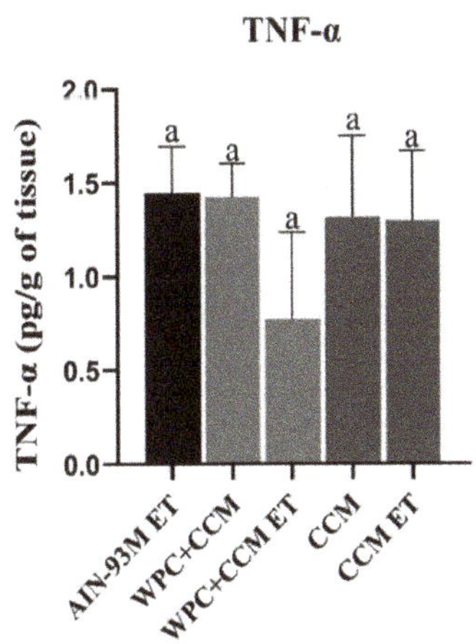

Figure 4. TNF-α profile in muscle using the cytometric bead array kit. AIN-93M: group not submitted to ET that received a standard diet; AIN-93M ET: group submitted to the exhaustion test that received a standard diet; WPC + CCM: group not submitted to ET that received curcumin-added whey protein concentrate; WPC + CCM ET: group submitted to the exhaustion test that received curcumin-added whey protein concentrate; CCM: group not submitted to ET that received curcumin; CCM ET: group submitted to the exhaustion test that received curcumin; ET: exhaustion test. According to ANOVA, different lowercase letters indicate significant mean differences among the groups, followed by the Newman–Keuls test, at 5% probability. Same lowercase letters indicate that there was no significant mean difference among the groups. Data expressed as mean ± standard deviation.

3.4. Histopathology

Among the animals not submitted to ET, more significant areas of inflammation were observed in the animals of the AIN-93M group ($p < 0.05$) when compared to the animals of the WPC + CCM and CCM groups (Figure 5). The same was observed among the groups submitted to TE, with the WPC + CCM ET and CCM ET groups showing lower inflammation when compared to the AIN-93M ET group ($p < 0.05$) (Table 4). The groups showed a CV range of 8.1 to 28.3% (Table 3).

Among the animals not submitted to ET, it was possible to observe a greater amount of muscle fiber in the WPC + CCM and CCM groups when compared to the AIN-93M group ($p < 0.05$). Among the animals submitted to ET, those in the WPC + CCM and CCM groups had a greater amount of muscle fiber per evaluated area when compared to the animals in the AIN-93M group ($p < 0.05$) (Figure 5). The groups showed a CV range of 1.8 to 2.4% (Table 3). In turn, the presence of connective tissue among the animals not submitted to ET was higher in the AIN-93M group when compared to the WPC + CCM and CCM groups ($p < 0.05$). The same was found among the animals submitted to the exhaustive test, with a greater presence of connective tissue in the muscle tissues of the animals of the AIN-93M ET group when compared to the animals of the WPC + CCM ET and CCM ET groups ($p < 0.05$) (Table 3). The groups showed a CV range of 7.8 to 18.3% (Table 3).

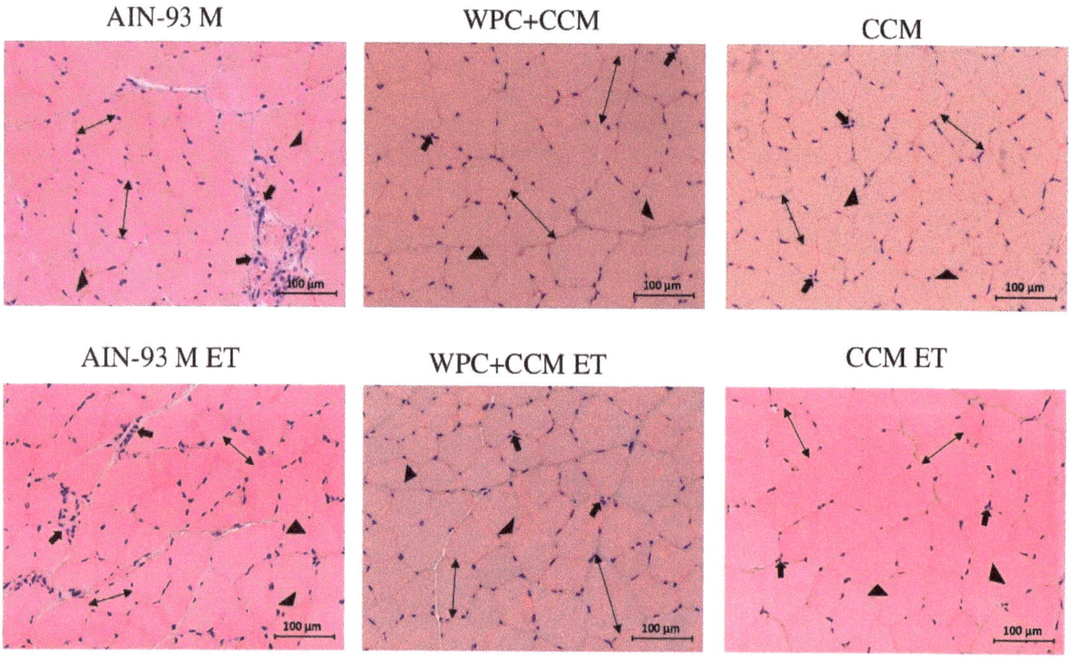

Figure 5. Cross-section of the muscle of rats that were or were not submitted to the exhaustive test. AIN-93M: group not submitted to ET that received a standard diet for rodents in the maintenance phase; AIN-93M ET: group submitted to the exhaustion test that received a standard diet for rodents in the maintenance phase; WPC + CCM: group not submitted to ET that received curcumin-added whey protein concentrate; WPC + CCM ET: group submitted to the exhaustion test that received curcumin-added whey protein concentrate; CCM: group not submitted to ET that received curcumin; CCM ET: group submitted to the exhaustion test that received curcumin. Inflammatory infiltrates (→); Muscle fiber (↔); Connective tissue (►).

Table 4. Histopathology of the muscle of rats that were submitted or not to an exhaustive swimming test.

Groups	Inflammation (%)	Count a Point on Muscle Fiber	Count a Point on Connective Tissue
AIN-93M	3.69 ± 0.53^{a}	243.38 ± 5.92^{b}	43.01 ± 5.88^{a}
WPC + CCM	2.03 ± 0.45^{b}	254.00 ± 5.72^{a}	33.70 ± 5.52^{b}
CCM	1.92 ± 0.28^{b}	258.60 ± 4.59^{a}	24.60 ± 1.93^{c}
AIN-93M ET	3.59 ± 0.29^{aa}	236.80 ± 4.43^{b}	39.31 ± 3.23^{a}
WPC + CCM ET	2.02 ± 0.29^{b}	259.15 ± 5.08^{a}	26.65 ± 4.88^{b}
CCM ET	1.73 ± 0.49^{b}	258.06 ± 4.88^{a}	26.46 ± 2.55^{b}

AIN-93M: group not submitted to ET that received a standard diet; AIN-93M ET: group submitted to the exhaustion test that received a standard diet; WPC + CCM: group not submitted to ET that received curcumin-added whey protein concentrate; WPC + CCM ET: group submitted to the exhaustion test that received curcumin-added whey protein concentrate; CCM: group not submitted to ET that received curcumin; CCM ET: group submitted to the exhaustion test that received curcumin; ET: exhaustion test. Means followed by the same letter in the column do not differ for the same condition (not submitted to ET × submitted to ET). According to ANOVA, different lowercase letters (**a–c**) indicate intra-group differences, followed by the Newman–Keuls test, at 5% probability. Same lowercase letters indicate that there was no significant mean difference among the groups. Data expressed as mean ± standard deviation.

Among the animals not submitted to ET, the area measurements remained similar. Moreover, no differences were observed within the same group (sedentary vs. ET). Among the animals submitted to ET, the animals of the WPC + CCM ET and CCM ET groups had a greater area of muscle fiber when compared to the animals of the AIN-93M ET group, who had muscle atrophy ($p < 0.05$) (Figure 6a). The groups showed a CV range of 2.7 to 14.4% (Table 3).

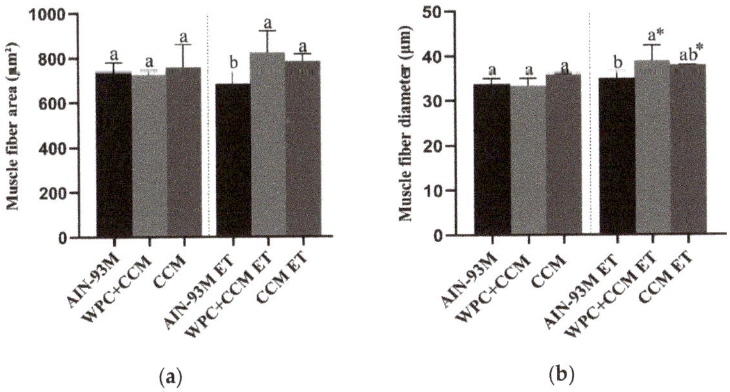

Figure 6. Histopathological analysis of muscle tissue. AIN-93M: group not submitted to ET that received a standard diet; AIN-93M ET: group submitted to the exhaustion test that received a standard diet; WPC + CCM: group not submitted to ET that received curcumin-added whey protein concentrate; WPC + CCM ET: group submitted to the exhaustion test that received curcumin-added whey protein concentrate; CCM: group not submitted to ET that received curcumin; CCM ET: group submitted to the exhaustion test that received curcumin; ET: exhaustion test. The graphs show (**a**) area of muscle fibers (µm^2), (**b**) diameter of muscle fibers (µm). * indicates significant differences within the same group (not submitted to ET × submitted to ET), according to the *t*-test ($p < 0.05$). According to ANOVA, different lowercase letters (**a**,**b**) indicate significant mean differences among the groups, followed by the Newman–Keuls test, at 5% probability. Same lowercase letters indicate that there was no significant mean difference among the groups. Data expressed as mean ± standard deviation.

Regarding the diameter of the muscle fibers, it was possible to observe a greater diameter among the animals in the WPC + CCM ET group when compared to the AIN-93M ET group ($p < 0.05$) (Figure 6b). Among the animals in the CCM ET group, the diameter of the muscle fibers remained similar to those in the WPC + CCM ET and AIN-93M ET groups. The groups showed a CV range of 0.1 to 9.2% (Table 3). Moreover, when comparing the diameters within the same group (not submitted to ET × submitted to ET), the animals of the WPC + CCM ET and CCM ET groups showed greater muscle fiber diameter when compared to the WPC + CCM and CCM groups not submitted to ET, respectively (Figure 6b). No differences were observed among the animals of the AIN-93M group (not submitted to ET × submitted to ET).

3.5. Blood Lactate and Swimming Time

Resting lactate levels and swimming time did not differ among groups. However, after performing the ET, the WPC + CCM group had lower lactate levels when compared to the AIN-93M group ($p < 0.05$). In addition, through comparisons among the groups, it was possible to observe that lactate levels after the end of the ET were higher in the AIN-93M group after the performance of the ET when compared to the same group before ET ($p < 0.05$). In the WPC + CCM and CCM groups, the lactate values before and after performing the ET remained similar (Figure 7). The groups showed a CV range of 16.7 to 41.9% for lactate and 30 to 41.8% for swimming time (Table 3).

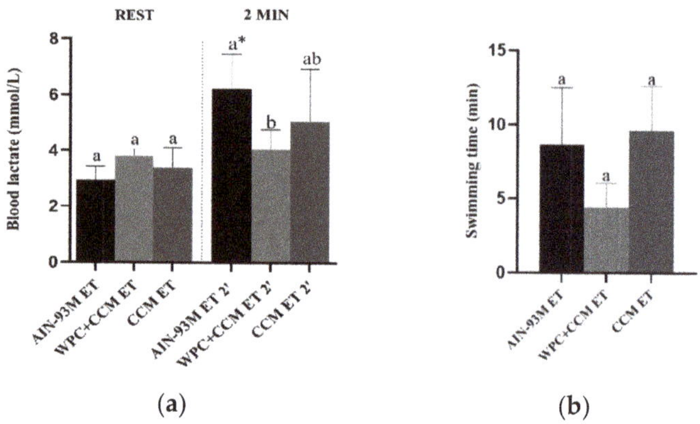

Figure 7. Blood lactate levels and swimming time. (**a**) Blood lactate levels collected at resting and 2 min after performing the exhaustive swimming test. (**b**) Swimming time. AIN-93M: group not submitted to ET that received a standard diet; AIN-93M ET: group submitted to the exhaustion test that received a standard diet; WPC + CCM: group not submitted to ET that received curcumin-added whey protein concentrate; WPC + CCM ET: group submitted to the exhaustion test that received curcumin-added whey protein concentrate; CCM: group not submitted to ET that received curcumin; CCM ET: group submitted to the exhaustion test that received curcumin; ET: exhaustion test. * indicates significant differences within the same treatment group (before ET × after ET), according to the *t*-test ($p < 0.05$). According to ANOVA, different lowercase letters (**a**,**b**) indicate significant mean differences among the groups, followed by the Newman–Keuls test, at 5% probability. Same lowercase letters indicate that there was no significant mean difference among the groups. Data expressed as mean ± standard deviation.

4. Discussion

Acute and exhaustive exercise is associated with increased tissue stress, especially inflammation, oxidative stress, and, consequently, changes in muscle morphology [2]. In this context, curcumin and whey proteins have been shown to be safe for consumption and exert antioxidant and anti-inflammatory effects [18,21,41]. Overall, our results showed that the consumption of curcumin-added whey protein concentrate had positive effects on oxidative balance and the development of inflammation in the muscle tissue, with a consequent change in muscle morphology. Thus, we can highlight a reduction in oxidative markers, an increase in antioxidant enzymes, and a reduction in inflammation and blood lactate after performing the exhaustive test.

Muscle damage resulting from unusual and exhaustive exercise is related to the production of free radicals and, consequently, ROS associated with inflammation [42], which can lead to cellular damage, such as membrane disorganization, protein oxidation, and the alteration of cellular functions [43]. Increased oxidative stress associated with an inability of the endogenous antioxidant system to remove excess free radicals impairs recovery and decreases exercise performance [42,44].

Substances reactive to thiobarbituric acid (TBARS) or MDA, and carbonyl protein levels are products of the oxidation of biomolecules and are commonly used as biomarkers and indicative of oxidative stress [45]. In turn, enzymes such as CAT, SOD, and GST are used as parameters to reflect the antioxidant status [46]. Our study demonstrated that SOD concentrations remained similar amonggroups, possibly because there was not enough free radical production to activate the antioxidant enzyme system beyond what the cell normally expresses. The curcumin-added whey protein concentrate, as with the exposure to isolated curcumin, provided higher antioxidant activity due to greater recruitment of the catalase enzyme by performing the exhaustive test. Our results show that WPC and CCM can significantly activate antioxidant enzymes in muscle tissue and, consequently, have a protective activity for the tissue. However, surprisingly, CAT concentrations among animals not submitted to ET were lower in the CCM group, probably due to lower H2O2 formation or an ability of curcumin itself to reduce H2O2 levels. It is worth mentioning that our data are based on the analysis of enzyme activity and not on their gene expression; therefore, we believe that future studies are necessary to clarify the possible mechanisms of action and consequent tissue alteration.

The administration of the protein concentrate mixed with curcumin allowed for a reduction in lipid peroxidation induced by free radicals and ROS, such as H_2O_2, provoked by the performance of the exhaustive swimming test. Similarly, a study with male rats undergoing resistance exercise training to induce muscle damage also demonstrated the positive modulatory effects of dietary curcumin, with attenuation in MDA concentrations [14]. In the current study, due to its antioxidant properties, CCM was able to reduce protein oxidation, one of the main processes responsible for the misfolding of proteins within the rough endoplasmic reticulum and, consequently, for their malfunction. A previous study by our research group [47] found that CCM consumption was responsible for reducing MDA concentrations, and that exposure to CCM, as well as WPC + CCM, was able to reduce concentrations of PCN in the liver. Our current findings partially corroborate the previous study. The differences found could possibly be related to the investigated tissue, since the liver is the main metabolism organ and, therefore, where the most relevant alterations are found.

Under conditions of oxidative stress, high levels of superoxide ion react with nitric oxide (NO), forming a reactive nitrogen species (RNS), consequently producing peroxynitrite [48]. It is already known that NO suppresses the installation of an inflammatory process by decreasing lymphocyte activation and by decreasing leukocyte adhesion, preventing the rolling and migration of cells and, consequently, inhibiting the installation of a tissue inflammatory process [49,50]. These reports corroborate our findings, since treatments with WPC + CCM and CCM showed increased NO levels and decreased inflammatory infiltrate foci in the tissue.

After unusual and/or prolonged exercise, muscle damage stimulates the inflammatory response through the recruitment of immune system cells, such as neutrophils [42], which send signals for the migration of macrophages that, in turn, mediate the expression of inflammatory cytokines, growth factors, ROS, and RNS directly involved in the progression of inflammation [43]. Thus, the inflammatory process must be controlled to avoid persistent tissue damage due to the continuous action of free radicals [43]. The WPC and CCM represent a promising therapy in controlling inflammation after exhaustive exercise, as reduced IL-6 expression levels and increased IL-10 levels directly reflect attenuation of the inflammatory response provoked by the exercise and improve muscular recovery.

Increased IL-6 in skeletal muscle is related to the incidence of muscle damage induced by strenuous and/or exhaustive exercise [51]. A potential explanation of the lower concentration of IL-6 in the groups that consumed the WPC + CCM or the isolated CCM is probably related to the high concentrations of WPC bioactive peptides and the concentrations of CCM in the blood, which were maintained during and after exercise. Previous studies have also reported that the inflammatory response is suppressed when curcumin is ingested several days before exercise [23,52]. Corroborating our results, Davis et al. (2007) [53] showed that curcumin could reduce IL-6 and TNF-α concentrations after eccentric exercise, controlling inflammation and protecting tissue. We also observed an increase in the levels of IL-10 expression in the muscle of the animals submitted to the exhaustive swimming test, which received the WPC + CCM or the isolated CCM.

IL-10 is an anti-inflammatory cytokine that acts by suppressing the activation of immune cells after the emergence of an initial inflammatory response [54]. The increase in IL-10 concentration can accelerate the inflammatory recovery process [6]. Curcumin has been shown to induce the expression and release of IL-10 [55], which partly explains its anti-inflammatory effects. In our previous study [47], we also observed increased IL-10 concentrations in the liver of Wistar rats that received WPC + CCM or isolated CCM.

The analysis of the inflammatory cytokine profile showed that, although without statistical significance ($p > 0.05$), the WPC + CCM ET group showed a tendency towards lower concentrations of TNF-α. However, levels of IL-10 in its protein form could not be detected. These results suggest that after the production of mRNA for IL-10 (as detected by RT-qPCR analysis), there was a deactivation of M1-type macrophages and an increase in M2-type macrophages reducing exercise-induced inflammation.

The anti-inflammatory role of WPC is related to its concentration of bioactive peptides, present in whey proteins, which can be generated by enzymatic hydrolysis, microbial fermentation, or through the gastrointestinal digestion of food proteins [21]. The ingestion of isolated CCM has been shown to be effective in the reduction in pro-inflammatory cytokines, since CCM can negatively regulate the inducible nitric oxide synthase (iNOS) enzymes, cyclooxygenase-2 (COX-2), lipoxygenase, and xanthine oxidase activity, suppressing the activation of NF-κB [56]. To corroborate our findings, histological evaluations also indicated less inflammation in both the animals that received WPC + CCM and the animals that only received CCM, demonstrating the ability to reduce inflammation. In addition, it was also possible to observe an increase in area and diameter muscle fibers in these groups, which may be related to the increase in muscle strength.

It was also possible to observe that the animals of the control group exhibited a more pronounced connective tissue when compared to the animals of the WPC + CCM and CCM groups. Increased connective tissue, formed mainly by type III collagen fibers, consequently implies decreased muscle fibers. In general, the thickening of collagen fibers is related to some type of injury, usually the beginning of the fibrotic process and cell death [57,58]. These findings are possibly associated with the chronicity of inflammation and the development of degenerative processes that culminate in cell death with consequent collagen deposition. The exhaustive swimming test caused morphological changes that may have led to cell death and increased tissue fibrosis. However, more studies are needed to investigate such effects.

Surprisingly, we observed that the groups that received the AIN93M diet showed increased inflammation in muscle tissue, as well as the development of increased connective tissue and muscle atrophy. The high carbohydrates content and low protein presented in this diet could be related to this result. In its formulation, the AIN93M diet contains 70% of carbohydrates and only 15% of protein impairing protein synthesis. Furthermore, it is known that an increased fats and carbohydrates' intake in the diet may be related to mechanisms that are involved in lipidic metabolism, such as a reduction in lipoprotein formation, inflammatory processes, and free radicals responsible for lipid chain peroxidation. A previous study by our research group also reported damage caused by the consumption of standard AIN-93M diet by showing that animals fed this diet showed increased fat in hepatocytes.

Exhaustion in swimming rats is a complex type of stress that generates central and peripheral fatigue, since the animals are highly motivated to avoid drowning, and thus, factors such as thermal and emotional stress are difficult to eliminate [30]. In our test of endurance swimming to exhaustion, rats supporting loads of 4% of their body weight usually reached maximal blood lactate stabilization around 4–6 mmol/L, which is indicative of peripheral fatigue [40,59]. More importantly, our results show that the group supplemented with WPC + CCM showed lower lactate levels after the exhaustive swimming test. Elevated levels of lactate in muscle and blood during exercise can provoke sensations of pain and discomfort, increasing the perception of exertion and, consequently, reducing performance [60]. Similar findings were demonstrated by Sahin et al. (2016) [14] in their study with male Wistar rats submitted to an exercise protocol, in which 20 mg of curcuminoids was provided per day for 6 weeks. The authors found that serum lactate levels in the supplemented group decreased compared to control group levels.

This study corroborates our previous findings that the consumption of WPC + CCM and isolated CCM contribute to the reduction in inflammation and oxidative damage caused by the exhaustive swimming test, despite the low dose of CCM offered (3 mg/kg/day). It is important to emphasize that these markers do not directly reflect muscle damage; however, they can be used with other variables to infer muscle changes resulting from exhaustive physical exercise. Future studies are needed to clarify the acute and chronic mechanisms related to the antioxidant and anti-inflammatory responses generated by the consumption of WPC + CCM and isolated CCM, and an evaluation for a longer time after performing an exhaustive test.

5. Conclusions

The consumption of WPC + CCM and/or isolated CCM for four weeks before performing an exhaustive swimming test was able to reduce the inflammatory response through the reduction in IL-6 and increase in IL-10. In addition, the consumption of WPC + CCM was able to reduce the lipid peroxidation caused by exhaustive exercise, and the consumption of CCM reduced the concentrations of carbonylated protein. Higher nitric oxide levels were observed by the consumption of WPC + CCM and CCM, evidencing a possible cardioprotective action. Furthermore, the consumption of WPC + CCM and CCM was able to reduce the areas of inflammatory infiltrate and connective tissue in the muscle caused by exhaustive exercise, as well as increase the area and diameter of muscle fibers. The association of WPC + CCM also allowed for lower lactate production during exhaustive exercise.

Author Contributions: Conceptualization, K.A.D., A.R.d.C., S.M.S.P., L.A.O., R.V.G., H.S.D.M. and C.M.D.L.; methodology, K.A.D., A.R.d.C., S.M.S.P., J.V.d.S.R. and R.S.D.; validation, R.S.D., A.J.N., H.S.D.M. and S.L.P.d.M.; formal analysis, K.A.D., A.R.d.C., S.M.S.P., L.A.O., J.V.d.S.R. and R.S.D.; investigation, K.A.D., A.R.d.C. and S.M.S.P.; resources, A.J.N., S.L.P.d.M., S.O.d.P., E.T., C.M.D.L., R.V.G. and H.S.D.M.; writing—original draft preparation, K.A.D., A.R.d.C., S.M.S.P. and L.A.O.; writing—review and editing, R.V.G., E.T. and E.T.; supervision, R.V.G., H.S.D.M., E.T. and C.M.D.L.; project administration, funding acquisition, K.A.D., A.R.d.C. and C.M.D.L. All authors have read and agreed to the published version of the manuscript.

Funding: This work was supported by the National Counsel for Technological and Scientific Development (CNPq, Brazil) (Finance Code 428701/2018-3), Foundation for Research Support of Minas Gerais (FAPEMIG, Brazil), and Coordination for the Improvement of Higher Education Personnel (CAPES, Brazil) (Finance Code 001).

Institutional Review Board Statement: Ethics Committee on the Use of Animals, Federal University of Viçosa (CEUA/UFV), under protocol number 72/2018.

Informed Consent Statement: Not applicable.

Data Availability Statement: Data can be made available on request.

Conflicts of Interest: The authors declare no conflict of interest.

References

1. Clarkson, P.M.; Nosaka, K.; Braun, B. Muscle Function after Exercise-Induced Muscle Damage and Rapid Adaptation. *Med. Sci. Sports Exerc.* **1992**, *24*, 512–520. [CrossRef] [PubMed]
2. González-Bartholin, R.; Mackay, K.; Valladares, D.; Zbinden-Foncea, H.; Nosaka, K.; Peñailillo, L. Changes in Oxidative Stress, Inflammation and Muscle Damage Markers Following Eccentric versus Concentric Cycling in Older Adults. *Eur. J. Appl. Physiol.* **2019**, *119*, 2301–2312. [CrossRef] [PubMed]
3. Tanabe, Y.; Chino, K.; Sagayama, H.; Lee, H.J.; Ozawa, H.; Maeda, S.; Takahashi, H. Effective Timing of Curcumin Ingestion to Attenuate Eccentric Exercise-Induced Muscle Soreness in Men. *J. Nutr. Sci. Vitaminol.* **2019**, *65*, 82–89. [CrossRef] [PubMed]
4. Howatson, G.; Van Someren, K.A. The Prevention and Treatment of Exercise-Induced Muscle Damage. *Sport. Med.* **2008**, *38*, 483–503. [CrossRef]
5. Huang, W.C.; Chiu, W.C.; Chuang, H.L.; Tang, D.W.; Lee, Z.M.; Li, W.; Chen, F.A.; Huang, C.C. Effect of Curcumin Supplementation on Physiological Fatigue and Physical Performance in Mice. *Nutrients* **2015**, *7*, 905–921. [CrossRef] [PubMed]
6. Faria, F.R.; Gomes, A.C.; Antunes, A.; Rezende, K.R.; Pimentel, G.D.; Oliveira, C.L.P.; Antunes, B.M.; Lira, F.S.; Aoki, M.S.; Mota, J.F. Effects of Turmeric Extract Supplementation on Inflammation and Muscle Damage after a Half-Marathon Race: A Randomized, Double-Blind, Placebo-Controlled Trial. *Eur. J. Appl. Physiol.* **2020**, *120*, 1531–1540. [CrossRef]
7. Fuster-Muñoz, E.; Roche, E.; Funes, L.; Martínez-Peinado, P.; Sempere, J.M.; Vicente-Salar, N. Effects of Pomegranate Juice in Circulating Parameters, Cytokines, and Oxidative Stress Markers in Endurance-Based Athletes: A Randomized Controlled Trial. *Nutrition* **2016**, *32*, 539–545. [CrossRef]

8. Howatson, G.; McHugh, M.P.; Hill, J.A.; Brouner, J.; Jewell, A.P.; Van Someren, K.A.; Shave, R.E.; Howatson, S.A. Influence of Tart Cherry Juice on Indices of Recovery Following Marathon Running. *Scand. J. Med. Sci. Sport.* **2010**, *20*, 843–852. [CrossRef]
9. Hewlings, S.J.; Kalman, D.S. Curcumin: A Review of Its Effects on Human Health. *Foods* **2017**, *6*, 92. [CrossRef]
10. Liu, F.; Gao, S.; Yang, Y.; Zhao, X.; Fan, Y.; Ma, W.; Yang, D.; Yang, A.; Yu, Y. Antitumor Activity of Curcumin by Modulation of Apoptosis and Autophagy in Human Lung Cancer A549 Cells through Inhibiting PI3K/Akt/MTOR Pathway. *Oncol. Rep.* **2018**, *39*, 1523–1531. [CrossRef]
11. Li, M.; Liu, F. Effect of Whey Protein Supplementation during Resistance Training Sessions on Body Mass and Muscular Strength: A Meta-Analysis. *Food Funct.* **2019**, *10*, 2766–2773. [CrossRef] [PubMed]
12. Tanabe, Y.; Maeda, S.; Akazawa, N.; Zempo-Miyaki, A.; Choi, Y.; Ra, S.G.; Imaizumi, A.; Otsuka, Y.; Nosaka, K. Attenuation of Indirect Markers of Eccentric Exercise-Induced Muscle Damage by Curcumin. *Eur. J. Appl. Physiol.* **2015**, *115*, 1949–1957. [CrossRef] [PubMed]
13. Nicol, L.M.; Rowlands, D.S.; Fazakerly, R.; Kellett, J. Curcumin Supplementation Likely Attenuates Delayed Onset Muscle Soreness (DOMS). *Eur. J. Appl. Physiol.* **2015**, *115*, 1769–1777. [CrossRef] [PubMed]
14. Sahin, K.; Pala, R.; Tuzcu, M.; Ozdemir, O.; Orhan, C.; Sahin, N.; Juturu, V. Curcumin Prevents Muscle Damage by Regulating NF-KB and Nrf2 Pathways and Improves Performance: An in Vivo Model. *J. Inflamm. Res.* **2016**, *9*, 147–154. [CrossRef]
15. Fridén, J.; Lieber, R.L. Structural and Mechanical Basis of Exercise-Induced Muscle Injury. *Med. Sci. Sports Exerc.* **1992**, *24*, 521–529. [CrossRef]
16. Davies, R.W.; Carson, B.P.; Jakeman, P.M. The Effect of Whey Protein Supplementation on the Temporal Recovery of Muscle Function Following Resistance Training: A Systematic Review and Meta-Analysis. *Nutrients* **2018**, *10*, 221. [CrossRef]
17. Atherton, P.J.; Smith, K. Muscle Protein Synthesis in Response to Nutrition and Exercise. *J. Physiol.* **2012**, *590*, 1049–1057. [CrossRef]
18. Teixeira, K.R.; Silva, M.E.; de Lima, W.G.; Pedrosa, M.L.; Haraguchi, F.K. Whey Protein Increases Muscle Weight Gain through Inhibition of Oxidative Effects Induced by Resistance Exercise in Rats. *Nutr. Res.* **2016**, *36*, 1081–1089. [CrossRef]
19. Athira, S.; Mann, B.; Sharma, R.; Kumar, R. Ameliorative Potential of Whey Protein Hydrolysate against Paracetamol-Induced Oxidative Stress. *J. Dairy Sci.* **2013**, *96*, 1431–1437. [CrossRef]
20. Kerasioti, E.; Stagos, D.; Jamurtas, A.; Kiskini, A.; Koutedakis, Y.; Goutzourelas, N.; Pournaras, S.; Tsatsakis, A.M.; Kouretas, D. Anti-Inflammatory Effects of a Special Carbohydrate-Whey Protein Cake after Exhaustive Cycling in Humans. *Food Chem. Toxicol.* **2013**, *61*, 42–46. [CrossRef]
21. Ma, Y.; Liu, J.; Shi, H.; Yu, L. Isolation and Characterization of Anti-Inflammatory Peptides Derived from Whey Protein. *J. Dairy Sci.* **2016**, *99*, 6902–6912. [CrossRef] [PubMed]
22. Combaret, L.; Dardevet, D.; Rieu, I.; Pouch, M.N.; Béchet, D.; Taillandier, D.; Grizard, J.; Attaix, D. A Leucine-Supplemented Diet Restores the Defective Postprandial Inhibition of Proteasome-Dependent Proteolysis in Aged Rat Skeletal Muscle. *J. Physiol.* **2005**, *569*, 489–499. [CrossRef]
23. Tanabe, Y.; Chino, K.; Ohnishi, T.; Ozawa, H.; Sagayama, H.; Maeda, S.; Takahashi, H. Effects of Oral Curcumin Ingested before or after Eccentric Exercise on Markers of Muscle Damage and Inflammation. *Scand. J. Med. Sci. Sport.* **2019**, *29*, 524–534. [CrossRef] [PubMed]
24. Wang, I.L.; Hsiao, C.Y.; Li, Y.H.; Meng, F.B.; Huang, C.C.; Chen, Y.M. Nanobubbles Water Curcumin Extract Reduces Injury Risks on Drop Jumps in Women: A Pilot Study. *Evidence-Based Complement. Altern. Med.* **2019**, *2019*, 8647587. [CrossRef] [PubMed]
25. Perrone, Í.T.; Simeão, M.; Rodrigues Júnior, P.H.; Stephani, R.; Carvalho, A.F. Influence of Pilot Spray Dryer Settings on Moisture and Water Activity of Whole Milk Powder. *Rev. Do Inst. Laticínios Cândido Tostes* **2013**, *68*, 5–9. [CrossRef]
26. Stankovic, I. Curcumin—Chemical and Technical Assessment. *FAO* **2004**, *61st JECFA*, 1–8.
27. Schoenfeld, B.J.; Aragon, A.A. How Much Protein Can the Body Use in a Single Meal for Muscle-Building? Implications for Daily Protein Distribution. *J. Int. Soc. Sports Nutr.* **2018**, *15*, 10. [CrossRef]
28. Moher, D.; Schulz, K.F.; Altman, D.G.; Lepage, L. The CONSORT Statement: Revised Recommendations for Improving the Quality of Reports of Parallel-Group Randomised Trials. *Lancet* **2001**, *357*, 1191–1194. [CrossRef]
29. Reeves, P.G.; Nielsen, F.H.; Fahey, G.C. AIN-93 Purified Diets for Laboratory Rodents: Final Report of the American Institute of Nutrition Ad Hoc Writing Committee on the Reformulation of the AIN-76A Rodent Diet. *J. Nutr.* **1993**, *123*, 1939–1951. [CrossRef]
30. Vendetti, P.; Di Meo, S. Antioxidants, Tissue Damage, and Endurance in Trained and Untrained Young Male Rats. *Arch. Biochem. Biophys.* **1996**, *331*, 63–68. [CrossRef]
31. Dawson, C.; Horvath, S.M. Swimming in Small Laboratory Animals. *Med. Sci. Sport. Exerc.* **1970**, *2*, 51–78. [CrossRef]
32. Aebi, H. Catalase in Vitro. *Methods Enzymol.* **1984**, *105*, 121–126. [CrossRef] [PubMed]
33. Dieterich, S.; Bieligk, U.; Beulich, K.; Hasenfuss, G.; Prestle, J. Gene Expression of Antioxidative Enzymes in the Human Heart: Increased Expression of Catalase in the End-Stage Failing Heart. *Circulation* **2000**, *101*, 33–39. [CrossRef] [PubMed]
34. Habig, W.H.; Jakoby, W.B. Glutathione S-Transferases (Rat and Human). *Methods Enzymol.* **1981**, *77*, 218–231. [CrossRef] [PubMed]
35. Buege, J.A.; Aust, S.D. Microsomal Lipid Peroxidation. *Methods Enzymol.* **1978**, *52*, 302–310. [CrossRef]
36. Tsikas, D. Analysis of Nitrite and Nitrate in Biological Fluids by Assays Based on the Griess Reaction: Appraisal of the Griess Reaction in the l-Arginine/Nitric Oxide Area of Research. *J. Chromatogr. B Anal. Technol. Biomed. Life Sci.* **2007**, *851*, 51–70. [CrossRef]
37. Levine, R.L.; Garland, D.; Oliver, C.N.; Amici, A.; Climent, I.; Lenz, A.G.; Ahn, B.W.; Shaltiel, S.; Stadtman, E.R. Determination of Carbonyl Content in Oxidatively Modified Proteins. *Methods Enzymol.* **1990**, *186*, 464–478. [CrossRef]

38. Lowry, O.H.; Rosebrough, N.J.; Farr, A.L. Protein Measurement with the Folin Phenol Reagent. *J. Biol. Chem.* **1951**, *193*, 265–275. [CrossRef]
39. Sabarense, C.M.; Rocha, K.S.O.; Rosa, D.D.; Martins, J.H.; da Pereira, M.M.L.S.; Silva, F.F.; Steward, B.L. A New Computational Method for Hepatic Fat Microvesicles Counting in Histological Study in Rats. *Biochem. Biophys. Res. Commun.* **2012**, *418*, 284–289. [CrossRef]
40. Gobatto, C.A.; De Mello, M.A.R.; Sibuya, C.Y.; De Azevedo, J.R.M.; Dos Santos, L.A.; Kokubun, E. Maximal Lactate Steady State in Rats Submitted to Swimming Exercise. *Comp. Biochem. Physiol. A Mol. Integr. Physiol.* **2001**, *130*, 21–27. [CrossRef]
41. Dias, K.A.; Da Conceição, A.R.; Oliveira, L.A.; Pereira, S.M.S.; Paes, S.D.S.; Monte, L.F.; Sarandy, M.M.; Novaes, R.D.; Gonçalves, R.V.; Della Lucia, C.M. Effects of Curcumin Supplementation on Inflammatory Markers, Muscle Damage, and Sports Performance during Acute Physical Exercise in Sedentary Individuals. *Oxid. Med. Cell. Longev.* **2021**, *2021*, 9264639. [CrossRef] [PubMed]
42. Campbell, M.S.; Carlini, N.A.; Fleenor, B.S. Influence of Curcumin on Performance and Post-Exercise Recovery. *Crit. Rev. Food Sci. Nutr.* **2021**, *61*, 1152–1162. [CrossRef] [PubMed]
43. Gonçalves, R.V.; Costa, A.M.A.; Grzeskowiak, L. Oxidative Stress and Tissue Repair: Mechanism, Biomarkers, and Therapeutics | Hindawi. *Oxid. Med. Cell. Longev.* **2021**, *2021*, 6204096. [CrossRef] [PubMed]
44. Radak, Z.; Taylor, A.W. Exercise and Hormesis. In *The Science of Hormesis in Health and Longevity*; Elsevier: Amsterdam, The Netherlands, 2018; pp. 63–73. ISBN 9780128142547.
45. Tsikas, D. Assessment of Lipid Peroxidation by Measuring Malondialdehyde (MDA) and Relatives in Biological Samples: Analytical and Biological Challenges. *Anal. Biochem.* **2017**, *524*, 13–30. [CrossRef]
46. Hendrix, J.; Nijs, J.; Ickmans, K.; Godderis, L.; Ghosh, M.; Polli, A. The Interplay between Oxidative Stress, Exercise, and Pain in Health and Disease: Potential Role of Autonomic Regulation and Epigenetic Mechanisms. *Antioxidants* **2020**, *9*, 1166. [CrossRef]
47. Da Conceição, A.R.; Dias, K.A.; Pereira, S.M.S.; Saraiva, L.C.; Oliveira, L.A.; De Souza, E.C.G.; Gonçalves, R.V.; Da Matta, S.L.P.; Natali, A.J.; Martino, H.S.D.; et al. Protective Effects of Whey Protein Concentrate Admixtured of Curcumin on Metabolic Control, Inflammation and Oxidative Stress in Wistar Rats Submitted to Exhaustive Exercise. *Br. J. Nutr.* **2021**, *127*, 526–539. [CrossRef]
48. Halliwell, B.; Gutteridge, J.M.C. *Free Radicals in Biology and Medicine*, 5th ed.; Oxford University Press: New York, NY, USA, 2015.
49. Robbins, R.A.; Grisham, M.B. Nitric Oxide. *Int. J. Biochem. Cell Biol.* **1997**, *29*, 857–860. [CrossRef]
50. Garcia, V.; Sessa, W.C. Endothelial NOS: Perspective and Recent Developments. *Br. J. Pharmacol.* **2019**, *176*, 189–196. [CrossRef]
51. Prestes, J.; Donatto, F.F.; Dias, R.; Frollini, A.B.; Cavaglieri, C.R. Papel Da Interleucina-6 Como Um Sinalizador Em Diferentes Tecidos Durante O Exercício Físico. *Fit. Perform. J.* **2006**, *5*, 348–353. [CrossRef]
52. Drobnic, F.; Riera, J.; Appendino, G.; Togni, S.; Franceschi, F.; Valle, X.; Pons, A.; Tur, J. Reduction of Delayed Onset Muscle Soreness by a Novel Curcumin Delivery System (Meriva®): A Randomised, Placebo-Controlled Trial. *J. Int. Soc. Sports Nutr.* **2014**, *11*, 31. [CrossRef]
53. Davis, J.M.; Murphy, E.A.; Carmichael, M.D.; Zielinski, M.R.; Groschwitz, C.M.; Brown, A.S.; Gangemi, J.D.; Ghaffar, A.; Mayer, E.P. Curcumin Effects on Inflammation and Performance Recovery Following Eccentric Exercise-Induced Muscle Damage. *Am. J. Physiol. Regul. Integr. Comp. Physiol.* **2007**, *292*, 2168–2173. [CrossRef] [PubMed]
54. Mosser, D.M.; Zhang, X. Interleukin-10: New Perspectives on an Old Cytokine. *Immunol. Rev.* **2008**, *226*, 205–218. [CrossRef] [PubMed]
55. Mollazadeh, H.; Cicero, A.F.G.; Blesso, C.N.; Pirro, M.; Majeed, M.; Sahebkar, A. Immune Modulation by Curcumin: The Role of Interleukin-10. *Crit. Rev. Food Sci. Nutr.* **2019**, *59*, 89–101. [CrossRef] [PubMed]
56. Kocaadam, B.; Şanlier, N. Curcumin, an Active Component of Turmeric (*Curcuma longa*), and Its Effects on Health. *Crit. Rev. Food Sci. Nutr.* **2017**, *57*, 2889–2895. [CrossRef]
57. Hargis, A.M.; Myers, S. The Integument. In *Pathologic Basis of Veterinary Disease*; Mosby: Maryland Heights, MI, USA, 2017; pp. 1009–1146.e1. [CrossRef]
58. Nogueira, B.C.F.; Campos, A.K.; Alves, R.S.; de Cássia Vieira Faria, R.; Sarandy, M.M.; Fonseca e Silva, F.; Gonçalves, R.V. Oxidative and Local Histopathological Response on Skin Wound of Horses Due to Amblyomma Sculptum Tick Parasitism. *Res. Vet. Sci.* **2021**, *136*, 550–560. [CrossRef]
59. Kostricky, I.M.; Basso, R.D.B.; Dos Santos, A.B.; Sulzbacher, M.M.; Goettems-Fiorin, P.B.; Frizzo, M.N.; Ludwig, M.S.; Heck, T.G. Effects of Water Temperature on Blood Lactate Concentration in Swimming Exercise Rat Model. *J. Exerc. Physiol. Online* **2021**, *24*, 13–21.
60. Kantanista, A.; Kusy, K.; Zarębska, E.; Włodarczyk, M.; Ciekot-Sołtysiak, M.; Zieliński, J. Blood Ammonia and Lactate Responses to Incremental Exercise in Highly-Trained Male Sprinters and Triathletes. *Biomed. Hum. Kinet.* **2016**, *8*, 32–38. [CrossRef]

nutrients

MDPI

Article

Intra-Amniotic Administration—An Emerging Method to Investigate Necrotizing Enterocolitis, In Vivo (*Gallus gallus*)

Nikolai Kolba, Jacquelyn Cheng, Cydney D. Jackson and Elad Tako *

Department of Food Science, Cornell University, Stocking Hall, Ithaca, NY 14850, USA
* Correspondence: et79@cornell.edu; Tel.: +1-607-255-0884

Abstract: Necrotizing enterocolitis (NEC) is a severe gastrointestinal disease in premature infants and a leading cause of death in neonates (1–7% in the US). NEC is caused by opportunistic bacteria, which cause gut dysbiosis and inflammation and ultimately result in intestinal necrosis. Previous studies have utilized the rodent and pig models to mimic NEC, whereas the current study uses the in vivo (*Gallus gallus*) intra-amniotic administration approach to investigate NEC. On incubation day 17, broiler chicken (*Gallus gallus*) viable embryos were injected intra-amniotically with 1 mL dextran sodium sulfate (DSS) in H_2O. Four treatment groups (0.1%, 0.25%, 0.5%, and 0.75% DSS) and two controls (H_2O/non-injected controls) were administered. We observed a significant increase in intestinal permeability and negative intestinal morphological changes, specifically, decreased villus surface area and goblet cell diameter in the 0.50% and 0.75% DSS groups. Furthermore, there was a significant increase in pathogenic bacterial (*E. coli* spp. and *Klebsiella* spp.) abundances in the 0.75% DSS group compared to the control groups, demonstrating cecal microbiota dysbiosis. These results demonstrate significant physiopathology of NEC and negative bacterial–host interactions within a premature gastrointestinal system. Our present study demonstrates a novel model of NEC through intra-amniotic administration to study the effects of NEC on intestinal functionality, morphology, and gut microbiota in vivo.

Keywords: necrotizing enterocolitis; NEC; dextran sodium sulfate; intraamniotic administration; *Gallus gallus*; gut microbiome; dysbiosis; intestinal immaturity

Citation: Kolba, N.; Cheng, J.; Jackson, C.D.; Tako, E. Intra-Amniotic Administration—An Emerging Method to Investigate Necrotizing Enterocolitis, In Vivo (*Gallus gallus*). *Nutrients* **2022**, *14*, 4795. https://doi.org/10.3390/nu14224795

Academic Editor: Rosa Casas

Received: 11 October 2022
Accepted: 8 November 2022
Published: 12 November 2022

1. Introduction

In premature infants, a leading gastrointestinal disease, necrotizing enterocolitis (NEC), accounts for approximately 2–13% of preterm and very-low-birth-weight (VLBW, <1500 g) infants in the United States [1–3]. Variations in incidences are attributed to different risk factor profiles, such as differing populations, detection rates, and inclusion and exclusion criteria for the disease [4–7]. Currently, there is no global incidental rate on NEC. Previous literature suggests that NEC is caused by intra-luminal pathogenic bacteria disrupting the intestinal villi, which upregulates inflammatory pathways, causing dysbiosis, and ultimately results in intestinal necrosis [8–10]. NEC is a multifactorial disease wherein symptoms start slowly, but decompensation occurs quickly, leading to fulminate NEC with pneumatosis intestinalis and portal gases [11–14]. The bacterial endotoxins released from opportunistic bacteria bind to Toll-like receptor four within epithelial cells, which activate pathogen-associated molecular pattern (PAMP) and release a complement and coagulation cascade effect within the immune system to break down gut mucosa [15–17]. Intestinal barrier disruption leads to bacteria entering intestinal cells and causes possible ischemia–reperfusion injury to the tissue [18].

NEC was initially investigated and induced in rodent and pig models through hypoxic/hypothermic and/or surgical interventions to mimic the multifactorial nature of the human disease [6,19–21]. In 1974, Barlow et al. demonstrated the first NEC model in rats, where gut flora and lack of immunoglobulin A (IgA) from breast milk were found to be essential factors contributing to NEC-like injury [22]. Currently, model organisms for NEC include rodents, pigs, and gnotobiotic quail, each with distinct strengths and weaknesses in modeling NEC [23,24]. Rats modeling NEC have practical benefits such as a low cost and high resilience to stress compared with mice; however, rats lack biomolecular reagents such as antibodies, meaning specific genetic techniques cannot be utilized to understand the mechanisms of pathophysiology, unlike mice [24]. The mouse model has been used to demonstrate NEC prevention mechanisms but is limited, as reproducible data is not necessarily obtained [19,25–27]. Another model, pigs, which have a closer resemblance in size, physiology, and anatomy when compared to a premature infant [28–30], is costly to maintain, and genetic techniques are limited [20,31–33]. Additionally, piglet models utilize intestinal injury to induce NEC, affecting the whole GI tract, while human NEC occurs primarily in the distal small intestine [32]. Lastly, there is the quail model, which is practical for NEC because of their modest size; rapid, productive maturation; resilience to research manipulation; transgenic lines; fully sequenced genome; and availability for molecular manipulation [34–37].

The current study suggests an alternative and novel model for NEC, the chicken (*Gallus gallus*). The chicken has been a well-studied model organism since the last century due to genetic analysis in developmental biology, virology, oncology, and immunology [38]. Chickens has been utilized for several human diseases, including muscular dystrophy [39–41], bacterial infections [42–45], autoimmunity [46–48], cancer [49–51], the microbiome [52–57], and micronutrient deficiencies [58–64]. The external embryology of the chicken has been a leading system investigating vertebrate development using functional genomics and biochemistry to study diseases similar to NEC. We hypothesize that the intra-amniotic administration [64–66] of dextran sulfate sodium (DSS, a compound previously demonstrated to induce NEC) will lead to NEC development, causing clinical symptoms within the brush border membrane functionality, tissue morphology, and dysbiosis of the intestinal microbial populations.

2. Materials and Methods

2.1. Sample Preparation

Dextran sulfate sodium (>98%) (Catalog #J62101.14, molecular weight 165.19 g/mol, ThermoFisher, Waltham, MA, USA) was used for the intra-amniotic administration experiment. In addition, 4 kDa fluorescein isothiocyanate-dextran (FITC-Dextran, Catalog #SIAL-46944-100M, Sigma-Aldrich, St. Louis, MO, USA) was used for the intestinal permeability assay.

2.2. Animals and Study Design

Cornish cross-fertile broiler eggs (n = 59) were purchased from a hatchery (Moyer's Chicks, Quakertown, PA, USA). The eggs were incubated under standard conditions at the Cornell University Animal Science poultry farm. All animal experiments were approved and performed in compliance with Cornell University IACUC (protocol code: 2020-0077).

Intra-Amniotic Administration

Pure DSS solutions were individually diluted in deionized (DI) water. As previously described [54,56,61–63,67], intra-amniotic administration was completed on day 17 of embryonic development with viable embryos (n = 60). Eggs were weighed and divided into six treatment groups of equal weight distribution (n = 10), using a random sequence generation [68]. For the intra-amniotic administration, all eggs were disinfected by spraying 70% ethanol. In the H_2O control and DSS-treated groups, a 21-gauge needle was inserted into the amniotic fluid, and 1 mL of sterile solution was injected. The site for intra-amniotic administration was identified via candling. After the administration, the injection sites were sprayed with 70% ethanol and sealed with transparent tape. The eggs were distributed into six groups: (1) no injection, (2) DI H_2O, (3) 0.10% DSS, (4) 0.25% DSS, (5) 0.50% DSS, and (6) 0.75% DSS. The eggs were equally distributed in each incubator to reduce possible allocation bias. Upon hatch, on incubation day 21, chicks were euthanized by CO_2. The blood was obtained via cardiac puncture and stored at 4 °C, then fractionated and stored at −20 °C. The proventriculus, gizzard, liver, pectoral muscle, duodenum, and cecum were obtained, flash frozen in liquid nitrogen, and stored at −20 °C until analysis.

2.3. Intestinal Permeability Test: Fluorescein Isothiocyanate Dextran (FITC-Dextran) Test

The intestinal permeability of the hatchlings was determined on day of hatch, as previously described by Barekatain et al., 2019 [69]. Briefly, on the day of hatch, each bird was orally gavaged with a 0.5 mL aqueous solution containing 1.1 mg of fluorescein isothiocyanate dextran (FITC-Dextran) before euthanization. A blood sample was taken from each bird after 4 h via myocardial puncture. Blood samples were fractionated via centrifugation at 1000× *g* for 15 min (Allegra X-30R, Beckman Coulter, Brea, CA, USA) and kept at −20 °C until analysis. Plasma samples and standards were analyzed in triplicate for FITC-Dextran concentration using a Biotek Epoch Microplate Spectrophotometer (Agilent Technologies, Santa Clara, CA, USA) with excitation and emission wavelengths set at 485 and 530 nm, respectively.

2.4. Glycogen Analysis as a Measurement of Energetic Status

All procedures were conducted as previously described [54,56,70]. A total of 20 mg of the liver was collected for glycogen analysis. Hepatic glycogen content was determined by multiplying the weight of the tissue by the amount of glycogen per 1 g of wet tissue.

2.5. Isolation of the Total RNA from the Duodenum Samples

As previously described [54,56,67], a RNeasy Mini Kit (Catalog #74106, Qiagen Inc., Valencia, CA, USA) utilized 30 mg of duodenal tissue (n = 5) to extract the total RNA according to the manufacturer's protocol. Total RNA was eluted in 50 µL of RNase-free water. All steps were carried out under RNase-free conditions. RNA was quantified by absorbance at 260/280 nm, and the integrity of the RNA was verified by 1.5% agarose gel electrophoresis followed by ethidium bromide staining. RNA was stored at −80 °C.

2.6. Real-Time Polymerase Chain Reaction (RT-PCR)

To create the cDNA, a 20 µL reverse transcriptase (RT) reaction was completed in a BioRad CFX1000 Touch thermocycler (BioRad, Hercules, CA, USA) using the Improm-II Reverse Transcriptase Kit (Catalog #A1250; Promega, Madison, WI, USA). The concentration of the cDNA obtained was determined by measuring the absorbance at 260/280 nm with an extinction coefficient of 33 (single-stranded DNA) by a NanoDrop 1000 Spectrophotometer (ThermoFisher Scientific, Waltham, MA USA). A RT-PCR assay assessed genomic DNA contamination for the genetic samples [67,71].

2.7. Intestinal Primer Design and Real-Time Quantitative PCR Design

The primers used in the RT-qPCR were designed based on ten gene sequences from the Genbank database, using Real-Time Primer Design Tool software (IDT DNA, Coralvilla, IA, USA). The sequences and the description of the primers used in this work are found in Table 1. The *Gallus gallus* 18s rRNA primer was designed as the reference gene, and the results obtained from the qPCR system were used to normalize the primers listed in Table 1. As previously described [62,63,72,73], all real-time quantitative PCR procedures were conducted with the specific primers listed in Table 1.

Table 1. Primer sequences used in the study.

Target Gene	Forward (5′-3′)	Reverse (3′-5′)	Amplicon Length (Base Pairs)	NCBI Accession	Ref.
		Inflammatory Genes			
NF-κβ	CACAGCTGGAGGGAAGTAAAT	TTGAGTAAGGAAGTGAGGTTGAG	100	2130627	
TNF-α	GACAGCCTATGCCAACAAGTA	TTACAGGAAGGGCAACTCATC	109	53854909	
IL-1β	TCATCCATCCCAAGTTCATTCA	GACACACTTCTCTGCCATCTT	105	395872	
IL-6	ACCTCATCCTCCGAGACTTTA	GCACTGAAACTCCTGGTCTT	105	302315692	
		Brush Border Membrane (BBM) Functionality Genes			
OCLN	GTCTGTGGGTTCCTCATCGT	GTTCTTCACCCACTCCTCCA	124	396026	[74]
MUC2	CCTGCTGCAAGGAAGTAGAA	GGAAGATCAGAGTGGTGCATAG	272	423101	
AP	CGTCAGCCAGTTTGACTATGTA	CTCTCAAAGAAGCTGAGGATGG	138	45382360	
SI	CCAGCAATGCCAGCATATTG	CGGTTTCTCCTTACCACTTCTT	95	2246388	
SGLT1	GCATCCTTACTCTGTGGTACTG	TATCCGCACATCACACATCC	106	8346783	
18S rRNA	GCAAGACGAACTAAAGCGAAAG	TCGGAACTACGACGGTATCT	100	7262899	

NF-κβ, nuclear factor kappa-light-chain-enhancer of activated B cells; TNF-α, tumor necrosis factor-alpha; IL-1β: interleukin one beta; IL-6: interleukin 6; OCLN: occludin; MUC2, mucin 2; AP: aminopeptidase; SI: sucrose isomaltase; SGLT1: sodium-glucose transporter 1; 18S rRNA: 18S ribosomal subunit. Target genes were created from accessions within National Center for Biotechnology Information (NCBI).

2.8. Intestinal Content DNA Isolation, Bacterial Primer Design, and PCR Amplification of Bacterial 16S rDNA

Frozen cecal contents were placed into a sterile tube containing 9 mL of phosphate-buffered saline (PBS) (Catalog#75800-998, VWR, Radnor, PA, USA) and homogenized with silicone bead-beating for 3 min [67,75,76]. All procedures were conducted as previously described [63,67,72].

As previously described [76–79], primers for *Lactobacillus*, *Bifidobacterium*, *Clostridium*, *Escherichia coli*, and *Klebsiella* were used with a universal primer variable region in bacterial 16S rRNA and were used as an internal standard. The PCR products were loaded on 2% agarose gel, stained with ethidium bromide, and quantified by Quantity-One 1D analysis software version 4.6.8 (BioRad, Hercules, CA, USA). The results were given by proportions of each bacterial group compared to the universal primer, giving relative abundance as previously conducted and demonstrated, with primers listed in Table 2 [62,63,72,73,80].

Table 2. Microbial primer sequences for bacteria within cecum.

Target Gene	Forward (5′-3′)	Reverse (3′-5′)	Ref.
Lactobacillus spp.	CATCCAGTGCAAACCTAAGAG	GATCCGCTTGCCTTCGCA	[77]
Bifidobacterium spp.	GGGTGGTAATGCCGGATG	CCACCGTTACACCGGGAA	[77]
E. coli spp.	GACCTCGGTTTAGTTCACAGA	CACACGCTGACGCTGACCA	[77]
Clostridium spp.	AAAGGAAGATTAATACCGCATAA	ATCTTGCGACCGTACTCCCC	[77]
Klebsiella spp.	CGCGTACTATACGCCATGAACGTA	ACCGTTGATCACTTCGGTCAGG	[78,79]
16S rRNA	CGTGCCAGCCGCGGTAATACG	GGGTTGCGCTCGTTGCGGGACTTAACCCAACAT	[77]

2.9. Morphological Examination

As previously described [54,65–67,81–85], intestinal samples (duodenum) were collected after the study and fixed in 4% (v/v) buffered formaldehyde. The samples were fixed further in 4% (v/v) buffered formaldehyde, dehydrated, cleared, and embedded in paraffin. The duodenum tissue was cut into 5 µm sections and placed on positively charged slides. Sections were: deparaffinized in xylene, rehydrated in different concentrations of alcohol, and stained. Periodic acid–Schiff and Alcian blue were used to distinguish neutral (magenta) and acidic (blue) mucins. Four sections of the duodenum per chick (n = 5 per treatment group) were examined. Villus height, villus width, crypt depth, goblet cell number, and goblet cell diameter were measured in each segment, using light microscopy with CellSens Standard version 3.2 (Olympus Corporation, Tokyo, Japan). Villi height was measured using the lamina propria as the base; villi width, the depth of the crypt, and the number of goblet cells were counted per side of a cross-sectional view through the villus; goblet cell size was measured as the diameter of the goblet cells (µm^2). Villi surface area was calculated from the villus height and width at half height according to Uni et al. [86] and calculated using the following equation:

$$Villus\ surface\ area = 2\prod \times \frac{VW}{2} \times VL \quad (1)$$

where VW is the average of three measurements of villus width, and VL is the villus length [87]. For the Alcian Blue and periodic acid–Schiff stain, the segments were counted for the types of goblet cells in the villi epithelium and goblet cells within the crypts. Goblet cells were counted in ten randomly selected villi or crypts per intestinal section (four intestinal sections per subject, 40 villi or crypts counted per subject). Goblet cell type was identified based on color, as periodic acid–Schiff and Alcian blue stains distinguishes between neutral (magenta), mixed (purple), and acidic (blue) mucins. Paneth cells were identified by their triangular shape within 10 randomly selected crypts per intestinal section and then counted and measured. The means were utilized for statistical analysis.

2.10. Statistical Analysis

Experimental treatments for the in ovo assay were arranged entirely randomly. The Shapiro–Wilk test was used to assess for normality. Statistical analyses were performed using one-way Analysis of Variance (ANOVA). Data is presented as means and standard deviations. Differences were considered significant at $p < 0.05$ using a post hoc Duncan or Tukey test was used to compare different NEC severity treatments, as described in figure or table legends. Statistical analysis was conducted using SPSS version 27.0 software (IBM, Armonk, NY, USA).

3. Results

3.1. Gross Physical Findings

There was a total hatchability rate of 95%. As shown in Table 3, there was no significant difference between body weight observed between DSS treatment groups and the controls. However, the cecum weight in the DSS treatment groups (0.1% and 0.5%) was significantly higher compared to the no-injection group ($p < 0.05$, Table 2).

Table 3. The effect of DSS on the body weight, cecum weight, and cecum-to-body-weight ratio.

Group	Body Weight (g)	Cecum Weight (g)	Cecum: Body Weight
No Injection	40.06 ± 4.06^{b}	0.42 ± 0.06^{b}	0.015 ± 0.005^{a}
H_2O Injection	47.49 ± 1.21^{a}	$0.47 \pm 0.03^{a,b}$	0.010 ± 0.001^{a}
0.1% DSS	$45.81 \pm 1.23^{a,b}$	0.62 ± 0.08^{a}	0.013 ± 0.002^{a}
0.25% DSS	$45.25 \pm 1.01^{a,b}$	$0.49 \pm 0.05^{a,b}$	0.011 ± 0.001^{a}
0.50% DSS	$45.24 \pm 1.11^{a,b}$	0.64 ± 0.08^{a}	0.014 ± 0.002^{a}
0.75% DSS	$45.78 \pm 0.86^{a,b}$	$0.60 \pm 0.06^{a,b}$	0.010 ± 0.000^{a}

Values are means ± stand error, n = 8–10. ^{a,b} within a column, means without a common letter are significantly different, $p < 0.05$ (Duncan's post-hoc test).

Additionally, intra-abdominal abscesses were found within the proventriculus and gizzard within only the 0.5% and 0.75% DSS treatments (depicted in Supplementary Figure S1).

3.2. Hb Concentration and Hepatic Glycogen Levels

The Hb value in the 0.75% DSS treatment group was significantly higher than the water injected, 0.1% DSS, and 0.25% DSS treatment groups (Table 4). Furthermore, there was a significant ($p < 0.05$) difference between the 0.75% DSS group and all other hepatic glycogen treatment groups.

Table 4. Blood hemoglobin (Hb) concentrations (g/dL) and hepatic glycogen levels (mg/mL).

Group	Hb (g/dL)	Hepatic Glycogen (mg/mL)
No Injection	10.48 ± 1.31 [a]	0.002 ± 0.001 [b]
H_2O Injection	9.82 ± 0.77 [a]	0.003 ± 0.001 [b]
0.1% DSS	10.70 ± 1.16 [a]	0.003 ± 0.001 [b]
0.25% DSS	10.22 ± 1.56 [a]	0.004 ± 0.001 [b]
0.50% DSS	10.13 ± 0.77 [a]	0.004 ± 0.001 [b]
0.75% DSS	10.94 ± 3.24 [a]	0.008 ± 0.002 [a]

Values are means ± standard error, $n = 5$. [a,b] within a column, means without a common letter are significantly different, $p < 0.05$ (Tukey's post-hoc test).

3.3. *Change of Intestinal Permeability across the Groups*

DSS treatments were not significantly different from the non-treated FITC-dextran birds. However, the 0.75% DSS treatment group was significantly ($p < 0.05$, Figure 1) different than the other treatment groups treated with FITC-dextran. Furthermore, no dose-response occurred from the titration of concentrations between the experimental groups and controls.

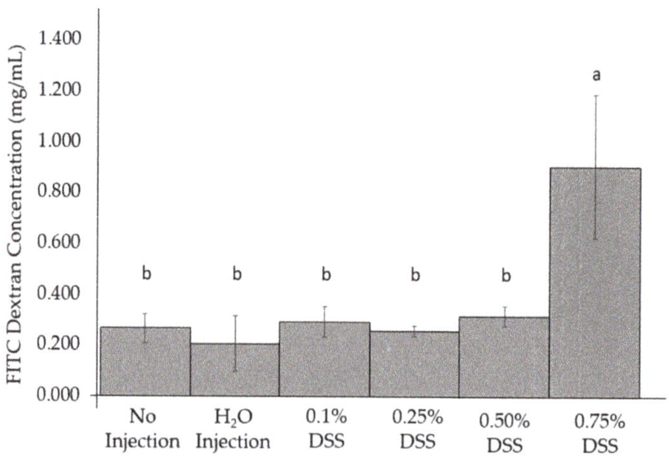

Figure 1. Comparison of the intra-amniotic administration of DSS to controls on the day of hatch within the small intestine (duodenum). Values are means ± stand error, $n = 3$. [a,b] within a column, means without a common letter are significantly different, $p < 0.05$ (Duncan's post-hoc test).

3.4. *Duodenal Gene Expression*

The gene expression of the inflammatory marker, NF-κβ1, was lower ($p < 0.05$) in the 0.50% and 0.75% DSS treatment groups compared to the control groups (no injection and H_2O injection) (Figure 2). However, other concentrations of DSS did not affect the expression of NF-κβ1 ($p < 0.05$). The relative expression of TNF-α was significantly ($p < 0.05$) decreased in all of the DSS experimental groups (0.1%, 0.25%, 0.5%, and 0.75%) compared to the controls. Similarly, IL-6 was lowered in all of the DSS-treated groups compared only to the H_2O injection group (Figure 2). However, no significant differences in IL-1β expression were found between any of the groups.

	NF-κB	TNF-α	IL-1β	IL-6	OCLN	MUC2	AP	SI	SGLT1
No Injection	^{ab} 11.685 ± 0.637	^{b} 3.158 ± 0.187	^{a} 1.017 ± 0.002	^{b} $1.197\times10^5 \pm 4.176\times10^4$	^{a} 1.094 ± 0.003	^{b} 11.697 ± 1.230	^{b} 1.493 ± 0.022	^{a} 2.632 ± 0.018	^{ab} 0.94 ± 0.003
H_2O Injection	^{a} 13.651 ± 1.050	^{a} 3.477 ± 0.066	^{a} 1.015 ± 0.005	^{a} $3.535\times10^5 \pm 1.188\times10^5$	^{a} 1.100 ± 0.007	^{a} 14.521 ± 0.775	^{a} 1.536 ± 0.016	^{a} 2.623 ± 0.020	^{b} 0.927 ± 0.008
0.1% DSS	^{bc} 10.899 ± 1.076	^{c} 2.846 ± 0.158	^{a} 1.023 ± 0.005	^{b} $5.959\times10^4 \pm 3.694\times10^4$	^{a} 1.096 ± 0.003	^{bc} 9.75 ± 1.076	^{bc} 1.46 ± 0.015	^{a} 2.673 ± 0.022	^{ab} 0.94 ± 0.007
0.25% DSS	^{bc} 10.065 ± 0.278	^{c} 2.692 ± 0.019	^{a} 1.02 ± 0.003	^{b} $2.654\times10^4 \pm 2.771\times10^3$	^{a} 1.091 ± 0.003	^{c} 8.92 ± 0.160	^{c} 1.432 ± 0.004	^{a} 2.648 ± 0.023	^{a} 0.944 ± 0.003
0.50% DSS	^{c} 9.463 ± 0.198	^{c} 2.65 ± 0.026	^{a} 1.023 ± 0.001	^{b} $2.796\times10^4 \pm 4.731\times10^3$	^{a} 1.091 ± 0.001	^{c} 8.707 ± 0.187	^{c} 1.431 ± 0.005	^{a} 2.678 ± 0.025	^{a} 0.948 ± 0.002
0.75% DSS	^{c} 9.014 ± 0.222	^{c} 2.582 ± 0.033	^{a} 1.026 ± 0.000	^{b} $2.048\times10^4 \pm 2.427\times10^3$	^{a} 1.092 ± 0.001	^{c} 8.319 ± 0.178	^{c} 1.425 ± 0.004	^{a} 2.650 ± 0.020	^{a} 0.951 ± 0.002

Low AU — High AU

Figure 2. Effect of the intra-amniotic administration of DSS on intestinal gene expression on the day of hatch within the small intestine (duodenum). Values are means ± stand error, $n = 5$. ^{a,b,c} within a column, means without a common letter are significantly different, $p < 0.05$ (Duncan's post-hoc test). NF-κβ, nuclear factor kappa-light-chain-enhancer of activated B cells; TNF-α, tumor necrosis factor-alpha; IL-1β: interleukin one beta; IL-6: interleukin 6; OCLN: occludin; MUC2, mucin 2; AP: aminopeptidase; SI: sucrose isomaltase; SGLT1: sodium-glucose transporter 1; 18s rRNA: 18S ribosomal subunit.

The gene expression of brush border membrane functionality proteins, sucrose isomaltase (SI), and occludin (OCLN) were not significantly different. Despite no significant difference in OCLN gene expression, there was a trend of decreased gene expression with increased DSS treatment concentration (Figure 2). There was a significant ($p < 0.05$) down-regulation of MUC2 and AP gene expression in the DSS treatment groups compared to the H_2O and no-injection control. There is a significant increase ($p < 0.05$) in the gene expression of SGLT1 in the 0.25%, 0.50%, and 0.75% DSS-treatment groups compared to the H_2O-injected group.

3.5. Microbial Dysbiosis

Figure 3 shows cecal bacterial populations. The relative abundance of *Bifidobacterium* spp. was significantly decreased in the DSS-treated groups ($p < 0.05$) relative to the control groups (non-injected and H_2O-injected groups). *Bifidobacterium* spp. and *Lactobacillus* spp. were significantly decreased in the 0.50% and 0.75% DSS groups compared to the non-injected and H_2O-injected groups. The highest relative abundances of *Lactobacillus* spp. were in 0.1% and 0.25% DSS following the exposure compared to the controls.

	Bifidobacterium	*Lactobacillus*	*E. coli*	*Clostridium*	*Klebsiella*
No Injection	0.873 ± 0.004 ^{a}	0.754 ± 0.014 ^{bc}	0.901 ± 0.009 ^{b}	1.197 ± 0.012 ^{a}	1.081 ± 0.054 ^{b}
H_2O Injection	0.839 ± 0.01 ^{b}	0.773 ± 0.015 ^{b}	0.885 ± 0.011 ^{bc}	1.23 ± 0.010 ^{a}	1.142 ± 0.057 ^{b}
0.1% DSS	0.789 ± 0.011 ^{c}	0.823 ± 0.015 ^{a}	0.843 ± 0.036 ^{c}	1.196 ± 0.022 ^{a}	1.124 ± 0.032 ^{b}
0.25% DSS	0.739 ± 0.007 ^{d}	0.838 ± 0.007 ^{a}	0.908 ± 0.005 ^{b}	1.072 ± 0.014 ^{b}	1.136 ± 0.026 ^{b}
0.50% DSS	0.753 ± 0.004 ^{d}	0.725 ± 0.007 ^{c}	1.013 ± 0.008 ^{a}	1.077 ± 0.008 ^{b}	1.386 ± 0.018 ^{a}
0.75% DSS	0.727 ± 0.017 ^{d}	0.689 ± 0.010 ^{d}	1.029 ± 0.013 ^{a}	1.104 ± 0.024 ^{b}	1.472 ± 0.033 ^{a}

Low AU — High AU

Figure 3. Effect of the intra-amniotic administration of DSS on cecal bacterial populations (day of hatch). Values are means ± SEM. Per bacterial category, $^{a–d}$ within a column, treatment groups that do not share letters are significantly different according to one-way ANOVA with Tukey's post-hoc test ($p < 0.05$).

E. coli and *Klebsiella* spp., opportunistic and possibly pathogenic bacteria, were significantly increased ($p < 0.05$) in the two highest concentrations of DSS (0.50% and 0.75%) compared with the non-injected and the H_2O-injected groups. However, the relative abundance of *Clostridium* spp. was significantly ($p < 0.05$) lowered in 0.25%, 0.50%, and 0.75% DSS compared to the control groups (non-injected and H_2O-injected groups) and 0.1% DSS group.

3.6. Intestinal Morphology

The villus surface area and crypt goblet cell diameter were significantly ($p < 0.05$) lowered in the 0.75% DSS group compared to the H_2O injection group (Table 5, images withing Supplemental Figure S2), indicating that DSS negatively impacted intestinal development. A significant ($p < 0.05$, Table 5) increase was found in the villi goblet cell diameters, Paneth cell number, and Paneth cell diameter of 0.1% and 0.75% DSS groups compared to the non-injected and H_2O-injected groups. There was no significant difference in crypt depth between experimental groups.

Table 5. Effects on intestinal villi and crypts of the duodenum after the intra-amniotic administration of experimental DSS.

Treatment	Villus Surface Area (μm^2)	Crypt Depth (μm)	Villi Goblet Diameter (μm)	Crypt Goblet Diameter (μm)	Paneth Cell #	Paneth Cell Diameter (μM)
No Injection	109.99 ± 3.06^{d}	$25.17 \pm 0.93^{a,b}$	3.57 ± 0.05^{d}	2.99 ± 0.05^{b}	1.09 ± 0.02^{c}	1.56 ± 0.03^{b}
H_2O Injection	205.15 ± 5.03^{a}	$26.35 \pm 0.98^{a,b}$	4.04 ± 0.06^{c}	3.16 ± 0.04^{a}	1.03 ± 0.01^{c}	1.47 ± 0.02^{c}
0.1% DSS	147.51 ± 3.28^{b}	27.89 ± 1.08^{a}	4.55 ± 0.07^{a}	2.99 ± 0.04^{b}	1.80 ± 0.05^{b}	1.69 ± 0.03^{a}
0.75% DSS	130.35 ± 0.03^{c}	24.32 ± 0.78^{b}	4.25 ± 0.05^{b}	2.73 ± 0.04^{c}	1.93 ± 0.06^{a}	1.67 ± 0.03^{a}

Values are means ± stand error, $n = 5$. $^{a–d}$ within a column means without a common letter are significantly different, $p < 0.05$ (Duncan's post-hoc test). # Number of cells.

A closer investigation of goblet cells within crypts and villi was viewed to determine differences (Table 6). Crypt goblet cell count was significantly increased in the 0.1% and 0.75% DSS groups compared to the non-injected and H_2O injection groups. Different goblet cell types were analyzed in crypts; acidic goblet cells were significantly higher ($p < 0.05$) in the 0.1% and 0.75% DSS groups compared to the control groups. There was a significantly ($p < 0.05$) higher amount of mixed goblet cells in crypts within the 0.1% DSS group compared to the control groups. Similarly, the villi goblet cell number and the acidic and mixed goblet cell number were significantly increased with 0.1% DSS exposure compared with the controls, as seen in Table 6.

Table 6. Effects on intestinal villi and crypt goblet cells of the duodenum after the intra-amniotic administration of experimental DSS.

Treatment	Crypt Goblet Cell #	Crypt Goblet Cell Type Number			Villi Goblet Cell #	Villi Goblet Cell Type Number		
		Acidic	Neutral	Mixed		Acidic	Neutral	Mixed
No Injection	8.57 ± 0.32^{c}	6.59 ± 0.26^{c}	0.00 ± 0.0	1.97 ± 0.18^{d}	15.78 ± 0.45^{c}	13.7 ± 0.42^{c}	0.00 ± 0.00	2.08 ± 0.13^{c}
H_2O Injection	7.96 ± 0.24^{c}	7.42 ± 0.22^{b}	0.00 ± 0.00	0.49 ± 0.06^{c}	22.93 ± 0.6^{b}	18.4 ± 0.53^{b}	0.00 ± 0.00	4.53 ± 0.23^{b}
0.1% DSS	14.7 ± 0.41^{a}	10.23 ± 0.31^{a}	0.00 ± 0.00	4.47 ± 0.19^{a}	30.37 ± 0.84^{a}	23.91 ± 0.68^{a}	0.00 ± 0.00	6.46 ± 0.36^{a}
0.75% DSS	13.48 ± 0.04^{b}	9.76 ± 0.32^{a}	0.00 ± 0.00	3.74 ± 0.19^{b}	16.64 ± 0.56^{c}	14.63 ± 0.49^{c}	0.00 ± 0.00	2.09 ± 0.14^{c}

Values are means ± stand error, $n = 5$. $^{a–d}$ within a column means without a common letter are significantly different, $p < 0.05$ (Duncan's post-hoc test). # Number of cells.

4. Discussion

NEC is an acute inflammatory disease that results in the intestinal necrosis of the bowels, systemic sepsis, and multiorgan failure from a complex combination of pathological events, including patchy inflammation of the small intestine and intestinal hypoxic/reperfusion injuries and bacterial dysbiosis [3,13,16,88]. In our present study, we investigated the effects of dextran sodium sulfate (DSS) utilizing the *Gallus gallus* intra-amniotic administration model to mimic necrotizing enterocolitis (NEC). The present study indicates that the intra-amniotic administration of DSS at the highest concentration (0.75%) had similar findings as NEC manifestations in humans, including but not limited to the inflammation of the small intestines, intra-abdominal abscesses of the gizzard, increased hemoglobin levels, increased permeability within the intestines, and the increased presence of potentially pathogenic bacteria. NEC has previously been induced and investigated in the rodent and pig models through hypoxic/hypothermic and/or surgical interventions to simulate the multifactorial nature of NEC in humans [6,19–21].

Current NEC in vivo models, such as rodents, pigs, and quails, have distinct strengths and weaknesses. Rodents (mice and rats) are induced to have NEC by cesarian section delivery before term, then gavage fed with the formula [19,89,90]. Mice have antibodies, and specific genetic techniques can be utilized to understand NEC pathophysiology mechanisms, whereas rats lack antibodies but are more resilient to stress [21]. Unfortunately, studies using the mouse model have shown inconsistent results, wherein data reproducibility presents potential issues [19,25–27]. While pigs have closer resemblance relative to rodents in physiology, there are drawbacks in which NEC is induced via intestinal injury to the whole intestine, and genetic techniques also present limitations. In another model, gnotobiotic quails, NEC is induced via the oral gavage of bacteria associated with NEC to affect its small intestine, which has aided in understanding the inducible nitric oxide synthase (iNOS) pathway before macroscopic lesions [34,35,91]. Quails have been shown to be an NEC model that balances practicality, resilience, and molecular manipulation [34–37]. Given these aforementioned factors and that the chicken model has been a leading system investigating vertebrate development using functional genomics and biochemistry to study diseases similar to NEC [38], we sought to develop another potential model for NEC using the embryonic stage of the *Gallus gallus*.

In our experimental trials, the intra-amniotic administration of DSS was utilized to induce intestinal inflammation to cause NEC physiopathology [92–94]. DSS is a sulfated polysaccharide with various molecular weights (5–1400 kDa), commonly used to induce enteric colitis in rodents by penetrating the intestinal mucosal membrane [19,93–96]. DSS-induced colitis is a widely used model because it is rapid, simple, reproducible, and controllable. Recent studies have shown that DSS added to DI H_2O induces clinical, gross, and histological factors associated with enteritis in broiler chickens, such as decreased body weight, bloody diarrhea, intestinal lesions, shortened villi height, and increased goblet cell density [92,94,97,98]. Furthermore, Zou et al. (2018) demonstrated that DSS exposure to broiler chickens increases gut leakiness and induces pro- and anti-inflammatory cytokine response elements in a dose-dependent manner [92]. Nevertheless, there is limited research characterizing DSS-induced NEC in *Gallus gallus*, and this study demonstrates the first-ever intra-amniotic administration of DSS to induce NEC.

During the initial necropsies, there was little to no inflammation within the internal organs of the chicks on the day of hatch in the 0.1, 0.25, and 0.5% DSS exposed groups. However, the 0.75% DSS treatment group had a few slightly patchy inflammation sites in the small intestines and showed intra-abdominal abscesses within the proventriculus and gizzard (Supplemental Figure S1); these observations are in agreement with previous studies that demonstrated the initial hypoxic and reperfusion injuries (typical in NEC cases from rodents and clinical models) and an increase in innate immune responses (cytokines and white bloods cells). [11,20,24,92,99]. The intra-abdominal abscesses and patchy inflammation in the distal digestive organs resulting from the treatment with the highest DSS concentration (0.75%) can be associated with hypoxic reperfusion injuries, which have previously been demonstrated with DSS exposure [92,94,95]. To further illustrate the effectiveness of DSS penetrating through the mucosal layers of the small intestine, a FITC-dextran assay was performed (Figure 1). Compared to all other groupings, there was a significant ($p < 0.05$) increase in the intestinal mucosal layer penetration in the 0.75% DSS treatment group. The increased intestinal permeability suggests that the DSS successfully breaks down the mucosal layer, potentially allowing pathogenic bacteria to invade the host's villi, as previously described [9,25,69,100,101]. However, occludin (OCLN) gene expression, a tight junction protein between the intestinal enterocytes, does not significantly differ between the groups (Figure 2). The lack of significance of the gene expression of the tight junctions on the basolateral surface could be the short duration of a DSS exposure time to allow for occludin degradation. Though no significant alterations in OCLN gene expression were found, our results supporting increased intestinal permeability with DSS exposure were further supported by our hemoglobin and hepatic glycogen results. The hemoglobin concentration was raised within the DSS groups compared to the non-injected group (Table 4). As intestinal irritability and instability increase, hemoglobin values were found to increase in clinical patients with inflammatory bowel disease [102]. Hepatic glycogen levels were increased only within the 0.75% DSS group compared to others (Table 4). However, Sodhi et al. (2009) demonstrated that enterocyte proliferation is inhibited in rat intestinal cell lines (IEC-6 cells enterocytes) and TLR4$^{-/-}$ mice and that glycogen synthase kinase decreases when under NEC conditions with lipopolysaccharides [103]. This observed difference could be due to the treatment to induce the condition and the difference between the models utilized.

Brush border membrane (BBM) functionality was investigated by measuring the gene expression of the functional proteins viewing the digestive capabilities, as seen in Figure 2. There was no significant difference between the treatment groups in sucrose isomaltase or sodium–glucose transporter 1. However, there was a substantial lowering of mucin 2 (MUC2) and aminopeptidase (AP) gene expression in the DSS treatment groups (0.25%, 0.50%, and 0.75%) compared to the no-injection and H_2O-injection groups. As previously mentioned, DSS is a sulfated polysaccharide that disrupts the luminal mucus layer, allowing mucosal thinning and opportunistic bacteria to penetrate the BBM, causing intestinal trauma [104,105]. This reasoning on MUC2 supports the findings on the lowering of aminopeptidase expression, as AP is primarily located near the apical side of the lumen in the intestinal epithelial cells [106]. Aminopeptidases are enzymes that catalyze the amino terminus of a protein within subcellular organelles, cytosol, and membrane components. Thus, it can be suggested that, if the BBM membrane is injured, the AP capacity would be significantly reduced, increasing the possibility of immune responses (pro-inflammation and apoptosis) [107].

Inflammation resulting from NEC is a primary identifier of injury from hypoxia/reperfusion conditions and an indicative marker for the disease [15,32,108,109]. Inflammation biomarkers, nuclear factor kappa-light-chain-enhancer of activated B cells (NF-κβ), tumor necrosis factor-alpha (TNF-α), interleukin one beta (IL-1β), and interleukin 6 (IL-6) were all analyzed via RT-qPCR (Figure 2) due to the use of these biomarkers in other NEC models and clinical trials [108–112]. NF-κβ and TNF-α were significantly ($p > 0.05$) lower in the 0.50% and 0.75% DSS groups compared to the control groups (non-injected and H_2O-injected). The downregulation of NF-κβ gene expression originates from the upstream signaling of TNF-α being downregulated since these proteins operate in tandem with pro-inflammation pathways triggered by microbial products (i.e., endotoxins, metabolites, amino acids, etc.) and signal transductions mechanisms in the innate immune system [113–117]. The downregulation of TNF-α is potentially derived from the upregulation of microbial byproducts within the duodenum in the DSS treatments. Krishnaveni and Jayachandran (2009) found that ethyl acetate extracts from two different marine bacteria caused the downregulation of TNF-α [118]. Similarly, Lou et al. (2018) found that *Brucella* caused the same downregulation of TNF-α within porcine and murine models [119].

There were significant changes within the bacterial profiles in the DSS treatment groups (Figure 3). Bacterial profiles of the 0.75% DSS group were significantly ($p < 0.05$) lower abundance of *Bifidobacterium, Lactobacillus,* and *Clostridium* spp. In contrast, the DSS-treatment groups demonstrated significantly higher abundance levels of *E. coli* and *Klebsiella* spp. ($p < 0.05$). The lower abundance of beneficial bacteria (Bifidobacterium and Lactobacillus) suggests an opportunity for dysbiosis via the proliferation of opportunistic bacteria such as *E. coli* and *Klebsiella*. Our results are similar to other NEC models that utilized different treatments to induce NEC [6,9,17,20,24,26,34,91,120]. One of the first NEC models used *Klebsiella* to create an NEC model (mice) because the genus produces hydrogen-sulfide-rich gas pockets of pneumatosis in vivo [22,121]. Similarly, it is theorized that gas produced by *E. coli* can invade the same intraluminal cavities as *Klebsiella,* which leads to pneumatosis intestinalis, which is a radiographic sign of NEC [122–125]. Additionally, Tarracchini et al. (2021) found that *E. coli* and other opportunistic bacteria are found within the next-generation sequencing of NEC clinical patients, suggesting that the bacterial abundance of *E. coli* could induce NEC pathology [126]. These invasive bacterial changes the grouping/profile of the gut and influences the intestinal BBM morphology [126–132].

Within this study, the duodenum was sectioned to investigate DSS's effect on its intestinal morphology (Supplemental Figure S2). Previous studies have shown that NEC results in various levels of intestinal degradation due to microbial dysbiosis effects on the brush border membrane morphology (i.e., the villi surface area, goblet cell number, type, size, Paneth cell production) [133–135]. The 0.75% DSS treatment group had significantly lower villus surface area, crypt goblet cell diameter, and villi goblet cell number and type. In parallel, there was a significant increase in the Paneth cell number and size (Table 5) and crypt goblet cell number and types (Table 6) populations. Since goblet cells produce mucin, which lubricates the passage of food through the intestines and protects the intestine from the potential damage from digestive enzymes, the DSS treatments at the highest concentration would be associated with damage at the apical side of the enterocyte while lowering the villus surface area, which aligns with previous DSS studies and is in agreement with the present study [136–138]. Additionally, on the enterocyte's basolateral side, the crypts' goblet cells would anticipate the loss of mucin and increase its mucin production to overcome the loss [139,140]. This anticipation can be further supported by the increase in Paneth cell findings within the crypts of the intestinal epithelial cells ($p < 0.05$, Table 5). Paneth cells within the small intestine synthesize and secrete antimicrobial enzymes as a part of the innate immune system [100,141,142]. The antimicrobial peptide secretion by Paneth cells is recognized by MyD88-dependent Toll-like receptor (TLR) activations, triggering the expression of multiple peptides and proteins [19,25]. The Paneth cells migrate towards the base of the villi after differentiation within the crypts to protect commensal bacteria from the opportunistic bacteria within the gut, which supports the findings of bacterial dysbiosis mentioned earlier in Figure 3.

5. Conclusions

This study is the first to demonstrate NEC symptoms via the intra-amniotic administration of DSS in vivo (*Gallus gallus*). The 0.75% DSS treatment group decreased BBM functionality and demonstrated microbiota dysbiosis within a premature gut, mimicking other models of NEC. Although we did not observe significant severe pathologies (gas-filled lesions or necrotic plaques in histological sectioning), there was a clear trend of opportunistic bacterial populations proliferation and overtaking the distal gastrointestinal tract. This transformation of untreated and DSS-treated individuals' microbial profiles can potentially affect several bacterial metabolic pathways related to bacterial, cellular, and metabolic processes. The results of this study are promising evidence to investigate increased concentrations of DSS to cause more severe NEC symptoms and identify potential novel biomarkers for less severe NEC cases. Furthermore, the suggested in vivo novel model and innovative approach will support the assessment of various potential interventions to ameliorate the pathophysiology of NEC.

Supplementary Materials: The following are available online at https://www.mdpi.com/article/10.3390/nu14224795/s1, Figure S1: Representative images of gross anatomical photos of dissections and proventriculus/gizzards of birds on the day of hatch; Figure S2. Representative histology (Alcian Blue and Pacific Acid Schiff staining) images of the duodenum of birds on the day of hatch.

Author Contributions: N.K. and E.T. conceived and designed the experiment. N.K., J.C. and C.D.J. collected and analyzed the data and wrote the initial draft. N.K., J.C., C.D.J. and E.T. reviewed and edited the manuscript. E.T. led the research. All authors have read and agreed to the published version of the manuscript.

Funding: This research received no external funding.

Institutional Review Board Statement: The animal protocol used in this study was conducted according to the guidelines of the Declaration of Helsinki and approved by the Cornell University Institutional Animal Care and Use Committee by ethic approval code 2020-0077.

Informed Consent Statement: Not applicable.

Conflicts of Interest: The authors declare no conflict of interest.

References

1. Alsaied, A.; Islam, N.; Thalib, L. Global Incidence of Necrotizing Enterocolitis: A Systematic Review and Meta-Analysis. *BMC Pediatr.* **2020**, *20*, 344. [CrossRef]
2. Kosloske, A. Epidemiology of Necrotizing Enterocolitis. *Acta Paediatr.* **1994**, *83*, 2–7. [CrossRef] [PubMed]
3. Eaton, S.; Rees, C.M.; Hall, N.J. Current Research on the Epidemiology, Pathogenesis, and Management of Necrotizing Enterocolitis. *Neonatology* **2017**, *111*, 423–430. [CrossRef] [PubMed]
4. Faraday, C.; Hamad, S.; Jones, K.D.; Sim, K.; Cherian, S.; James, A.; Godambe, S.; New, H.V.; Kroll, J.S.; Clarke, P. Characteristics and Incidence of Transfusion-Associated Necrotizing Enterocolitis in the UK. *J. Matern. Fetal. Neonatal. Med.* **2020**, *33*, 398–403. [CrossRef] [PubMed]
5. Chauhan, N.; Tiwari, S.; Jain, U. Potential Biomarkers for Effective Screening of Neonatal Sepsis Infections: An Overview. *Microb. Pathog.* **2017**, *107*, 234–242. [CrossRef]
6. Doğan, G.; İpek, H. The Development of Necrotizing Enterocolitis Publications: A Holistic Evolution of Global Literature with Bibliometric Analysis. *Eur. J. Pediatr. Surg.* **2020**, *30*, 293–303. [CrossRef] [PubMed]
7. Battersby, C.; Santhalingam, T.; Costeloe, K.; Modi, N. Incidence of Neonatal Necrotising Enterocolitis in High-Income Countries: A Systematic Review. *Arch. Dis. Child. Fetal. Neonatal Ed.* **2018**, *103*, F182–F189. [CrossRef] [PubMed]
8. De Plaen, I.G. Inflammatory Signaling in Necrotizing Enterocolitis. *Clin. Perinat.* **2013**, *40*, 109–124. [CrossRef]
9. Zhang, C.; Sherman, M.P.; Prince, L.S.; Bader, D.; Weitkamp, J.-H.; Slaughter, J.C.; McElroy, S.J. Paneth Cell Ablation in the Presence of Klebsiella Pneumoniae Induces Necrotizing Enterocolitis (NEC)-like Injury in Immature Murine Small Intestine. *Dis. Models Mech.* **2012**, *5*, 522–532. [CrossRef]
10. Bazacliu, C.; Neu, J. Pathophysiology of Necrotizing Enterocolitis: An Update. *Curr. Pediatr. Rev.* **2019**, *15*, 68–87. [CrossRef]
11. Meister, A.L.; Doheny, K.K.; Travagli, R.A. Necrotizing Enterocolitis: It's Not All in the Gut. *Exp. Biol. Med.* **2020**, *245*, 85–95. [CrossRef] [PubMed]
12. Rich, B.S.; Dolgin, S.E. Necrotizing Enterocolitis. *Pediatr. Rev.* **2017**, *38*, 552–559. [CrossRef] [PubMed]
13. Müller, M.J.; Paul, T.; Seeliger, S. Necrotizing Enterocolitis in Premature Infants and Newborns. *NPM* **2016**, *9*, 233–242. [CrossRef]
14. Adams-Chapman, I. Necrotizing Enterocolitis and Neurodevelopmental Outcome. *Clin. Perinatol.* **2018**, *45*, 453–466. [CrossRef]
15. Molteni, M.; Gemma, S.; Rossetti, C. The Role of Toll-Like Receptor 4 in Infectious and Noninfectious Inflammation. *Mediat. Inflamm.* **2016**, *2016*, 6978936. [CrossRef]
16. Tirone, C.; Pezza, L.; Paladini, A.; Tana, M.; Aurilia, C.; Lio, A.; D'Ippolito, S.; Tersigni, C.; Posteraro, B.; Sanguinetti, M.; et al. Gut and Lung Microbiota in Preterm Infants: Immunological Modulation and Implication in Neonatal Outcomes. *Front. Immunol.* **2019**, *10*, 2910. [CrossRef]
17. Schnabl, K.-L.; Van Aerde, J.-E.; Thomson, A.-B.; Clandinin, M.-T. Necrotizing Enterocolitis: A Multifactorial Disease with No Cure. *World J. Gastroenterol.* **2008**, *14*, 2142–2161. [CrossRef] [PubMed]
18. Alganabi, M.; Lee, C.; Bindi, E.; Li, B.; Pierro, A. Recent Advances in Understanding Necrotizing Enterocolitis. *F1000Research* **2019**, *8*, 107. [CrossRef] [PubMed]
19. Jilling, T.; Simon, D.; Lu, J.; Meng, F.J.; Li, D.; Schy, R.; Thomson, R.B.; Soliman, A.; Arditi, M.; Caplan, M.S. The Roles of Bacteria and TLR4 in Rat and Murine Models of Necrotizing Enterocolitis. *J. Immunol.* **2006**, *177*, 3273–3282. [CrossRef]
20. Di Lorenzo, M.; Bass, J.; Krantis, A. An Intraluminal Model of Necrotizing Enterocolitis in the Developing Neonatal Piglet. *J. Pediatr. Surg.* **1995**, *30*, 1138–1142. [CrossRef]
21. Kappel, S.S.; Sangild, P.T.; Hilsted, L.; Hartmann, B.; Thymann, T.; Aunsholt, L. Gastric Residual to Predict Necrotizing Enterocolitis in Preterm Piglets As Models for Infants. *J. Parenter. Enter. Nutr.* **2021**, *45*, 87–93. [CrossRef] [PubMed]
22. Barlow, B.; Santulli, T.V.; Heird, W.C.; Pitt, J.; Blanc, W.A.; Schullinger, J.N. An Experimental Study of Acute Neonatal Enterocolitis—The Importance of Breast Milk. *J. Pediatr. Surg.* **1974**, *9*, 587–595. [CrossRef]
23. Sodhi, C.; Richardson, W.; Gribar, S.; Hackam, D.J. The Development of Animal Models for the Study of Necrotizing Enterocolitis. *Dis. Model. Mech.* **2008**, *1*, 94–98. [CrossRef]
24. Sulistyo, A.; Rahman, A.; Biouss, G.; Antounians, L.; Zani, A. Animal Models of Necrotizing Enterocolitis: Review of the Literature and State of the Art. *Innov. Surg. Sci.* **2018**, *3*, 87–92. [CrossRef] [PubMed]
25. White, J.R.; Gong, H.; Pope, B.; Schlievert, P.; McElroy, S.J. Paneth Cell Disruption-Induced Necrotizing Enterocolitis Requires Live Bacteria and Occurs Independent of TLR4 Signaling. *Dis. Model. Mech.* **2017**, *10*, 727–736. [CrossRef]
26. Lyu, C.; Jiang, S.; Kong, M.; Chen, X.; Zhang, L. Vitamin D Protects against Necrotising Enterocolitis in Newborn Mice by Activating the ERK Signalling Pathway. *Mol. Med. Rep.* **2020**, *22*, 2107–2114. [CrossRef]

27. MohanKumar, K.; Kaza, N.; Jagadeeswaran, R.; Garzon, S.A.; Bansal, A.; Kurundkar, A.R.; Namachivayam, K.; Remon, J.I.; Bandepalli, C.R.; Feng, X.; et al. Gut Mucosal Injury in Neonates Is Marked by Macrophage Infiltration in Contrast to Pleomorphic Infiltrates in Adult: Evidence from an Animal Model. *Am. J. Physiol. Gastrointest. Liver Physiol.* **2012**, *303*, G93–G102. [CrossRef]
28. Roura, E.; Koopmans, S.-J.; Lallès, J.-P.; Le Huerou-Luron, I.; de Jager, N.; Schuurman, T.; Val-Laillet, D. Critical Review Evaluating the Pig as a Model for Human Nutritional Physiology. *Nutr. Res. Rev.* **2016**, *29*, 60–90. [CrossRef]
29. Lunney, J.K.; Van Goor, A.; Walker, K.E.; Hailstock, T.; Franklin, J.; Dai, C. Importance of the Pig as a Human Biomedical Model. *Sci. Transl. Med.* **2021**, *13*, eabd5758. [CrossRef]
30. Merrifield, C.A.; Lewis, M.; Claus, S.P.; Beckonert, O.P.; Dumas, M.-E.; Duncker, S.; Kochhar, S.; Rezzi, S.; Lindon, J.C.; Bailey, M.; et al. A Metabolic System-Wide Characterisation of the Pig: A Model for Human Physiology. *Mol. BioSyst.* **2011**, *7*, 2577. [CrossRef]
31. Holgersen, K.; Gao, X.; Narayanan, R.; Gaur, T.; Carey, G.; Barton, N.; Pan, X.; Muk, T.; Thymann, T.; Sangild, P.T. Supplemental Insulin-Like Growth Factor-1 and Necrotizing Enterocolitis in Preterm Pigs. *Front. Pediatr* **2020**, *8*, 602047. [CrossRef] [PubMed]
32. Nguyen, D.N.; Thymann, T.; Goericke-Pesch, S.K.; Ren, S.; Wei, W.; Skovgaard, K.; Damborg, P.; Brunse, A.; van Gorp, C.; Kramer, B.W.; et al. Prenatal Intra-Amniotic Endotoxin Induces Fetal Gut and Lung Immune Responses and Postnatal Systemic Inflammation in Preterm Pigs. *Am. J. Pathol.* **2018**, *188*, 2629–2643. [CrossRef] [PubMed]
33. Di Lorenzo, M.; Bass, J.; Krantis, A. Use of L-Arginine in the Treatment of Experimental Necrotizing Enterocolitis. *J. Pediatr. Surg.* **1995**, *30*, 235–241. [CrossRef]
34. Waligora-Dupriet, A.J.; Dugay, A.; Auzeil, N.; Nicolis, I.; Rabot, S.; Huerre, M.R.; Butel, M.J. Short-Chain Fatty Acids and Polyamines in the Pathogenesis of Necrotizing Enterocolitis: Kinetics Aspects in Gnotobiotic Quails. *Anaerobe* **2009**, *15*, 138–144. [CrossRef] [PubMed]
35. Waligora-Dupriet, A.-J.; Dugay, A.; Auzeil, N.; Huerre, M.; Butel, M.-J. Evidence for Clostridial Implication in Necrotizing Enterocolitis through Bacterial Fermentation in a Gnotobiotic Quail Model. *Pediatr. Res.* **2005**, *58*, 629–635. [CrossRef]
36. Baer, J.; Lansford, R.; Cheng, K. Japanese Quail as a Laboratory Animal Model. In *Laboratory Animal Medicine*; Elsevier: Amsterdam, The Netherlands, 2015; pp. 1087–1108. ISBN 978-0-12-409527-4.
37. Ares, G.J.; McElroy, S.J.; Hunter, C.J. The Science and Necessity of Using Animal Models in the Study of Necrotizing Enterocolitis. *Semin. Pediatr. Surg.* **2018**, *27*, 29–33. [CrossRef]
38. Dodgson, J.B.; Romanov, M.N. Use of Chicken Models for the Analysis of Human Disease. *Curr. Protoc. Hum. Genet.* **2004**, *40*, 15.5.1–15.5.12. [CrossRef]
39. Huang, H.; Liu, L.; Li, C.; Liang, Z.; Huang, Z.; Wang, Q.; Li, S.; Zhao, Z. Fat Mass- and Obesity-Associated (FTO) Gene Promoted Myoblast Differentiation through the Focal Adhesion Pathway in Chicken. *3 Biotech* **2020**, *10*, 403. [CrossRef]
40. Cui, C.; Han, S.; Tang, S.; He, H.; Shen, X.; Zhao, J.; Chen, Y.; Wei, Y.; Wang, Y.; Zhu, Q.; et al. The Autophagy Regulatory Molecule CSRP3 Interacts with LC3 and Protects Against Muscular Dystrophy. *Int. J. Mol. Sci.* **2020**, *21*, 749. [CrossRef]
41. Sundekilde, U.K.; Rasmussen, M.K.; Young, J.F.; Bertram, H.C. High Resolution Magic Angle Spinning NMR Spectroscopy Reveals That Pectoralis Muscle Dystrophy in Chicken Is Associated with Reduced Muscle Content of Anserine and Carnosine. *Food Chem.* **2017**, *217*, 151–154. [CrossRef]
42. Thu, H.M.; Myat, T.W.; Win, M.M.; Thant, K.Z.; Rahman, S.; Umeda, K.; Nguyen, S.V.; Icatlo, F.C., Jr.; Higo-Moriguchi, K.; Taniguchi, K.; et al. Chicken Egg Yolk Antibodies (IgY) for Prophylaxis and Treatment of Rotavirus Diarrhea in Human and Animal Neonates: A Concise Review. *Korean J. Food Sci. Anim. Resour.* **2017**, *37*, 1–9. [CrossRef] [PubMed]
43. Sood, U.; Gupta, V.; Kumar, R.; Lal, S.; Fawcett, D.; Rattan, S.; Poinern, G.E.J.; Lal, R. Chicken Gut Microbiome and Human Health: Past Scenarios, Current Perspectives, and Futuristic Applications. *Indian J. Microbiol.* **2020**, *60*, 2–11. [CrossRef] [PubMed]
44. Lacharme-Lora, L.; Owen, S.V.; Blundell, R.; Canals, R.; Wenner, N.; Perez-Sepulveda, B.; Fong, W.Y.; Gilroy, R.; Wigley, P.; Hinton, J.C.D. The Use of Chicken and Insect Infection Models to Assess the Virulence of African Salmonella Typhimurium ST313. *PLoS Negl. Trop. Dis.* **2019**, *13*, e0007540. [CrossRef] [PubMed]
45. Leigh, S.A.; Branton, S.L.; Evans, J.D.; Collier, S.D. Fluorescent Microspheres as a Positive Indicator in an Intratracheal Infection Model. *J. Microbiol. Methods* **2020**, *172*, 105886. [CrossRef]
46. Ochoa-Repáraz, J.; Magori, K.; Kasper, L.H. The Chicken or the Egg Dilemma: Intestinal Dysbiosis in Multiple Sclerosis. *Ann. Transl. Med.* **2017**, *5*, 145. [CrossRef]
47. Sundick, R.S.; Bagchi, N.; Brown, T.R. The Obese Strain Chicken as a Model for Human Hashimoto's Thyroiditis. *Exp Clin Endocrinol. Diabetes* **2009**, *104*, 4–6. [CrossRef]
48. Teixeira, A.R.L.; Nitz, N.; Bernal, F.M.; Hecht, M.M. Parasite Induced Genetically Driven Autoimmune Chagas Heart Disease in the Chicken Model. *J. Vis. Exp.* **2012**, e3716. [CrossRef]
49. Hu, J.; Ishihara, M.; Chin, A.I.; Wu, L. Establishment of Xenografts of Urological Cancers on Chicken Chorioallantoic Membrane (CAM) to Study Metastasis. *Precis. Clin. Med.* **2019**, *2*, 140–151. [CrossRef]
50. Kunz, P.; Schenker, A.; Sähr, H.; Lehner, B.; Fellenberg, J. Optimization of the Chicken Chorioallantoic Membrane Assay as Reliable in Vivo Model for the Analysis of Osteosarcoma. *PLoS ONE* **2019**, *14*, e0215312. [CrossRef]
51. Hawkridge, A.M. The Chicken Model of Spontaneous Ovarian Cancer. *Prot. Clin. Appl.* **2014**, *8*, 689–699. [CrossRef]
52. Reed, S.; Neuman, H.; Glahn, R.P.; Koren, O.; Tako, E. Characterizing the Gut (*Gallus gallus*) Microbiota Following the Consumption of an Iron Biofortified Rwandan Cream Seeded Carioca (*Phaseolus vulgaris* L.) Bean-Based Diet. *PLoS ONE* **2017**, *12*, e0182431. [CrossRef]

53. Reed, S.; Knez, M.; Uzan, A.; Stangoulis, J.; Glahn, R.P.; Koren, O.; Tako, E. Alterations in the Gut (*Gallus gallus*) Microbiota Following the Consumption of Zinc Biofortified Wheat (*Triticum aestivum*)-Based Diet. *J. Agric. Food Chem.* **2018**, *66*, 6291–6300. [CrossRef] [PubMed]
54. Carboni, J.; Reed, S.; Kolba, N.; Eshel, A.; Koren, O.; Tako, E. Alterations in the Intestinal Morphology, Gut Microbiota, and Trace Mineral Status Following Intra-Amniotic Administration (*Gallus gallus*) of Teff (Eragrostis Tef) Seed Extracts. *Nutrients* **2020**, *12*, 3020. [CrossRef] [PubMed]
55. Shterzer, N.; Rothschild, N.; Sbehat, Y.; Stern, E.; Nazarov, A.; Mills, E. Large Overlap Between the Intestinal and Reproductive Tract Microbiomes of Chickens. *Front. Microbiol.* **2020**, *11*, 1508. [CrossRef] [PubMed]
56. Juste Contin Gomes, M.; Stampini Duarte Martino, H.; Tako, E. Effects of Iron and Zinc Biofortified Foods on Gut Microbiota In Vivo (*Gallus gallus*): A Systematic Review. *Nutrients* **2021**, *13*, 189. [CrossRef] [PubMed]
57. Beasley, J.T.; Johnson, A.A.T.; Kolba, N.; Bonneau, J.P.; Glahn, R.P.; Ozeri, L.; Koren, O.; Tako, E. Nicotianamine-Chelated Iron Positively Affects Iron Status, Intestinal Morphology and Microbial Populations in Vivo (*Gallus gallus*). *Sci. Rep.* **2020**, *10*, 2297. [CrossRef]
58. Reed, S.; Neuman, H.; Moscovich, S.; Glahn, R.; Koren, O.; Tako, E. Chronic Zinc Deficiency Alters Chick Gut Microbiota Composition and Function. *Nutrients* **2015**, *7*, 9768–9784. [CrossRef]
59. Tako, E.; Glahn, R.P. Iron Status of the Late Term Broiler (*Gallus gallus*) Embryo and Hatchling. *Int. J. Poult. Sci.* **2011**, *10*, 42–48. [CrossRef]
60. Warkentin, T.; Kolba, N.; Tako, E. Low Phytate Peas (*Pisum sativum* L.) Improve Iron Status, Gut Microbiome, and Brush Border Membrane Functionality In Vivo (Gallus gallus). *Nutrients* **2020**, *12*, 2563–2581. [CrossRef]
61. Martino, H.S.D.; Kolba, N.; Tako, E. Yacon (Smallanthus sonchifolius) Flour Soluble Extract Improve Intestinal Bacterial Populations, Brush Border Membrane Functionality and Morphology in Vivo (Gallus gallus). 2020, 137, 109705-109716. *Food Res. Int.* **2020**, *137*, 109705–109716. [CrossRef]
62. Dias, D.M.; Kolba, N.; Hart, J.J.; Ma, M.; Sha, S.T.; Lakshmanan, N.; Nutti, M.R.; Martino, H.S.D.; Glahn, R.P.; Tako, E. Soluble Extracts from Carioca Beans (*Phaseolus vulgaris* L.) Affect the Gut Microbiota and Iron Related Brush Border Membrane Protein Expression in Vivo (*Gallus gallus*). *Food Res. Int.* **2019**, *123*, 172–180. [CrossRef] [PubMed]
63. Wang, X.; Kolba, N.; Liang, J.; Tako, E. Alterations in Gut Microflora Populations and Brush Border Functionality Following Intra-Amniotic Administration (*Gallus gallus*) of Wheat Bran Prebiotic Extracts. *Food Funct.* **2019**, *10*, 4834–4843. [CrossRef] [PubMed]
64. Hou, T.; Tako, E. The In Ovo Feeding Administration (*Gallus gallus*)—An Emerging In Vivo Approach to Assess Bioactive Compounds with Potential Nutritional Benefits. *Nutrients* **2018**, *10*, 418–435. [CrossRef] [PubMed]
65. Agarwal, N.; Kolba, N.; Jung, Y.; Cheng, J.; Tako, E. Saffron (*Crocus sativus* L.) Flower Water Extract Disrupts the Cecal Microbiome, Brush Border Membrane Functionality, and Morphology In Vivo (*Gallus gallus*). *Nutrients* **2022**, *141*, 220–233. [CrossRef]
66. Agarwal, N.; Kolba, N.; Khen, N.; Even, C.; Turjeman, S.; Koren, O.; Tako, E. Quinoa Soluble Fiber and Quercetin Alter the Composition of the Gut Microbiome and Improve Brush Border Membrane Morphology In Vivo (*Gallus gallus*). *Nutrients* **2022**, *14*, 448–462. [CrossRef]
67. Pereira da Silva, B.; Kolba, N.; Duarte Martino, H.S.; Hart, J.J.; Tako, E. Soluble Extracts from Chia Seed (*Salvia hispanica* L.) Affect Brush Border Membrane Functionality, Morphology and Intestinal Bacterial Populations In Vivo (*Gallus gallus*). *Nutrients* **2019**, *11*, 2457–2474. [CrossRef]
68. Haahr, M.; Haahr, S. *Random Sequence Generator*; Dublin, Ireland, 2022.
69. Barekatain, R.; Chrystal, P.V.; Howarth, G.S.; McLaughlan, C.J.; Gilani, S.; Nattrass, G.S. Performance, Intestinal Permeability, and Gene Expression of Selected Tight Junction Proteins in Broiler Chickens Fed Reduced Protein Diets Supplemented with Arginine, Glutamine, and Glycine Subjected to a Leaky Gut Model. *Poult. Sci.* **2019**, *98*, 6761–6771. [CrossRef]
70. Dreiling, C.E.; Brown, D.E.; Casale, L.; Kelly, L. Muscle Glycogen: Comparison of Iodine Binding and Enzyme Digestion Assays and Application to Meat Samples. *Meat Sci.* **1987**, *20*, 167–177. [CrossRef]
71. Morais Dias, D.; Kolba, N.; Binyamin, D.; Ziv, O.; Regini Nutti, M.; Stampini Duarte Martino, H.; Glahn, R.P.; Koren, O.; Tako, E. Iron Biofortified Carioca Bean (*Phaseolus vulgaris* L.)-Based Brazilian Diet Delivers More Absorbable Iron and Affects the Gut Microbiota In Vivo (*Gallus gallus*). *Nutrients* **2018**, *10*, 1970–1990. [CrossRef]
72. Pacifici, S.; Song, J.; Zhang, C.; Wang, Q.; Glahn, R.; Kolba, N.; Tako, E. Intra Amniotic Administration of Raffinose and Stachyose Affects the Intestinal Brush Border Functionality and Alters Gut Microflora Populations. *Nutrients* **2017**, *9*, 304. [CrossRef]
73. Hou, T.; Kolba, N.; Glahn, R.; Tako, E. Intra-Amniotic Administration (*Gallus gallus*) of Cicer Arietinum and Lens Culinaris Prebiotics Extracts and Duck Egg White Peptides Affects Calcium Status and Intestinal Functionality. *Nutrients* **2017**, *9*, 785. [CrossRef] [PubMed]
74. Stephens, C.S.; Johnson, P.A. Occludin Expression and Regulation in Small Follicles of the Layer and Broiler Breeder Hen. *Gen. Comp. Endocrinol.* **2017**, *248*, 106–113. [CrossRef] [PubMed]
75. Hartono, K.; Reed, S.; Ankrah, N.A.; Glahn, R.P.; Tako, E. Alterations in Gut Microflora Populations and Brush Border Functionality Following Intra-Amniotic Daidzein Administration. *RSC Adv.* **2015**, *5*, 6407–6412. [CrossRef]
76. Tako, E.; Glahn, R.P.; Welch, R.M.; Lei, X.; Yasuda, K.; Miller, D.D. Dietary Inulin Affects the Expression of Intestinal Enterocyte Iron Transporters, Receptors and Storage Protein and Alters the Microbiota in the Pig Intestine. *Br. J. Nutr.* **2008**, *99*, 472–480. [CrossRef]

77. Zhu, X.Y.; Zhong, T.; Pandya, Y.; Joerger, R.D. 16S RRNA-Based Analysis of Microbiota from the Cecum of Broiler Chickens. *Appl. Environ. Microbiol.* **2002**, *68*, 124–137. [CrossRef]
78. Brisse, S.; Verhoef, J. Phylogenetic Diversity of Klebsiella Pneumoniae and Klebsiella Oxytoca Clinical Isolates Revealed by Randomly Amplified Polymorphic DNA, GyrA and ParC Genes Sequencing and Automated Ribotyping. *Int. J. Syst. Evol. Microbiol.* **2001**, *51*, 915–924. [CrossRef]
79. Younis, A.I.; Elbialy, A.I. Molecular Detection of Genus Klebsiella and Genotypic Identification of Klebsiella Pneumoniae and Klebsiella Oxytoca by Duplex Polymerase Chain Reaction in Poultry. *Glob. Vet.* **2017**, *18*, 234–241.
80. Tako, E.; Glahn, R.P.; Knez, M.; Stangoulis, J.C. The Effect of Wheat Prebiotics on the Gut Bacterial Population and Iron Status of Iron Deficient Broiler Chickens. *Nutr. J.* **2014**, *13*, 1. [CrossRef]
81. Gomes, M.J.C.; Kolba, N.; Agarwal, N.; Kim, D.; Eshel, A.; Koren, O.; Tako, E. Modifications in the Intestinal Functionality, Morphology and Microbiome Following Intra-Amniotic Administration (*Gallus gallus*) of Grape (*Vitis vinifera*) Stilbenes (Resveratrol and Pterostilbene). *Nutrients* **2021**, *13*, 3247. [CrossRef]
82. Gomes, M.J.C.; Martino, H.S.D.; Kolba, N.; Cheng, J.; Agarwal, N.; De Moura Rocha, M.; Tako, E. Zinc Biofortified Cowpea (*Vigna unguiculata* L. Walp.) Soluble Extracts Modulate Assessed Cecal Bacterial Populations and Gut Morphology in Vivo (*Gallus gallus*). *Front. BioScience Landmark* **2022**, *27*, 140–153. [CrossRef]
83. Agarwal, N.; Shukla, V.; Kolba, N.; Jackson, C.; Cheng, J.; Padilla-Zakour, O.I.; Tako, E. Comparing the Effects of Concord Grape (*Vitis labrusca* L.) Puree, Juice, and Pomace on Intestinal Morphology, Functionality, and Bacterial Populations In Vivo (*Gallus gallus*). *Nutrients* **2022**, *14*, 3539. [CrossRef] [PubMed]
84. Agrizzi Verediano, T.; Stampini Duarte Martino, H.; Kolba, N.; Fu, Y.; Cristina Dias Paes, M.; Tako, E. Black Corn (*Zea mays* L.) Soluble Extract Showed Anti-Inflammatory Effects and Improved the Intestinal Barrier Integrity in Vivo (*Gallus gallus*). *Food Res. Int.* **2022**, *157*, 111227–111239. [CrossRef] [PubMed]
85. Kolba, N.; Zarei, A.; Cheng, J.; Agarwal, N.; Dadmohammadi, Y.; Khazdooz, L.; Abbaspourrad, A.; Tako, E. Alterations in Intestinal Brush Border Membrane Functionality and Bacterial Populations Following Intra-Amniotic Administration (*Gallus gallus*) of Nicotinamide Riboside and Its Derivatives. *Nutrients* **2022**, *14*, 3130–3150. [CrossRef]
86. Uni, Z.; Ganot, S.; Sklan, D. Posthatch Development of Mucosal Function in the Broiler Small Intestine. *Poult. Sci.* **1998**, *77*, 75–82. [CrossRef] [PubMed]
87. Cheng, J.; Kolba, N.; Sisser, P.; Turjeman, S.; Even, C.; Koren, O.; Tako, E. Intraamniotic Administration (*Gallus gallus*) of Genistein Alters Mineral Transport, Intestinal Morphology, and Gut Microbiota. *Nutrients* **2022**, *14*, 3473. [CrossRef]
88. Spinner, J.A.; Morris, S.A.; Nandi, D.; Costarino, A.T.; Marino, B.S.; Rossano, J.W.; Shamszad, P. Necrotizing Enterocolitis and Associated Mortality in Neonates With Congenital Heart Disease: A Multi-Institutional Study. *Pediatr. Crit. Care Med.* **2020**, *21*, 228–234. [CrossRef]
89. Buffie, C.G.; Bucci, V.; Stein, R.R.; McKenney, P.T.; Ling, L.; Gobourne, A.; No, D.; Liu, H.; Kinnebrew, M.; Viale, A.; et al. Precision Microbiome Reconstitution Restores Bile Acid Mediated Resistance to Clostridium Difficile. *Nature* **2015**, *517*, 205–208. [CrossRef]
90. Korbecki, J.; Bajdak-Rusinek, K. The Effect of Palmitic Acid on Inflammatory Response in Macrophages: An Overview of Molecular Mechanisms. *Inflamm. Res.* **2019**, *68*, 915–932. [CrossRef]
91. Bousseboua, H.; Le Coz, Y.; Dabard, J.; Szylit, O.; Raibaud, P.; Popoff, M.R.; Ravisse, P. Experimental Cecitis in Gnotobiotic Quails Monoassociated with Clostridium Butyricum Strains Isolated from Patients with Neonatal Necrotizing Enterocolitis and from Healthy Newborns. *Infect. Immun.* **1989**, *57*, 932–936. [CrossRef]
92. Zou, X.; Ji, J.; Wang, J.; Qu, H.; Shu, D.M.; Guo, F.Y.; Luo, C.L. Dextran Sulphate Sodium (DSS) Causes Intestinal Histopathology and Inflammatory Changes Consistent with Increased Gut Leakiness in Chickens. *Br. Poult. Sci.* **2018**, *59*, 166–172. [CrossRef]
93. Eichele, D.D.; Kharbanda, K.K. Dextran Sodium Sulfate Colitis Murine Model: An Indispensable Tool for Advancing Our Understanding of Inflammatory Bowel Diseases Pathogenesis. *World J. Gastroenterol.* **2017**, *23*, 6016–6029. [CrossRef]
94. Perše, M.; Cerar, A. Dextran Sodium Sulphate Colitis Mouse Model: Traps and Tricks. *J. Biomed. Biotechnol.* **2012**, *2012*, 1–13. [CrossRef] [PubMed]
95. Ginzel, M.; Feng, X.; Kuebler, J.F.; Klemann, C.; Yu, Y.; von Wasielewski, R.; Park, J.-K.; Hornef, M.W.; Vieten, G.; Ure, B.M.; et al. Dextran Sodium Sulfate (DSS) Induces Necrotizing Enterocolitis-like Lesions in Neonatal Mice. *PLoS ONE* **2017**, *12*, e0182732. [CrossRef] [PubMed]
96. Denning, T.L.; Bhatia, A.M.; Kane, A.F.; Patel, R.M.; Denning, P.W. Pathogenesis of NEC: Role of the Innate and Adaptive Immune Response. *Semin. Perinatol.* **2017**, *41*, 15–28. [CrossRef] [PubMed]
97. Menconi, A.; Hernandez-Velasco, X.; Vicuña, E.A.; Kuttappan, V.A.; Faulkner, O.B.; Tellez, G.; Hargis, B.M.; Bielke, L.R. Histopathological and Morphometric Changes Induced by a Dextran Sodium Sulfate (DSS) Model in Broilers. *Poult. Sci.* **2015**, *94*, 906–911. [CrossRef]
98. Simon, K.; Arts, J.A.J.; de Vries Reilingh, G.; Kemp, B.; Lammers, A. Effects of Early Life Dextran Sulfate Sodium Administration on Pathology and Immune Response in Broilers and Layers. *Poult. Sci.* **2016**, *95*, 1529–1542. [CrossRef]
99. Gonzalez, L.M.; Moeser, A.J.; Blikslager, A.T. Animal Models of Ischemia-Reperfusion-Induced Intestinal Injury: Progress and Promise for Translational Research. *Am. J. Physiol. Gastrointest. Liver Physiol.* **2015**, *308*, G63–G75. [CrossRef]
100. Joldrichsen, M.R.; Kim, E.; Cormet-Boyaka, E.; Boyaka, P.N. Paneth Cells Regulate Diet-Induced Obesity and Trafficking of Inflammatory Immune Cells into Adipose Tissues. *J. Immunol.* **2020**, *204*, 83.3.

101. Bergstrom, K.S.B.; Kissoon-Singh, V.; Gibson, D.L.; Ma, C.; Montero, M.; Sham, H.P.; Ryz, N.; Huang, T.; Velcich, A.; Finlay, B.B.; et al. Muc2 Protects against Lethal Infectious Colitis by Disassociating Pathogenic and Commensal Bacteria from the Colonic Mucosa. *PLoS Pathog.* **2010**, *6*, e1000902. [CrossRef]
102. Song, C.S.; Park, D.I.; Yoon, M.Y.; Seok, H.S.; Park, J.H.; Kim, H.J.; Cho, Y.K.; Sohn, C.I.; Jeon, W.K.; Kim, B.I. Association Between Red Cell Distribution Width and Disease Activity in Patients with Inflammatory Bowel Disease. *Dig. Dis. Sci.* **2012**, *57*, 1033–1038. [CrossRef]
103. Sodhi, C.P.; Shi, X.; Richardson, W.M.; Grant, Z.S.; Shapiro, R.A.; Prindle, T.; Branca, M.; Russo, A.; Gribar, S.C.; Ma, C.; et al. Toll-Like Receptor-4 Inhibits Enterocyte Proliferation via Impaired β-Catenin Signaling in Necrotizing Enterocolitis. *Gastroenterology* **2010**, *138*, 185–196. [CrossRef] [PubMed]
104. Morgan, M.E.; Zheng, B.; Koelink, P.J.; van de Kant, H.J.G.; Haazen, L.C.J.M.; van Roest, M.; Garssen, J.; Folkerts, G.; Kraneveld, A.D. New Perspective on Dextran Sodium Sulfate Colitis: Antigen-Specific T Cell Development during Intestinal Inflammation. *PLoS ONE* **2013**, *8*, e69936. [CrossRef] [PubMed]
105. Johansson, M.E.V.; Gustafsson, J.K.; Holmén-Larsson, J.; Jabbar, K.S.; Xia, L.; Xu, H.; Ghishan, F.K.; Carvalho, F.A.; Gewirtz, A.T.; Sjövall, H.; et al. Bacteria Penetrate the Normally Impenetrable Inner Colon Mucus Layer in Both Murine Colitis Models and Patients with Ulcerative Colitis. *Gut* **2014**, *63*, 281–291. [CrossRef] [PubMed]
106. Bröer, S. Amino Acid Transport Across Mammalian Intestinal and Renal Epithelia. *Physiol. Rev.* **2008**, *88*, 249–286. [CrossRef]
107. Kong, S.; Zhang, Y.H.; Zhang, W. Regulation of Intestinal Epithelial Cells Properties and Functions by Amino Acids. *BioMed Res. Int.* **2018**, *2018*, 2819154. [CrossRef]
108. Gephart, S.M.; Gordon, P.V.; Penn, A.H.; Gregory, K.E.; Swanson, J.R.; Maheshwari, A.; Sylvester, K. Changing the Paradigm of Defining, Detecting, and Diagnosing NEC: Perspectives on Bell's Stages and Biomarkers for NEC. *Semin. Pediatr. Surg.* **2018**, *27*, 3–10. [CrossRef]
109. Caplan, M.S.; Sun, X.-M.; Hsueh, W.; Hageman, J.R. Role of Platelet Activating Factor and Tumor Necrosis Factor-Alpha in Neonatal Necrotizing Enterocolitis. *J. Pediatr.* **1990**, *116*, 960–964. [CrossRef]
110. Klinke, M.; Wiskemann, H.; Bay, B.; Schäfer, H.-J.; Pagerols Raluy, L.; Reinshagen, K.; Vincent, D.; Boettcher, M. Cardiac and Inflammatory Necrotizing Enterocolitis in Newborns Are Not the Same Entity. *Front. Pediatr.* **2021**, *8*, 593926. [CrossRef]
111. Miyake, H.; Li, B.; Lee, C.; Koike, Y.; Chen, Y.; Seo, S.; Pierro, A. Liver Damage, Proliferation, and Progenitor Cell Markers in Experimental Necrotizing Enterocolitis. *J. Pediatr. Surg.* **2018**, *53*, 909–913. [CrossRef]
112. Zmora, O.; Gutzeit, O.; Segal, L.; Boulos, S.; Millo, Z.; Ginsberg, Y.; Khatib, N.; Dabbah-Assad, F.; Fainaru, O.; Weiner, Z.; et al. Prophylactic Antenatal N-Acetyl Cysteine Administration Combined with Postnatal Administration Can Decrease Mortality and Injury Markers Associated with Necrotizing Enterocolitis in a Rat Model. *PLoS ONE* **2020**, *15*, e0233612. [CrossRef]
113. Karin, M.; Lawrence, T.; Nizet, V. Innate Immunity Gone Awry: Linking Microbial Infections to Chronic Inflammation and Cancer. *Cell* **2006**, *124*, 823–835. [CrossRef] [PubMed]
114. Karin, M.; Yamamoto, Y.; Wang, Q.M. The IKK NF-KB System: A Treasure Trove for Drug Development. *Nat. Rev. Drug Discov.* **2004**, *3*, 17–26. [CrossRef] [PubMed]
115. Akira, S.; Uematsu, S.; Takeuchi, O. Pathogen Recognition and Innate Immunity. *Cell* **2006**, *124*, 783–801. [CrossRef]
116. Dobrovolskaia, M.; Kozlov, S. Inflammation and Cancer: When NF-κB Amalgamates the Perilous Partnership. *Curr. Cancer Drug Targets* **2005**, *5*, 325–344. [CrossRef] [PubMed]
117. Lawrence, T. The Nuclear Factor NF-B Pathway in Inflammation. *Cold Spring Harb. Perspect. Biol.* **2009**, *1*, a001651. [CrossRef]
118. Krishnaveni, M.; Jayachandran, S. Inhibition of MAP Kinases and down Regulation of TNF-α, IL-β and COX-2 Genes by the Crude Extracts from Marine Bacteria. *Biomed. Pharmacother.* **2009**, *63*, 469–476. [CrossRef]
119. Luo, X.; Zhang, X.; Wu, X.; Yang, X.; Han, C.; Wang, Z.; Du, Q.; Zhao, X.; Liu, S.-L.; Tong, D.; et al. Brucella Downregulates Tumor Necrosis Factor-α to Promote Intracellular Survival via Omp25 Regulation of Different MicroRNAs in Porcine and Murine Macrophages. *Front. Immunol.* **2018**, *8*, 2013. [CrossRef]
120. Claud, E.C.; Keegan, K.P.; Brulc, J.M.; Lu, L.; Bartels, D.; Glass, E.; Chang, E.B.; Meyer, F.; Antonopoulos, D.A. Bacterial Community Structure and Functional Contributions to Emergence of Health or Necrotizing Enterocolitis in Preterm Infants. *Microbiome* **2013**, *1*, 20. [CrossRef]
121. Hill, H.R.; Hunt, C.E.; Matsen, J.M. Nosocomial Colonization with Klebsiella, Type 26, in a Neonatal Intensive-Care Unit Associated with an Outbreak of Sepsis, Meningitis, and Necrotizing Enterocolitis. *J. Pediatr.* **1974**, *85*, 415–419. [CrossRef]
122. Azzaroli, F.; Turco, L.; Ceroni, L.; Sartoni Galloni, S.; Buonfiglioi, F.; Calvanese, C.; Mazzella, G. Pneumatosis Cystoides Intestinalis. *WJG* **2011**, *17*, 4932. [CrossRef]
123. Ling, F.; Guo, D.; Zhu, L. Pneumatosis Cystoides Intestinalis: A Case Report and Literature Review. *BMC Gastroenterol.* **2019**, *19*, 176. [CrossRef] [PubMed]
124. Meini, S.; Zini, C.; Passaleva, M.T.; Frullini, A.; Fusco, F.; Carpi, R.; Piani, F. Pneumatosis Intestinalis in COVID-19. *BMJ Open Gastroenterol.* **2020**, *7*, e000434. [CrossRef] [PubMed]
125. Im, J.; Anjum, F. Pneumatosis Intestinalis. In *StatPearls [Internet]*; StatPearls Publishing: Treasure Island, FL, USA; St. Petersburg, Russia, 2021; pp. 1–9.
126. Tarracchini, C.; Milani, C.; Longhi, G.; Fontana, F.; Mancabelli, L.; Pintus, R.; Lugli, G.A.; Alessandri, G.; Anzalone, R.; Viappiani, A.; et al. Unraveling the Microbiome of Necrotizing Enterocolitis: Insights in Novel Microbial and Metabolomic Biomarkers. *Microbiol. Spectr.* **2021**, *9*, e01176-21. [CrossRef] [PubMed]

127. Hunter, C.J.; Bean, J.F. Cronobacter: An Emerging Opportunistic Pathogen Associated with Neonatal Meningitis, Sepsis and Necrotizing Enterocolitis. *J. Perinatol.* **2013**, *33*, 581–585. [CrossRef] [PubMed]
128. Zhou, Y.; Shan, G.; Sodergren, E.; Weinstock, G.; Walker, W.A.; Gregory, K.E. Longitudinal Analysis of the Premature Infant Intestinal Microbiome Prior to Necrotizing Enterocolitis: A Case-Control Study. *PLoS ONE* **2015**, *10*, e0118632. [CrossRef]
129. Schönherr-Hellec, S.; Aires, J. Clostridia and Necrotizing Enterocolitis in Preterm Neonates. *Anaerobe* **2019**, *58*, 6–12. [CrossRef] [PubMed]
130. Hosny, M.; Baptiste, E.; Levasseur, A.; La Scola, B. Molecular Epidemiology of Clostridium Neonatale and Its Relationship with the Occurrence of Necrotizing Enterocolitis in Preterm Neonates. *New Microbes New Infect.* **2019**, *32*, 100612. [CrossRef] [PubMed]
131. Baranowski, J.R.; Claud, E.C. Necrotizing Enterocolitis and the Preterm Infant Microbiome. In *Probiotics and Child Gastrointestinal Health*; Guandalini, S., Indrio, F., Eds.; Advances in Experimental Medicine and Biology; Springer International Publishing: Cham, Switzerland, 2019; Volume 1125, pp. 25–36. ISBN 978-3-030-14635-1.
132. Neu, J.; Pammi, M. Necrotizing Enterocolitis: The Intestinal Microbiome, Metabolome and Inflammatory Mediators. *Semin. Fetal Neonatal Med.* **2018**, *23*, 400–405. [CrossRef]
133. Garcia, M.I.; Ghiani, M.; Lefort, A.; Libert, F.; Strollo, S.; Vassart, G. LGR5 Deficiency Deregulates Wnt Signaling and Leads to Precocious Paneth Cell Differentiation in the Fetal Intestine. *Dev. Biol.* **2009**, *331*, 58–67. [CrossRef]
134. Juber, B.A.; McElroy, S.J. The Paneth Cell and Its Role in the Development of NEC. In *Necrotizing Enterocolitis*, 1st ed.; Hackam, D.J., Ed.; CRC Press: Boca Raton, FL, USA, 2021; pp. 242–247. ISBN 978-0-429-28830-2.
135. Liu, D.; Xu, Y.; Feng, J.; Yu, J.; Huang, J.; Li, Z. Mucins and Tight Junctions Are Severely Altered in Necrotizing Enterocolitis Neonates. *Am. J. Perinatol.* **2021**, *38*, 1174–1180. [CrossRef]
136. Zhao, B.; Wu, J.; Li, J.; Bai, Y.; Luo, Y.; Ji, B.; Xia, B.; Liu, Z.; Tan, X.; Lv, J.; et al. Lycopene Alleviates DSS-Induced Colitis and Behavioral Disorders via Mediating Microbes-Gut–Brain Axis Balance. *J. Agric. Food Chem.* **2020**, *68*, 3963–3975. [CrossRef] [PubMed]
137. Wu, H.; Chen, Q.-Y.; Wang, W.-Z.; Chu, S.; Liu, X.-X.; Liu, Y.-J.; Tan, C.; Zhu, F.; Deng, S.-J.; Dong, Y.-L.; et al. Compound Sophorae Decoction Enhances Intestinal Barrier Function of Dextran Sodium Sulfate Induced Colitis via Regulating Notch Signaling Pathway in Mice. *Biomed. Pharmacother.* **2021**, *133*, 110937. [CrossRef] [PubMed]
138. Li, L.; Cheng, L.; Li, Z.; Li, C.; Hong, Y.; Gu, Z. Butyrylated Starch Protects Mice from DSS-Induced Colitis: Combined Effects of Butyrate Release and Prebiotic Supply. *Food Funct.* **2021**, *12*, 11290–11302. [CrossRef] [PubMed]
139. Gehart, H.; Clevers, H. Tales from the Crypt: New Insights into Intestinal Stem Cells. *Nat. Rev. Gastroenterol. Hepatol.* **2019**, *16*, 19–34. [CrossRef] [PubMed]
140. Rodríguez-Colman, M.J.; Schewe, M.; Meerlo, M.; Stigter, E.; Gerrits, J.; Pras-Raves, M.; Sacchetti, A.; Hornsveld, M.; Oost, K.C.; Snippert, H.J.; et al. Interplay between Metabolic Identities in the Intestinal Crypt Supports Stem Cell Function. *Nature* **2017**, *543*, 424–427. [CrossRef]
141. Bel, S.; Pendse, M.; Wang, Y.; Li, Y.; Ruhn, K.A.; Hassell, B.; Leal, T.; Winter, S.E.; Xavier, R.J.; Hooper, L.V. Paneth Cells Secrete Lysozyme via Secretory Autophagy during Bacterial Infection of the Intestine. *Science* **2017**, *357*, 1047–1052. [CrossRef]
142. Bevins, C.L.; Salzman, N.H. Paneth Cells, Antimicrobial Peptides and Maintenance of Intestinal Homeostasis. *Nat. Rev. Microbiol.* **2011**, *9*, 356–368. [CrossRef]

nutrients

MDPI

Article

Effect of Black Corn Anthocyanin-Rich Extract (*Zea mays* L.) on Cecal Microbial Populations In Vivo (*Gallus gallus*)

Thaisa Agrizzi Verediano [1,†], Nikita Agarwal [2,†], Hércia Stampini Duarte Martino [1], Nikolai Kolba [2], Mariana Grancieri [1], Maria Cristina Dias Paes [3] and Elad Tako [2,*]

1 Nutrition and Health Department, Universidade Federal de Viçosa, Vicosa 36571-000, Minas Gerais, Brazil
2 Department of Food Science, Cornell University, Stocking Hall, Ithaca, NY 14853, USA
3 Empresa Brasileira de Pesquisa e Agropecuária (EMBRAPA), Sete Lagoas 35701-970, Minas Gerais, Brazil
* Correspondence: et79@cornell.edu; Tel.: +1-607-255-0884
† These authors contributed equally to this work.

Abstract: Black corn has been attracting attention to investigate its biological properties due to its anthocyanin composition, mainly cyanidin-3-glucoside. Our study evaluated the effects of black corn extract (BCE) on intestinal morphology, gene expression, and the cecal microbiome. The BCE intra-amniotic administration was evaluated by an animal model in *Gallus gallus*. The eggs (n = 8 per group) were divided into: (1) no injection; (2) 18 MΩ H_2O; (3) 5% black corn extract (BCE); and (4) 0.38% cyanidin-3-glucoside (C3G). A total of 1 mL of each component was injected intra-amniotic on day 17 of incubation. On day 21, the animals were euthanized after hatching, and the duodenum and cecum content were collected. The cecal microbiome changes were attributed to BCE administration, increasing the population of *Bifidobacterium* and *Clostridium*, and decreasing *E. coli*. The BCE did not change the gene expression of intestinal inflammation and functionality. The BCE administration maintained the villi height, Paneth cell number, and goblet cell diameter (in the villi and crypt), similar to the H_2O injection but smaller than the C3G. Moreover, a positive correlation was observed between *Bifidobacterium*, *Clostridium*, *E. coli*, and villi GC diameter. The BCE promoted positive changes in the cecum microbiome and maintained intestinal morphology and functionality.

Keywords: cyanidin; intestinal barrier; phenolic components; goblet cells

Citation: Agrizzi Verediano, T.; Agarwal, N.; Stampini Duarte Martino, H.; Kolba, N.; Grancieri, M.; Dias Paes, M.C.; Tako, E. Effect of Black Corn Anthocyanin-Rich Extract (*Zea mays* L.) on Cecal Microbial Populations In Vivo (*Gallus gallus*). *Nutrients* **2022**, *14*, 4679. https://doi.org/10.3390/nu14214679

Academic Editor: Dennis Savaiano

Received: 12 October 2022
Accepted: 1 November 2022
Published: 4 November 2022

1. Introduction

Corn, also known as maize (*Zea mays* L.), is one of the most produced cereals and one of the major food sources worldwide [1]. In recent decades, scientific research has focused on pigmented corn varieties due to their beneficial health properties [2]. Among them, black corn (*Zea mays* spp.) is a variety traditionally cultivated in South and Central America that has an affinity for warm and dry climates [1]. Black and purple corn can accumulate anthocyanin in different tissues; thus, these varieties have a significant concentration of these flavonoids [3].

Anthocyanins are bioactive water-soluble pigments observed in nature, mainly in the form of glycosides, providing color in plants, fruits, vegetables, and flowers [4]. Extract from anthocyanin-rich foods has been reported to have health properties [5], as anti-inflammatory [6], antioxidant [7], gut microbiota modulation [8], improving cholesterol [9], and glucose metabolism [10]. Seven hundred anthocyanin structures have been identified in nature; however, some of them are verified in higher concentrations in plants, as cyanidin, delphinidin, malvidin, pelargonidin, peonidin, and petunidin [4,11]. Among them, cyanidin-3-glucoside (C3G) is the most predominant anthocyanin naturally observed in plants [12]. C3G is the main anthocyanin observed in the black corn flour composition (30.40 mg/100 g); however, a high concentration of phenolic components (614.30 mg GAE—gallic acid equivalent/100 g) is also observed in the food matrix [13].

C3G has an *O*-glycosylated anthocyanin with two hydroxyls on the third aromatic ring, which confers vigorous antioxidant activity [14]. The leading site of C3G catabolism is the small intestine, in which the C3G molecules are hydrolyzed to aglycones and degraded to specific phenolic compounds such as protocatechuic acid, phloroglucinaldehyde, vanillic acid, and ferulic acid by the gut microbiota [12]. Intestinal microbiota are able to utilize phenolic compounds as a substrate to obtain energy and to create fermentable metabolites with biological functions [15]. The metabolites produced from the intestinal metabolism of C3G inhibit inflammatory pathways such as nuclear factor-kappa B (Nf-κB) [14]. Since the intestinal tract acts as a barrier against external pathogens [16], C3G and its metabolites contribute to maintaining the intestinal barrier integrity, mucosal barrier, and microbiota composition [12,17].

Recently, we demonstrated that black corn soluble extract, composed of soluble fiber and phenolic compounds, promoted goblet cell proliferation and upregulated biomarkers related to the epithelial intestinal integrity pathway such as AMP—activated protein kinase (AMPK) and caudal-related homeobox transcriptional factor 2 (CDX2) [18]. Previously, it was suggested that fermented soluble extract promotes the proliferation of beneficial gut bacteria, which affects intestinal brush border membrane morphology, including the growth of villus and crypt and goblet cell proliferation [19]. However, the effects of the isolated phenolic extract on the morphology and gut microbiota might differ [20]. Saffron flower, a source of polyphenols, showed an unfavorable effect on the microbial population, brush border morphology, and functionality [21]. Depending on their dosage, polyphenols can have negative impacts due to interference with nutrient metabolism [20]. Resveratrol and pterostilbene (5%) did not promote modification in the taxonomy of the cecal microbiota but increased morphological changes [22]. For this reason, phenolics and other bioactive compounds in food sources and byproducts need to be quantified and their biological effects and safety validated in living organisms.

The intra-amniotic (in ovo) approach is widely accepted for assessing potential effects of bioactive components [18,21,23–25]. The in ovo technique in *Gallus gallus* allows the administration of components into the amniotic fluid. Therefore, as the embryo consumes the amniotic fluid before hatching, the biological changes after hatching are a predictor of the effects of the bioactive component administered [19]. Considering the possible consumption of bioactive compound-rich extracts, it is relevant to assess the biological effects of dried extracts to validate their application. Since anthocyanin-rich extract from black corn has not been explored for intestinal health so far, this experiment was carried out to investigate the impact of intra-amniotic administration of black corn anthocyanin-rich extract on the intestinal brush border membrane functionality, morphology, and cecal microbial populations.

2. Materials and Methods

2.1. Materials

Black corn grains (TO002) were provided by the Brazilian Agriculture Research Corporation (EMBRAPA) from the Maize Germplasm Bank of the Maize and Sorghum National Research Center (Sete Lagoas, MG, Brazil). Cyanidin-3-glucoside chloride (>98%) was obtained from Sigma-Aldrich® (Cat # PHL89616, St. Louis, MO, USA).

2.2. Black Corn Extract Procedure

Prior to extraction, black corn grains were ground with a 1.0 mm stainless steel sieve (Willy, Solab®, Piracicaba, Brazil) to prepare the corn flour. The production of the extract was performed at room temperature without any light. The black corn flour was added to ethanol 50% (1:10 *v*/*v*), then submitted to a magnetic stir plate (100 rpm/60 min/room temperature). After the allotted time had passed, the suspension was vacuum-filtered via filter paper. The ethanol in the extract was evaporated in a rotatory evaporator (40 °C) [26]. Then, the resulting concentrate was lyophilized, resulting in a dried extract (Figure 1), whose weight was quantified to calculate the final yield considering the initial flour mass.

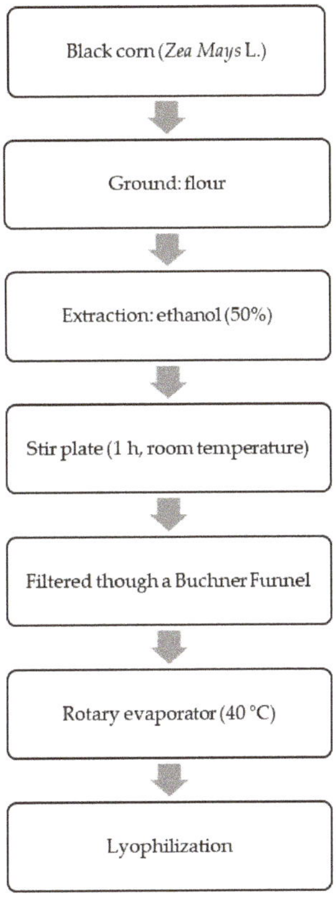

Figure 1. Flowchart of the black corn extract procedure.

2.3. *Extract Chemical Characterization*

2.3.1. Total Polyphenols and Antioxidant Capacity

The analysis of total polyphenols was determined in the dried extract by the Single-Ciocalteau assay [27]. The absorbance was measured (760 nm) and total polyphenols were expressed as grams of gallic acid equivalent (GAE) per 100 g of wet weight sample.

The antioxidant capacity was determined by the radical scavenging activity assay using DPPH (1,1-diphenyl-2-picrylhydrazyl) [28]. Briefly, lyophilized black corn extract (100 μL) was added to an ethanolic DPPH solution and stirred by vortex (3000 rpm/30 s). After incubation (30 min), the absorbance was measured (517 nm) and the DPPH radical scavenging activity was calculated as:

$$\text{Scavenging } \% = \frac{\text{Ablank} - \text{Asample}}{\text{Ablank}} \tag{1}$$

2.3.2. Anthocyanin Profile Analysis

The black corn anthocyanin-rich extract was analyzed by High Performance Liquid Chromatography (HPLC) Alliance Waters® model 2690/5, with a Waters® photodiode array detector model 2996 (scanning from 210 to 600 nm with quantification at 520 nm). The chromatographic separation was performed using a Thermo Hypersil BDS (Thermo Fisher Scientific, Waltham, MA, USA) C_{18} column (100 mm × 4.6 mm × 2.4 μm) at 40 °C, an injection volume of 20 μL, a total run time of 20 min, and a 1.0 mL min^{-1} flow rate. The mobile phase used was an aqueous solution of formic acid (Phase A) and acetonitrile (Phase B). The quantification was performed by using external standards. The gradient elution was 20% solvent B over 3 min, followed by a linear gradient up to 30% solvent

B within 15 min and held there for 2 min, and then a linear gradient up to 60% solvent B in 13 min and held there for 2 min. Returning to initial conditions, 20% solvent B in 5 min and held there for 8 min for column rinse and re-equilibration. Mobile phase 2 consisted of an aqueous solution of formic acid as solvent A and acetonitrile as solvent B. The gradient was linear from 5 to 10.5% solvent B over 7 min 30 s and held there for 4 min 30 s, then a linear gradient up to 12% solvent B over 1 min, then another linear gradient up to 14% solvent B over 1 min, and then reduced back to 5% solvent B at 2 min 30 s and held there for 3 min 30 s for column rinse and re-equilibration [29]. Cyanidin-3-glucoside and pelaronidin-3-O-glucoside were used as standards.

2.4. Intra-Amniotic Experiment

Forty Cornish-cross fertile eggs from a commercial hatchery (Moyer's chicks, Quakertown, PA, USA) were incubated under controlled temperature (37 ± 2 °C) and humidity (89 ± 2% humidity) in a poultry farm incubator at Cornell University Animal Science. All experimental procedures were carried out in accordance with the Cornell University International Animal Care and Use Committee (IACUC, protocol code: 2020-0077).

The black corn extract and the cyanidin-3-glucoside (C3G) were diluted in 18 MΩ H_2O to verify the concentration to achieve an osmolarity value (Osm) of <320 Osm [18,22], in order to certify that the viable embryos would not be dehydrated upon the administration of the amniotic fluid. During the embryonic development (a total of 21 days), on day 17 of incubation, eggs with viable embryos (n = 36) were distributed by randomization into four groups with a similar weight frequency distribution. The groups were distributed as follows: No-injection (n = 7); H_2O injection, 18 MΩ H_2O (n = 8); BCE, 5% black corn extract in 18 MΩ H_2O (n = 8); and C3G, 0.38% cyanidin-3-glucoside in 18 MΩ H_2O (n = 8).

The intra-amniotic administration of black corn extract (1 mL/animal) was prepared at a concentration of 5% in accordance with our previous study [18]. The C3G was administered at a concentration of 0.38%, as this compound has yet to be tested intra-amniotically; hence, we chose to proceed with a lower dosage. A 1 mL solution was administered using a 21-gauge needle into amniotic fluid following candling [22,25]. Afterward, cellophane tape was used to seal the injection holes, and all the eggs were allocated to hatching baskets to minimize bias related to allocation. On day 21, after hatching, chickens were weighed and then euthanized by CO_2 exposure, and the blood was collected by cardiac puncture. The duodenum and cecum were immediately collected, and part of the duodenum and cecum were immersed in liquid nitrogen and then kept at −80 °C until further analysis. Meanwhile, the other portion of the duodenum was fixed in a 10% (v/v) formalin solution for histological analysis.

2.5. Total RNA Extraction from Duodenum

Total RNA extraction from the proximal duodenum (n = 5 animals/group) was performed with a RNeasy Mini Kit, Qiagen Inc. (Cat # 74004, Valencia, CA, USA), as suggested by the manufacturer's protocol. The procedures were performed under RNase-free conditions, and RNA was quantified by absorbance (260/280 nm). The integrity of the 18S ribosomal RNAs was carried out using agarose gel electrophoresis (1.5%) and staining with ethidium bromide. Extracted RNA samples were frozen at (−80 °C) until further analysis.

2.6. Gene Expression Analysis

The gene expression of the duodenum was determined by real-time polymerase chain reaction (RT-PCR) as described earlier [18,25]. Briefly, cDNA was created with a total of 20 μL of reverse transcriptase (RT) reaction completed in a BioRad C1000 touch thermocycler using the Improm-II Reverse Transcriptase Kit (Ca # A1250; Promega, Madison, WI, USA). The cDNA obtained was analyzed by Nanodrop (Thermo Fisher Scientific, Waltham, MA, USA). The concentration of cDNA was verified by measuring the absorbance (260/280 nm) with an extinction coefficient of 33 (for single-stranded DNA). The forward and reverse primers and the tested genes' descriptions were designed based on the Genbank database, using Real-Time primer Design Tool software (IDT DNA, Coralvilla, IA, USA) (Table 1).

Table 1. Sequence and description of experimental primers.

Analyte	Forward P. (5′-3′)	Reverse P. (5′-3′)	Base Pairs Length	GI Identifier
		Inflammatory Response		
TNFα	GACAGCCTATGCCAACAAGTA	TTACAGGAAGGGCAACTCATC	109	53,854,909
NF-κB1	CACAGCTGGAGGGAAGTAAAT	TTGAGTAAGGAAGTGAGGTTGAG	100	2,130,627
IL-1β	CTCACAGTCCTTCGACATCTTC	TGTTGAGCCTCACTTTCTGG	119	88,702,685
		Intestinal Functionality		
MUC2	CCTGCTGCAAGGAAGTAGAA	GGAAGATCAGAGTGGTGCATAG	272	423,101
OCLN	GTCTGTGGGTTCCTCATCGT	GTTCTTCACCCACTCCTCCA	124	396,026
AMPK	CTCCACTTCCAGAAGGTTACTT	GCAGTAGCTATCGTTCATCCTATC	140	427,185
CDX2	ACCAGGACGAAGGACAAATAC	CTTTCCTCCGGATGGTGATATAG	103	374,205
VDAC2	CAGCACTCGCTTTGGAATTG	GTGTAACCCACTCCAACTAGAC	99	395,498
18S rRNA	GCAAGACGAACTAAAGCGAAAG	TCGGAACTACGACGGTATCT	100	7,262,899

P: primers; TNFα: tumor necrosis factor-alpha; NF-κB1: nuclear factor kappa beta 1; IL-1β: interleukin 1 beta; MUC2: mucin 2; OCLN: occludin; AMPK: AMP-activated protein kinase; CDX2: caudal-related homeobox transcriptional factor 2; VDAC: voltage-dependent anion channel.

Real-time PCR amplifications were carried out under specific conditions: 95 °C (30 s followed by 40 cycles (95 °C, 15 s), annealing temperature for 30 s, and elongation at 60 °C for 30 s in the Bio-Rad CFX96 Touch (Hercules, CA, USA). The gene expression data was obtained as the lowest cyclic product (Cp) values based on the "second derivative maximum" as computed by Bio-Rad CFX Maestro 1.1 (Version 4.1.2433.1219, Hercules, CA, USA). The assays were quantified through a standard curve in the real-time qPCR analysis, and a 1:10 dilution prepared a standard curve with four points. The software procedure a Cp vs. log 10 concentration graph, and the efficiencies were calculated as 10 (1/slope). The specificity of the amplified real-time RT-PCR procedures was verified by melting curve analysis (60–95 °C) after 40 cycles, resulting in several different specific products with specific melting temperatures.

2.7. Intestinal Content and DNA Isolation

The cecum (n = 5 animals/group) from a separate chicken was aseptically removed and treated as shown elsewhere [18,30]. In short, the cecum content (200 mg) was placed into a plastic tube with phosphate-buffered saline (PBS) solution and homogenized through a vortex with glass beads (3 mm in diameter) for 3 min. To remove the debris, it was centrifuged, and the supernatant was collected. Before DNA extraction, the pellet was washed twice with PBS and stored at −20 °C. In order to perform the purification of DNA, the pellet was re-suspended in 50 mM ethylenediaminetetraacetic acid (EDTA) and treated with lysozyme (Sigma Aldrich Co., St. Louis, MO, USA). The bacterial genomic DNA was isolated using the Wizard Genomic DNA purification kit (Cat # A1120, Promega Corp., Madison, WI, USA).

2.8. Primers Design and PCR Amplification of Bacterial 16S rDNA

Primers for *Lactobacillus*, *Bifidobacterium*, *Clostridium*, *Escherichia coli*, and *L. planetarium* were used. The universal primers were designed based on prior research [25,30,31]. PCR products were separated by electrophoresis on a 2% agarose gel, stained with ethidium bromide, and quantified by Quantity One 1-D analysis software (Version 4.6.8, Bio-Rad, Hercules, CA, USA). All products were expressed relative to the content of the universal 16s rRNA primer product and the proportions of each examined bacterial group.

2.9. Histological Analysis

Duodenal morphology was performed as previously described [18,32]. Briefly, duodenum sections were fixed using buffered formaldehyde solution 4% (v/v), dehydrated, cleared, and embedded in paraffin. Sections (5 μm) were added to glass slices, deparaffinized in xylene, rehydrated in ethanol, and stained with Alcian blue/Periodic acid–Schiff. The morphometric measurements of villus height (μM), villus surface (μM), depth of crypts (μM), goblet cell number, and goblet cell diameter (μM) in the crypt and the villi, Paneth cell number, and Paneth cell diameter were assessed using a light microscope (CellSens Standard software, Olympus, Waltham, MA, USA). Five segments of each biological sample (n = 3/treatment group) were assessed, and ten randomly selected villi and crypts were analyzed per segment (50 replicates per biological sample). Villus surface area was obtained by the equation:

$$\text{Villus surface area} = 2\,\frac{VW}{2} \times VL \tag{2}$$

where VW = villus width average of three measurements, and VL = villus length.

2.10. Statistical Analysis

Experimental groups were completely randomized. Statistically significant differences between experimental groups were conducted by a one-way Analysis of Variance (ANOVA) and a post-hoc Duncan test for those with a normal distribution. The mean for a normal distribution is tested using the Shapiro–Wilk normality test. The means without normal distribution were analyzed using Kruskal–Wallis and a post-hoc Dunn's test. Data were expressed as mean ± standard error deviation (SED) and differences were considered significant when $p < 0.05$. The association and significance between intestinal biomarkers, bacterial population, and histological parameters were analyzed by Spearman's rank correlation coefficient. GraphPad Prism® version 8.0 software packages (GraphPad Software Inc., San Diego, CA, USA) were used for graphing and data analysis.

3. Results

3.1. Black Corn Extract Characterization

The cyanidin-3-glucoside (C3G) was identified as the principal anthocyanin constituent of black corn extract (BCE), followed by pelargonidin-3-O-glucoside. The BCE showed a high concentration of total phenolic compounds (555 mg GAE/100 g), and the antioxidant capacity was 70.79% (Table 2).

Table 2. Characterization of black corn anthocyanin-rich extract (BCE).

Components	Amount	Retention Time (min)
Cyanidin-3-glucoside (mg/100 g)	283.91	6.5
Pelargonidin-3-O-glucoside (mg/100 g)	39.57	8.7
Total phenolic compounds (mg GAE/100 g sample)	555.00	-
DPPH (%)	70.79	-

GAE: gallic acid equivalent. Cyanidin-3-glucoside and pelargonidin-3-O-glucoside were quantified by High Performance Liquid Chromatography; total phenolic compounds and DPPH were analyzed by spectrophotometry.

3.2. Effect of BCE on the Bacterial Population on Cecum Content

The BCE promoted significant changes in the cecum bacterial populations. Specifically, the BCE and the G3G increased ($p < 0.05$) *Bifidobacterium* and decreased ($p < 0.05$) *E. coli* populations compared to No injection and H_2O injection. The BCE group had the highest abundance of *Clostridium* compared to the other treatment groups. Further, the abundance of *Lactobacillus* significantly ($p < 0.05$) decreased after the C3G intra-amniotic administration compared to the control and BCE groups. The abundance of *L. plantarum* was similar ($p > 0.05$) among all experimental groups (Figure 2).

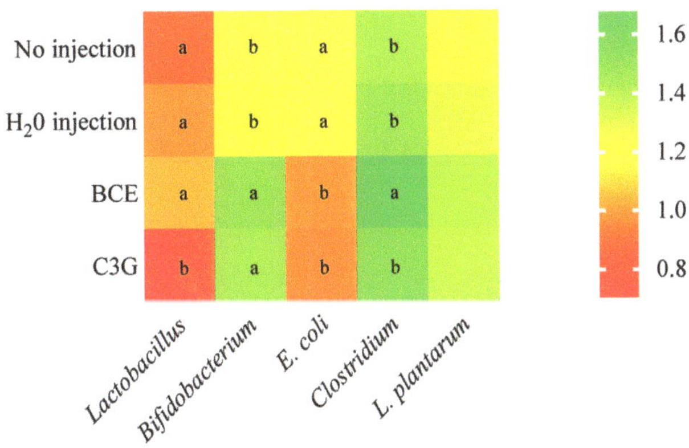

Figure 2. The effect of intra-amniotic black corn extract administration on the bacterial population from cecal content. The relative abundance is expressed in arbitrary units (AU). Values are means ± SED, $n = 5$ animals/group. BCE: black corn extract; C3G: cyanidin-3-glucoside. The treatment groups not indicated by the same letter are different ($p < 0.05$) by the post-hoc Duncan test. Squares without any letters: no difference among the treatments ($p > 0.05$).

3.3. *Effect of BCE on Duodenal Gene Expression*

The gene expression of duodenal interleukin one beta (IL-1β) and nuclear factor kappa beta (NF-κβ) was similar ($p > 0.05$) among the experimental groups. The pro-inflammatory cytokine tumor necrosis factor-alpha (TNFα) was downregulated ($p < 0.05$) in the C3G group compared to BCE and the H_2O injection (Figure 3A). Furthermore, to evaluate the intestinal physical barrier integrity, the mRNA expression of AMP-activated protein kinase (AMPK), occludin (OCLN), and voltage-dependent anion channel (VDAC) were determined, but no significant difference ($p > 0.05$) was observed among the groups for these variables. On the other hand, the caudal-related homeobox transcriptional factor 2 (CDX2) gene expression was downregulated ($p < 0.05$) after the intra-amniotic administration of H_2O, BCE, and C3G compared with No injection (Figure 3B).

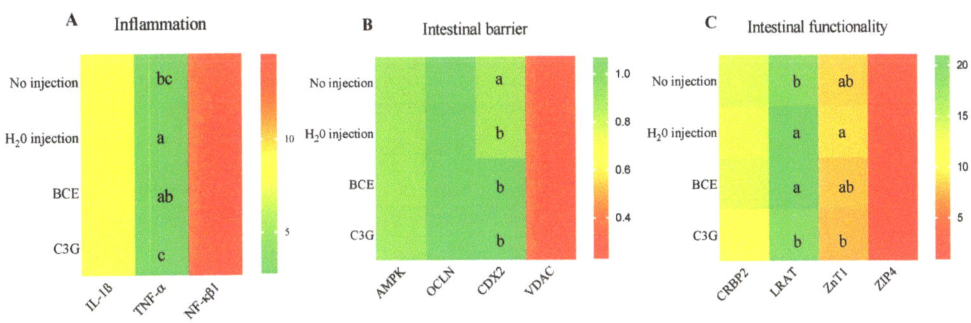

Figure 3. Effect of intra-amniotic administration of black corn extract on duodenal gene expression related to (**A**): intestinal inflammation biomarkers; (**B**): intestinal barrier biomarkers; (**C**): intestinal functionality biomarkers. Values are means (AU: arbitrary units) ± SED, $n = 5$ animals/group. BCE: black corn extract; C3G: cyanidin-3-glucoside. The treatment groups not indicated by the same letter are different ($p < 0.05$) by the post-hoc Duncan test. Squares without any letters: no difference among the treatments ($p > 0.05$). TNFα: tumor necrosis factor-alpha; NF-κβ1: nuclear factor kappa beta-1; IL-1β: interleukin 1 beta; AMPK: AMP-activated protein kinase; OCLN: occludin; CDX2: caudal-related homeobox transcriptional factor 2; VDAC: voltage-dependent anion channel; CRBP2: cellular retinol-binding protein-2; LRAT: lecithin retinol acyltransferase; ZnT1: Zinc transporter 1.

Intestinal functionality was assessed through intestinal transporters. The mRNA expression of cellular retinol-binding protein-2 (CRBP2) and ZIP 4 was similar among all experimental groups ($p > 0.05$). However, lecithin: retinol acyltransferase (LRAT) and zinc transporter 1 (ZnT1) downregulated ($p < 0.05$) in the C3G group compared to the H_2O injection group, but there was no difference between the intra-amniotic administration of BCE and H_2O for these markers (Figure 3C).

3.4. Effect of BCE on Duodenal Morphology

A morphological analysis of the duodenum was performed to observe the intra-amniotic effects of BCE in the duodenal mucosa. The animals that received the BCE had no changes in the villi height compared to the H_2O injection ($p > 0.05$). The C3G group showed the highest villi height among all experimental groups ($p < 0.05$). Further, the duodenal depth crypt and the Paneth cell number were higher in the C3G compared to the BCE group ($p < 0.05$). The Paneth number was higher ($p < 0.05$) in the BCE when compared to the No injection (Table 3).

Table 3. Effect of intra-amniotic administration of black corn extract on villi height, surface, and depth crypt.

	No Injection	H_2O Injection	BCE	C3G
Villi height (μM)	193.12 ± 3.75 ^{b}	171.50 ± 4.01 ^{c}	169.69 ± 2.10 ^{c}	202.43 ± 2.81 ^{a}
Villi surface (μM^2)	12,324.31 ± 344.23 ^{b}	11,740.52 ± 336.59 ^{b}	11,181.40 ± 224.95 ^{b}	15,250.89 ± 390.56 ^{a}
Depth crypt (μM)	24.98 ± 1.01 ^{b}	37.30 ± 1.08 ^{a}	24.78 ± 0.62 ^{b}	35.35 ± 0.99 ^{a}
Paneth cell number	0.95 ± 0.07 ^{c}	1.30 ± 0.04 ^{b}	1.27 ± 0.04 ^{b}	2.49 ± 0.14 ^{a}
Paneth cell diameter	1.69 ± 0.03 ^{a}	1.55 ± 0.02 ^{b}	1.50 ± 0.02 ^{b}	1.58 ± 0.02 ^{a,b}

Values are means ± SED, $n = 3$ animals/group. BCE: black corn extract; C3G: cyanidin-3-glucoside. Treatment group means for specific variables followed by the same letter are not significantly different ($p > 0.05$) by Kruskal–Wallis and a post-hoc Dunn's test.

Moreover, goblet cell (GC) morphological analysis was performed in the villi and the crypt. In the villi, the GC diameter (Figure 4A) and number (Figure 4B) were higher ($p < 0.05$) after the C3G administration intra-amniotically compared to the BCE group, which had similar values to the control groups. Furthermore, the BCE promoted a decrease ($p < 0.05$) of acid GC compared to the C3G, H_2O injection, and No injection (Figure 4C). The villi mixed GC was higher in the BCE and C3G than in the H_2O injection and No injection (Figure 4D). In the same way, in the crypt, the C3G increased ($p < 0.05$) the GC diameter compared to the other experimental groups, and BCE was similar to the control groups (Figure 4E). Further, the BCE and the C3G promoted a decrease in the GC number compared to the other groups (Figure 4F). After classifying the GC, we observed that the BCE and C3G have the lowest number of villi acid GC compared to the control groups (Figure 4G). There was no difference in the crypt mixed GC in the BCE group compared to the H_2O injection and C3G (Figure 4H).

In our results, significant intestinal correlations were observed between the intestinal parameters investigated (Figure 5). Positive correlations were observed between *Bifidobacterium* and *Clostridium*, *E. coli* and villi GC diameter, and CDX2 and OCLU. Furthermore, villi height, TNFα, NF-κB1, and CDX2 showed a negative correlation.

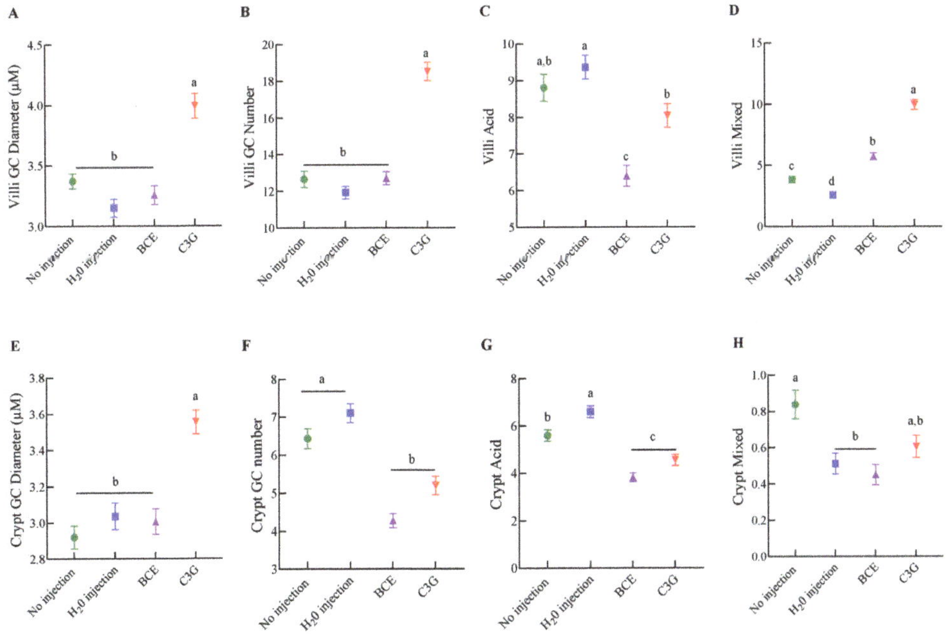

Figure 4. Effect of intra-amniotic administration of black corn extract on goblet cells: (**A–D**): villi goblet cell characteristics; (**E–H**) crypt goblet cell characteristics. Values are means ± SED, n = 3 animals/group. Treatment groups not indicated by the same letter are different ($p < 0.05$) by Kruskal–Wallis and a post-hoc of Dunn's test. Mixed goblet cells are acidic and neutral. BCE: black corn extract; C3G: cyanidin-3-glucoside; GC: goblet cell.

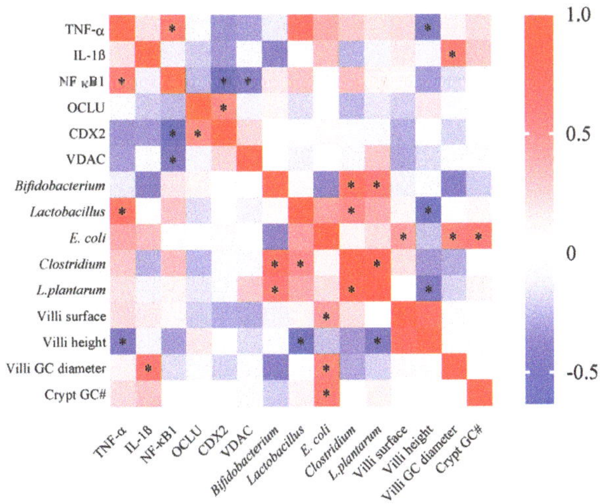

Figure 5. Correlation between intestinal biomarkers, bacterial population, and histological parameters. Colors from blue to red represented the p-value of the Spearman's rank correlation coefficient. Blue: negative and red: positive correlation. * Significant correlation $p < 0.05$. TNF−α: tumor necrosis factor-alpha; NF−κB: nuclear factor kappa beta; IL−1β: interleukin 1 beta; MUC2: mucin 2; OCLN: occludin; CDX2: caudal-related homeobox transcriptional factor 2; VDAC: voltage-dependent anion channel. GC: goblet cells; #: number.

4. Discussion

The current scientific literature suggests that the dietary intake of bioactive components offers significant health-promoting benefits [5,15]. Bioactive components include a range of phenolic components, in which each subgroup exerts different tissue and/or cellular effects and promotes beneficial responses in the organism [15,20]. Our study focused on the effects of black corn (*Zea mays*) anthocyanin-rich extract on intestinal functionality, morphology, and microbial populations in an intraamniotic approach. The intraamniotic administration of black corn extract (BCE) promoted a significant improvement in cecal *Bifidobacterium*, *Clostridium*, and reduced *E. coli*. populations. BCE did not change the duodenal brush border membrane morphology and functionality compared to the control groups.

The BCE composition showed significant levels of C3G and total phenolic compounds. Purple corn flour has shown an amount of anthocyanin (mg cyanidin-3-glucoside/100 g) varying from 220 [33] to 310.04 mg [24]. A wide variation has also been observed among different genotypes, from 12.8 to 93.5 mg C3G/g in 20 different genotypes [34]. In this context, considering our previous study with the same food source but as a flour (black corn flour) [13], the values of C3G in the extract (283.91 mg/100 g) were almost ten-fold higher relative to the flour (30.40 mg/100 g). Total phenolic compounds were similar in the extract (555 mg GAE/100 g sample) compared to the flour (614.30 mg GAE/100 g). Solid-liquid extraction with solvents is the simplest and most common method for extracting phenolic compounds, which is performed to achieve higher yields of the required compounds [35]. Polyphenolic compounds are secondary metabolites of plants that have an effect on plant adaptation to the environment [36], as well as potential bioactivities in animal organisms [15]. According to their chemical structure, phenolic compounds are classified into categories, in which the largest group is the flavonoids, with anthocyanin as a subgroup [37].

Furthermore, the 16s rDNA analysis investigated five bacterial populations and revealed that BCE and C3G increased *Bifidobacterium* and reduced *E. coli* populations in comparison to the other experimental groups (No injection and H_2O injection). The C3G metabolism promotes the proliferation of the genus *Bifidobacterium* in the cecum [38]. Species of *Bifidobacterium* can produce a β-glucosidase enzyme, which supports the hydrolysis of C3G into aglycones and phenolic compounds, which in turn promotes the growth of these beneficial bacteria [17]. In a study with berries, the bifidogenic effect was attributed to the content of anthocyanin but also of polyphenols, as polyphenols contribute to creating a redox environment beneficial to the *Bifidobacteria* selection, which is favorable by a low oxidation-reduction potential [39,40]. Furthermore, the *E. coli* genus contains diverse pathogenic strains that may impair the epithelial barrier by disrupting tight junction proteins [41]. The protective effect of anthocyanin on pathogenic bacteria might be through its intestinal metabolite protocatechuic acid [12], which has been shown to inhibit the growth of *E. coli* [42], which agrees with the observed reduction of *E. coli* abundance in the current study. In agreement, the inhibition of *E. coli* might be associated with the villi goblet cell (GC) diameter and crypt GC number, as indicated by the positive correlation between these variables. GC produces the most important substance in the mucus layer: mucin, which forms a gel barrier against pathogenic bacteria [43]. Therefore, we speculate that a reduction in *E. coli* due to the BCE contributes to maintaining the GC number as the control.

Moreover, the BCE administration increased the *Clostridium* and *Lactobacillus* populations in cecal content compared to the C3G administration. In addition to C3G, other phenolic compounds are found in the black corn extract, which might explain these findings. Several *Lactobacillus* strains use phenolic compounds as a carbon source, thus maintaining their growth besides being involved in the hydrolysis of phenolic compounds due to fermentation [44]. Polyphenols are suggested to exert a prebiotic-like effect by increasing the *Lactobacillus* populations [15], and strains of *Lactobacillus* are considered probiotics due to their immunomodulatory and anti-inflammatory actions, inhibition of bacterial toxins, and competition with pathogens [45]. Therefore, for further investigation, and in addition to the anthocyanin profile, the focus should also be on phenolic characterization.

In the present study, the BCE did not affect the intestinal brush border membrane (BBM) biomarkers: interleukin one beta (IL-1 β), tumor necrosis factor-alpha (TNFα), and nuclear factor kappa beta (NF-κβ). However, the isolated C3G administration downregulated TNFα expression compared to H_2O injection and BCE. Considering the chemical composition of the BCE, it provided a higher administration of cyanidin-3-glucoside (0.014 mg/mL) compared to the isolated C3G, which provided an administration of 0.003 mg/mL. It was previously demonstrated that the effect of polyphenols to downregulate TNFα gene expression was concentration dependent, as 2% saffron extract downregulated TNFα expression, but 5 and 10% did not have this effect, as tested in vivo via intra-amniotic administration [21]. Additionally, we highlight that even with a high dosage of cyanidin-3-glucoside, the black corn extract did not exert any detrimental effect on the investigated inflammatory pathway, as there was no difference in the biomarkers in the BCE group versus the controls. Further investigation and other biomarkers are required to address phenolic compounds' dosage and profile to exert an anti-inflammatory effect.

The BCE administration did not alter villi height and GC diameter (crypt and villi), relative to the H_2O injection. However, these variables were lower in the BCE compared to the C3G. Therefore, we hypothesize that this result is not attributed to the anthocyanin level but probably to other phenolic components that might be present in the extract, such as protocatechuic, vanillic, p-hydroxycinnamic, and ferulic acid [46]. In agreement, the administration of saffron flower extract, a source of phenolic components, showed a dose-dependent effect on decreasing villus surface area, goblet cell number, and diameter [21]. The duodenal morphometric observations in the current study may indicate that depending on the polyphenolic components can exert distinct effects on the brush border development and absorptive capacity [21,23,34]. The variation of polyphenol composition in four distinct types of beans contributed to different results in intestinal morphology and functionality [23]. Interestingly, regarding the type of goblet cells, the BCE group showed the lowest number of acidic GC (in the villi and crypt). A luminal acidic pH facilitates the growth of beneficial bacteria over detrimental bacteria [47]. Therefore, the decrease of acidic GC might be associated with the increase in the *Clostridium* population verified in the BCE group [48]. However, even with the growth of *Clostridium* bacteria, the BCE did not affect Paneth cell number relative to the H_2O injection groups. These cells indicate an early state of inflammation, infection, and toxicity due to the secretion of antimicrobial peptides [49].

Finally, in a prior experiment, we showed the beneficial effects of black corn soluble extract (composed of 6.33 g of total dietary fiber/100 g) on intestinal inflammation parameters, morphology, and BBM barrier function [18]. On the other hand, in the present study, the black corn extract (5%) is composed mainly of phenolic components without any dietary fiber. It modulates the cecal microbiome by changing specific bacterial populations and maintaining intestinal morphology and functionality without detrimental effects. Thus, we highlight the positive effects of black corn anthocyanin-rich extract without any soluble dietary fiber, which was able to improve the cecal microbial populations and maintain intestinal morphology and functionality without any detrimental effects in vivo.

5. Conclusions

The black corn anthocyanin-rich extract improved the cecal microbiome by increasing *Bifidobacterium* and *Clostridium*, reducing the *E. coli* population while maintaining intestinal morphology and functionality. Further, the C3G group showed additional effects on improving intestinal morphology versus the BCE, suggesting that the combination and dosage of phenolic compounds might interfere with intestinal morphology development. Therefore, our results suggest that black corn anthocyanin-rich extract is a promising target matrix to be used as a functional extract to improve intestinal microbial populations, and further studies in terms of dosage and profile of phenolic compounds in this food matrix are now warranted.

Author Contributions: Conceptualization, E.T., T.A.V. and H.S.D.M.; methodology, T.A.V., N.A., N.K. and M.G.; formal analysis, T.A.V., N.A. and N.K.; investigation, T.A.V., N.A., H.S.D.M., N.K. and E.T.; resources, H.S.D.M. and E.T.; data curation, T.A.V., N.K. and E.T.; writing—original draft preparation, T.A.V.; writing—review and editing, T.A.V., H.S.D.M., M.G., M.C.D.P. and E.T.; supervision, H.S.D.M., M.C.D.P. and E.T.; project administration, E.T. and H.S.D.M.; funding acquisition, E.T. All authors have read and agreed to the published version of the manuscript.

Funding: This research received no external funding.

Institutional Review Board Statement: The animal protocol used in this study was conducted according to the guidelines of the Declaration of Helsinki and approved by the Cornell University Institutional Animal Care and Use Committee by ethic approval code 2020-0077.

Informed Consent Statement: Not applicable.

Data Availability Statement: Data available upon reasonable request.

Acknowledgments: To the Coordination for the Improvement of Higher Educational Personnel (CAPES), Brazil, for providing the scholarship support for the doctor in the Capes-Print Program (process number 88887.569929/2020-00) and the Brazilian Agricultural Research Corporation (EMBRAPA, Brazil) for providing the black corn grains and the anthocyanin profile analysis. All authors have consented to all acknowledgement included.

Conflicts of Interest: The authors declare no conflict of interest.

References

1. Salvador-Reyes, R.; Clerici, M.T.P.S. Peruvian Andean Maize: General Characteristics, Nutritional Properties, Bioactive Compounds, and Culinary Uses. *Food Res. Int.* **2020**, *130*, 108934. [CrossRef] [PubMed]
2. Colombo, R.; Ferron, L.; Papetti, A. Colored Corn: An Up-Date on Metabolites Extraction, Health Implication, and Potential Use. *Molecules* **2021**, *26*, 199. [CrossRef] [PubMed]
3. Petroni, K.; Pilu, R.; Tonelli, C. Anthocyanins in Corn: A Wealth of Genes for Human Health. *Planta* **2014**, *240*, 901–911. [CrossRef]
4. Sui, X.; Zhang, Y.; Jiang, L.; Zhou, W. Anthocyanins in Food. *Encycl. Food Chem.* **2019**, *2*, 10–17.
5. Verediano, T.A.; Stampini Duarte Martino, H.; Dias Paes, M.C.; Tako, E. Effects of Anthocyanin on Intestinal Health: A Systematic Review. *Nutrients* **2021**, *13*, 1331. [CrossRef] [PubMed]
6. Zhang, Q.; Luna-Vital, D.; de Mejia, E.G. Anthocyanins from Colored Maize Ameliorated the Inflammatory Paracrine Interplay between Macrophages and Adipocytes through Regulation of NF- KB and JNK-Dependent MAPK Pathways. *J. Funct. Foods* **2019**, *54*, 175–186. [CrossRef]
7. Gowd, V.; Bao, T.; Wang, L.; Huang, Y.; Chen, S.; Zheng, X.; Cui, S.; Chen, W. Antioxidant and Antidiabetic Activity of Blackberry after Gastrointestinal Digestion and Human Gut Microbiota Fermentation. *Food Chem.* **2018**, *269*, 618–627. [CrossRef]
8. Peng, Y.; Yan, Y.; Wan, P.; Chen, D.; Ding, Y.; Ran, L.; Mi, J.; Lu, L.; Zhang, Z.; Li, X.; et al. Gut Microbiota Modulation and Anti-Inflammatory Properties of Anthocyanins from the Fruits of Lycium Ruthenicum Murray in Dextran Sodium Sulfate-Induced Colitis in Mice. *Free Radic. Biol. Med.* **2019**, *136*, 96–108. [CrossRef]
9. Wang, H.; Liu, D.; Ji, Y.; Liu, Y.; Xu, L.; Guo, Y. Dietary Supplementation of Black Rice Anthocyanin Extract Regulates Cholesterol Metabolism and Improves Gut Microbiota Dysbiosis in C57BL/6J Mice Fed a High-Fat and Cholesterol Diet. *Mol. Nutr. Food Res.* **2020**, *64*, e1900876. [CrossRef]
10. Jurgoński, A.; Juśkiewicz, J.; Zduńczyk, Z. An Anthocyanin-Rich Extract from Kamchatka Honeysuckle Increases Enzymatic Activity within the Gut and Ameliorates Abnormal Lipid and Glucose Metabolism in Rats. *Nutrition* **2013**, *29*, 898–902. [CrossRef]
11. Hribar, U.; Ulrih, P. The Metabolism of Anthocyanins. *Curr. Drug Metab.* **2014**, *15*, 3–13. [CrossRef] [PubMed]
12. Tan, J.; Li, Y.; Hou, D.X.; Wu, S. The Effects and Mechanisms of Cyanidin-3-Glucoside and Its Phenolic Metabolites in Maintaining Intestinal Integrity. *Antioxidants* **2019**, *8*, 479. [CrossRef] [PubMed]
13. Verediano, T.A.; Sant' Ana, C.T.; Grancieri, M.; Parzanini Brilhante de São José, V.; Toledo, R.C.L.; Dias Paes, M.C.; Duarte Martino, H.S. Black Corn (*Zea Mays* L.) Whole Flour Improved the Antioxidant Capacity and Prevented Adipogenesis in Mice Fed a High-Fat Diet. *Food Funct.* **2022**, *13*, 5590–5601. [CrossRef]
14. Rahman, S.; Mathew, S.; Nair, P.; Ramadan, W.S.; Vazhappilly, C.G. Health Benefits of Cyanidin-3-Glucoside as a Potent Modulator of Nrf2-Mediated Oxidative Stress. *Inflammopharmacology* **2021**, *29*, 907–923. [CrossRef]
15. Domínguez-Avila, J.A.; Villa-Rodriguez, J.A.; Montiel-Herrera, M.; Pacheco-Ordaz, R.; Roopchand, D.E.; Venema, K.; González-Aguilar, G.A. Phenolic Compounds Promote Diversity of Gut Microbiota and Maintain Colonic Health. *Dig. Dis. Sci.* **2021**, *66*, 3270–3289. [CrossRef]

16. Vancamelbeke, M.; Vermeire, S. The Intestinal Barrier: A Fundamental Role in Health and Disease. *Expert Rev. Gastroenterol. Hepatol.* **2017**, *11*, 821–834. [CrossRef]
17. Tian, L.; Tan, Y.; Chen, G.; Wang, G.; Sun, J.; Ou, S.; Chen, W.; Bai, W. Metabolism of Anthocyanins and Consequent Effects on the Gut Microbiota. *Crit. Rev. Food Sci. Nutr.* **2018**, *59*, 982–991. [CrossRef]
18. Verediano, T.A.; Stampini Duarte Martino, H.; Kolba, N.; Fu, Y.; Cristina Dias Paes, M.; Tako, E. Black Corn (*Zea Mays* L.) Soluble Extract Showed Anti-Inflammatory Effects and Improved the Intestinal Barrier Integrity in Vivo (*Gallus gallus*). *Food Res. Int.* **2022**, *157*, 111227. [CrossRef]
19. Hou, T.; Tako, E. The in Ovo Feeding Administration (*Gallus gallus*)—An Emerging in Vivo Approach to Assess Bioactive Compounds with Potential Nutritional Benefits. *Nutrients* **2018**, *10*, 418. [CrossRef]
20. Abdel-Moneim, A.M.E.; Shehata, A.M.; Alzahrani, S.O.; Shafi, M.E.; Mesalam, N.M.; Taha, A.E.; Swelum, A.A.; Arif, M.; Fayyaz, M.; Abd El-Hack, M.E. The Role of Polyphenols in Poultry Nutrition. *J. Anim. Physiol. Anim. Nutr.* **2020**, *104*, 1851–1866. [CrossRef]
21. Agarwal, N.; Kolba, N.; Jung, Y.; Cheng, J.; Tako, E. Saffron (*Crocus Sativus* L.) Flower Water Extract Disrupts the Cecal Microbiome, Brush Border Membrane Functionality, and Morphology In Vivo (*Gallus gallus*). *Nutrients* **2022**, *14*, 220. [CrossRef]
22. Gomes, M.J.C.; Kolba, N.; Agarwal, N.; Kim, D.; Eshel, A.; Koren, O.; Tako, E. Modifications in the Intestinal Functionality, Morphology and Microbiome Following Intra-Amniotic Administration (*Gallus gallus*) of Grape (*Vitis vinifera*) Stilbenes (Resveratrol and Pterostilbene). *Nutrients* **2021**, *13*, 3247. [CrossRef]
23. Gomes, M.J.C.; Martino, H.S.D.; Kolba, N.; Cheng, J.; Agarwal, N.; de Moura Rocha, M.; Tako, E. Zinc Biofortified Cowpea (*Vigna Unguiculata* L. Walp.) Soluble Extracts Modulate Assessed Cecal Bacterial Populations and Gut Morphology In Vivo (*Gallus gallus*). *Front. Biosci.* **2022**, *27*, 140. [CrossRef]
24. Ranilla, L.G.; Christopher, A.; Sarkar, D.; Shetty, K.; Chirinos, R.; Campos, D. Phenolic Composition and Evaluation of the Antimicrobial Activity of Free and Bound Phenolic Fractions from a Peruvian Purple Corn (*Zea mays* L.) Accession. *J. Food Sci.* **2017**, *82*, 2968–2976. [CrossRef]
25. Martino, H.S.D.; Kolba, N.; Tako, E. Yacon (*Smallanthus sonchifolius*) Flour Soluble Extract Improve Intestinal Bacterial Populations, Brush Border Membrane Functionality and Morphology in Vivo (*Gallus gallus*). *Food Res. Int.* **2020**, *137*, 109705. [CrossRef]
26. Lao, F.; Giusti, M.M. Extraction of Purple Corn (*Zea Mays* L.) Cob Pigments and Phenolic Compounds Using Food-Friendly Solvents. *J. Cereal Sci.* **2018**, *80*, 87–93. [CrossRef]
27. Singleton, V.L.; Rossi, J.A.J. Colorimetry of Total Phenolics with Phosphomolybdic Phosphotungstic Acid Reagents. *Am. J. Enol. Vitic.* **1965**, *16*, 144–158.
28. Grancieri, M.; Martino, H.S.D.; de Mejia, E.G. Digested Total Protein and Protein Fractions from Chia Seed (*Salvia Hispanica* L.) Had High Scavenging Capacity and Inhibited 5-LOX, COX-1-2, and INOS Enzymes. *Food Chem.* **2019**, *289*, 204–214. [CrossRef]
29. de Araujo Santiago, M.C.P.; Galhardo Borguini, R.; da Silva de Mattos do Nascimento, L.; de Oliveira Braga, E.C.; de Carvalho Martins, V.; Senna Gouvêa, A.C.M.; Marques Peixoto, F.; Pacheco, S.; Nogueira, R.I.; de Oliveira Godoy, R.L. Jabuticaba (Myrciaria Jaboticaba (Vell.) O. Berg) Peel Powder Produced by Convective Drying Process: A Rich Anthocyanin Product. *Fruits* **2018**, *73*, 201–208. [CrossRef]
30. Dias, D.M.; Kolba, N.; Hart, J.J.; Ma, M.; Sha, S.T.; Lakshmanan, N.; Nutti, M.R.; Martino, H.S.D.; Glahn, R.P.; Tako, E. Soluble Extracts from Carioca Beans (*Phaseolus Vulgaris* L.) Affect the Gut Microbiota and Iron Related Brush Border Membrane Protein Expression in Vivo (*Gallus gallus*). *Food Res. Int.* **2019**, *123*, 172–180. [CrossRef]
31. Tako, E.; Glahn, R.P.; Welch, R.M.; Lei, X.; Yasuda, K.; Miller, D.D. Dietary Inulin Affects the Expression of Intestinal Enterocyte Iron Transporters, Receptors and Storage Protein and Alters the Microbiota in the Pig Intestine. *Br. J. Nutr.* **2008**, *99*, 472–480. [CrossRef]
32. Cheng, J.; Kolba, N.; Sisser, P.; Turjeman, S.; Even, C.; Koren, O.; Tako, E. Intraamniotic Administration (*Gallus gallus*) of Genistein Alters Mineral Transport, Intestinal Morphology, and Gut Microbiota. *Nutrients* **2022**, *14*, 3473. [CrossRef]
33. Bhaswant, M.; Shafie, S.R.; Mathai, M.L.; Mouatt, P.; Brown, L. Anthocyanins in chokeberry and purple maize attenuate diet-induced metabolic syndrome in rats. *Nutrition* **2017**, *41*, 24–31. [CrossRef]
34. Zhang, Q.; Mejia, E.G.; Luna-Vital, D.; Tao, T.; Chandrasekaran, S.; Chatham, L.; Juvik, J.; Singh, V.; Kumar, D. Relationship of phenolic composition of selected purple maize (*Zea mays* L.) genotypes with their anti-inflammatory, anti-adipogenic and anti-diabetic potential. *Food Chem.* **2019**, *289*, 739–750. [CrossRef]
35. Cong-Cong, X.U.; Wang, B.; Yi-Qiong, P.U.; Jian-Sheng, T.; Tong, Z. Advances in Extraction and Analysis of Phenolic Compounds from Plant Materials. *Chin. J. Nat. Med.* **2017**, *15*, 721–731.
36. Brglez Mojzer, E.; Knez Hrnčič, M.; Škerget, M.; Knez, Ž.; Bren, U. Polyphenols: Extraction Methods, Antioxidative Action, Bioavailability and Anticarcinogenic Effects. *Molecules* **2016**, *21*, 901. [CrossRef]
37. Kumar, S.; Pandey, A.K. Chemistry and Biological Activities of Flavonoids. *Hindawi Sci. World J.* **2013**, *2013*, 533–548. [CrossRef]
38. Hanske, L.; Engst, W.; Loh, G.; Sczesny, S.; Blaut, M.; Braune, A. Contribution of Gut Bacteria to the Metabolism of Cyanidin 3-Glucoside in Human Microbiota-Associated Rats. *Br. J. Nutr.* **2013**, *109*, 1433–1441. [CrossRef]
39. Jayamanne, V.S.; Adams, M.R. Modelling the Effects of PH, Storage Temperature and Redox Potential (Eh) on the Survival of Bifidobacteria in Fermented Milk. *Int. J. Food Sci. Technol.* **2009**, *44*, 1131–1138. [CrossRef]
40. Vendrame, S.; Klimis-Zacas, D. Anti-Inflammatory Effect of Anthocyanins via Modulation of Nuclear Factor- KB and Mitogen-Activated Protein Kinase Signaling Cascades. *Nutr. Rev.* **2015**, *73*, 348–358. [CrossRef]

41. Pawłowska, B.; Sobieszczańska, B.M. Intestinal Epithelial Barrier: The Target for Pathogenic Escherichia Coli. *Adv. Clin. Exp. Med.* **2017**, *26*, 1437–1445. [CrossRef]
42. Ajiboye, T.O.; Habibu, R.S.; Saidu, K.; Haliru, F.Z.; Ajiboye, H.O.; Aliyu, N.O.; Ibitoye, O.B.; Uwazie, J.N.; Muritala, H.F.; Bello, S.A.; et al. Involvement of Oxidative Stress in Protocatechuic Acid-Mediated Bacterial Lethality. *Microbiologyopen* **2017**, *6*, e00472. [CrossRef]
43. Tarabova, L.; Makova, Z.; Piesova, E.; Szaboova, R.; Faixova, Z. Intestinal Mucus Layer and Mucins (A Review). *Folia Vet.* **2016**, *60*, 21–25. [CrossRef]
44. Liu, L.; Zhang, C.; Zhang, H.; Qu, G.; Li, C.; Liu, L. Biotransformation of Polyphenols in Apple Pomace Fermented by β-Glucosidase-Producing Lactobacillus Rhamnosus L08. *Foods* **2021**, *10*, 1343. [CrossRef]
45. Markowiak, P.; Ślizewska, K. Effects of Probiotics, Prebiotics, and Synbiotics on Human Health. *Nutrients* **2017**, *9*, 1021. [CrossRef]
46. Huang, B.; Wang, Z.; Park, J.H.; Ryu, O.H.; Choi, M.K.; Lee, J.Y.; Kang, Y.H.; Lim, S.S. Anti-Diabetic Effect of Purple Corn Extract on C57BL/KsJ Db/Db Mice. *Nutr. Res. Pract.* **2015**, *9*, 17–21. [CrossRef]
47. Witten, J.; Samad, T.; Ribbeck, K. Selective Permeability of Mucus Barriers. *Curr. Opin. Biotechnol.* **2018**, *52*, 124–133. [CrossRef]
48. Fachi, J.L.; de Souza Felipe, J.; Pral, L.P.; da Silva, B.K.; Corrêa, R.O.; de Andrade, M.C.P.; da Fonseca, D.M.; Basso, P.J.; Câmara, N.O.S.; e Souza, É.L.d.S.; et al. Butyrate Protects Mice from Clostridium Difficile-Induced Colitis through an HIF-1-Dependent Mechanism. *Cell Rep.* **2019**, *27*, 750–761.e7. [CrossRef]
49. Gassler, N. Paneth Cells in Intestinal Physiology and Pathophysiology. *World J. Gastrointest. Pathophysiol.* **2017**, *8*, 150–160. [CrossRef]

MDPI
St. Alban-Anlage 66
4052 Basel
Switzerland
www.mdpi.com

Nutrients Editorial Office
E-mail: nutrients@mdpi.com
www.mdpi.com/journal/nutrients